TWO-DEGREE-OF-FREEDOM CONTROL SYSTEMS

TWO-DEGREE-OF-FREEDOM CONTROL SYSTEMS

The Youla Parameterization Approach

"WHO KNOWS WRITES, WHO READS UNDERSTANDS"

LÁSZLÓ KEVICZKY AND CSILLA BÁNYÁSZ

ELSEVIER

Amsterdam • Boston • Heidelberg • London
New York • Oxford • Paris • San Diego
San Francisco • Singapore • Sydney • Tokyo
Academic Press is an imprint of Elsevier

AP

Academic Press is an imprint of Elsevier
125 London Wall, London EC2Y 5AS, UK
525 B Street, Suite 1800, San Diego, CA 92101-4495, USA
225 Wyman Street, Waltham, MA 02451, USA
The Boulevard, Langford Lane, Kidlington, Oxford OX5 1GB, UK

ISBN: 978-0-12-803310-4

British Library Cataloguing-in-Publication Data
A catalogue record for this book is available from the British Library

Library of Congress Cataloging-in-Publication Data
A catalog record for this book is available from the Library of Congress

Publisher: Andre Wolff
Acquisition Editor: Sonnini Yura
Editorial Project Manager: Mariana Kuhl
Production Project Manager: Jason Mitchell
Designer: Matthew Limbert

Typeset by TNQ Books and Journals
www.tnq.co.in

Printed and bound in the United States of America

DEDICATION

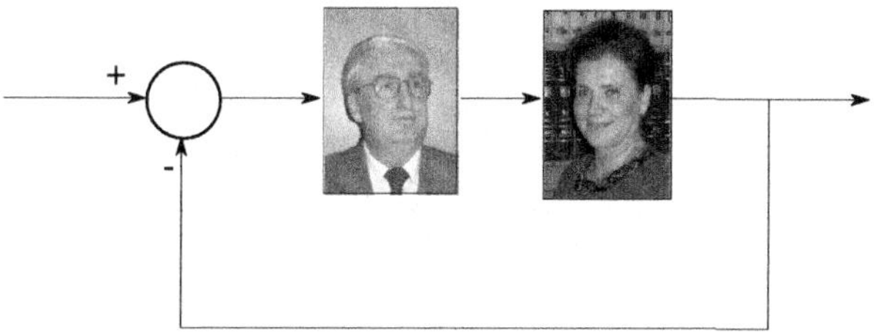

We would like to express our thanks to our children, Zoltán and Tamás, who have always supported our work. Their love and patience have allowed us to achieve the results presented in this book.

ISBN: 978-0-12-803310-4

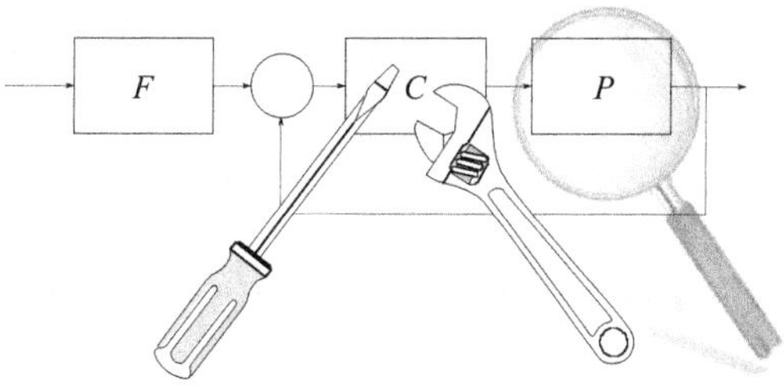

THE BEST STUDENTS

József Bokor
Jenő Hetthéssy
Róbert Haber
István Vajk

Our students are carrying the torch further. Their cooperation helped and inspired the work of the teachers. All of them have successful university carriers and they are good colleagues and friends. Apart from the relationship of parents and children, this continuity of human life is perhaps the most important one.

THE CLOSEST COWORKERS

We express our gratitude to Ruth Bars and Jenő Hetthéssy, who were our coworkers writing a former university lecture notes and they gave their permission to shortly repeat some parts of that work in this book for ease of understanding.

NOTATION

Transfer function of continuous-time systems	H
Transfer function of discrete-time systems	G
Regulator transfer function	C
Process transfer function	P
Discrete-time process pulse transfer function	G (or P_d)
Sensitivity function	S
Complementary sensitivity function	T
Transfer function of open control loop	L
Gain of control loop	K
Transfer coefficient of control loop	k
Youla parameter	Q (or $\mathbf{Q}$)
KB parameter	Q_{KB}
Continuous-time	(t)
Discrete-time	$[k]$
Laplace transform	$\mathcal{L}\{\ldots\}$
z-transform	$\mathcal{Z}\{\ldots\}$
Complex variable argument ($\mathcal{L}$ transf.)	s
Complex variable argument ($\mathcal{Z}$-transf.)	z
Reference signal	r (or y_r)
Controlled variable	y
Error signal	e
Actuating signal (or output of the regulator)	u
Input noise	y_{ni}
Output noise	y_n (or y_{no})
Vector	$\boldsymbol{a}, \boldsymbol{b}, \boldsymbol{c}, \ldots$
Row vector	$\boldsymbol{a}^T, \boldsymbol{b}^T, \boldsymbol{c}^T, \ldots$
Matrix	$\boldsymbol{A}, \boldsymbol{B}, \boldsymbol{C}, \ldots$
Transpose of a matrix	$\boldsymbol{A}^T$

Adjunct of a matrix	$\mathbf{adj}(\boldsymbol{A})$
Determinant of a matrix	$\det(\boldsymbol{A})$ (or $\lvert\boldsymbol{A}\rvert$)
Trace of a matrix	$\mathrm{tr}(\boldsymbol{A})$
State variable	$\boldsymbol{x}$
Parameters of state equation (continuous)	$\boldsymbol{A}$, $\boldsymbol{b}$, $\boldsymbol{c}$, d (or $\mathbf{A}$, $\mathbf{B}$, $\mathbf{C}$, $\mathbf{D}$)
Parameters of state equation (discrete)	$\boldsymbol{F}$, $\boldsymbol{g}$, $\boldsymbol{c}$, d
Diagonal matrix	$\mathbf{diag}[a_{11},a_{22},\ldots,a_{nn}]$
Identity matrix	$\boldsymbol{I}=\mathbf{diag}[1,1,\ldots,1]$
Sampling time	T_s
Dead-time (continuous)	T_d
Time delay (discrete)	d
Step response function	$v(t)$
Weighting function	$w(t)$
Frequency	ω
Frequency spectrum of a continuous signal	$F(j\omega)$
Frequency spectrum of a sampled signal	$F^*(j\omega)$
Frequency spectrum of a discrete time model	$G(j\omega)$ (or $P_d(j\omega)$)
Polynomials	$\mathcal{A}$, $\mathcal{B}$, $\mathcal{C}$, $\mathcal{D}$, $\mathcal{G}$, $\mathcal{F}$, $\mathcal{R}$, $\mathcal{X}$, $\mathcal{Y}$, $\mathcal{V}$
Matrix polynomials	$\boldsymbol{\mathcal{A}}$, $\boldsymbol{\mathcal{B}}$, $\boldsymbol{\mathcal{C}}$, $\boldsymbol{\mathcal{D}}$, $\boldsymbol{\mathcal{U}}$, $\boldsymbol{\mathcal{V}}$
Parameter matrices of matrix polynomials	$\mathbf{N}_L^i, \mathbf{D}_L^i, \mathbf{T}_i, \mathbf{P}_i$
Order of a polynomial	$\deg\{\mathcal{A}\}$
Characteristic equation	$\mathcal{A}(s)=0$
Characteristic equation (discrete)	$\mathcal{F}(z)=0$
Limit of control output	$\mathbb{U}$
Gradient vector	$\mathbf{grad}[f(\boldsymbol{x})]$
For all ω	$\forall\omega$
Angle of a complex number or function	$\angle$ (or $\mathrm{arc}(\ldots)$)
Exponential function	$e^{(\ldots)}$ (or $\exp(\ldots)$)
Natural logarithm	$\ln(\ldots)$
Common (base 10)	$\lg(\ldots)$
Expected value	$E\{\ldots\}$
Probability limit value	$\mathrm{plim}\{\ldots\}$
Matrix exponential	$e^{\boldsymbol{A}}$
Matrix logarithm	$\ln(\boldsymbol{A})$
Set of stable linear DT processes	$\mathcal{S}_{DT}$
Set of stable linear CT processes	$\mathcal{S}_{CT}$ (or $\mathcal{S}$)

ABBREVIATIONS

CAD	Computer-aided design
CT	Continuous-time
DT	Discrete-time
ID	Process identification
DE	Diophantine equation
ELS method	Extended least squares method
GMV	Generalized minimum variance
IV	Instrumental variables
IS	Inverse stable
IU	Inverse unstable
LS method	Least squares method
MFD	Matrix fraction description
ML	Maximum likelihood
MSE	Mean square error
MV	Minimum variance
MIMO	Multiple input multiple output
ODOF	One-degree-of-freedom
ST	Self-tuning
SISO	Single input single output
SRE	Step response equivalent
SA method	Stochastic approximation method
TDOF	Two-degree-of-freedom
TFM	Transfer function matrix
YP	Youla-parameterized
PFE	Partial fractional expansion

CONTENTS

PREFACE

From earliest times humankind has aimed to preserve its experiences and opinions and pass them on to succeeding generations. This is reflected in cave drawings, tally marks, clay boards, and papyrus rolls, as well as in books and Gutenberg technologies. The process is continued by contemporary digital methods. A person can share knowledge by describing her/his experiences, results, and opinions. This process together with genetic heritage ensures the future of humankind but the latter is much slower and is not transparent yet in practice.

This is why we as authors decided to publish our results, methods, experiences, and problem-solving approaches. This book is recommended for those working in and outside the subject area, students, colleagues, interested persons, and particularly, our successors. The reader is very important in this process, perhaps the most important. In general it is true that "Who knows writes, who reads understands." This is the most effective situation since the time book writing and publishing exists.

Mountain peaks can be scaled in many different ways. There are well-designed incline rails, pedestrian path, and even roads, but the real friends of nature prefer to go on trails. This book tries to offer solutions via hidden trails, hoping that the reader will understand and appreciate these methods rather more.

The majority of people rarely come across the concept of automatic control, despite the fact that they operate machines by pushing buttons, tripping switches, or using instrument panels. While being largely unaware of it, they almost always apply some kind of *Control* when using common everyday devices. In the modern technologies of the twenty-first century the basic processing, evaluating, and decision-making tasks are executed by digital *Computation*, i.e., by computers. Observation of the signals and characteristics of the real-time processes and the transfers of executive commands are made possible by digital *Communication*. These three areas (*Control–Computation–Communication* = C^3) are often considered in close synergy, sometimes extended by a fourth, not easily defined, subject area, namely built-in machine *Intelligence*. This constitutes the famous C^3I complex. Our book deals specifically with the *Control* area of this complex.

"*Navigare necesse est.*" i.e., a ship must be navigated, as the ancient Romans recognized. "*Controlare necesse est.*" i.e., a system must be controlled, something we have been saying since the technical revolution of the nineteenth century. In fact, our everyday life cannot be imagined without automatic equipment. Sometimes this fact is not transparent because everyone is used to pushing the buttons on their modern or less modern devices, switching on and adjusting screens etc., and the technology that completes the demands is hidden, invisible. One therefore does not think about how many control loops or regulatory processes are in operation when a radio, television, or CD or DVD system is activated or simple equipment such as irons, hot water boilers, or central heating systems is used. It is indeed hard to find equipment that does not contain at least one or more control tasks solved by automation for our comfort. This subject area also includes the investigation of control systems operating in different life forms, regulate the supply, and demand ratio in economics, or the environmental processes of the globe.

In an iron the temperature control system is operated by a relay, in the gas heating system the temperature is controlled, but in more sophisticated systems the temperature of the environment is also taken into consideration. In the home, modern audio–visual systems contain dozens of control tasks, e.g., the regulation of the speed of tape recorders, the start and stop operation of the equipment; CD and DVD systems use similar operation modes, not forgetting the temperature control of the processor in the PC, the positioning of the hard disks' heads, etc. In cars the quantity of petrol used and the harmonized operation of the brakes are all governed by automatic regulators. Modern aircraft typically could not fly without controllers, since their operation is a typical example of unstable systems. The number of control tasks in modern aircraft is over a hundred. The universe could not be investigated without automatic control and guidance systems for launching rockets and satellites. In Mars explorers, sophisticated high-level, so-called autonomous intelligent, equipments have been applied.

In complex, industrial processes the number of tasks to be solved is over a thousand or even ten thousand. The quantity and quality of the products and the safety of the environment cannot be guaranteed without these automatically operated systems. Launching products onto the market requires the accurate control of a number of variables (in many cases within a prescribed fidelity limit).

In almost all factory assemblies, from simple production beltways to robots, automatic control is applied.

With the development of systems biology it was discovered that in any organ, or the entire body, dozens of basic control processes are at work (e.g., governing blood pressure, body temperature, the level of blood sugar content, the level of hormones, etc.), and current technologies are approaching the point when some of these tasks can be taken over by devices in the case of illness or other problems.

Several basic economic processes (e.g., supply and demand, storage inventory, macro- and microbalance, etc.) offer possibilities for automatic control.

The discussion of control engineering methods requires many formulas. This does not mean pure mathematics, though the results and methods of mathematics are intensively used. Control engineering belongs to the technical sciences, but its results are applied from biology to economics.

Let us ascend via the trails we have found. On the way we will try to produce results that cannot be seen from anywhere else. The new view is worth the trek.

László Keviczky

Csilla Bányász

CHAPTER 1

Introduction

A majority of people rarely come across the concept of automatic control, despite the fact that they operate machines by pushing buttons, tripping switches, or using instrument panels. That is why control is often considered as a *hidden* technology. This phenomenon used to be the reason for the regrettable opinion that there is no need to study the theory of control and regulation, since it comes with the equipment embedded. But do not forget that such equipment needs to be designed and produced before being launched onto the market. The most advanced countries are at the frontline of the development of these kinds of instruments and processes.

Even at an early age we see that there are machines that do not work without control, for example, the bicycle. It is unstable by itself, it needs to be driven and steered, otherwise it falls down. From the perspective of dynamic behavior, very similar problems assail the case of the umbrella or a rod balanced on the finger-tip or the flight of planes, which we encounter as adults. Very special capabilities are required to balance two umbrellas or rods on top of each other or ride a unicycle. The controls of helicopters are perhaps the most complex of all flying instruments. Conventional courses in control engineering offer a solution which can be performed easily by computers.

Contrary to the popularizing preface, our book is intended to be used by those who have studied control engineering at BSc and MSc levels and would like to learn further methods for the design and realization of more precise and reliable control algorithms, that is, tasks requiring high-performance solutions.

A control is a specific action required to reach the desired behavior of a system. In control of industrial processes generally technological processes are considered, but control is required to keep any physical, chemical, biological, communicational, economical, social process in a desired manner.

Control methods should be used whenever some quantity must be kept at a desired value. For example, control is used to maintain the temperature of a flat at a comfortable specific value in both winter and summer. When controlling an aircraft the pilot (or the autopilot) has to execute extremely diverse control tasks to keep the speed, the direction, the altitude of the aircraft at the desired values. Control systems are all around us, in the household (e.g., setting the program of a washing machine, ironing by means of an on–off temperature control, air conditioning, etc.), and in transportation, space research, communication, industrial manufacturing, economics, medicine, etc. A lot of control systems operate in living organisms as well.

Control engineering is an interdisciplinary area of science. The operation of the process needs to be understood and consequently there is a need for knowledge of physical, chemical, biological, and other phenomena. To investigate the operation of control systems, knowledge is needed about signals, systems, and the behavior of systems with negative feedback. During the design process rational considerations and basic restrictions also have to be taken into account. The design has to cover economic, safety, environmental protection, and other aspects as well. To solve a more complex control task, the coordinated work of different professionals is needed.

Interdisciplinarity also means that several scientific subject areas can use the concepts of control engineering and explain its vital phenomena. The control of human organs, blood pressure, blood sugar, or the level of certain hormones is a complex multivariable system, whose settings are very difficult tasks similar to technical processes. Who would imagine that basic economic processes (e.g., supply and demand, storage inventory, macro- and micro-balance, etc.) are problems analogous to the level control in a tank.

This kind of reasoning naturally in control and joint systems engineering—(according to which processes are dynamic, have input and output signals, and inner state variables)—provides useful methodology and basically rational approaches for different subject areas.

Contrary to the optimistic picture above, however, not all tasks can be solved by control engineering. In many cases the process to be controlled is sluggish, inert, its dynamics are slow and it needs to be speeded up. This can be resolved only by adding extra power (overshoot) to the process (for a longer or shorter time) by the commands computed from the closed-loop control. In many cases, however, they cannot be performed because of technical limitations. The jumbo valve of an oil pipeline, for example, cannot be switched on and off in seconds. This is, of course, an extreme example, but many similar ones can be found where the actuator signal is limited. In practice, this fundamentally influences the amount of speedup. In the case of fighter aircraft, however, the limits are not only technical, but usually influenced by the effects tolerable for the pilots (e.g., the value of the "g-load" acceptable for the pilot).

In certain cases the solution does not depend on us, but on the process to be controlled. It is interesting that the real problem is not the stabilization of an unstable process (see the case of the balanced rod, i.e., in our terminology the inverted pendulum) because stabilization is a routine task. The other dynamic attributes of the process, however, are of basic importance and they cannot be changed: these are called invariant behaviors, and their effect cannot be eliminated by any theoretical solution. It is a very important question whether given the invariant process behaviors and the physical limitations of the intervention there is any chance of optimizing control.

The basic feature of control is feedback. Feedback can be found in nature everywhere: in the balance of the species, atmospheric phenomena, organs, inner human processes, for example, blood pressure, etc.

Very simple examples are presented, which can be repeated even by simple paper and pencil methods. This is to ease understanding. "Nothing is more theoretical than a good example." In this context, we preempt the criticism that the book is a tutorial. We do not want to prove that we are very clever; our goal is for readers to find the content of the book useful and valuable. Valuable for those who want to see through the jungle of available methods using the unified approach of our view; valuable for educators who want to explain the background of certain methods in the class; valuable for PhD students who want to apply new approaches; and even valuable for those who want to prepare a computer code with a given algorithm.

The title of our book is derived from feedback, since the closed-loop control is known as feedback, and the two-degrees-of-freedom (*TDOF*) control loops are the most general and high-performance ones. These control loops make it possible to prescribe simultaneously the demands of the control and meet requirements in all circumstances.

The book includes 11 chapters, an appendix, and references. The outline of each chapter is as follows:

Chapter 1, after a short introduction, describes the models of the processes used, and then the concepts of feedback control systems. A special section is devoted to the stability of closed-loop control and the possible parameterization methods known from the literature.

Chapter 2 discusses the control of stable processes. The concept of Youla parameterization (*YP*), of vital importance in our book, is introduced, and then it is extended for *TDOF* control loops. The Keviczky–Bányász parameterization (*KB*) method, developed for closed-loop systems, is presented. The core of our book is formed by *YP* and *KB* parameterization, since even the classical control design methods are discussed or compared through them. This chapter also presents other classical parameterization methods. The chapter concludes with the discussion of deadbeat and predictive regulators.

Chapter 3 deals with feedback regulator design methods. These methods can be applied to arbitrary, and thus unstable, processes, as well. First, the classical methods of state feedback and then the *linear-quadratic* regulators are discussed. Finally, a general polynomial method is given for the direct design of the feedback. Comparison with the *YP* is given for all methods.

Chapter 4 presents a new method for the decomposition of the sensitivity function of the closed loop to three components. These components give the main losses, fundamentally determining the quality of control, namely the design, the realizability, and modeling losses. Very detailed mathematical analysis is given for all components, and their optimization possibilities by different norms are also presented.

Chapter 5 briefly discusses conventional *proportional-integral-derivative* regulators, devoting special attention to the design methods applying dominant pole cancellation. These methods can be compared to the regulators obtained by *YP*.

Chapter 6 deals with the design of closed-loop controls under stochastic output disturbances.

Chapter 7 treats the problem of multivariable process control and extends the results of *YP* and *KB* parameterization to this case.

Chapter 8 investigates the applicability of the results of *YP* and *KB* parameterization for a special class of nonlinear dynamic processes, namely processes represented by cascade models.

Chapter 9 deals with robust control. First the case of *YP* regulators is investigated, then the robustness of these control systems is discussed. Besides the classical Vinnicombe gap metrics, other similar gap metrics are introduced. A special chapter is devoted to the dialectic and relationships between the quality and the robustness of control. Finally, the auspicious features of *YP* and *KB* parameterization are demonstrated via the so-called product inequality.

Chapter 10 discusses process identification methods. It presents off-line identification methods, primarily important from the perspective of the methods applied in the book, and then treats recursive identification methods. Special attention is devoted to identification in closed loops, as this is the most frequently applied method in practice. This part of the book demonstrates that the identification method using *KB* parameterization is the best one.

Chapter 11 treats adaptive control and iterative tuning. First, adaptive learning algorithms are demonstrated and then the methods of iterative joint identification and control are investigated. Next, the method of triple control in a new structure is introduced, which besides the traditional joint identification and control design steps contains a third loop, which optimizes the modeling loss by computing the optimal reference signal.

The Appendix contains a mathematical summary of three subject areas: state-space methods, sampled systems, and optimization. A special section is devoted to the norms of signals and operators, and derivation of the complex algorithms applied in identification methods.

Throughout the book, when it is possible and reasonable, discussion of the *TDOF* control always appears, and the methods, when possible, are presented by *YP* and *KB* parameterization. This methodology affords general comparisons, and shows the limits of a given method, or the differences between the different methods, or their special character. The details supplied are enough to write their computer algorithms.

The university booklet [105] provides the basic knowledge for this book (to facilitate understanding, some parts are shortly included here).

The book basically follows the general English notations, although there is no unified system. Abbreviations common in the literature are also used.

1.1 PROCESS MODELS

The design and implementation of a control system is an iterative task. First the requirements for control have to be formulated, that is, the mathematical model of the process needs to be postulated based on the physical operation of the process, then the parameters of the model have to be determined by measurements or via identification procedures. Based on the requirements set and the process model, the control system needs to be designed, and then the operation of the control system needs to be checked by simulation. If necessary, the control system needs to be redesigned. During the implementation, the tuning of the controller can be further refined. (The simpler controllers are called regulators hereafter.)

Three tasks appear during the solution of a control problem [105].

The model P of the system to be controlled needs to be constructed from its signal transfer properties on the basis of the physical relationships describing the process or from the measurements of its input and output signals. This latter is called *process identification* (*ID*) (Figure 1.1.1).

If the input signal and the process P are known then the output signal can be calculated, and the properties of the system can be analyzed in the time and frequency domain or in the parameter domain (Figure 1.1.2.). This method is termed *analysis*.

If the process P is given and the output signal is prescribed (Figure 1.1.3), then the task is to determine the necessary input signal. The input signal is provided by a regulator or an algorithm, that is, finally by a control circuit. This step is called regulator design or *synthesis*.

Control engineering is an interdisciplinary subject area. To understand the operation of the process wide ranging knowledge of physics, chemistry, biology, etc. may be required. Extensive mathematical knowledge is needed for analysis, synthesis, and modeling. To investigate control systems, basic knowledge on the signals and systems, and on the properties of the feedback control systems is essential. During the design, rational considerations and basic restrictions also need to be taken into account. The design procedure includes, at the same time, economy, safety, environmental protection, etc., considerations as well. To solve a more complex control task joint efforts of several experts working on different fields are required.

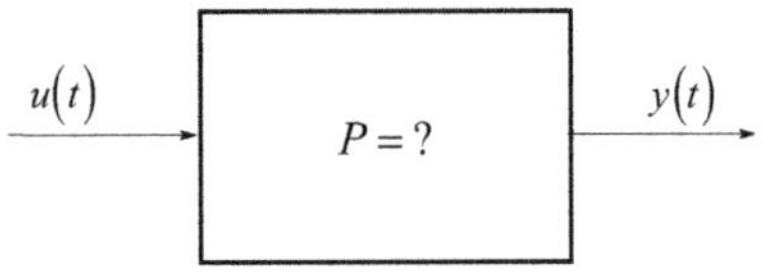

Figure 1.1.1 The identification paradigm.

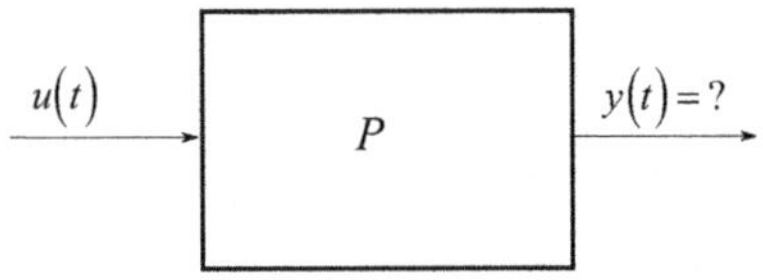

Figure 1.1.2 The analysis paradigm.

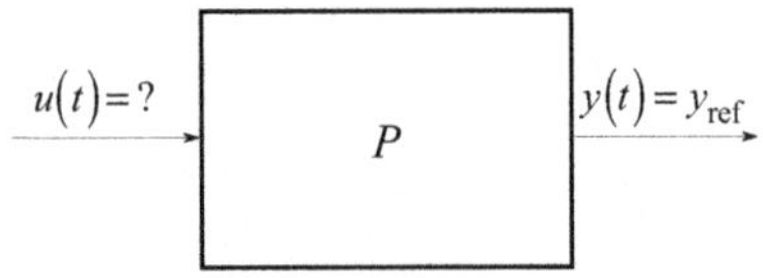

Figure 1.1.3 The synthesis paradigm.

During the implementation, the state of the process needs to be observed, the considered output needs to be measured with appropriate measuring equipment, and an intervention in the process with a proper actuator is required. The measurement noises of the sensors, the signal ranges of the actuators, the limits of the actuating effects produced must be taken into consideration. In many cases the measured data need to be transferred for longer distances, and thus data transfer must be ensured. There are standards, the so-called protocols, for data transfer. The control signal must be determined by a corresponding algorithm and forwarded to the input of the process. During the design of the control algorithm the disturbances acting on the process, the uncertainties in the process parameters, and the restrictions owed to practical realization must also be considered. The control system works with real data and a real-time signal transfer. During the signal transfer nondeterministic signal delays appear and may distort the operation. The connection and information exchange of the individual elements of the control system need to be solved with the proper interfaces.

Besides the continuous-time (CT) control systems the computer control systems are of increasing importance, as the process and the computer controller (or regulator) are connected via A/D (*analogous/digital*) and D/A (*digital/analogous*) converters. The relevant functions of the regulator are performed by the computer in real time, repeated at every sampling instance. Distributed control systems are preferred in the industrial process control where the control system elements distributed in the ground communicate in alignment with each other.

Model building is a very important stage in the design of control systems. The model describes the signal transfer properties of a system in mathematical form. With a model the static and dynamic behavior of a system can be analyzed without the need for experiments on the real system. On the basis of the model, calculations can be executed and the behavior of the system can be simulated numerically. The model can also be used for regulator design.

A model is *static* if its output depends only on the actual value of its input signal. A static system is, for example, a resistance when the input signal is the voltage and the output signal is the current. A model is *dynamic* if its output depends on previous signal values, as well. An electrical circuit consisting of a serially connected resistor and capacitor is a dynamical system as the voltage drop in the capacitance depends on the charge, and thus on the previous current values.

A model can be *linear* or *nonlinear.* The static characteristics embody the steady values of the output signal versus the steady values of the input signal. If the static characteristics are on a straight line, the system is linear; otherwise it is nonlinear.

A model can be *deterministic* or *stochastic.* The signals of a deterministic model can be described by analytical relationships (formulas, explicit, or implicit equations). In a stochastic model the signals can be given by probabilistic variables and can contain uncertainties.

In the space a model can be of *lumped* or *distributed parameters.* Lumped parameter systems can be described by ordinary differential equations, whereas distributed parameter systems can be described by partial differential equations.

A model can be a *continuous-time* (CT) or a *discrete-time* (DT) model. The CT model generally describes the relationship between its continuous input and output in the form of a differential equation. If the input and the output signals are sampled, the system is a DT or sampled data system, wherein the relationship between the input and the output signals is described by a difference equation.

Considering the number of the input and the output signals the model can be *single-input single-output* (*SISO*), *multi-input multi-output* (*MIMO*), *single-input multi-output* (*SIMO*), or *multi-input single-output* (*MISO*). Besides the input and the output signals, *state variables* of the system can also be defined. The state variables are the internal variables of the system, whose current values have evolved by means of the previous signal changes in the system. Their values cannot be changed abruptly when the input signals change abruptly. The instant value of the input signals and that of the state variables determine the further motion of the system.

The so-called state equations are widely used in the scientific and engineering fields for the description of dynamic systems. The necessity for this kind of description is explained in different ways. Perhaps the best is the recognition that the operation of a wide class of complex dynamic systems can be modeled with relatively high precision by the first-order vector differential equation:

$$\begin{aligned} \frac{\mathrm{d}\boldsymbol{x}(t)}{\mathrm{d}t} &= \dot{\boldsymbol{x}}(t) = \boldsymbol{f}[\boldsymbol{x}(t), u(t)] \\ y(t) &= g[\boldsymbol{x}(t), u(t)] \end{aligned} \tag{1.1.1}$$

The state variables of the system as scalar components, x_i, are collected in a vector $\boldsymbol{x}$ called a state vector. The system input is u, and the output is y. The dimension of $\boldsymbol{x}$ is called the degree or the order of the system. The function $\boldsymbol{f}(\boldsymbol{x},u)$ means the varying "speed" of the state vector as the function of the states and the input signal. The function $g(\boldsymbol{x},u)$ is called the sensor or measurement function since it provides the output of the system. Note that here $\boldsymbol{f}(\boldsymbol{x},u)$ and $g(\boldsymbol{x},u)$ do not depend on time in an explicit way. (We emphasize, that nevertheless the signals of the state equations obviously depend on time!) These kinds of systems are called *time-invariant* (*TI*) systems. The state variables contain information on the history of the system, and the future values of the signals can be predicted; therefore the state vector behaves like the memory of the system.

This book basically concentrates on *SISO* processes, but in Chapter 7 the extension of certain control systems to *MIMO*, that is, multivariable cases, is also discussed.

In engineering systems, state vectors are often related to the basic physical processes, whereby the relations necessary to describe the storage of the mass, flow, impulse, and power need to be determined. (It needs to be noted, however, that in certain fields, e.g., chemistry, the definition of the state vector is different from the above general system theoretical concept, as it mostly reflects other variables—like pressure, temperature, composition, etc.—representing the physicochemical state of the investigated material, mixture, compound, etc.)

State variables as coordinates define a space (*state space*). The state vector $\boldsymbol{x}(t)$ is interpreted in this space. The motion of the end point of the vector means the motion of the system. The curve described by the motion of the end point of the state vector gives the state trajectory.

A special class of the nonlinear dynamic systems is given by Equations (1.1.1), whose possible equilibrium state is $(\boldsymbol{x}_o,u_o)$ (where $\dot{\boldsymbol{x}} = \boldsymbol{0}$) is obtained from the equation

$$\boldsymbol{f}(\boldsymbol{x}_o, u_o) = \boldsymbol{0} \tag{1.1.2}$$

(Remark: in general cases, several equilibrium states can be obtained. These equilibrium states can provide different stable states. The performance of these states requires the investigation of the second-order derivatives of $\boldsymbol{f}(\boldsymbol{x},u)$.)

Static systems can be described by degenerated state equations, since they do not have memory, or corresponding states, so they can be described only by the second equation of (1.1.1)

$$y = g(u) \tag{1.1.3}$$

By Taylor series expansion in the point u_o, we get

$$y = g(u_o) + \frac{dg(u_o)}{du}(u - u_o) + \ldots = g(u_o) + g'(u_o)(u - u_o) + \ldots \tag{1.1.4}$$

and the linearized model

$$y - y_o = \Delta y = y - g(u_o) = g'(u_o)(u - u_o) = g'(u_o)\Delta u \tag{1.1.5}$$

can be obtained from the first-order term of (1.1.4).

The linearized model substitutes the original curve with its tangent in the working point u_o and establishes a static linear connection between the changes $(\Delta y, \Delta u)$ around the working point. (Naturally, an arbitrary point can be chosen as working point u_o beyond the equilibrium point.)

Actually, the linearization of the state-space equation (1.1.1) can also be given in a very similar way. Using the following notation valid for the changes around the equilibrium state $(\boldsymbol{x}_o, u_o)$, that is,

$$\boldsymbol{x} = \boldsymbol{x}_o + \Delta\boldsymbol{x}; \quad u = u_o + \Delta u; \quad y = y_o + \Delta y \tag{1.1.6}$$

let us calculate the first-order linearized approach of (1.1.1)

$$\begin{aligned} \frac{d\boldsymbol{x}}{dt} &= \boldsymbol{f}(\boldsymbol{x}_o + \Delta\boldsymbol{x}, u_o + \Delta u) \approx \boldsymbol{f}(\boldsymbol{x}_o, u_o) + \frac{d\boldsymbol{f}(\boldsymbol{x}_o, u_o)}{d\boldsymbol{x}^T}\Delta\boldsymbol{x} + \frac{d\boldsymbol{f}(\boldsymbol{x}_o, u_o)}{du}\Delta u \\ y &= g(\boldsymbol{x}_o + \Delta\boldsymbol{x}, u_o + \Delta u) \approx g(\boldsymbol{x}_o, u_o) + \frac{dg(\boldsymbol{x}_o, u_o)}{d\boldsymbol{x}^T}\Delta\boldsymbol{x} + \frac{dg(\boldsymbol{x}_o, u_o)}{du}\Delta u \end{aligned} \tag{1.1.7}$$

Use the fact that in the equilibrium point $\boldsymbol{f}(\boldsymbol{x}_o, u_o) = \boldsymbol{0}$ and introduce the notation $y_o = g(\boldsymbol{x}_o, u_o)$, so the linearized model valid for small changes obtains the form

$$\begin{aligned} \frac{d(\boldsymbol{x} - \boldsymbol{x}_o)}{dt} &= \frac{d\Delta\boldsymbol{x}}{dt} = \boldsymbol{A}(\boldsymbol{x} - \boldsymbol{x}_o) + \boldsymbol{b}(u - u_o) = \boldsymbol{A}\Delta\boldsymbol{x} + \boldsymbol{b}\Delta u \\ y - y_o &= \Delta y = \boldsymbol{c}^T(\boldsymbol{x} - \boldsymbol{x}_o) + d(u - u_o) = \boldsymbol{c}^T\Delta\boldsymbol{x} + d_c\Delta u \end{aligned} \tag{1.1.8}$$

where the following notations are introduced

$$A = \frac{\mathrm{d}f(x_o, u_o)}{\mathrm{d}x^T}; \quad b = \frac{\mathrm{d}f(x_o, u_o)}{\mathrm{d}u}$$
$$c^T = \frac{\mathrm{d}g(x_o, u_o)}{\mathrm{d}x^T}; \quad d_c = \frac{\mathrm{d}g(x_o, u_o)}{\mathrm{d}u} \qquad (1.1.9)$$

The obtained model is a *linear time-invariant* (*LTI*) system, that is, the parameters do not change in time. It is a widely applied practice: the original variables x,u,y are used instead of the small changes (Δx,Δu,Δy) for simplicity, but they are considered as the changes around the working point. In this way we arrive at the *LTI* state-space equation of the system generally applied in systems and control theory

$$\frac{\mathrm{d}x(t)}{\mathrm{d}t} = Ax(t) + bu(t) \qquad \text{or simply} \qquad \frac{\mathrm{d}x}{\mathrm{d}t} = Ax + bu$$
$$y(t) = c^T x(t) + d_c u(t) \qquad\qquad y = c^T x + d_c u \qquad (1.1.10)$$

Here the parameter matrices of the system are A,b,c^T,d. Since this book largely treats *SISO* systems, in n-order case, matrix A means a $(n \times n)$ square matrix, which is the so-called state matrix, b is a column vector of $(n \times 1)$ size, c^T is a row vector of $(1 \times n)$ size, and d_c is scalar. The block diagram of the state equations (1.1.10) can be seen in Figure 1.1.4.

The most important relationships of the state equation description are summarized in Appendix A.2.

This book deals mainly with linear processes, but the extension of control systems for special nonlinear processes is discussed in Chapter 8.

The classical model of the dynamic *LTI* processes, the transfer function, $P(s)$ defined by the ratio of the Laplace transforms of the output and input signals, can be easily derived from the state equation (1.1.10) (see Appendix A.2)

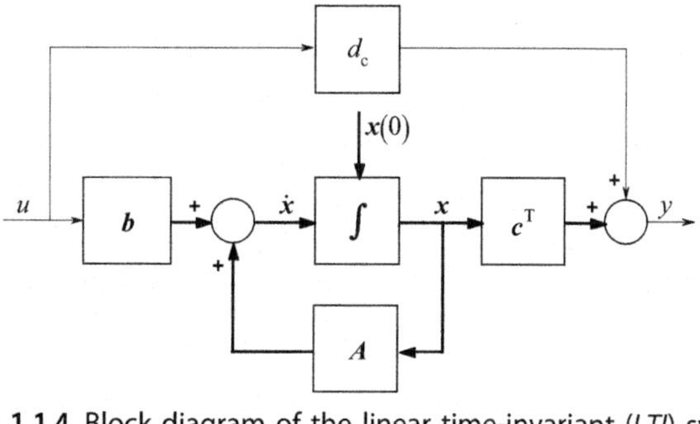

Figure 1.1.4 Block diagram of the linear time-invariant (*LTI*) system.

$$P(s) = \frac{Y(s)}{U(s)} = \mathbf{c}^{\mathrm{T}}(s\mathbf{I} - \mathbf{A})^{-1}\mathbf{b} + d_{\mathrm{c}} = \frac{\mathcal{B}(s)}{\mathcal{A}(s)} \tag{1.1.11}$$

where

$$\begin{aligned} \mathcal{A}(s) &= \det(s\mathbf{I} - \mathbf{A}) = s^n + a_1 s^{n-1} + \ldots + a_n \\ \mathcal{B}(s) &= b_{\mathrm{o}} s^m + b_1 s^{m-1} + \ldots + b_m \end{aligned} \tag{1.1.12}$$

The roots of the equation $\mathcal{A}(s) = 0$ are called poles; the roots of $\mathcal{B}(s) = 0$ are called zeros. A CT linear process is stable, if all roots of the polynomial $\mathcal{A}(s)$ are located on the left-hand side of the complex plane. Let us note the set of the CT linear stable processes by $\mathcal{S}$. Concerning the order of the polynomials $\mathcal{A}(s)$ and $\mathcal{B}(s)$ it should be noted that the number of the state variables is n, m is the order of the polynomial $\mathcal{B}(s)$, and the relation $m \leq n$ exists. The difference between the order of the numerator and denominator $p_{\mathrm{T}} = n - m$ is called *pole access.* If $p_{\mathrm{T}} > 0$, then $P(s)$ is strictly proper, if $p_{\mathrm{T}} = 0$, then the transfer function is proper. In the practice arbitrary relation $0 \leq p_{\mathrm{T}} \leq n$ might occur.

The first term of $P(s)$ (if $d_{\mathrm{c}} = 0$) is strictly proper since it consists of the linear combination of only proper elements (see (A.2.13), that is, the order of the adjunct is always lower than that of the determinant). If $d_{\mathrm{c}} \neq 0$, then $P(s)$ is proper, that is, the order of the numerator is equal to that of the denominator. The physical meaning of d_{c} is how the input directly influences the output without any dynamics. Note that this effect does not disappear even for very high frequencies, and thus $P(j\omega \to \infty) = d_{\mathrm{c}}$. This means, at the same time, that the jump of the transfer function at time $t = 0$ is $v(t = 0) = d_{\mathrm{c}}$. In practice, the case $d_{\mathrm{c}} \neq 0$ is usually traced back to the case $d_{\mathrm{c}} = 0$ by introducing a new output $\tilde{y} = y - d_{\mathrm{c}}u$. The case $d_{\mathrm{c}} \neq 0$ can also be considered as a nonperfect linearization, which needs certain correction. In the case $p_{\mathrm{T}} = 0$ the initial value of the step response is d_{c}.

Note that if the transfer function of the system is given, then first the common divisors of the numerator and denominator polynomials must be investigated. The common factor can only be a common root. It is reasonable to continue the simplification until there is no more common divisor. These polynomials are called relative prime. A transfer function $P(s) = \mathcal{B}(s)/\mathcal{A}(s)$ is called nonreducible (cannot be simplified) if the polynomials $\mathcal{A}(s)$ and $\mathcal{B}(s)$ in (1.1.11) are relative prime, whose algebraic condition is such that the special Diophantine (or Bezout) equation

$$\mathcal{A}(s)\mathcal{X}(s) + \mathcal{B}(s)\mathcal{Y}(s) = 1 \tag{1.1.13}$$

must have a solution, that is, the corresponding Sylsvester matrix must be regular [38,120] (see more details in Chapter 4).

In practice, there are *LTI* processes which have a special property that cannot be represented by rational fractional functions. This is the pure dead-time. The ideal dead-time lag delays its input signal by dead-time T_d. The transfer function of the dead-time lag is e^{-sT_d}, which is not a rational fractional function. In this case the transfer function of the process has the form

$$P(s) = \frac{\mathcal{B}(s)}{\mathcal{A}(s)} e^{-sT_d} \tag{1.1.14}$$

A right-side (unstable) pole modifies the Bode phase–frequency diagram by positive, and a right-side zero modifies it by negative phase angles; thus the zero does not decrease but increases the negative phase angle. (This property motivated the denomination of nonminimum-phase systems.)

To illustrate the nonminimum-phase property, let us consider the following two transfer functions:

$$H_a(s) = \frac{1 + sT}{1 + sT_1} \quad \text{and} \quad H_b(s) = \frac{1 - sT}{1 + sT_1} \tag{1.1.15}$$

For positive T_1 and T values both systems are stable. H_a has a stable zero, whereas H_b has an unstable zero. The amplitude–frequency functions of the two systems are the same

$$a(\omega) = \sqrt{\frac{1 + (\omega T)^2}{1 + (\omega T_1)^2}} \tag{1.1.16}$$

but their phase angles differ:

$$\varphi_a(\omega) = -\arctan\frac{\omega(T_1 - T)}{1 + \omega^2 T_1 T} \quad \text{and} \quad \varphi_b(\omega) = -\arctan\frac{\omega(T_1 + T)}{1 + \omega^2 T_1 T}$$

where $|\varphi_a(\omega)|$ is less than $|\varphi_b(\omega)|$ in each frequency range.

Every nonminimum-phase system H_{nmp} can be converted to the product of a so-called all-pass H_{ap} and a minimum-phase H_{mp} element.

$$H_{nmp}(s) = \frac{1 - sT}{1 + sT_1} = \frac{1 - sT}{1 + sT}\frac{1 + sT}{1 + sT_1} = H_{ap}(s)H_{mp}(s) \tag{1.1.17}$$

These systems are called nonminimum-phase systems and their zeros are located on the right-hand side of the complex plane. If a system is of minimum phase, that is, the zeros of its transfer function are on the left-hand side of the complex plane, then the phase angle belonging to the poles is negative, and the phase angle belonging to the zeros is positive. Thus, the phase angle curve can be unambiguously assigned to the asymptotic amplitude curve.

The property of an all-pass nonminimum-phase element is that its absolute frequency function is a unity constant at all frequencies. The transfer function of the n-order all-pass element in the case of real poles is

$$H_{\mathrm{ap}}(s) = \prod_{i=1}^{n} \frac{1 - sT_i}{1 + sT_i} = \prod_{i=1}^{n} \frac{s - s_i}{s + s_i} \tag{1.1.18}$$

Nonminimum-phase systems show unusual behavior in the time domain. For example, in the case of one right-hand side zero the unit step response starts in the direction opposite to its steady state value, finally changing direction and reaching its steady state.

The dead-time which appears very often in CT systems can be expressed by the form $e^{-sT_{\mathrm{d}}}$ which is a nonrational fractional function. This causes problems for simulation and design methods, which is why the Pade approximation is usually used, which approximates the transfer function of the dead-time by nonminimum-phase rational fractions, whereby the first elements of the Taylor expansion are equal to the first elements of the Taylor expansion of the exponential expression of the transfer function of the dead-time element.

The nth order Pade approximation gives a rational fraction where there are n number of zeros and n number of poles differing only in their sign.

$$H_{\mathrm{H}}(s) = e^{-sT_{\mathrm{d}}} \approx \frac{(s - s_1)\ldots(s - s_n)}{(s + s_1)\ldots(s + s_n)} = H_{\mathrm{Pade}}(s) \tag{1.1.19}$$

The absolute value of the frequency function of the rational fraction for any $s = j\omega$ is one; similarly to the frequency function of the dead-time element only its phase angle changes with the frequency. Such elements are called all-pass filters. The Pade approximation is a nonminimum-phase rational fraction. The s_i poles or the coefficients of the numerator and the denominator can be determined by taking the first $N+M+1$ terms of the Taylor expansion of the transfer function $H_{\mathrm{H}}(s)$ equal to the first $N+M+1$ terms of the rational fraction transfer function $H_{\mathrm{Pade}}(s)$. Here M is the degree of the numerator and N is the degree of the denominator of the rational fraction ($M \leq N$).

$$e^{-x} = \sum_{i=0}^{\infty} b_i x^i \approx \frac{\sum_{k=0}^{M} d_k x^k}{\sum_{j=0}^{N} c_j x^j} \tag{1.1.20}$$

In the equation, there are $N+M+2$ unknown coefficients. Therefore choosing $c_0=1$, for the remaining $N+M+1$ parameters, considering the condition mentioned above, we can write $N+M+1$ linear equations. With this method in the case $N=M=3$, the following form is obtained:

$$e^{-x} \approx \frac{1-\frac{1}{2}x+\frac{1}{10}x^2-\frac{1}{120}x^3}{1+\frac{1}{2}x+\frac{1}{10}x^2+\frac{1}{120}x^3} \tag{1.1.21}$$

Expressions of the first- and second-order Pade approximations according to similar calculations are as follows:

$$e^{-x} \approx \frac{1-\frac{1}{2}x}{1+\frac{1}{2}x}; \quad e^{-x} \approx \frac{1-\frac{1}{2}x+\frac{1}{12}x^2}{1+\frac{1}{2}x+\frac{1}{12}x^2} \tag{1.1.22}$$

The approximations are summarized in Table 1.1. The higher the degree of the rational fraction, the more the terms are identical in the Taylor approximations of the two functions.

Figure 1.1.5 shows the unit step responses of the Pade approximation for the first-, second-, and third-order cases. It is seen that the approximations do not fit the initial point, but approximate the steady state well. With higher-order approximation the step responses fit the step response of the dead-time element better.

The basic property of stable linear systems is that sinusoidal input signals in *steady state* respond with sinusoidal output signals of the same frequency as that of the input signal, after the deceasing of the transients. The amplitude and the phase angle of the output signal, however, depend on that frequency.

Table 1.1 First-, second-, and third-order Pade approximations of the dead-time element e^{-sT_d}

Dead-time element	$H(s) = e^{-sT_d}$
First-order Pade approximation	$H(s) \approx \frac{2-sT_d}{2+sT_d}$
Second-order Pade approximation	$H(s) \approx \frac{12-6sT_d+(sT_d)^2}{12+6sT_d+(sT_d)^2}$
Third-order Pade approximation	$H(s) \approx \frac{120-60sT_d+12(sT_d)^2-(sT_d)^3}{120+60sT_d+12(sT_d)^2+(sT_d)^3}$

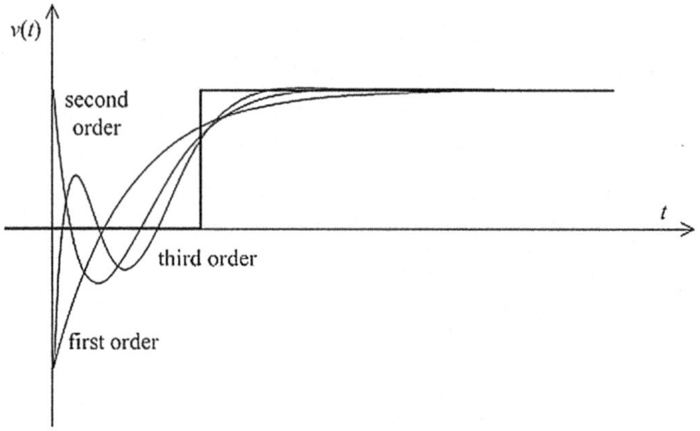

Figure 1.1.5 Unit step responses of first-, second-, and third-order Pade approximations of a dead-time element.

Let the input signal of the system be $u(t) = A_u \sin(\omega t + \varphi_u)$, $t \geq 0$. The output signal is

$$y(t) = y_{steady}(t) + y_{tranzient}(t) \tag{1.1.23}$$

The output signal in steady (quasi-stationary) state is

$$y_{steady}(t) = A_y \sin(\omega t + \varphi_y)$$

The *frequency function* is a complex function representing the frequency dependence of two system properties, the *amplitude ratio* A_y/A_u and the *phase difference* $(\varphi_y - \varphi_u)$. It can be proven that *formally* the frequency function can be derived from the transfer function by substituting $s = j\omega$.

$$H(j\omega) = H(s)|_{s=j\omega} = |H(j\omega)|e^{j\varphi(\omega)} = a(\omega)e^{j\varphi(\omega)} \tag{1.1.24}$$

In the frequency function the expressions of the *amplitude function* $a(\omega)$ (the absolute value of the frequency function) and the *phase function* $\varphi(\omega)$ (the phase angle of the frequency function) are

$$a(\omega) = |H(j\omega)| = \frac{A_y(\omega)}{A_u(\omega)} \quad \text{and}$$
$$\varphi(\omega) = \arg\{H(j\omega)\} = \varphi_y(\omega) - \varphi_u(\omega) \tag{1.1.25}$$

The frequency function is the Fourier transform of the impulse response if it exists.

The frequency function can be plotted in several forms. A Nyquist diagram draws the frequency function in the complex plane as a polar diagram. For each value of the frequency function in the selected frequency range, a point can be given in the complex plane corresponding to the pair of values $a(\omega)$ and $\varphi(\omega)$. Connecting these points by a contour gives us the Nyquist diagram. For plotting the Nyquist diagram, the frequency is generally taken to be between zero and infinity (Figure 1.1.6). The arrow shows the direction of the increasing frequency parameter. Often the curve is supplemented by values calculated for negative frequencies. In this case the diagram is called a complete Nyquist diagram. The part of the diagram given for the frequency range $-\infty < \omega < 0$ (indicated by a dotted line in Figure 1.1.6.) is the mirror image of the curve plotted for positive frequencies related to the real axis. The Nyquist diagram can also be considered as conform mapping of the straight line $s = j\omega$, $-\infty < \omega < \infty$ according to function $H(s)$.

The shape of the Nyquist diagram always characterizes the system. Analysis of the Nyquist diagram means a qualitative picture of the system. Investigating it important system properties (e.g., stability) can be derived.

The Bode diagram simultaneously plots the absolute value $a(\omega)$ and the phase angle $\varphi(\omega)$ of the frequency function versus the frequency in a given frequency range (Figure 1.1.7). Generally the frequency scale is logarithmic

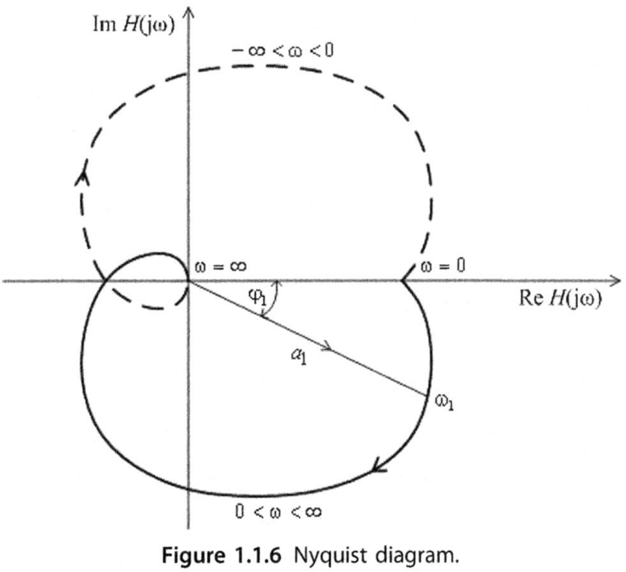

Figure 1.1.6 Nyquist diagram.

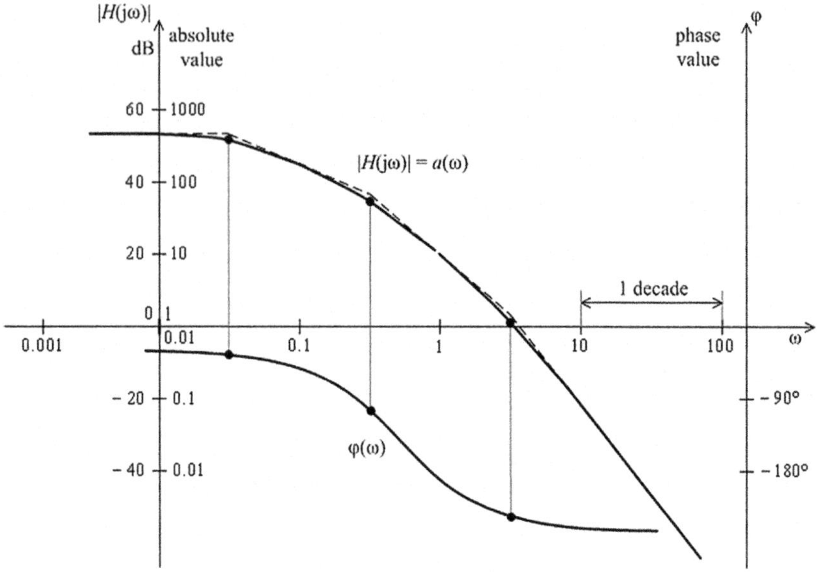

Figure 1.1.7 Bode diagram.

in order to cover a wide frequency range. The frequency range when the frequency is changed by a factor of 10 is called a decade. (In music an octave is used which gives a frequency band when the frequency is changed by a factor of two.) The absolute value—following telecommunication tradition—is scaled in decibels. The decibel (dB) constitutes the decimal logarithm of a number multiplied by 20. The phase angle is drawn in linear scale.

The advantage of the Bode diagram, on the one hand, is that multiplying individual frequency function components—because of the logarithmic scale—the Bode diagrams of the individual components are simply added. On the other hand, a further advantage is that generally the Bode amplitude–frequency diagram can be approximated by its asymptotes. The course and breakpoints of the asymptotic amplitude–frequency curves allow quick evaluation of the fundamental system properties.

Sampled Data Systems

The book discusses both the CT and the sampled DT systems in terms of their advantages and disadvantages. One advantage is that the repetition of all relationships can be avoided since they differ from each other for CT or DT systems only in the applied notation or formalism. Joint discussion shows the vital differences. A disadvantage is, however, that the reading and understanding of the relationships requires more effort, but the unambiguous notation helps considerably. The most important properties of the sampled data systems are summarized in Section A.3 of the Appendix.

The model of the process, giving the relationship between the input $u[k]$ and output $y[k]$ signals sampled by sampling time T_s, can be given by the so-called pulse transfer operator

$$G(q) = \mathbf{c}^{\mathrm{T}}(q\mathbf{I} - \mathbf{F})^{-1}\mathbf{g} + d_{\mathrm{c}} = \frac{\mathcal{G}(q)}{\mathcal{F}(q)} \tag{1.1.26}$$

(where q is the generalized "shift" operator) or by the equivalent pulse transfer function

$$G(z) = \mathbf{c}^{\mathrm{T}}(z\mathbf{I} - \mathbf{F})^{-1}\mathbf{g} + d_{\mathrm{c}} = \frac{\mathcal{G}(z)}{\mathcal{F}(z)} \tag{1.1.27}$$

The essence of the DT models is that the following difference equation gives the connection between $u[k]$ and $y[k]$

$$\begin{aligned} y[k] = g_1 u[k-1] + g_2 u[k-2] + \ldots + g_n u[k-n] - f_1 y[k-1] - \ldots \\ - f_n y[k-n] \end{aligned} \tag{1.1.28}$$

where $k = 0,1,2,\ldots$ means the serial number of the sampling, that is, the DT. For correctness, it should be mentioned that although the form $G(z)$ gives the connections between the Dirac delta ($\delta[k]$) series of values $u[k]$ and $y[k]$, the $G(q)$ does the same but between the signal series $u[k]$ and $y[k]$. For simplicity, the form $G(z)$ will be used. The roots of the equation $\mathcal{F}(z) = 0$ are called poles; the roots of $\mathcal{G}(z) = 0$ are called zeros.

Comparing the variables of the z and Laplace transforms, we obtain the following relation

$$z = e^{sT_s} \tag{1.1.29}$$

It can easily be seen that the stability region of the CT systems in the complex frequency domain $s = \sigma + j\omega$ (left-hand side plane) is mapped by $z = e^{sT_s}$ in the unit disc in the z complex plane, since the line $s = j\omega$ $(-\infty < \omega < \infty)$ is mapped in the unit disc $z = e^{j\omega T_s}$. Furthermore, even for very high frequencies the stability region will also stay on the left-hand side. More precisely, the sampling maps the main region $-\pi/T_s < \omega < \pi/T_s$ of the left-hand side of the s plane in the unit disc (in the other regions outside this main band, because of the property of the exponential function, the values are repeated periodically). The two lines parallel to the negative real axis in the s plane at $\omega = \pm j\pi/T_s$ are transformed into one single line, that is, into the negative real axis of the z plane. Thus the poles according to the limit of the Shannon sampling law are transformed to the negative real axis of the z plane.

The lines belonging to the complex conjugated roots with given constant attenuating factor of a CT system are mapped into a special "heart form curve" (see Figure 2.3.3). A DT linear process is stable, if all roots of the polynomial $\mathcal{F}(z)$ are inside the unit disc of the complex plane. Next, the set of linear stable processes is marked by $\mathcal{S}_{\mathrm{DT}}$.

The dead-time e^{-sT_d} of the CT system can be given in the form (A.3.41)

$$z^{-d}; \quad q^{-d} \tag{1.1.30}$$

where

$$d = \mathrm{int}\left\{\frac{T_\mathrm{d}}{T_\mathrm{s}}\right\} \tag{1.1.31}$$

Here int{...} means the separation of the integer part. z^{-d} corresponds to the general delay relationship by (A.3.17). The z^{-1} and q^{-1} terms in the expression of $G(z^{-1})$ and $G(q^{-1})$, respectively, are not part of the z^{-d} and q^{-d} terms representing the time delay. Thus for a process containing actual time delay it is reasonable to use the pulse transfer function

$$G(z^{-1}) = \frac{g_1 + g_2 z^{-1} + \ldots + g_n z^{-(n-1)}}{1 + f_1 z^{-1} + f_2 z^{-2} + \ldots + f_n z^{-n}} z^{-(d+1)} \tag{1.1.32}$$

The function $G(z)$ or the equivalent pulse transfer function $G(z^{-1})$ can be obtained by *step response equivalent* (*SRE*) transformation from the transfer function $P(s)$ of a CT process, if a zero-order holding is applied in the DT system.

The *SRE* transformation of the first-order lag $H(s) = K/(1 + sT)$ provides the first-order pulse transfer function $G(z) = g_1/(z + f_1) = g_1 z^{-1}/(1 + f_1 z^{-1})$ (see A.3.45). Note that the *SRE* transformation always gives pole excess $p_\mathrm{T} = 1$ for $G(z)$ in DT case for any arbitrary pole excess of $P(s)$ in CT case.

What the allocation of poles and zeros of an nth order pulse transfer function $G(z)$ looks like in the case of *SRE* transformation is an interesting issue. The transformation of the poles follows the result obtained for first-order lag (see Section A.3 of the Appendix (A.3.45) and (A.3.46)), that is, in the case of real poles $p_i^\mathrm{d} = e^{-x_i}$, where $x_i = T_s/T_i$. The pole of the CT system is $-1/T_i$. This completely corresponds to (1.1.29). (The relationship is valid also for complex poles; however, although it is reasonable to handle those poles as complex conjugate pairs the transformation relationship is much more complex.) The exponential relationship maps the stable CT poles to stable DT poles; thus the complete CT stability region (left-hand side half plane) is transformed into the stable DT region (inside the unit disc). Similarly, the unstable CT poles (right-hand side half plane) are transformed into the unstable DT poles (outside the unit disc). The transformation of the zeros, however, does not follow this pattern. $G(z)$ always has $(n-1)$ zeros (see above about the pole excess). Provided that the CT system has only m stable zeros, then m out of the $(n-1)$ DT zeros are transformed approximately by $z_i^{\mathrm{d}} \approx e^{-x_i}$ (the exact relationship is far more complicated). The rest of the zeros—there are $(n-m-1)$ of them—follow a different law, as those zeros have no CT counterparts. Typically, these (unmatched) zeros are located on the real axis.

Let the transfer function of a CT system be

$$H(s) = K\frac{\prod_{i=1}^{m}(1+s\tau_i)}{\prod_{i=1}^{n}(1+sT_i)} \tag{1.1.33}$$

whose high frequency approximation is

$$H(\infty) \approx K\frac{\prod_{i=1}^{m}\tau_i}{\prod_{i=1}^{n}T_i}\frac{1}{s^{n-m}} = K\frac{\prod_{i=1}^{m}\tau_i}{\prod_{i=1}^{n}T_i}\frac{1}{s^{p_{\mathrm{T}}}} \tag{1.1.34}$$

This model, valid for small sampling time ($T_{\mathrm{s}} \to 0$), is an integrator of the order $p_{\mathrm{T}} = n - m$ [9,53,72]. The general relationship (A.3.33) of the *SRE* transformation is now [72]

$$\frac{z}{z-1}G(z) = \mathcal{Z}\left\{\mathcal{L}^{-1}\left[\frac{1}{s^{p_{\mathrm{T}}+1}}\right]_{t=kT_{\mathrm{s}}}\right\} = \frac{(-1)^{p_{\mathrm{T}}}}{p_{\mathrm{T}}!}\frac{\partial^{p_{\mathrm{T}}}}{\partial a^{p_{\mathrm{T}}}}\left[\frac{z}{z-e^{-aT_{\mathrm{s}}}}\right]_{a=0} \tag{1.1.35}$$

(The z-transform expression applied here can be obtained from the well-known tables.) The pulse transfer function of the DT system from (1.1.35) is

$$G(z) = \frac{(-1)^{p_{\mathrm{T}}}(z-1)}{p_{\mathrm{T}}!z}\frac{\partial^{p_{\mathrm{T}}}}{\partial a^{p_{\mathrm{T}}}}\left[\frac{z}{z-e^{-aT_{\mathrm{s}}}}\right]_{a=0} \tag{1.1.36}$$

The obtained transfer functions $G(z)$ and their zeros are shown in Table 1.2.

Table 1.2 The obtained trasfer functions and their zeros

Δ	G(z)	Zeros
1	$\frac{T_s}{z-1}$	—
2	$\frac{T_s^2(z+1)}{2(z-1)^2}$	-1
3	$\frac{T_s^3(z^2+4z+1)}{6(z-1)^3}$	$-2+\sqrt{3}$ $-2-\sqrt{3}$
4	$\frac{T_s^4(z+1)^3}{24(z-1)^4}$	$-1, -1, -1$

It can be seen from the table that in the case of $T_s \to 0$ and pole excess $p_T = 3$, we obtain an unstable zero.

If the pole excess in the transfer function of the CT system is higher than two, then we have to assume, even in the case of minimum-phase systems (stable zeros), that among the $(n-m-1)$ zeros of the DT systems will be an unstable zero. (For detailed calculations see the derivation of the numerator for the general case in Section A.3 of the Appendix.)

Similarly, an important observation is that for fractional time delay unstable zeros may appear in $G(z)$. Consider now the transfer function of a first-order CT lag with dead-time

$$H(s) = \frac{K}{1+sT}e^{-sT_d} \tag{1.1.37}$$

where the dead-time T_d is not an integer multiple of the sampling time T_s. Let $T_d = dT_s + \varepsilon T_s$, where d is the integer DT time delay. Now the modified z-transform must be applied for the *SRE* transformation. Using the result of Åström and coworkers [9] and Vajk [152], we obtain the DT form

$$\begin{aligned} G(z) &= Kz^{-d}\frac{1-p}{z-p}\frac{(1-\alpha)z+\alpha}{z} = K(1-p)(1-\alpha)\frac{z-q}{z-p}z^{-d+1} \\ &= k\frac{z-q}{z-p}z^{-d'} \end{aligned} \tag{1.1.38}$$

where

$$p = e^{-T_s/T_d}; \quad \alpha = p\frac{p^{-\varepsilon}-1}{1-p} \cdot q = -\frac{\alpha}{1-\alpha} \tag{1.1.39}$$

$$k = K(1-p)(1-\alpha); \quad d' = d+1$$

It can be checked easily whether the zero of $G(z)$

$$q = -\frac{\alpha}{1-\alpha} = \frac{(1-p^{-\varepsilon})p}{1-p^{1-\varepsilon}} \tag{1.1.40}$$

is unstable, since it can have the value $0 \le q \le \infty$ in the region $0 \le \varepsilon \le 1$.

As unstable zeros (inverse instability) play an important role in the DT regulator design procedure, it should be noted that whereas for CT systems only very special technologies, processes have nonminimum-phase character, for DT processes it is rather common, so proper regulator design methods have to be chosen.

On the Uncertainty of Process Models

The knowledge of a process is never exact, independent of the method—whether measurement-based ID or physicochemical theoretical considerations—by which its model is determined. The uncertainty of the plant can be expressed by the absolute model error

$$\Delta P = P - \hat{P} \tag{1.1.41}$$

and the relative model error

$$\ell = \frac{\Delta P}{\hat{P}} = \frac{P-\hat{P}}{\hat{P}} \tag{1.1.42}$$

where $\hat{P}$ is the available nominal model intended for regulator design and P is the real plant (see Figure 1.1.8(a)). Besides these classical error categories another two preferred uncertainty models are applied in the references. The application of these models was probably inspired by design purposes to weight the real errors in certain frequency regions.

Additive Uncertainty Model

Let us decompose the classical model error Δ to the product of two factors: $\Delta = \Delta_a W_a$, where Δ_a is the normalized additive error ($|\Delta_a| \le 1$) and the

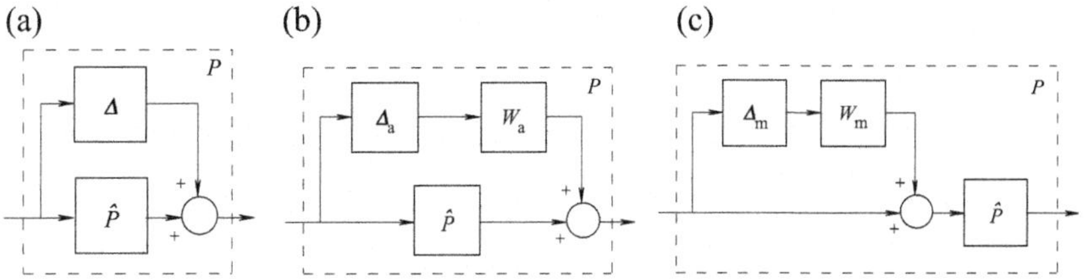

Figure 1.1.8 Uncertainty models.

weight W_a covers the design requirements. The resulting additive uncertainty model is shown in Figure 1.1.8(b).

Multiplicative Uncertainty Model

The multiplicative model shown in Figure 1.1.8(c) can easily be obtained if the Δ_m is the normalized multiplicative error ($|\Delta_m| \leq 1$) and W_m contains the design considerations.

The relationship between the three model types can be given by

$$P = \hat{P} + \Delta = \hat{P} + \Delta_a W_a = \hat{P}(1 + \Delta_m W_m) \tag{1.1.43}$$

Consequently the additive error is

$$\Delta = P - \hat{P} = \Delta_a W_a = \hat{P} \Delta_m W_m \tag{1.1.44}$$

and the relative model error is

$$\ell = \frac{\Delta}{\hat{P}} = \Delta_m W_m = \frac{\Delta_a W_a}{\hat{P}} \tag{1.1.45}$$

It should be mentioned that further inverse additive (a) and inverse multiplicative (b) models can be constructed as shown in Figure 1.1.9.

In these models $|\Delta_{ia}| \leq 1$ and $|\Delta_{im}| \leq 1$ are the normalized errors and W_{ia}, W_{im} are the design weights.

The goal of the ID is always to approximate the process P through its model $\hat{P}$. The process is never known completely; only a close model class P^* can be applied what is known. In practice, it is almost always assumed that $P \approx P^* = g(u,\boldsymbol{p})$, and thus the process belongs to an applied model class, where the parameter vector is denoted by $\boldsymbol{p}$.

As a consequence the model $\hat{P}$ used in the control design is not equal to the theoretical model P because our knowledge is incomplete or a simplified model is used. It is assumed that $P = B\hat{P}$ where B is termed as bias factor. It is clear from the comparison of the bias factor and the

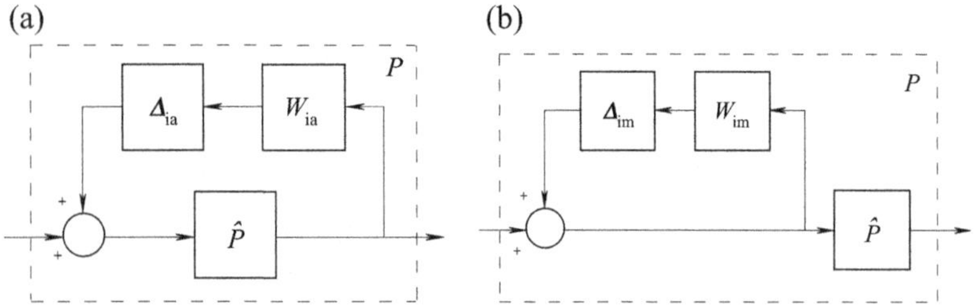

Figure 1.1.9 Further uncertainty models.

multiplicative model that the expression $\Delta_m W_m = B - 1$ means the relationship between the two types of approaches.

1.2 CLOSED-LOOP CONTROL

The application of a negative feedback is not necessary to realize a control system. If there is enough a priori knowledge about the process then the open control loop can be used as shown in Figure 1.2.1, where P means the transfer function of the process and C is the transfer function of the regulator. In the figure, the reference signal is r, y is the output signal (controlled signal), u is the actuating signal, and y_{ni} and y_{no} are the input and output noises, respectively. This scheme is called open-loop control.

The reference signal tracking would be ideal if the regulator could perform the inverse of the process transfer function, but the perfect inverse usually cannot be realized. The control can provide the reference signal tracking, but it cannot eliminate the effects of disturbances.

The control is performed via negative feedback, if the input signal (manipulated variable or actuating signal) of the process is influenced by the so-called error signal e, which is the deviation between the measured and desired value of the process output signal. Based on the error signal e, the regulator C generates the manipulated variable u, which modifies the output of the process P. The output signal of the process changes according to the dynamics of the process until the output signal reaches its prescribed value. The negative feedback-based regulation is called closed-loop control. The block scheme of such a control system is given in Figure 1.2.2. In many cases the reference signal is filtered via a term of transfer function F (denoted by the dashed line in the figure).

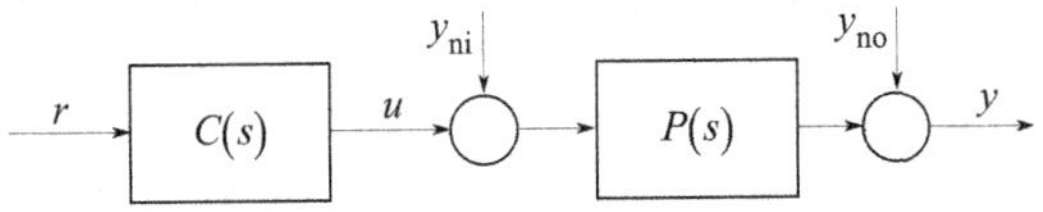

Figure 1.2.1 The block scheme of open-loop control.

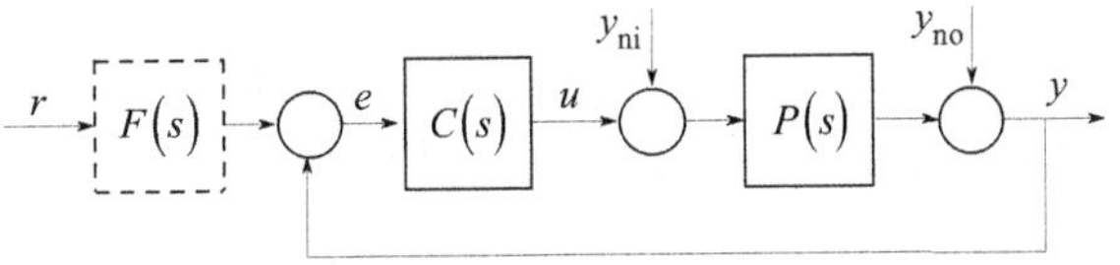

Figure 1.2.2 The block scheme of the closed-loop control system.

The relationships between the signals in Figure 1.2.2 can be given by the following expressions:

$$Y(s) = \frac{F(s)C(s)P(s)}{1+C(s)P(s)}R(s) + \frac{1}{1+C(s)P(s)}Y_{no}(s) + \frac{P(s)}{1+C(s)P(s)}Y_{ni}(s) \tag{1.2.1}$$

$$E(s) = \frac{F(s)}{1+C(s)P(s)}R(s) - \frac{1}{1+C(s)P(s)}Y_{no}(s) - \frac{P(s)}{1+C(s)P(s)}Y_{ni}(s) \tag{1.2.2}$$

$$U(s) = \frac{F(s)C(s)}{1+C(s)P(s)}R(s) - \frac{C(s)}{1+C(s)P(s)}Y_{no}(s) - \frac{C(s)P(s)}{1+C(s)P(s)}Y_{ni}(s) \tag{1.2.3}$$

If $F(s) = 1$, it is called the *one-degree-of-freedom* (*ODOF*) control; if $F(s)$ is a given transfer function, it is called the *TDOF* control. In the case of *ODOF*, four basic transfer functions and in the case of *TDOF*, six basic transfer functions determine the signal transmissions between the output signals y and u, and the input signals: the reference signal and the output noise. Since the input noise y_{ni} can always be transferred to an equivalent output disturbance y_n (see Figure 1.2.3), so hereafter it is sufficient to investigate the following four transfer functions (the signs can be disregarded):

$$\frac{Y}{R} = \frac{FCP}{1+CP} \quad ; \quad \frac{Y}{Y_n} = \frac{P}{1+CP}$$
$$\frac{U}{R} = \frac{FC}{1+CP} \quad ; \quad \frac{Y}{Y_n} = \frac{1}{1+CP} \tag{1.2.4}$$

In the case of $F = 1$ the transfer function of the system is

$$T = \frac{CP}{1+CP} = \frac{L}{1+L} \tag{1.2.5}$$

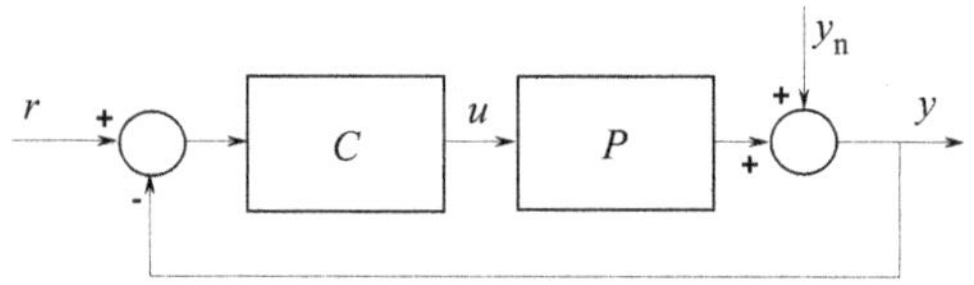

Figure 1.2.3 *One-degree-of-freedom* closed-loop control system.

which is the *complementary sensitivity function*. Here $L = CP$ is the so-called loop transfer function of the open-loop system. The error transfer function regarding the reference signal is

$$S = \frac{1}{1+CP} = \frac{1}{1+L} \tag{1.2.6}$$

and it is also called the *sensitivity function*. This terminology can be interpreted if the sensitivity of T is calculated for the change of the process.

Let us investigate the behavior of the control loop if the transfer function of the process changes from a nominal value P to $\hat{P} = P + \Delta P$. This means, in addition, that instead of the ideal process only its model $\hat{P}$ is known. Let the relative error of L be

$$\ell_{\rm L} = \frac{\Delta L}{L} = \frac{C\Delta P}{CP} = \frac{\Delta P}{P} = \ell \tag{1.2.7}$$

that is, it is equal to the relative error ℓ of the process model. Then the relative error of T can be obtained by simple computation as

$$\frac{\Delta T}{T} = \frac{1}{1+CP}\frac{\Delta P}{P} = S\frac{\Delta P}{P} = S\ell \tag{1.2.8}$$

since

$$\Delta T = \frac{\partial T}{\partial P}\Delta P = \frac{C}{(1+CP)^2}\Delta P \tag{1.2.9}$$

Thus the sensitivity function S shows how the relative change of the process influences the relative change of the overall transfer function. The complementary sensitivity function can be explained by the apparent relationship

$$T + S = 1 \tag{1.2.10}$$

The *TDOF* control systems can also be given in the form shown in Figure 1.2.4.

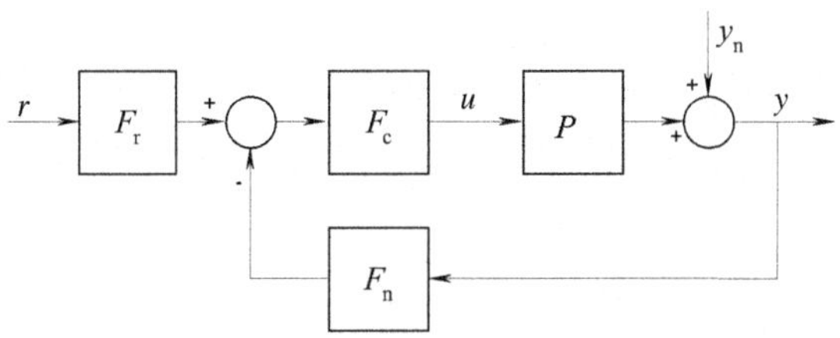

Figure 1.2.4 *Two-degrees-of-freedom* closed-loop control.

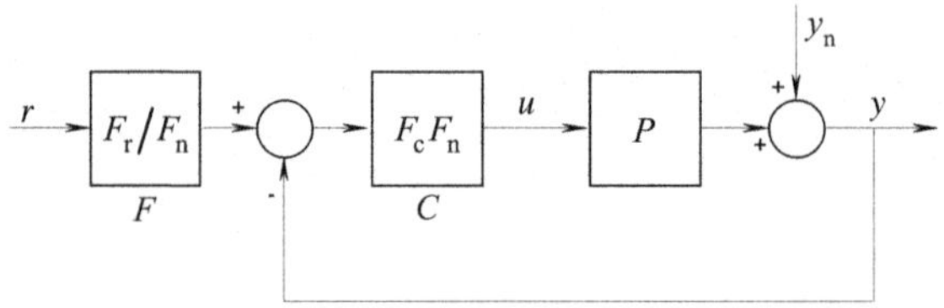

Figure 1.2.5 Equivalences of the *two-degrees-of-freedom* closed-loop controls.

This form can be rearranged to the equivalent form given in Figure 1.2.2 with the equivalences of Figure 1.2.5.

M − α and *E* − β Curves

For deeper analysis of the relationship between the frequency functions of the open- and closed-loop systems, let us analyze the following. The complementary sensitivity function of the closed system can be calculated by the expression (1.2.5). The frequency function of the closed-loop system can be expressed by its amplitude and phase angle:

$$T(j\omega) = \frac{L(j\omega)}{1+L(j\omega)} = M(\omega)e^{j\alpha(\omega)} \tag{1.2.11}$$

The gain of the closed system approaches the constant value one for high amplification value of the open system. This conclusion can be drawn from the expression

$$|T(j\omega)| = \left|\frac{L(j\omega)}{L(j\omega)+1}\right|_{|L|\gg 1} \approx 1 \tag{1.2.12}$$

Since the control, in general, ensures high gain in the low frequency region, the gain of the closed loop is close to one. Low gain values of the closed loop belongs to the low amplification values of the open loop

$$|T(j\omega)| = \left|\frac{L(j\omega)}{L(j\omega)+1}\right|_{|L|\ll 1} \approx |L(j\omega)|$$

Since the amplification of physical systems decreases at high frequencies, this expression can be interpreted as indicating that the amplification values of the open- and closed-loop systems are almost the same at high frequencies, that is, the negative feedback does not change the open loop.

During the control design, the relationship between the transfer functions of the closed loop $T(s)$ and of the open loop $L(s) = C(s)P(s)$ is taken into account. This relatively simple relationship actually means conform nonlinear mapping from a complex plane $L(s)$ to $T(s)$. This nonlinearity is the reason why the control design cannot always be performed unambiguously with simple methods.

Every mapping point (complex vector) can be determined for each point of the complex plane by the expression (1.2.5). First, investigate what kinds of curves belong to the values $|T| = M$, where M is a constant. The points of closed systems with the same amplitude correspond to circles on the complex plane. This can easily be verified by solving the equation

$$M = \left|\frac{L(j\omega)}{1 + L(j\omega)}\right| = \left|\frac{u + jv}{1 + u + jv}\right| = \sqrt{\frac{u^2 + v^2}{1 + 2u + u^2 + v^2}} \tag{1.2.13}$$

The equation of the curves belonging to the constant amplitude M can be obtained by rearranging the above equation as

$$\left(u - \frac{M^2}{1 - M^2}\right)^2 + v^2 = \left(\frac{M}{1 - M^2}\right)^2 \tag{1.2.14}$$

This gives the equation of a circle with radius r and center point (u_o, v_o):

$$r = \left|\frac{M}{1 - M^2}\right| \quad ; \quad u_o = \frac{M^2}{1 - M^2} \quad ; \quad v_o = 0 \tag{1.2.15}$$

The curves belonging to different constant, closed-loop amplitudes M are shown in Figure 1.2.6.

The curve at $M = 1$ is a vertical line at $u = -0.5$. For $M > 1$, the curves are to the left of the line; for $M < 1$, they are to the right of the line. If M tends to infinity, the curves shrink to the point $-1 + 0j$, and if M tends to zero, the curve shrinks to the origin.

The phase angles can be represented in a similar way. Besides the so-called Archimedes circles [105] belonging to constant M values, those curves can also be given what belong to constant α values, which are also Archimedes circles. The joint curve system is usually called $M - \alpha$ curves [105].

If the Nyquist diagram of the open loop is drawn on the complex plane containing the constant M curves then the amplitude–frequency diagram of the closed loop can be obtained. The highest amplitude of the closed loop is determined by how closely the Nyquist diagram of the open loop approaches the point $(-1 + 0j)$, that is, which circle of highest M value is tangential to the Nyquist diagram.

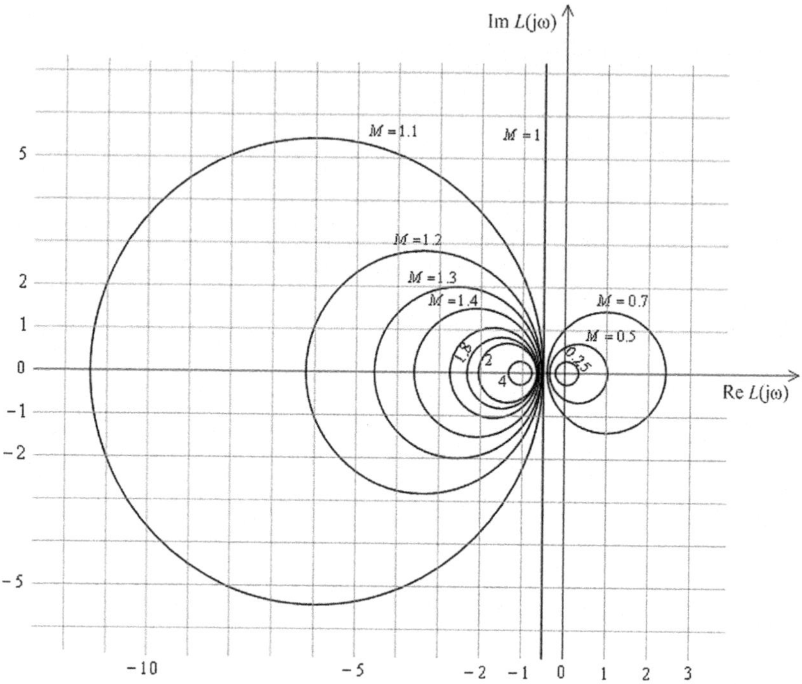

Figure 1.2.6 Constant *M* curves.

Some characteristic features of the M curves are shown on the complex plane in Figure 1.2.7. The frequency ω_b, where the frequency function $L(j\omega)$ intercepts the circle of $M = 1/\sqrt{2}$, gives the so-called bandwidth of the closed-loop system. In the figure, the cut-off frequency ω_c and the frequency ω_m, to which the maximum $M = \sqrt{2}$ belongs, are also indicated. The maximum value is the highest M value belonging to the circle tangential to the curve $L(j\omega)$. The amplitude–frequency diagram of the closed loop has amplification only if the Nyquist diagram of the open loop intersects the vertical line corresponding to $M = 1$; thus there is a frequency region where the Nyquist curve is to the left of this line. At the frequency of the intersection, $|T(j\omega)| = 1$.

Approximating relationships can also be applied between the maximum amplification $M_m = M_{max}$ of the closed-loop amplitude–frequency curve and the maximum value ν_m of the step response function [105] (see Figure 1.2.7):

$$\begin{array}{ll} M_m \geq 1.5 & \nu_m \leq M_m - 0.1 \\ 1.25 \leq M_m \leq 1.5 & \nu_m \approx M_m \\ M_m \leq 1.25 & \nu_m < M_m \end{array} \qquad (1.2.16)$$

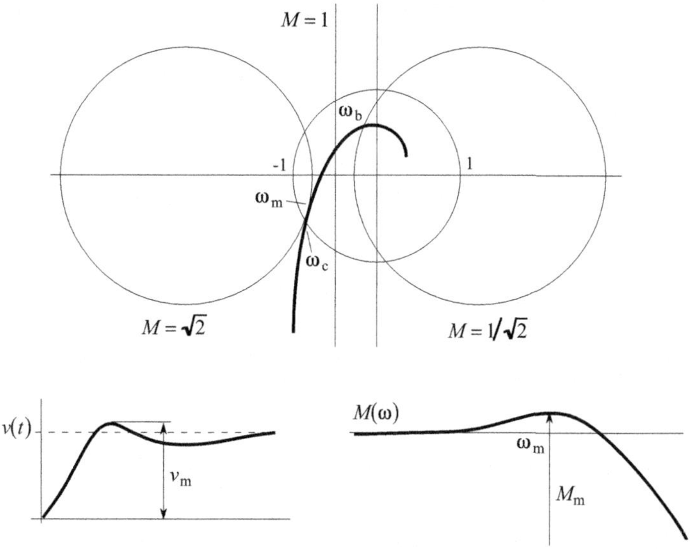

Figure 1.2.7 Some characteristic features of *M* curves on the complex plane: the shape of the unit step response and the amplitude–frequency characteristics.

To avoid oscillations and big overshoot in the time response, high amplifications must not be allowed in the amplitude–frequency diagram of the closed loop. The ideal and actual frequency curves of the closed loop are shown in Figure 1.2.8.

Similarly to the $M-\alpha$ curves of the frequency function $T(j\omega)$, the so-called $E-\beta$ curves can be constructed based on the overall error transfer function $S(j\omega)$ (sensitivity function)

$$S(j\omega) = \frac{1}{1+L(j\omega)} = E(\omega)e^{j\beta(\omega)} \tag{1.2.17}$$

The curves belonging to constant E can easily be given as

$$E = |S(j\omega)| = \frac{1}{|1+L(j\omega)|} \tag{1.2.18}$$

and the expression $|1+L(j\omega)|$ in the denominator means the distance from point $-1+j0$. Thus these curves are concentric circles around $(-1+j0)$ with radius $1/E$. Among them the circle belonging to $E=1$ has special significance.

In Figure 1.2.9 the curves corresponding to values $M=1$, $E=1$, and $|L|=1$, and also the distance $|1+L|$ are indicated. It is also shown how to determine the maximum value M_m in the open loop Nyquist diagram. The frequencies ω_1 and ω_2, belonging to the intersection of $L(j\omega)$ and the circle $E=1$, give the region where $|S(j\omega)|>1$.

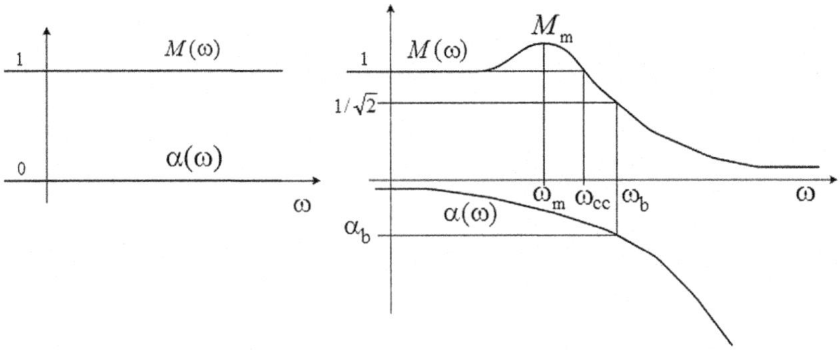

Figure 1.2.8 The ideal and actual frequency curves of the closed loop.

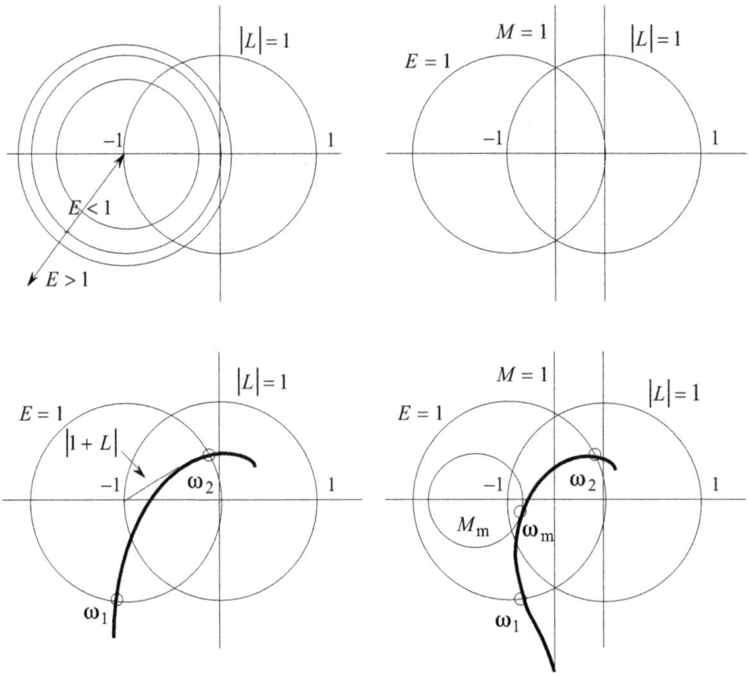

Figure 1.2.9 Curves corresponding to values $M=1$, $E=1$, and $|L|=1$.

1.3 STABILITY OF THE CLOSED-LOOP CONTROL

In practical applications the stability of control systems has significant importance. None of the disturbance-rejecting or reference signal-tracking controls can be performed by unstable control loops. The stability of a control system needs to be distinguished from the stability of the process itself, however. There are cases when an unstable process needs to be stabilized by the control loop. There are processes which would not even operate without control and the closed-loop control stabilizes the process. The best-known examples of these systems are the control of an airplane or the everyday activity of riding a bicycle.

Stability is a system property which shows whether the system, moved from its equilibrium state and left to itself, can return back to its equilibrium or not. If the system is nonlinear, then its stability depends also on its input signal and working point. In this case, stability is the property only of a state of the system and not of the whole system. In the case of linear systems, stability is the property of the system; it depends on the structure and parameters of the system, but does not depend on the input signal. A number of various formulations exist for the determination of stability.

Stability of a System Without Excitation

A system is stable if after being removed it from its equilibrium state and allowed free motion it returns to its original state. If the system moves away from its original state, its behavior is unstable. In the boundary case the system neither returns to its steady state nor moves away from it, but remains in the close vicinity of the steady state which depends on the extent of the removal (e.g., it makes undamped oscillations with bounded amplitude around the initial state). A nonlinear system is even considered stable when, in a boundary case, it is removed from its steady state and it returns to the arbitrarily prescribed small vicinity of the steady state. The system is *asymptotically stable* if after being removed from its equilibrium state it returns to its original state. A stable linear system is *asymptotically stable*, and in this case its weighting function $w(t)$ decreases, if

$$\lim_{t \to \infty} w(t) = 0 \tag{1.3.1}$$

or it is absolutely integrable, that is,

$$\int_0^\infty |w(t)| \mathrm{d}t < \infty \tag{1.3.2}$$

Stability of an Excited System

The system is stable if it responds to a bounded input signal with a bounded output signal under any initial conditions. The stability of an excited system is called *BIBO* (*bounded-input–bounded output*) stability [31].

For linear systems, stability is a system property. Stability does not depend on the magnitude of the excitation. In the case of linear systems, if the nonexcited system is stable, then the excited system is also stable. The stability can be unambiguously checked from the system response given for a simple input signal.

Internal Stability

A closed-loop control system fulfills the requirement of *internal stability* if its output signal and all of its inner signals respond in a stable way to any outer excitation signal. Consider the control system shown in Figure 1.3.1. Besides the reference signal r, let us investigate the noise rejection effect for the input y_{ni} and output y_{no} disturbances of the process P, and effect of the measurement noise y_n to the output. The system is stable, if for all the considered bounded input signals, the controlled output signal y, the manipulated control signal u, and the error signal e are bounded. It can be seen that for the structure shown in Figure 1.3.1, it is always sufficient to choose two arbitrary external and two arbitrary internal signals. The internal stability requires the investigation of the stability of the following four overall transfer functions: $CP/(1+CP)$, $1/(1+CP)$, $P/(1+CP)$, $C/(1+CP)$. This can be given by the so-called *transfer matrix* of the closed-loop control system [33,105,143]:

$$\mathbf{T}_t = \begin{bmatrix} \dfrac{CP}{1+CP} & \dfrac{P}{1+CP} \\ \dfrac{C}{1+CP} & \dfrac{1}{1+CP} \end{bmatrix} \tag{1.3.3}$$

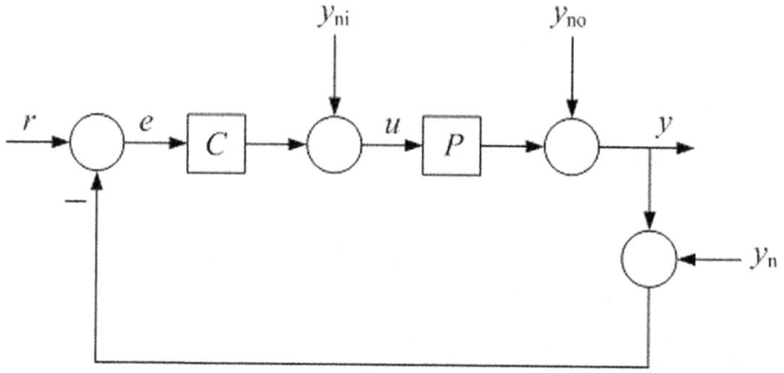

Figure 1.3.1 Block scheme of a closed-loop control system.

A closed-loop control system has the property of internal stability if matrix $\boldsymbol{T}_t$ is stable, that is, all its elements are stable. Internal stability is equivalent to the stability of the excited system if the open-loop system has no nonobservable or noncontrollable right-side poles (i.e., the zeros of the regulator do not cancel the right-side poles of the plant). (Note that the unstable pole of the plant must not be canceled by the right-side zero of the regulator, because the unstable pole becomes unseen only in the relationship between the controlled signal and the reference signal, but the unstable pole remains in the relationship between the disturbance acting at the input of the plant and the controlled signal.)

Lyapunov Stability

According to the Lagrange energy theorem, a system is in balance if its potential energy is minimal. Lyapunov prescribes the determination of a scalar function of energy property (the so-called Lyapunov function) belonging to the differential equation or state equation of a general nonlinear system with constant coefficients [60,64]. If in the considered range of the state variables this function is positive and its derivative is negative, the system is asymptotically stable. Investigation methods of Lyapunov provide sufficient conditions for determination of stability properties of nonlinear systems. Choosing the Lyapunov function is not always a simple task. Lyapunov suggests first the stability investigation of the linearized system in individual working points. Of course, the Lyapunov method can also be applied to investigate stability of linear systems, but for linear systems it is expedient to use simpler direct methods.

Stability Criteria

The negative feedback, which is the basic structure of a closed-loop control system, also involves the risk of instability. If a nonexcited closed-loop control system is asymptotically stable then the time function describing its transients contains components of decreasing time functions. The transient time function is the combination of such exponential components whose exponents are the roots of the characteristic equation of the system.

In a controllable and observable control system (when the zeros of the regulator do not cancel the poles of the plant) the roots of the characteristic equation are identical to the poles of the overall transfer function of the closed loop. Formally, the characteristic equation of the differential equation describing the system is equivalent to the denominator of the overall transfer function of the closed-loop system.

The overall transfer function of the closed loop between the output signal y and the reference signal r is

$$T(s) = \frac{Y(s)}{R(s)} = \frac{C(s)P(s)}{1 + C(s)P(s)} = \frac{C(s)P(s)}{1 + L(s)} \tag{1.3.4}$$

The differential equation of the system is the inverse Laplace transform of the expression below:

$$[1 + L(s)]Y(s) = C(s)P(s)R(s) \tag{1.3.5}$$

The characteristic equation is formally

$$1 + L(s) = 0 \tag{1.3.6}$$

Thus the roots of the characteristic equation are the same as the poles of the overall transfer function of the closed-loop system.

If the loop transfer function is a rational fraction, $L(s) = \mathcal{N}(s)/\mathcal{D}(s)$ where $\mathcal{N}(s)$ and $\mathcal{D}(s)$ are polynomials, and so the characteristic equation can also be given in the following forms:

$$\mathcal{A}(s) = \mathcal{D}(s) + \mathcal{N}(s) = 0 \tag{1.3.7}$$

or

$$a_n s^n + a_{n-1} s^{n-1} + \ldots + a_1 s + a_o = a_n(s - p_1)(s - p_2)\ldots(s - p_n) = 0 \tag{1.3.8}$$

If the system is described by its state equation with state matrix $\boldsymbol{A}$, then the characteristic equation can be given by the relationship (see also (A.2.4.))

$$\det(s\boldsymbol{I} - \boldsymbol{A}) = 0 \tag{1.3.9}$$

If the coefficients of the characteristic equation are real numbers, then the roots of the equation are real or conjugate complex numbers. The stability can be decided from the location of the roots of the characteristic equation which are the poles of the closed-loop control system.

The condition for asymptotic stability is that the real part of p_i, the poles of the closed loop, be of negative real part, as this condition ensures decreasing time functions of the transients. This condition can also be formulated as follows: the closed-loop control system is asymptotically stable if all of its poles lie on the left side of the complex plane.

If any of the poles lies on the right-side plane, the system is unstable. If besides the left-side poles there are poles on the imaginary axis in the origin, then there is an integrating effect in the system; its transient does not decrease, but goes to infinity. If there are single conjugate complex poles on the imaginary axis, then steady oscillations appear in the transients. In the case of multiple poles the amplitudes of the oscillations increase. In practice, only asymptotic stability can be accepted.

If there is no dead-time, the characteristic equation is an algebraic equation, whose roots can be given analytically with a so-called solution formula provided for the case when the degree is below five (Galois theorem). For higher degrees, numerical root-searching methods can be applied which determine the roots with a given accuracy.

If the system contains dead-time, then the characteristic equation is a transcendent equation $\mathcal{D}(s)+\mathcal{N}(s)e^{-sT_d}=0$, whose solution is not simple, and in the case of instability it is difficult to decide how to stabilize the system. In this case the characteristic equation can be approximated with a rational fractional approximation of the dead-time, or investigation needs to be done in the frequency domain.

Several procedures have been elaborated to decide the stability without the solution of the characteristic equation. These procedures are called stability criteria. If there is no dead-time, then based on the relationships between the roots and the coefficients of the algebraic equation, it can be checked with analytical stability criteria whether all the roots are on the left side of the complex plane, that is, whether the system is stable or not. A necessary condition of stability is that all the coefficients of the characteristic equation must be of the same sign and none of the coefficients must be zero.

The best-known analytical methods are the Routh scheme and the Hurwitz determinant method [37,105].

With the Nyquist stability criterion, the stability of the closed-loop control system can be determined based on the frequency diagram of the open loop. Assume that the transfer function of the open-loop control system has no poles on the right side of the complex plane, and thus the open-loop system is stable. Draw the frequency function on the complex plane for the region $-\infty<\omega<\infty$ (complete Nyquist diagram). Go through the Nyquist diagram in the direction along increasing frequencies:

If the Nyquist diagram does not encircle the $-1+0j$ point, the closed-loop control system is stable.

If the Nyquist diagram crosses the $-1+0j$ point, the system is on the stability limit.

If the Nyquist diagram encircles the $-1+0j$ point, the system is unstable.

In a simpler formulation it is sufficient to draw the Nyquist diagram only for positive ω values. If we go through the diagram from $\omega = 0$ to ∞, and point $-1 + 0j$ is to the left of the curve, the closed-loop control system is stable. If the curve crosses the $-1 + 0j$ point, the system is on the stability limit. If the $-1 + 0j$ point is to the right of the curve, the system is unstable.

The generalized Nyquist stability criterion gives the condition of stability even for the case when the open loop has right-side poles, that is, the open loop is unstable. The question is whether the closed loop can be stabilized with negative feedback. The generalized Nyquist stability criterion can be formulated as follows: if the open loop is unstable and the number of its right-side poles is n_{rP}, then the closed-loop control system is asymptotically stable if the complete Nyquist diagram of the open loop encircles point $-1 + 0j$ on the complex plane counterclockwise (considered as the positive direction) as many times as the number of the right-side poles of the open loop is (i.e., n_{rP} times).

Practical Stability Measures

In the case of a stable open loop, the closed loop is stable if the Nyquist diagram of the open loop does not encircle the $-1 + 0j$ point. It is fair to say that the system has a certain amount of stability reserve, if the Nyquist diagram is kept "adequately far" from the $-1 + 0j$ point.

Some measures can be defined which indicate how far the Nyquist diagram of the open loop is from the $-1 + 0j$ point. Such measures are the *phase margin*, the *gain margin*, the *modulus margin*, and the *delay margin* [117].

Phase Margin

Let us draw the Nyquist diagram of the open loop for positive frequencies. Let us then determine the intersection point of the Nyquist diagram with the circle of unity radius. The frequency belonging to this point is called *cut-off frequency* and is denoted by ω_c. Let us connect the origin and the intersection point with a straight line. The angle of this straight line formed with the negative real axis is called the phase margin (Figure 1.3.2).

$$\varphi_t = \varphi(\omega_c) + 180^\circ = \arg L(j\omega_c) + 180^\circ \tag{1.3.10}$$

If the phase margin is positive, the system is stable. If the phase margin is zero, the system is on the stability limit. If the phase margin is negative, the system is unstable.

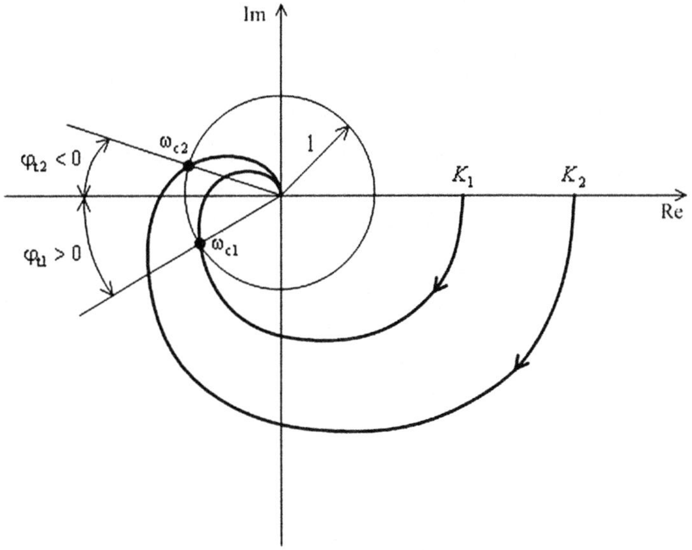

Figure 1.3.2 Interpretation of the phase margin.

Thus for the stability of the control system the following statements can be made:

$$
\begin{aligned}
&\varphi_t > 0 \quad \text{Stable system} \\
&\varphi_t = 0 \quad \text{Boundary of stability} \\
&\varphi_t < 0 \quad \text{Unstable system}
\end{aligned}
\tag{1.3.11}
$$

The stability of the system can be evaluated based on the phase margin as a single measure only if the Nyquist diagram of the open loop crosses the unit circle only once.

Gain Margin

Let us determine the intersection point of the Nyquist diagram with the negative real axis and also the distance $\kappa = |1 + L(j\omega_{180})|$ of this point from point $-1 + 0j$ (Figure 1.3.3). κ is called the gain margin. It is apparent that for $\kappa > 0$ the stability domain of the simple Nyquist criterion is obtained. The stability of the system can be evaluated based on the gain margin as a single measure only if the Nyquist diagram of the open loop crosses the negative real axis only once.

The κ'—the so-called modified gain margin—is defined by the intercept $\kappa' = L(j\omega_{180}) = 1 - \kappa$ as seen in Figure 1.3.3. If $\kappa' < 1$, the system is

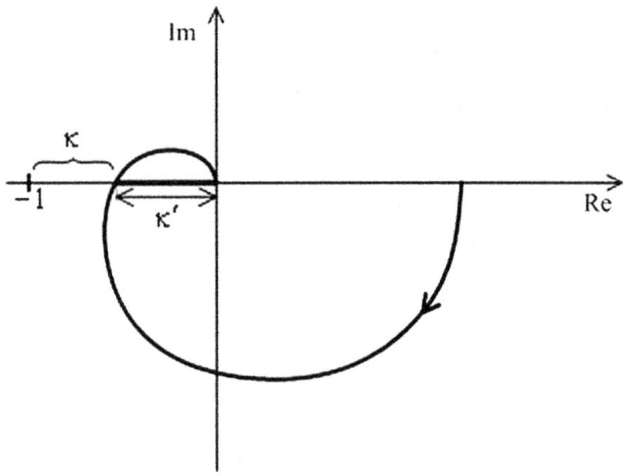

Figure 1.3.3 Interpretation of the gain margins.

stable. If $\kappa' = 1$, the system is on the stability limit. If $\kappa' > 1$, the system is unstable. Thus for the stability of the control system the following statements can be made:

$$\begin{array}{ll} \kappa' < 1 & \text{Stable system} \\ \kappa' = 1 & \text{Boundary of stability} \\ \kappa' > 1 & \text{Unstable system} \end{array} \tag{1.3.12}$$

The meaning of κ is more expressive than that of κ'; however, the reciprocal of κ' specifies the factor by which the system multiplies the actual loop gain to reach the stability limit. Therefore it is straightforward to use the measure $g_t = 1/\kappa' = 1/|L(j\omega_{180})|$ as the relative gain margin. Multiplying the loop gain by g_t, the value of the critical gain is obtained. The inequalities $g_t \geq M_m/(M_m - 1)$ and $\varphi_t \geq 2\arcsin(1/M_m)$ can be derived quite simply. (See Section 1.2 for a definition of M_m.)

Besides stability, the appropriate transient performance is also required. To ensure an overshoot of less than 10% in the step response of the closed-loop system, the desired phase margin is about 60° and the desired relative gain margin g_t is about $2(\kappa \approx \kappa' \approx 0.5)$. These values can be considered characteristic, if there are no resonant frequencies in the loop frequency function. The appropriate phase margin and gain margin do not give reliable information on the stability margin of the system in every case. The phase margin and the gain margin characterize the stability properties of the system only if the Nyquist diagram does not come too close to the unit circle before and after the cut-off frequency.

Modulus Margin

The modulus margin ρ_m is the minimum of the distance of the $-1+0j$ point from the Nyquist diagram, that is, it is the radius of the smallest circle tangential to the diagram and centered at $-1+0j$ (Figure 1.3.4). The modulus margin demonstrates how far the most sensitive point of the system is from the stability limit. We can reasonably let the modulus margin be $\rho_m > 0.5$.

The modulus margin ρ_m is also called the Nyquist stability margin. An important relationship is that ρ_m can be expressed as the reciprocal of the maximum of the absolute value of the sensitivity function (see Section 1.2):

$$\rho_m = \frac{1}{\max_{\omega}|S(j\omega)|} = \min_{\omega}\left|S^{-1}(j\omega)\right| = \min_{\omega}|1+L(j\omega)| \qquad (1.3.13)$$

The margins φ_t, κ, ρ_m are analog conceptions, as each of them aims to guarantee the distance from the $-1+0j$ point.

Delay Margin

The delay margin gives the smallest value of the dead-time T_{min}, which—inserted serially as an extra dead-time into the loop—allows the closed-loop control system to reach the stability limit. The delay margin can be calculated from the phase margin measured in radians with the following relationship:

$$T_{min} = \frac{\varphi_t}{\omega_c} \qquad (1.3.14)$$

where ω_c denotes the cut-off frequency.

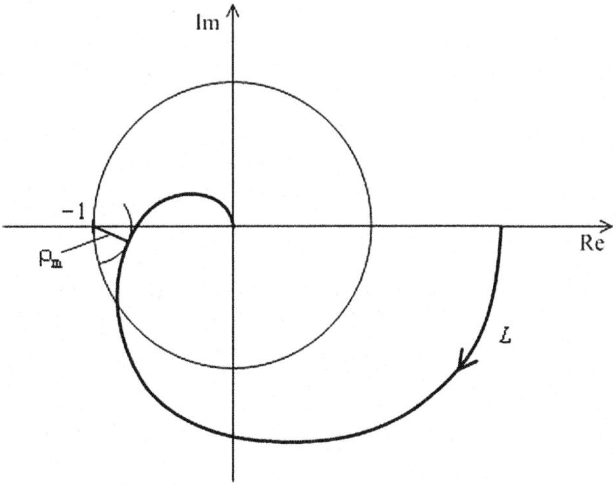

Figure 1.3.4 Interpretation of the modulus margin.

With the stability margins, the stability can not only be evaluated but it can also be established, "how far" the system is from the stability limit.

There are also systems, however, which—because of their structure—remain stable at any value of the loop gain. Such systems are called structurally stable systems. For example, a first-order or a second-order lag element or a pure integrator or an integrator serially connected to a first-order lag with negative constant feedback has this property, as its Nyquist diagram does not encircle point $-1+0j$ even if the loop gain is arbitrarily increased. By increasing the loop gain, the system will not become unstable, but its stability margins will decrease. There are systems, however, which are stable in a given region of the loop gain and in other regions become unstable. Therefore it is extremely important to set the loop gain correctly for this kind of system. These systems are called conditionally stable systems.

Robust Stability

Generally, the parameters of the plant are determined from measurement data. The parameters may change in terms of their nominal values in a given range. The closed-loop control system needs to be stable under the given uncertainty ranges of the parameters.

Suppose that the open loop is stable. The regulator designed for the nominal plant ensures the stability of the nominal closed-loop control system. Let us analyze whether the system remains stable with the parameter uncertainties of the open loop. Stability is maintained if the Nyquist diagram of the modified open loop does not encircle the $-1+0j$ point.

The absolute and relative model errors, and the different possible uncertainty models of a process model were discussed in Section 1.1.

If there is an uncertainty ΔP (or parameter change) in the transfer function of the plant, then if we apply the same regulator this uncertainty appears in the absolute error $\Delta L = C\Delta P$ of the loop transfer function, whereas its relative model error is (see (1.2.7))

$$\ell_L = \frac{\Delta L}{\hat{L}} = \frac{L-\hat{L}}{\hat{L}} = \frac{CP - C\hat{P}}{C\hat{P}} = \frac{P-\hat{P}}{\hat{P}} = \ell \qquad (1.3.15)$$

Here $\hat{L}$ denotes the nominal and L denotes the real loop transfer function.

Robust stability means that the closed-loop control system should not display unstable behavior even in the "worst case" parameter changes. The bound for ΔL can be formulated based on Figure 1.3.5 by taking the simple

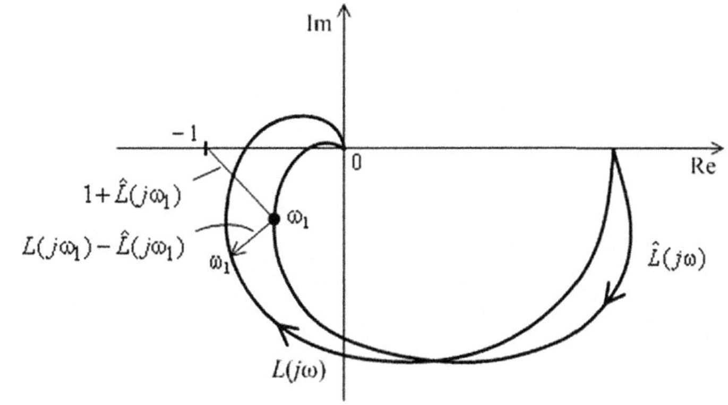

Figure 1.3.5 Change in the Nyquist diagram of an uncertain system.

geometrical considerations into account: the Nyquist diagram will not encircle the $-1+0j$ point, if the following relationship is satisfied for all frequencies:

$$|\Delta L(j\omega)| = |\ell(j\omega)||\hat{L}(j\omega)| < |1+\hat{L}(j\omega)| \quad \forall\omega \tag{1.3.16}$$

With further straightforward manipulations the necessary and sufficient condition for robust stability can be obtained as

$$|\ell(j\omega)| < \left|\frac{1+\hat{L}(j\omega)}{\hat{L}(j\omega)}\right| = \frac{1}{|\hat{T}(j\omega)|} \quad \forall\omega \tag{1.3.17}$$

where $\hat{T} = \hat{L}/(1+\hat{L})$ is the nominal complementary sensitivity function. Condition (1.3.17) can also be expressed as

$$|\hat{T}(j\omega)| < \frac{1}{|\ell(j\omega)|} \le \frac{1}{|W_{\mathrm{m}}(j\omega)|} \quad \forall\omega \tag{1.3.18}$$

because the normalization of the multiplicative error is $|\ell| = |\Delta_{\mathrm{m}} W_{\mathrm{m}}| \le |W_{\mathrm{m}}|$. It should be noted that considering the normalization $|\Delta_{\mathrm{a}}| \le 1$, we can obtain the condition

$$\left|\frac{C(j\omega)}{1+\hat{L}(j\omega)}\right| \le \frac{1}{|W_{\mathrm{a}}(j\omega)|} \quad \forall\omega \tag{1.3.19}$$

It is common practice to express the above inequalities for robust stability in the following form:

$$|\hat{T}(j\omega)||\ell(j\omega)| < 1 \quad \forall\omega \tag{1.3.20}$$

This form is also called the dialectic relationship of robust stability. In the design the first factor $|\hat{T}(j\omega)|$ is calculated for the supposed (known) nominal parameters of the plant, and thus it depends on the designer. The second factor $|\ell|$ does not (or only partly) depend on the designer, as it contains the uncertainties in the knowledge of the plant or its unexpected parameter changes. In those frequency ranges where the uncertainty is large, unfortunately only small transfer gain can be designed for the closed loop. Where $|\hat{T}(j\omega)|$ is high, very accurate information is needed to reach a small error. The higher the absolute value of the complementary sensitivity function, the smaller the permissible parameter uncertainty.

The condition (1.3.17), which considers the whole frequency range, is fairly strict, and therefore it is generally substituted by a more practical condition if the maximum value of $|\hat{T}(j\omega)|$ is known. Assume

$$\hat{T}_{\mathrm{m}} = \max_{\omega}|\hat{T}(j\omega)| \tag{1.3.21}$$

With this value, the relationship (1.3.17) can be simplified to the following satisfactory condition:

$$|\ell(j\omega)| < \frac{1}{\hat{T}_{\mathrm{m}}} \quad \forall\omega \tag{1.3.22}$$

(Let us refer to Section 1.2, where $M(\omega)$ is defined as $M(\omega) = |T(j\omega)|$.)

If the open loop is unstable, and the feedback stabilizes the nominal system, then the closed-loop system remains stable with the parameter uncertainties, and if the number of the right-hand side poles of the open loop does not change, and the number of the encirclements of the Nyquist diagram around the $-1+0j$ point does not change either.

The definition of the modulus margin and the formulation of the robust stability show that the $|1+L(j\omega)|$ distance of the Nyquist diagram of the loop transfer function $L(j\omega)$ from the point $(-1+j0)$ has a crucial role in the performance and stability of a closed-loop control system. In the definition of ρ_{m} it was obvious that the performance of the closed loop can be unambiguously connected to the restriction of the absolute value of the sensitivity function. The usual limiting inequality (*robust performance*) is $|W_{\mathrm{p}}(j\omega)S(j\omega)| \leq 1$, where $|W_{\mathrm{p}}(j\omega)|$ draws a circle (disc) around $-1+j0$ for all ω. This condition is completely the same as the geometrical condition

$$|W_{\mathrm{p}}(j\omega)| \leq |1+L(j\omega)| \quad \text{or} \quad |W_{\mathrm{p}}(j\omega)| \leq \rho_{\mathrm{m}} \quad \forall\omega \tag{1.3.23}$$

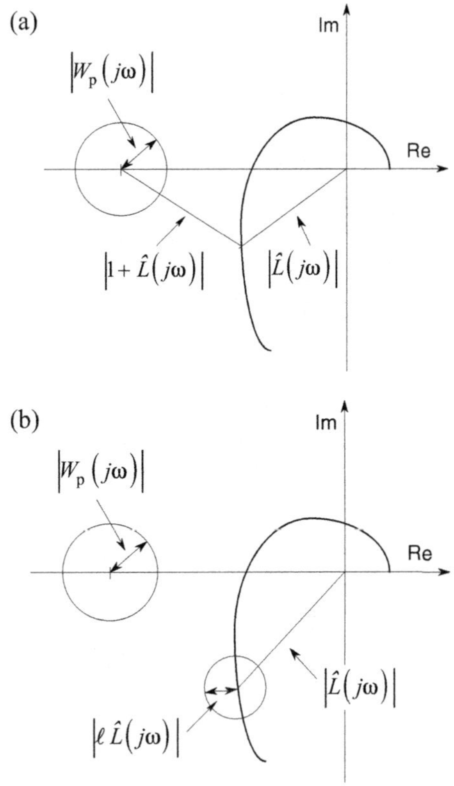

Figure 1.3.6 Representation of the robustness of performance and stability.

(see Figure 1.3.6(a)). Thus the frequency function of the open loop cannot step into the region of the disc $|W_p(j\omega)|$ for any single frequency. Figure 1.3.6(b) shows the classical interpretation of the robust stability discussed above.

These two conditions can be ensured simultaneously if the two circles (discs) have no intersections for any frequencies, that is,

$$|1 + L(j\omega)| > |W_p(j\omega)| + |\ell\hat{L}(j\omega)| \geq |W_p(j\omega)| + |W_m(j\omega)\hat{L}(j\omega)| \quad \forall\omega \tag{1.3.24}$$

Here it has been taken into account that, because $|\ell| \leq |W_m|$, ℓ can be substituted by W_m. This condition can be simply rewritten as

$$\left|W_p(j\omega)\hat{S}(j\omega)\right| + \left|W_m(j\omega)\hat{T}(j\omega)\right| < 1 \quad \forall\omega \tag{1.3.25}$$

Sampled Data Systems

In a controllable and observable control system (when the zeros of the regulator do not cancel the poles of the plant) the roots of the characteristic equation are identical to the poles of the overall transfer function of the closed loop. Formally the characteristic equation of the differential equation describing the system is equivalent to the denominator of the overall transfer function of the closed-loop system. The overall transfer function of the closed-loop system between the output $y[k]$ and the reference signal $r[k]$ is

$$T(z) = \frac{Y(z)}{R(z)} = \frac{C(z)G(z)}{1 + C(z)G(z)} = \frac{L(z)}{1 + L(z)} \tag{1.3.26}$$

where $L(z) = C(z)G(z)$. The differential equation of the system is the inverse z-transform of the expression below

$$[1 + L(z)]Y(z) = C(z)G(z)R(z) \tag{1.3.27}$$

The characteristic equation, however, is formally equivalent to the equation

$$1 + L(z) = 0 \tag{1.3.28}$$

Thus the roots of the characteristic equation are equal to poles of the overall pulse transfer function of the closed-loop system.

In sampled data systems, the pulse transfer function of the open loop is always a rational fraction function $L(z) = \mathcal{N}(z)/\mathcal{D}(z)$ (even for dead-time systems), where $\mathcal{N}(z)$ and $\mathcal{D}(z)$ are polynomials. The characteristic equation can also be given as

$$\mathcal{F}(z) = \mathcal{D}(z) + \mathcal{N}(z) = 0 \tag{1.3.29}$$

or

$$f_n z^n + f_{n-1} z^{n-1} + \ldots + f_1 z + f_o = a_n(z - p_1)(z - p_2)\ldots(z - p_n) = 0 \tag{1.3.30}$$

If the system is given by its state equation whose state matrix is $\boldsymbol{F}$, then the characteristic equation can be given as (see Appendix 3)

$$\det(z\boldsymbol{I} - \boldsymbol{F}) = 0 \tag{1.3.31}$$

If the coefficients of the characteristic equation are real numbers then the roots of the equation are real numbers or complex conjugate numbers. The stability can be defined on the basis of the roots of the characteristic equation, the poles of the control system.

The condition of the asymptotic stability $(T(z) \in \mathcal{S}_{\mathrm{DT}})$ is that the poles p_i^{d} of the DT closed system must lie inside the unit circle. The condition can also be formulated as the closed-control system is asymptotically stable if all its poles lie inside the unit circle.

If any of the poles lies outside the unit circle then the system is unstable. If the system has a pole equal to one, then the system has an integrating effect, its transient is not decreasing, and it goes to infinity. If single complex poles are on the unit circle, undamped oscillations occur. In the case of multiple poles the oscillations have increasing amplitude. In practice, only asymptotic stability can be accepted for the closed system.

Considering only the pulse transfer function itself the question arises where those poles are which result in a stable well-attenuated step response, that is, for unit step input they do not generate oscillating output. In case of CT systems the region of the well and slightly damped poles are separated by lines belonging to constant attenuations (damping factors) in the complex frequency region s. These lines are at a constant angle φ, depending on the damping, to the negative real axis $\cos(\varphi) = \xi$. Now the mapping of these lines needs to be found in the z plane. On the s plane the points $s = \sigma + j\omega$ are on the lines of the constant damping with the condition $\omega = \sigma\sqrt{1 - \xi^2}/\xi$. For a given damping the mapping of $z = e^{sT_s}$ to the $z = e^{\sigma + j\sigma T_s\sqrt{1-\xi^2}/\xi}$ can be calculated and drawn for different σ values. For example, the corresponding form in the z plane of the constant line $\xi = 0.4$ can be seen in Figure 1.3.7. This well-damped region ($\xi > 0.4$), as shown

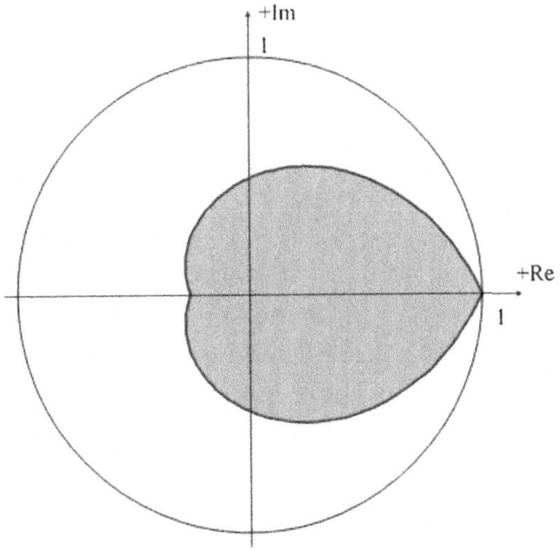

Figure 1.3.7 Region of the well-damped poles in the z plane.

in the figure, is also called the "heart form curve" in the literature. The formula $\xi = 1\Big/\sqrt{1+\pi^2/\left[\ln\left(|p_1|\right)\right]^2}$ is obtained after extensive calculations showing how ξ depends on a root p_1 falling to the negative real axis.

The allocation of these poles is interesting in terms of designing deadbeat regulators (Chapter 2.3) and in general, any DT regulator design method canceling underdamped zeros.

1.4 PARAMETERIZATION OF THE CLOSED-LOOP CONTROL

The most important element of a closed-loop control system is the regulator C, which needs to be determined during the design procedure. Direct and indirect parameterization methods are reviewed below which can help to make this task easier.

The closed system is called internally stable, if, given a bounded additional excitation at the arbitrary point of the system, the generated signals in any point remain bounded. Thus stable transfer functions must be obtained between any two inner points. The mathematical condition of this property can be formalized in the simplest way by introducing the transfer matrix $\boldsymbol{T}_t$ of the closed system (1.3.3), which represents the relationships between any two independent outer and two inner signals. A suitable choice for $\boldsymbol{T}_t$ is [33,91,143]

$$\boldsymbol{T}_t(P,C) = \frac{1}{1+CP}\begin{bmatrix} P \\ 1 \end{bmatrix}\begin{bmatrix} C & 1 \end{bmatrix} = \boldsymbol{s}\boldsymbol{D}^{-1}\boldsymbol{r}^{\mathrm{T}} = \boldsymbol{D}^{-1}\boldsymbol{s}\boldsymbol{r}^{\mathrm{T}} = \boldsymbol{s}\boldsymbol{r}^{\mathrm{T}}\boldsymbol{D}^{-1}$$
$$= \begin{bmatrix} \dfrac{CP}{1+CP} & \dfrac{P}{1+CP} \\ \dfrac{C}{1+CP} & \dfrac{1}{1+CP} \end{bmatrix} = \frac{1}{1+CP}\begin{bmatrix} CP & P \\ C & 1 \end{bmatrix} \tag{1.4.1}$$

where $\boldsymbol{s} = \begin{bmatrix} P & 1 \end{bmatrix}^{\mathrm{T}}$, $\boldsymbol{r} = \begin{bmatrix} C & 1 \end{bmatrix}^{\mathrm{T}}$, and $\boldsymbol{D} = (1+CP)\boldsymbol{I}$. It can be seen that the control loop is internally stable, if and only if $\boldsymbol{T}_t(P,C) \in \mathcal{S}$, where $\mathcal{S}$ is the set of the CT stable linear processes. Using simple algebraic rearrangements

$$\boldsymbol{T}_t(P,C) = \begin{bmatrix} -1 & 0 \\ 0 & 1 \end{bmatrix}\boldsymbol{H}_t(P,C) + \begin{bmatrix} 1 & 0 \\ 0 & 0 \end{bmatrix} \tag{1.4.2}$$

is obtained, where

$$\boldsymbol{H}_{\mathrm{t}}(P,C) = \begin{bmatrix} 1 & P \\ -C & 1 \end{bmatrix}^{-1} = \boldsymbol{D}^{-1}\begin{bmatrix} 1 & -P \\ C & 1 \end{bmatrix} = \frac{1}{1+CP}\begin{bmatrix} 1 & -P \\ C & 1 \end{bmatrix} \tag{1.4.3}$$

The stability of the transfer matrix $\boldsymbol{T}_{\mathrm{t}}(P,C)$ can be investigated via the stability of matrix $\boldsymbol{H}_{\mathrm{t}}(P,C)$. A similar statement can be made for the dual transfer function $\boldsymbol{T}_{\mathrm{t}}(C,P) = \boldsymbol{r}\boldsymbol{D}^{-1}\boldsymbol{s}^{\mathrm{T}} = \boldsymbol{D}^{-1}\boldsymbol{r}\boldsymbol{s}^{\mathrm{T}} = \boldsymbol{r}\boldsymbol{s}^{\mathrm{T}}\boldsymbol{D}^{-1}$ by introducing the matrix [91]

$$\boldsymbol{H}_{\mathrm{t}}(C,P) = \begin{bmatrix} 1 & C \\ -P & 1 \end{bmatrix}^{-1} = \boldsymbol{D}^{-1}\begin{bmatrix} 1 & -C \\ P & 1 \end{bmatrix} = \frac{1}{1+CP}\begin{bmatrix} 1 & -C \\ P & 1 \end{bmatrix} \tag{1.4.4}$$

Youla Parameterization

Let us find parameterization for the closed loop which makes all elements of $\boldsymbol{T}_{\mathrm{t}}(P,C)$ stable. Introduce the following transfer function as parameter

$$Q = \frac{C}{1+CP} \tag{1.4.5}$$

by means of which the new form of the transfer matrix is

$$\begin{aligned} \boldsymbol{T}_{\mathrm{t}}(P,C) &= \boldsymbol{T}_{\mathrm{t}}(P,Q) = \begin{bmatrix} P \\ 1 \end{bmatrix}\begin{bmatrix} Q & 1-QP \end{bmatrix} = \boldsymbol{s}\boldsymbol{q}^{\mathrm{T}} \\ &= \begin{bmatrix} QP & P(1-QP) \\ Q & 1-QP \end{bmatrix} \end{aligned} \tag{1.4.6}$$

It is clear that for the stable process $P \in \mathcal{S}$ and stable parameter $Q \in \mathcal{S}$, all elements of $\boldsymbol{T}_{\mathrm{t}}$ are stable; thus the internal stability of the closed loop is ensured (here $Q(\omega = \infty)$ is finite and regular). Otherwise Q represents the transfer function of a one-degree-of-freedom control loop between the reference signal r and the actuating signal u. All elements of $\boldsymbol{T}_{\mathrm{t}}$ are linear (therefore convex) in Q. Similarly, the sensitivity functions are also linear

$$T = \frac{CP}{1+CP} = QP; \quad S = \frac{1}{1+CP} = 1 - QP \tag{1.4.7}$$

The above procedure is called Youla parameterization (*YP*), where Q is the Youla parameter [127,162]. From (1.4.5) the Youla-parameterized regulator is

$$C = \frac{Q}{1-QP} \tag{1.4.8}$$

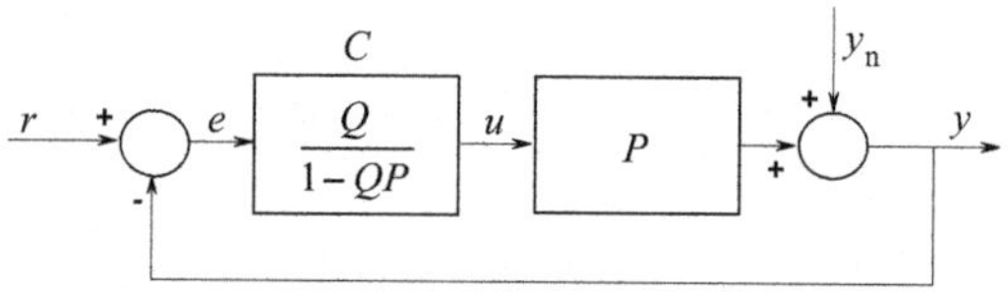

Figure 1.4.1 Youla-parameterized control loop.

The Youla-parameterized control loop is shown in Figure 1.4.1, where the above notations are used.

Assume that the regulator

$$C^* = \frac{Q^*}{1 - Q^* P} \tag{1.4.9}$$

is obtained by using Q^*

$$Q^* = \frac{C^*}{1 + C^* P} \tag{1.4.10}$$

It seems reasonable to express all stabilizing regulator in the form

$$q = Q/Q^* \tag{1.4.11}$$

$$C = C^* \frac{Q}{Q + C^*(1 - q)} \tag{1.4.12}$$

Here Equations (1.4.5) and (1.4.9) are used and P is eliminated. The condition for applying this parameterization is that the inverse of Q^* (or q) must be stable.

Another kind of parameterization can also be derived from the procedure discussed above using (1.4.12). Assume that

$$Q_f = \frac{C_f}{1 + C_f P}; \quad C_f = C^*(1 - q); \quad C_1 = \frac{Q_1}{1 - Q_1} = \frac{q}{1 - q}; \quad Q_1 = q \tag{1.4.13}$$

The new form of the regulator is

$$C = \frac{[C^*(1 - q)]}{1 + [C^*(1 - q)]P} \frac{q}{1 - q} = \frac{C_f}{1 + C_f P} \frac{q}{1 - q} = Q_f C_1 = C(q, P) \tag{1.4.14}$$

Here Q_f is the Y parameter of a nominal stabilizing regulator; furthermore C_1 is the first element, or the so-called basic regulator of a closed loop, where $P = 1$ and $Q_1 = q$. This simple basic system is shown in Figure 1.4.2.

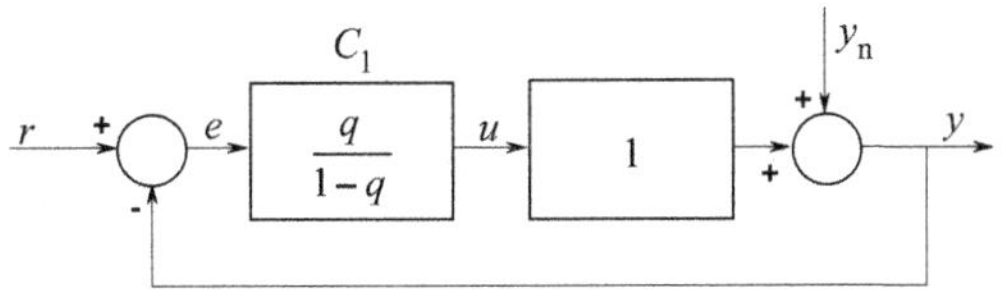

Figure 1.4.2 Block diagram of a simple basic closed-loop control system.

It is clear from Figure 1.4.1 that the order of C and P can be changed without causing any change in the overall transfer function. Consequently, the parameterization of the process transfer function can be introduced similarly to (1.4.5) in order to ensure the inner stability of the closed system. Since this paradigm is dual to the original Y parameterization, this procedure is called dual-Youla (*d-Y* or *H*) parameterization. The duality means that now the process can be written in the form

$$P = \frac{H}{1 - HC} \tag{1.4.15}$$

where with stable $H \in \mathcal{S}$ parameters, that is,

$$H = \frac{P}{1 + PC} \tag{1.4.16}$$

all the elements of $\boldsymbol{T}_t$ become stable, and thus the inner stability of the closed-control system is ensured (here $H(\omega = \infty)$ is finite and regular).

The *d-Y*-parameterized control loop is shown in Figure 1.4.3.

All the elements of $\boldsymbol{T}_t$ are linear (therefore convex) in H. Similarly, the sensitivity functions are also linear

$$T = \frac{CP}{1 + CP} = CH; \quad S = \frac{1}{1 + CP} = 1 - CH \tag{1.4.17}$$

Now investigate the relationship between the Youla parameter Q and the *d-Y* parameter H:

$$Q = C(1 - CH) = \frac{P - H}{P^2} \tag{1.4.18}$$

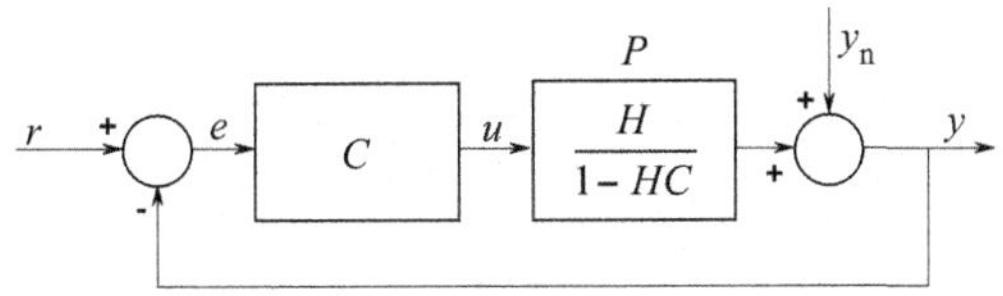

Figure 1.4.3 Block diagram of the dual-Youla-parameterized control loop.

or

$$H = P(1 - QP) = (C - Q)/C^2 \tag{1.4.19}$$

that is, the dependence between the parameters is also linear.

Assume that the process

$$P^* = \frac{H^*}{1 - H^* C} \tag{1.4.20}$$

is obtained by the d-Y parameterization H^*

$$H^* = \frac{P^*}{1 + P^* C} \tag{1.4.21}$$

Let us express the stabilizing processes by using the quotient

$$h = H/H^* \tag{1.4.22}$$

$$P = P^* \frac{H}{H + P^*(1 - h)} \tag{1.4.23}$$

Here Equations (1.4.15) and (1.4.20) are applied and C is eliminated. The condition for this parameterization is that the inverse of H^* (or h) must be stable.

Another kind of parameterization can also be derived from the above procedure using (1.4.23). Assume

$$H_f = \frac{P_f}{1 + P_f C}; \quad P_f = P^*(1 - h); \quad P_1 = \frac{H_1}{1 - H_1} = \frac{h}{1 - h}; \quad H_1 = h \tag{1.4.24}$$

The new form of the regulator is

$$P = \frac{[P^*(1 - h)]}{1 + [P^*(1 - h)]C} \frac{h}{1 - h} = \frac{P_f}{1 + P_f C} \frac{h}{1 - h} = H_f P_1 = P(h, C) \tag{1.4.25}$$

Here H_f is the d-Y parameter of a nominal stabilizing regulator; furthermore, P_1 is the first element, or the so-called basic process, of a closed-loop system, where $C = 1$ and $H_1 = h$. This basic system is shown in Figure 1.4.4.

Note the obvious duality in all quantities and variables of the above two parameterizations.

Of course, nothing prevents us applying both parameterizations simultaneously. This method is called Double-Youla (D-Y) parameterization

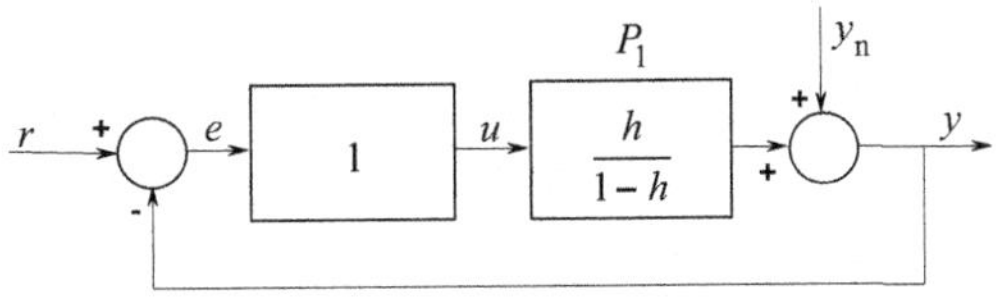

Figure 1.4.4 Block diagram of the basic closed-loop control system.

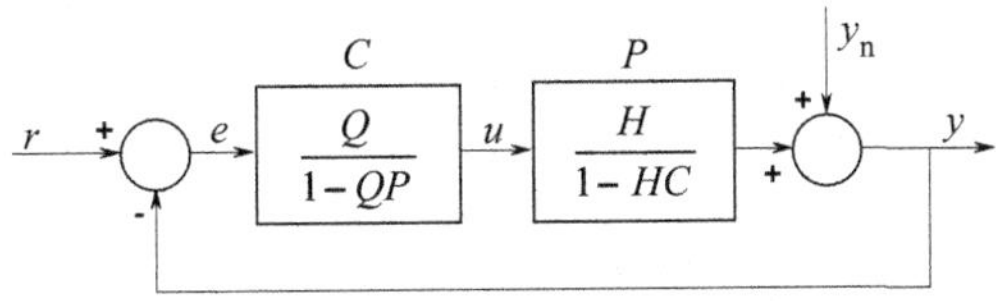

Figure 1.4.5 Block diagram of the Double-Youla parameterization.

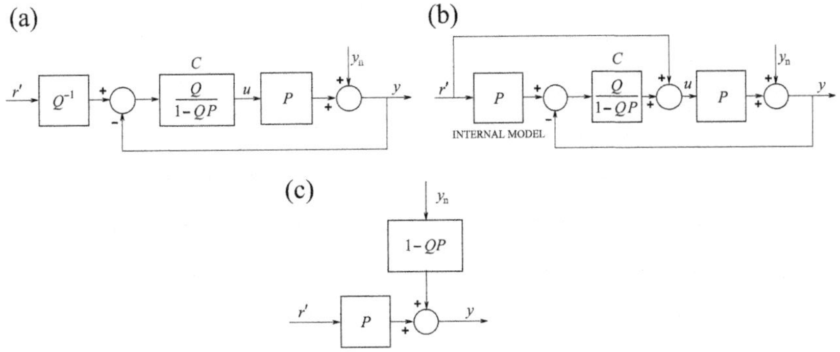

Figure 1.4.6 Block diagrams for opening the closed-loop control systems.

and its block diagram is shown in Figure 1.4.5. This block scheme has only theoretical importance.

Now apply the inverse of Q, connected serial as in Figure 1.4.6(a), in the block diagram (Figure 1.4.1) of the Youla-parameterized control loop. In this way the tracking property becomes independent of Q, that is, Pr', and thus the closed loop seems to be formally opened. It can be easily checked that this block diagram is equivalent to Figure 1.4.6(b).

The *YP* extended by Q^{-1}, which formally opens the closed loop, is called *KB* parameterization (after the authors Keviczky and Bányász). As regards the operation of the *KB*-parameterized loop it should be noted that here the reference signal has a direct effect on the input of the process, and thus it does not go through the regulator and the whole closed loop. All further control effects (regarding the reference signal) are in operation only when the inner model is not equal to the real process. It can be seen from

Figure 1.4.6(b) that the *KB* parameterization is independent of the *Y* parameterization concerning the reference signal effect, since it opens the closed loop for any arbitrary regulator *C*, and thus the overall tracking transfer function is always *P*. The effect of the disturbance signal, however, can be compensated only by the *Y* parameterization via a simple linear transfer function $(1 - QP)$ linear in *Q*. Thus, in the case of *TDOF* systems the two principles have to be jointly applied.

Youla–Kučera Parameterization

Youla [161] and Kučera [118] suggested independently the extension of the *YP* for processes that were not completely stable. This theory is discussed only insofar as it relates to the topic at hand. Let the process be given by its regular transfer function $P = \mathcal{B}/\mathcal{A}$ using the notations introduced in (1.1.11), where $\mathcal{B}$ and $\mathcal{A}$ are relative primes. Suppose that the process can be stabilized by a regulator $C^* = \mathcal{Y}/\mathcal{X}$, where $\mathcal{Y}$ and $\mathcal{X}$ are also relative primes. (Note that now the process is not supposed to be stable, which is substantially different from the *Y* parameterization.) All regulators stabilizing *P* can be given as

$$C = \frac{\mathcal{Y} + \mathcal{A}Q_C}{\mathcal{X} - \mathcal{B}Q_C} = C^*\left(1 + \frac{\mathcal{A}}{\mathcal{Y}}Q_C\right)\Big/\left(1 - \frac{\mathcal{B}}{\mathcal{X}}Q_C\right) \tag{1.4.26}$$

where $Q_C \neq Q$ is the so-called Youla–Kučera (*Y–K*) parameter, which is a regular and stable transfer function. (Here the index refers to the fact that the parameterization relates to the regulator.) The condition of joint relative prime means that $\mathcal{B}\mathcal{Y} + \mathcal{A}\mathcal{X} = 1$ must be fulfilled. In the case of $Q_C = 0$, the initial stabilizing regulator is retrieved, that is, $C(Q_C = 0) = C^*$. The complementary sensitivity function belonging to the initial regulator is

$$T^* = \frac{C^* P}{1 + C^* P} = \mathcal{B}\mathcal{Y} = \mathcal{Y}\mathcal{A}P = Q^* P \tag{1.4.27}$$

where Q^* is the Y-parameter belonging to C^*, that is,

$$Q^* = \frac{C^*}{1 + C^* P} = \mathcal{Y}\mathcal{A} \tag{1.4.28}$$

The initial sensitivity function can be obtained similarly

$$S^* = \frac{1}{1 + C^* P} = \mathcal{A}\mathcal{X} \tag{1.4.29}$$

Thus, in the general case, the overall transfer function of the closed system is

$$T = T^*\left(1 + \frac{\mathcal{A}}{\mathcal{Y}}Q_C\right) = \mathcal{A}(\mathcal{Y} + \mathcal{A}Q_C) = QP = \frac{CP}{1 + CP} \tag{1.4.30}$$

The error transfer function is

$$S = \mathcal{A}(\mathcal{X} - \mathcal{B}Q_C) = \frac{1}{1 + CP} = 1 - QP \tag{1.4.31}$$

It is important to note that Q and Q_C depend on each other. By simple computations

$$Q = Q^*\left(1 + \frac{\mathcal{A}}{\mathcal{Y}}Q_C\right) = \mathcal{A}(\mathcal{Y} + \mathcal{A}Q_C) \tag{1.4.32}$$

is obtained and vice versa

$$Q_C = \frac{\mathcal{Y}}{\mathcal{A}}\left(\frac{Q}{Q^*} - 1\right) = -\frac{\mathcal{Y}}{\mathcal{A}} + \frac{1}{\mathcal{A}^2}Q \tag{1.4.33}$$

Here it is quite obvious that $Q^* = Q(Q_C = 0)$. The Y and Y–K parameters differ and they are equal in only very special cases, that is, when $Q = Q_C = \overline{Q} = \mathcal{Y}Q^*/(1 - Q^*\mathcal{A})$. It is very easy to check that the Y–K parameterization returns the Y parameterization if Q_C is replaced by $Q_C(Q)$.

Note that all transfer functions are linear in Q and affine in Q_C (closest to linearity). Note also that each transfer function of the closed system can be rewritten in the quite general form $W_1^C + W_2^C Q_C W_3^C$. (This form is also well fitted to the different theoretical optimization methods.) The block diagram of the closed-loop control system containing the Y–K-parameterized and all P-stabilizing regulators is shown in Figure 1.4.7.

Y–K parameterization can also be applied in the generalization of d-Y parameterization [143]. Let the process be $P^* = \mathcal{B}/\mathcal{A}$ with regular relative prime polynomials $\mathcal{B}$ and $\mathcal{A}$, stabilized by the regulator $C = \mathcal{Y}/\mathcal{X}$, where $\mathcal{Y}$ and $\mathcal{X}$ are also relative primes. All processes stabilized by C can be given as

$$P = \frac{\mathcal{B} + \mathcal{X}Q_P}{\mathcal{A} - \mathcal{Y}Q_P} = P^*\left(1 + \frac{\mathcal{X}}{\mathcal{B}}Q_P\right) \Big/ \left(1 - \frac{\mathcal{Y}}{\mathcal{A}}Q_P\right) \tag{1.4.34}$$

where $Q_P \neq Q_C \neq Q$ is the dual-Youla–Kučera (d-Y–K) parameter, which is a regular and stable transfer function. (Here the index refers to the fact that parameterization is involved in the process now.) The condition for

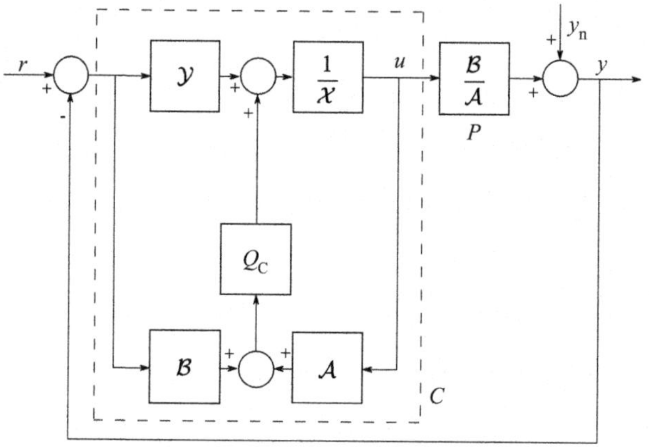

Figure 1.4.7 Block diagram of closed-loop control system containing the *Y–K*-parameterized and all *P*-stabilizing regulators.

joint relative prime means that $\mathcal{BY} + \mathcal{AX} = 1$ must be fulfilled. In the case of $Q_P = 0$ the initial stabilized process is retrieved, that is, $P(Q_P = 0) = P^*$. The complementary sensitivity function belonging to the initial regulator is

$$T^* = \mathcal{BY} = \mathcal{BX}C = H^*C \tag{1.4.35}$$

where H^* is the *d-Y–K* parameter belonging to P^*

$$H^* = P^*/(1 + P^*C) = \mathcal{BX} \tag{1.4.36}$$

It is interesting to note that the initial sensitivity function $S^* = \mathcal{BX}$ is the same as in (1.4.29), as a consequence of the duality.

Thus, in the general case, the overall transfer function of the closed system is

$$T = T^*\left(1 + \frac{\mathcal{X}}{\mathcal{B}}Q_P\right) = \mathcal{X}(\mathcal{B} + \mathcal{X}Q_P) = HC = \frac{CP}{1 + CP} \tag{1.4.37}$$

The error transfer function is

$$S = \mathcal{X}(\mathcal{A} - \mathcal{Y}Q_P) = 1/(1 + CP) = 1 - HC \tag{1.4.38}$$

It is important to note that H and Q_P depend on each other. By simple computation

$$H = H^*\left(1 + \frac{\mathcal{X}}{\mathcal{B}}Q_P\right) = \mathcal{X}(\mathcal{B} + \mathcal{Y}Q_P) \tag{1.4.39}$$

is obtained and vice versa

$$H_P = \frac{\mathcal{B}}{\mathcal{X}}\left(\frac{H}{H^*} - 1\right) = -\frac{\mathcal{B}}{\mathcal{X}} + \frac{1}{\mathcal{X}^2}H \tag{1.4.40}$$

$H^* = H(Q_P = 0)$ is obviously fulfilled. The *d-Y* and *d-Y–K* parameters are different and they are equal only in a very special case, that is, $H = Q_P = \overline{H} = \mathcal{B}H^*/(1 - H^*\mathcal{X})$. It is easy to check that the *Y–K* parameterization gives back the *Y* parameterization if Q_P is replaced by $Q_P(H)$. Note that all transfer functions are linear in *H* and affine in Q_P (closest to linearity). Note also that each transfer function of the closed system can be rewritten in the quite general form $W_1^P + W_2^P Q_P W_3^P$. (This form is also well fitted to the different theoretical optimization methods.) A block diagram of the *d-Y–K*-parameterized and all process-stabilizing closed-control systems is shown in Figure 1.4.8. The embedded open-loop subsystem with input signal *x*, output signal *z*, and the original disturbance signal y_n, shown in the figure, provides good structure for several closed-loop process identification (ID) methods (see also Section 10.3).

Following the principle of the Double-Youla parameterization, it is also possible to apply the Double-Youla–Kučera (*D-Y–K*) parameterization. This principle is shown in Figure 1.4.9, where Q_C and Q_P are applied. The stability matrix of *D-Y–K*-parameterized closed-control system is

$$\boldsymbol{H}_t(Q_C, Q_P) = \begin{bmatrix} 1 & -Q_C \\ Q_P & 1 \end{bmatrix}^{-1} = \frac{1}{1 + Q_C Q_P}\begin{bmatrix} 1 & Q_C \\ Q_P & 1 \end{bmatrix} \tag{1.4.41}$$

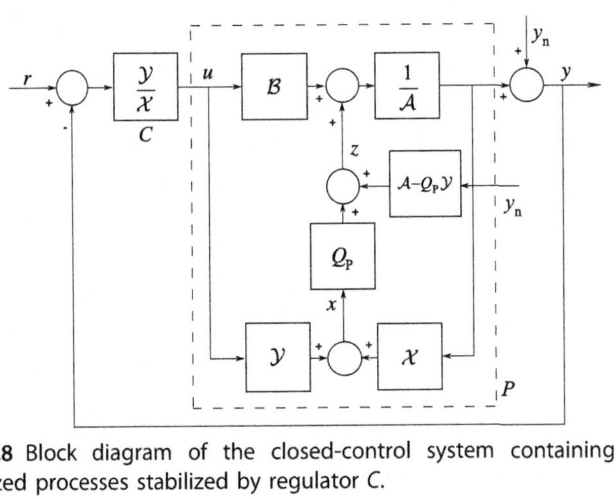

Figure 1.4.8 Block diagram of the closed-control system containing all *d-Y-K*-parameterized processes stabilized by regulator *C*.

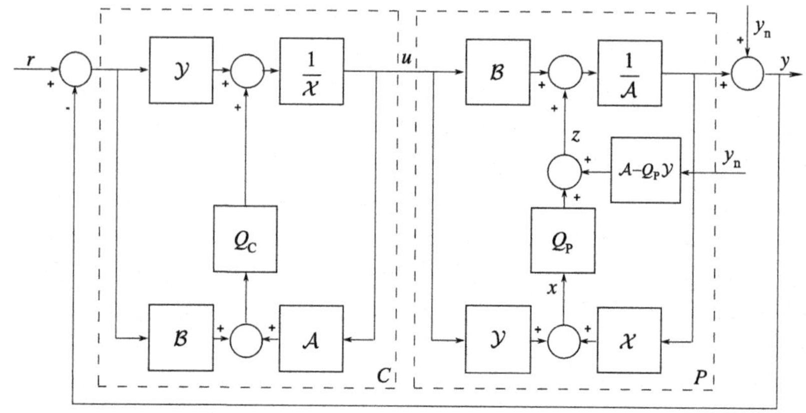

Figure 1.4.9 *Double-Youla–Kučera*-parameterized closed-loop control.

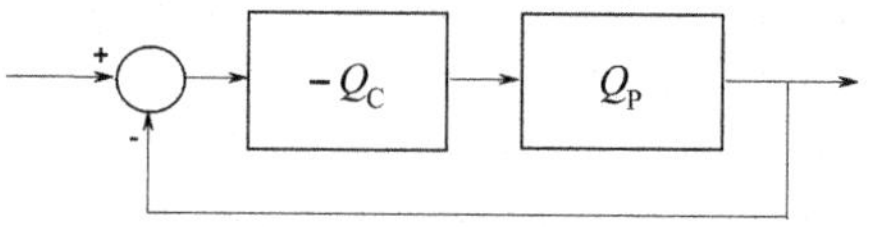

Figure 1.4.10 Equivalent block diagram for the stability investigation of the *Double-Youla–Kučera*-parameterized closed-loop control.

and its stability can be determined via the stability of the equivalent closed-control system shown in Figure 1.4.10. The analogy is apparent considering (1.4.4).

The *Y–K* parameterization has great importance besides its usefulness in ID tasks. This property can be connected to topological observations. Express the *d-Y–K* parameter Q_P from (1.4.34)

$$Q_P = \frac{\mathcal{A}}{\mathcal{X}(1+CP)}(P-P^*) = \frac{\mathcal{A}(1-QP)}{\mathcal{X}}(P-P^*)$$
$$= \frac{\mathcal{A}(1-HC)}{\mathcal{X}}(P-P^*) \tag{1.4.42}$$

and from this let us take C -t

$$C = \frac{\mathcal{A}(P-P^*) - Q_P\mathcal{X}}{Q_P\mathcal{X}P} = \frac{\mathcal{A}(P-P^*)}{Q_P\mathcal{X}P} - \frac{1}{P} \tag{1.4.43}$$

From this expression, formally $C(P=P^*) = -1/P$ is obtained, which is obviously not the solution. The reason for this misleading result is that the form 0/0, that is, the factor $(P-P^*)/Q_P$ in the first term of (1.4.43), is neglected. After detailed analysis

$$\frac{\mathcal{A}(P-P^*)}{Q_P\mathcal{X}} = 1 + CP = \frac{1}{1-QP} \tag{1.4.44}$$

is obtained, then, substituting it into (1.4.30) we obtain the limit value for the regulator, in case of convergence,

$$C = \frac{1}{1-QP}\frac{1}{P} - \frac{1}{P} = \frac{Q}{1-QP} \tag{1.4.45}$$

which is the well-known Y-parameterized form.

Modification of the *YP*

The special modification of the classical Y parameterization with two-step design is suitable for designing regulators for unstable processes. Using (1.4.26) and after simple rearrangements, we obtain a special additive term [127]

$$C = C^* + Q''/[1 - Q''(\mathcal{B}\mathcal{X})] = C^* + C'' \tag{1.4.46}$$

Here C'' means a regulator serially connected to C^*. This parallel regulator uses the special

$$Q'' = Q_C/\mathcal{X}^2 \tag{1.4.47}$$

Y parameter and assumes a virtual process

$$P'' = \mathcal{B}\mathcal{X} = P/(1 + C^*P) = T_{uy} \tag{1.4.48}$$

which needs to be stabilized. Observing the Figures 1.4.1 and 1.4.2, we can see how the Y parameterization operates: the regulator puts a transfer function $-P$ parallel to the process and an arbitrary stable transfer function into Q. The additive form of the regulator (1.4.46) obtained from the Y–K parameterization can be more complex, if the process is unstable. The interpretation of the second term C'' is very interesting, because this regulator stabilizes a virtual process $P'' = T_{uy}$. This virtual process is actually the overall transfer function of the initial closed system between the arbitrary outer disturbances in the input signal u of the process and the output y of the closed system (i.e., controlled signal). Since P is stabilized by C^*, T_{uy} must be stable. This two-step regulator design procedure can be seen in Figure 1.4.11, where in the first step the process is stabilized, then, in the second step, the required dynamics is modified. This kind of two-step design procedure, obtained from the methodology, also occurs in the next sections.

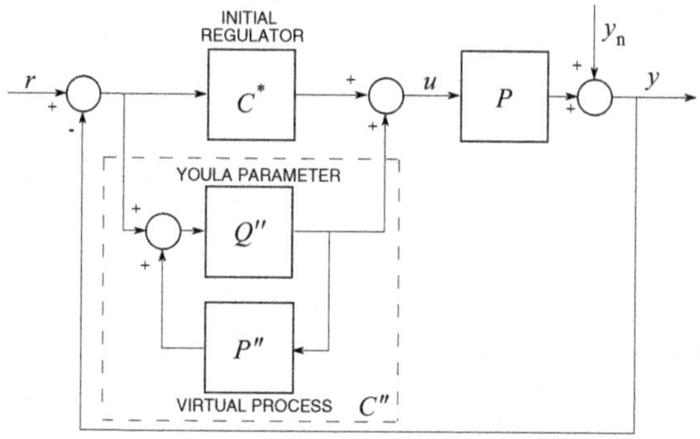

Figure 1.4.11 The two-step design scheme in the *Y–K*-parameterized closed-loop control system.

Additive terms can be obtained for the sensitivity and complementary sensitivity functions, as well

$$S = \mathcal{A}\mathcal{X} - \mathcal{A}\mathcal{B}Q_C = S^* - (\mathcal{A}\mathcal{X})(\mathcal{B}\mathcal{X})Q'' = S^* - S^*P''Q''$$
$$= S^*(1 - P''Q'') \tag{1.4.49}$$

and

$$T = T^*\left(1 + \frac{\mathcal{A}}{\mathcal{Y}}\right)Q_C = T^* + \frac{C^*P}{1 + C^*P}\frac{\mathcal{A}Q_C}{\mathcal{Y}} = T^* + (\mathcal{A}\mathcal{X})(\mathcal{B}\mathcal{X})Q''$$
$$= T^* + S^*P''Q'' \tag{1.4.50}$$

which are affine in Q'' and in the applied design method; the second step depends on the first one.

KB Parameterization

As mentioned previously, the *YP* extended by Q^{-1}, which formally opens the closed loop, is called *KB* parameterization after the authors (Keviczky and Bányász) who suggested the method. The basic scheme of the *KB* parameterization is shown in Figure 1.4.12.

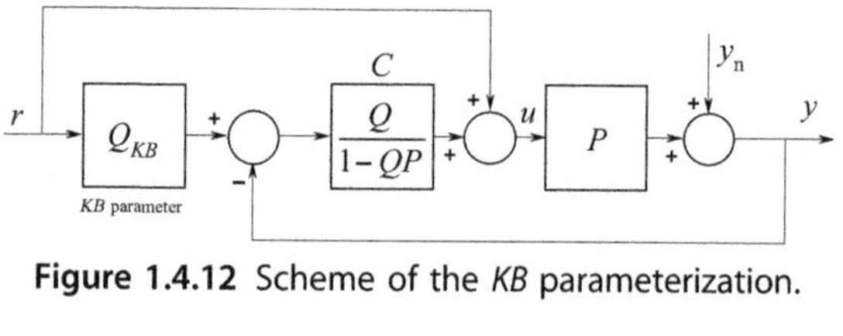

Figure 1.4.12 Scheme of the *KB* parameterization.

The basic relationships of the closed loop are

$$y=\left(1+Q_{KB}\frac{Q}{1-QP}\right)\frac{P}{1+\dfrac{QP}{1-QP}}r+\frac{1}{1+\dfrac{QP}{1-QP}}y_{\rm n}$$
$$=(1-QP+QQ_{KB})Pr+(1-QP)y_{\rm n}\Big|_{Q_{KB}=P}=Pr+(1-QP)y_{\rm n} \qquad (1.4.51)$$

It is worth noting that the reference signal takes effect directly on the input of the process, and thus it does not go through the regulator and the closed loop. The controlling effect regarding the reference signal operates only if the internal model is not equal to the real process. Considering the reference signal effect the *KB* parameterization shown in Figure 1.4.12 is independent of the *Y* parameterization, since it opens the closed loop even for the arbitrary regulator *C*, that is, the overall tracking transfer function is always *P* if $Q_{KB}=P$ is chosen. The effect of the disturbance signal is compensated only via the simple transfer function $(1-QP)$ linear in *Q*. Thus in the case of *TDOF* systems the two principles have to be applied simultaneously. If $Q_{KB}=P$ is chosen the equivalent closed loop corresponds to the open loop in Figure 1.4.13.

Although the invention of the *KB* parameterization is inspired by the *YP*, its validity is more general, since it opens the closed-control loop for any regulator

$$y=(1+Q_{KB}C)\frac{P}{1+CP}r+\frac{1}{1+CP}y_{\rm n}$$
$$=\frac{(1+Q_{KB}C)P}{1+CP}r+\frac{1}{1+CP}y_{\rm n}\Bigg|_{Q_{KB}=P}=Pr+\frac{1}{1+CP}y_{\rm n} \qquad (1.4.52)$$

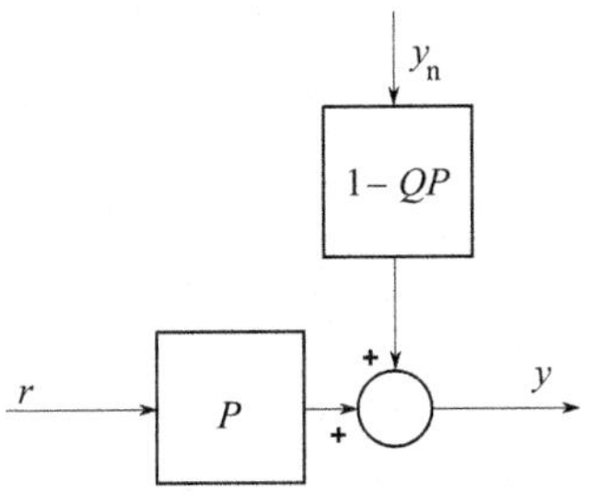

Figure 1.4.13 The ideal choice of the *KB* parameter $Q_{KB}=P$ opens the system.

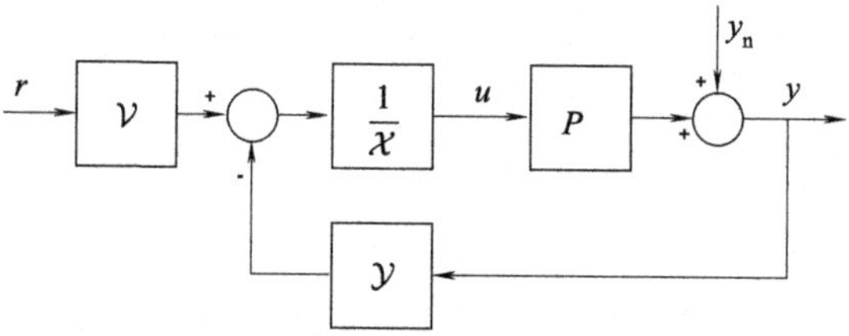

Figure 1.4.14 Conventional form of the $\mathcal{V}-\mathcal{X}-\mathcal{Y}$ parameterization.

$\mathcal{V}-\mathcal{X}-\mathcal{Y}$ Parameterization

$\mathcal{V}-\mathcal{X}-\mathcal{Y}$ parameterization (unattributed), appeared first in discussion of the pole placement self-tuning regulators in the early days of the Lund School by Professor Åström (the original name was $\mathcal{R}-\mathcal{S}-\mathcal{T}$ parameterization). In this approach the *TDOF* closed-loop control had the form shown in Figure 1.4.14.

The parameterization gives the polynomial forms of the filters of the regulator and the feedforward in a simplified way. It is clear that three polynomials are enough to parameterize the closed loop, which means that in this case it actually parameterizes the regulator.

The general schemes in Figures 1.2.2 and 1.2.5 of the *TDOF* control loops are connected to the above parameterization via the equalities $F=\mathcal{V}/\mathcal{Y}$ and $C=\mathcal{Y}/\mathcal{X}$. In the scheme shown in Figure 1.2.4, this parameterization means the following equalities $F_e=\mathcal{V}$, $F_c=1/\mathcal{X}$, and $F_n=\mathcal{Y}$.

The basic equation of the closed loop is now

$$
\begin{aligned}
y &= \mathcal{V}\frac{\frac{1}{\mathcal{X}}P}{1+\frac{\mathcal{Y}}{\mathcal{X}}P}r+\frac{1}{1+\frac{\mathcal{Y}}{\mathcal{X}}P}y_n=\frac{\mathcal{V}\frac{\mathcal{B}}{\mathcal{A}}}{\mathcal{X}+\mathcal{Y}\frac{\mathcal{B}}{\mathcal{A}}}r+\frac{\mathcal{X}}{\mathcal{X}+\mathcal{Y}\frac{\mathcal{B}}{\mathcal{A}}}y_n \\
&= \frac{\mathcal{B}\mathcal{V}}{\mathcal{A}\mathcal{X}+\mathcal{B}\mathcal{Y}}r+\frac{\mathcal{A}\mathcal{X}}{\mathcal{A}\mathcal{X}+\mathcal{B}\mathcal{Y}}y_n
\end{aligned}
\tag{1.4.53}
$$

where it is assumed that P has the transfer function (1.1.11). The polynomial $\mathcal{A}\mathcal{X}+\mathcal{B}\mathcal{Y}$ in the denominator is the characteristic polynomial of the closed system and thus the role of $\mathcal{X}$ and $\mathcal{Y}$ can be seen in the pole placement. The main advantage of this parameterization is that the actuating signal u can be easily given in the simple form of

$$
u=\frac{\mathcal{V}r-\mathcal{Y}y}{\mathcal{X}} \tag{1.4.54}
$$

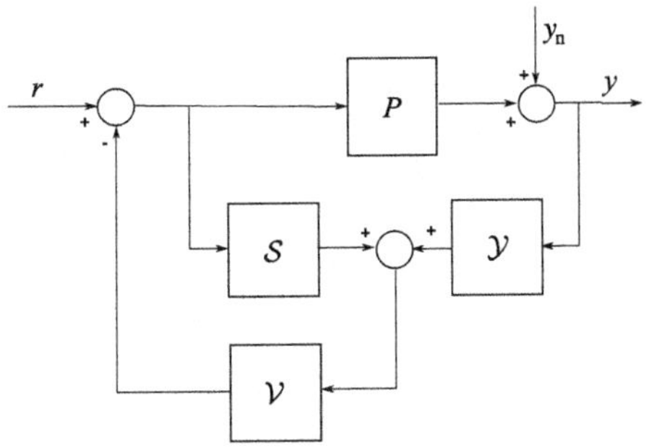

Figure 1.4.15 Conventional form of the parameterization.

Most of the regulators discussed in this book can be written in the above form, which is especially useful for the solution of DT tasks, because the difference equation of the regulator can be directly written. Note that the polynomials $\mathcal{V}$, $\mathcal{X}$, and $\mathcal{Y}$ are redundant as regards the necessary minimum number of parameters.

$\mathcal{Y}-\mathcal{S}-\mathcal{V}$ Parameterization

A further interesting form of the closed system shown in Figure 1.4.15 can be observed in the discussion of several modern control methods.

By equivalent rearrangements, this scheme can be translated to the form shown in Figure 1.2.2 with the expressions $F = \mathcal{V}/\mathcal{S}$ and $C = \mathcal{Y}/(\mathcal{V} - \mathcal{S})$. Note that this method practically also means the parameterization of the regulator.

The basic equation of the closed loop is now

$$y = \frac{\mathcal{B}\mathcal{V}}{\mathcal{A}(\mathcal{V} - \mathcal{S}) + \mathcal{B}\mathcal{Y}} r + \frac{\mathcal{A}(\mathcal{V} - \mathcal{S})}{\mathcal{A}(\mathcal{V} - \mathcal{S}) + \mathcal{B}\mathcal{Y}} y_n \tag{1.4.55}$$

In our experience, this parameterization does not have any advantage over the previous ones.

CHAPTER 2

Control of Stable Processes

In the early days, heuristic trial and error methods based on rule of thumb were widely used for regulator design. At the same time, great effort was made to elaborate a general mathematical methodology for the theoretical approach of design methods. The transfer function of an m-order regulator has $(2m+1)$ unknown parameters. During tuning of the higher-order regulators, it was noticed that a certain design goal can be reached by many parameter sets; moreover, as a consequence, in many cases the parameters were not independent, so the parameterization of the regulator was redundant. The main question is how to parameterize a general stable regulator to solve the basic design tasks of a closed-loop control system with a minimum number of nonredundant parameters. One of the most important theoretical solutions is provided by the so-called *Youla parameterization* (*YP*) discussed in Section 1.4.

The *YP* turns the transfer matrix $\mathbf{T}_t(P,C)$ of the closed system into the matrix $\mathbf{T}_t(P,Q)$, which is linear in the Y parameter Q (see (1.4.6)). In this case, the inner stability can be ensured only if $P \in \mathcal{S}$ and $Q \in \mathcal{S}$ are fulfilled. This requirement is not very restricting in practice, since a high percentage of the processes is stable. Therefore the techniques of *YP* and *Keviczky–Bányász* (*KB*) parameterization are chosen as the guiding principle in this book. *KB* parameterization gives the chance to extend and generalize the *YP* for the case of *two-degree-of-freedom* (*TDOF*) control systems. This section concentrates on stable processes.

The question arises why it is advantageous if the *KB* parameterization opens the closed system. Its most important advantages can be seen in the design methods discussed later in this section. It has special importance for closed-loop process identification methods, as well. In the case of real processes, in industry, only rarely it is used to open the operating control loops for any, especially experimental, purposes. The most beneficial advantage of the closed-loop identification is that it parameterizes the closed-control system in the model of the process, which is the ultimate goal of the identification. Several parameterization methods have been discussed in Section 1.4; however, they have only theoretical importance, since not all of them can be connected to practical solutions. The only exception is the Youla–Kučera parameterization which gives strong theoretical results for the design of stabilizing regulators. Though this method does not seem to be very useful for modeling, it has been used in some quarters [5,33,153].

Techniques prior to Y parameterization have also been discussed in which similar approaches have appeared: e.g., the methods of Truxal–Guillemin and the Smith regulator. It will be also shown that the Smith regulator can be considered as a subcase of the Youla-parameterized regulator, so nowadays there is no need for this kind of distinction.

In the more modern control engineering techniques, the design methods of finite settling time and predictive regulators can be included in this methodology. These regulators will be discussed later.

In the next sections, all the regulator design methods will be discussed concerning YP and its results. It will be seen that formal comparison is appropriate even in the case when this parameterization cannot be directly applied. This kind of comparison is especially important when it is not obvious that a case formally corresponds to the Y parameterization and the basic approach is hidden behind equations.

Of course, there are processes which are unstable because of their basic behavior. The control design methods which can be applied for arbitrary, and therefore unstable, processes will also be discussed in the next sections.

2.1 REGULATORS BASED ON *YP*

The Youla parameter, as a matter of fact, is a stable (by definition), regular transfer function

$$Q(s) = \frac{C(s)}{1 + C(s)P(s)} \quad \text{or shortly} \quad Q = \frac{C}{1 + CP} \tag{2.1.1}$$

where $C(s)$ is a stabilizing regulator, and $P(s)$ is the transfer function of the stable process.

It follows from the definition of the Youla parameter that the structure of the realizable and stabilizing regulator in the Youla-parameterized control loop is fixed:

$$C(s) = \frac{Q(s)}{1 - Q(s)P(s)} \quad \text{or shortly} \quad C = \frac{Q}{1 - QP} \tag{2.1.2}$$

The Youla-parameterized control loop was shown in Figure 1.4.1. The sensitivity and complementary sensitivity functions linear in Q of the

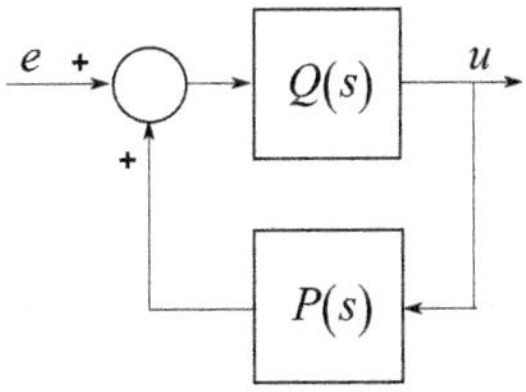

Figure 2.1.1 Realization of *Youla parameterization* regulator.

closed-control systems were defined by (1.3.7). It is interesting to observe that the *YP* regulator of (2.1.2) can be realized by a simple control loop with positive feedback as shown in Figure 2.1.1.

The relationships between the most important signals of the closed system can be obtained with simple calculations

$$
\begin{aligned}
u &= Qr - Qy_n \\
e &= (1 - QP)r - (1 - QP)y_n = Sr - Sy_n \\
y &= QPr + (1 - QP)y_n = Tr + Sy_n
\end{aligned}
\tag{2.1.3}
$$

The effect of r and y_n on u and e is completely symmetrical (not considering the sign). Thus the input of the process depends only on the external signals and $Q(s)$.

Using the scheme in Figure 2.1.1, the Youla-parameterized closed-loop control shown in Figure 1.4.1 can be redrawn to the equivalent form of Figure 2.1.2. This classical internal model-based scheme is called the *internal model control* (*IMC*) [131]. The basic principle of this control is that it has feedback only from the deviation (ε) between the process output and the model output to create the error signal of the control. This error signal is zero in the ideal case when the internal model is completely equal to the process. This case is shown. But in reality the internal model $\widehat{P}(s)$ is only a

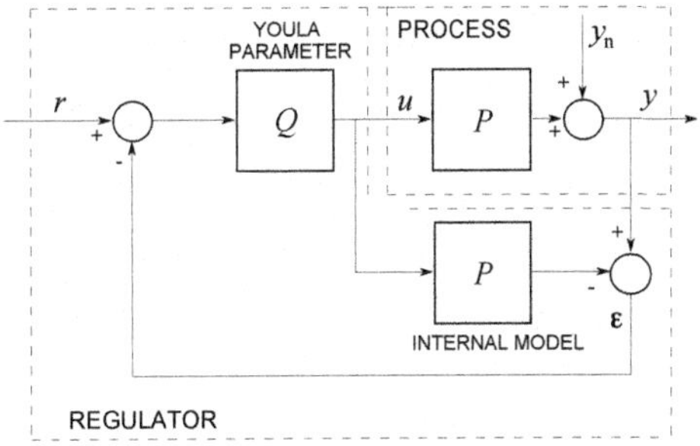

Figure 2.1.2 Equivalent *internal model control* loop.

good approximation of the true process $P(s)$, since the original system is not known exactly. For the sake of simplicity, only the ideal case is discussed here.

From the last equation in (2.1.3), it can be seen that both the *YP* and the *IMC* have the transfer function *QPr* concerning the reference signal tracking. If the *KB* parameterization introduced on the figures of Figure 1.4.6 is applied, then the *YP* can be simply extended for *TDOF* control systems. To do this, let us simply apply a parameter Q_r for the design of the tracking properties, and connect it in serial to the *KB*-parameterized loop of Figure 1.4.6, so the block diagram of Figure 2.1.3 is obtained.

The overall transfer characteristics for this system are

$$\begin{aligned} u &= Q_r y_r - Q y_n \\ e &= (1 - Q_r P) y_r - (1 - QP) y_n = (1 - T_r) y_r - S y_n \\ y &= Q_r P y_r + (1 - QP) y_n = T_r y_r + (1 - T) y_n = T_r y_r + S y_n \end{aligned} \tag{2.1.4}$$

where the tracking properties can be designed by choosing Q_r in $T_r = Q_r P$, and the noise rejection properties by choosing Q in $T = QP$. These two properties can be handled separately. The reference signal of the whole system is denoted by y_r. The conditions for Q_r are the same as for Q. The meaning of T_r is analogous to the meaning of the complementary sensitivity function T of the one-degree-of-freedom (ODOF) control loop for tracking.

The *IMC* of Figure 2.1.2 can be further developed according to Figure 2.1.4. Here the predicted value $\widehat{y}_n$ of the output disturbance y_n is constructed from the difference ε between the outputs of the process and the model by the predictor R_n. Similarly, the predictor R_r provides the predicted value $\widehat{y}_r$ of the reference signal y_r. The noise rejection is performed by giving the predicted value $-\widehat{y}_n$ of the disturbance to the process input via the inverse of the process model, thus in the case of accurate estimation the disturbance is eliminated. The reference signal tracking

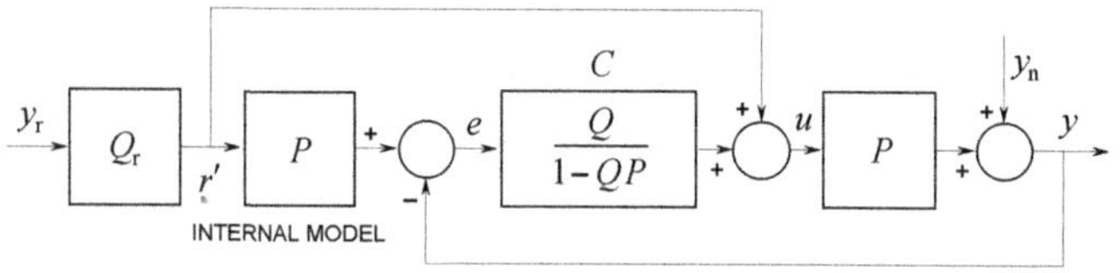

Figure 2.1.3 Two-degree-of-freedom version of the *Youla parameterization* control loop.

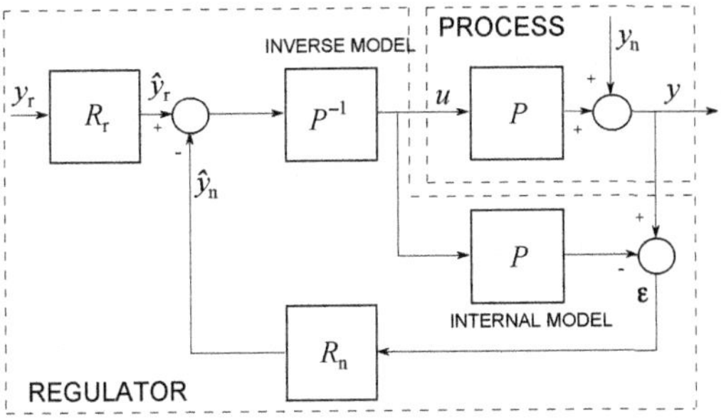

Figure 2.1.4 Extension of the *internal model control* (*IMC*)-based ideal control loop with reference models.

operates in a similar way. Here the operation of R_r can be considered a reference model (desired system dynamics), and therefore the introduced predictors are also called reference models. It is generally required that these predictors are strictly proper with unit static gain, i.e., $R_n(\omega = 0) = 1$ and $R_r(\omega = 0) = 1$.

The best operation of the *TDOF* control loop is reached under the special conditions $R_r = R_n = 1$ or $1 - R_n = 0$, but—as will be shown later—in general, it cannot be realized in most practical cases.

The block diagram of Figure 2.1.4 can be redrawn to the equivalent form of Figure 2.1.5. The most important observation is that the transfer

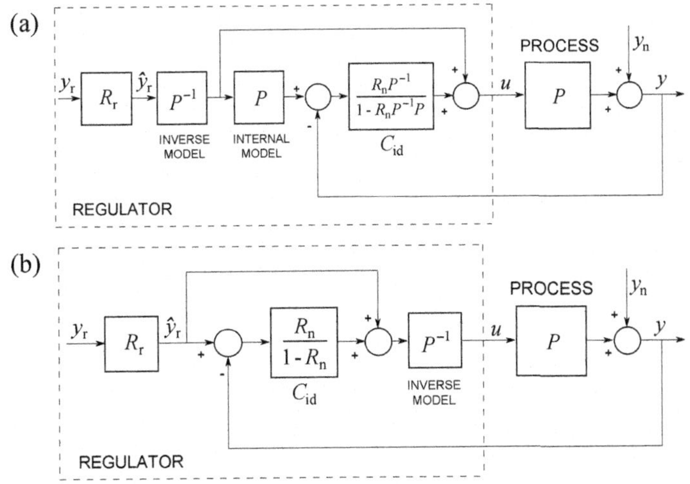

Figure 2.1.5 Extended *internal model control* principle-based equivalent ideal control loops.

function of the regulator—in an ideal case, i.e., when the inverse of the process is realizable and stable—can be given in the form

$$C_{id} = \frac{(R_n P^{-1})}{1-(R_n P^{-1})P} = \frac{Q}{1-QP} = \frac{R_n}{1-R_n} P^{-1} \tag{2.1.5}$$

which is a *YP* regulator with Youla parameter

$$Q = R_n P^{-1} \tag{2.1.6}$$

For the tracking, however, the parameter is

$$Q_r = R_r P^{-1} \tag{2.1.7}$$

(It can be clearly seen that the regulator C_{id} is realizable, if the pole access of R_n is greater or equal to that of the process. For example, the pole access j can easily be ensured by the reference model $R_n = 1/(1+sT_n)^j$.)

The most important signals of the closed loop for the ideal case are

$$\begin{aligned} u_{id} &= R_r P^{-1} y_r - R_n P^{-1} y_n \\ e_{id} &= (1-R_r)y_r - (1-R_n)y_n = \left(1-T_r^{id}\right)y_r - S_{id} y_n \\ y_{id} &= R_r y_r + (1-R_n)y_n = T_r^{id} y_r + (1-T_{id})y_n = T_r^{id} y_r + S_{id} y_n \end{aligned} \tag{2.1.8}$$

and thus the equalities $T_r^{id} = R_r$ and $T_{id} = R_n$ are fulfilled, which is our design goal.

The feed-forward from the reference signal shown in Figure 2.1.5 can also be found in earlier publications. The reason was simply that the change of the reference signal has to have a direct effect on the actuating signal and it is not a good solution if it goes through the closed loop. It can be seen from the detailed comparisons that earlier, mostly engineering schemes do not correspond to the topology of the *KB* parameterization and the generic (see later) *TDOF* control loops but have significant differences.

Note that the previously followed thoughts presented only the main idea of the *YP* and its equivalence to the control based on *IMC*. There is, however, a very critical point regarding the realizability of the resulting schemes, namely the realizability of the inverse of the process *P*. Unfortunately, with a few rare exceptions—this is not true for continuous-time (CT) systems. For practical applications versions of the above approaches have to be found where all elements of the *TDOF* system are realizable.

To introduce a generally applicable regulator, let us assume the transfer function of the process in the following factorized form:

$$P(s) = P_+(s)\overline{P}(s)_- = P_+(s)P_-(s)e^{-sT_d} \quad \text{or shortly} \quad P = P_+\overline{P}_- = P_+P_-e^{-sT_d} \tag{2.1.9}$$

where P_+ is stable, and its inverse is also stable (*inverse stable*) and *inverse stable realizable* (*ISR*). The inverse of $\overline{P}_-$ is unstable (*inverse unstable* (*IU*)) and not realizable (*non realizable*), i.e., *inverse unstable non realizable* (*IUNR*). P_- is *IU*. Here, in general, the inverse of the dead-time part e^{-sT_d} is not realizable, because it would be an ideal predictor. The generalized *IMC* principle can also be applied for the introduced general process structure shown in Figure 2.1.6.

The block scheme of Figure 2.1.6 can be redrawn into the equivalent form of Figure 2.1.7, where the realizable *YP*-regulator of the optimal structure obtained for the general case has the form

$$C_{\text{opt}} = \frac{Q_{\text{opt}}}{1 - Q_{\text{opt}}P} = \frac{R_n G_n P_+^{-1}}{1 - R_n G_n P_- e^{-sT_d}} = \frac{R_n K_n}{1 - R_n K_n P} = R_n G_n C'_{\text{opt}} \tag{2.1.10}$$

where the optimal Youla parameter is

$$Q_{\text{opt}} = R_n G_n P_+^{-1} = R_n K_n, \quad \text{where} \quad K_n = G_n P_+^{-1} \tag{2.1.11}$$

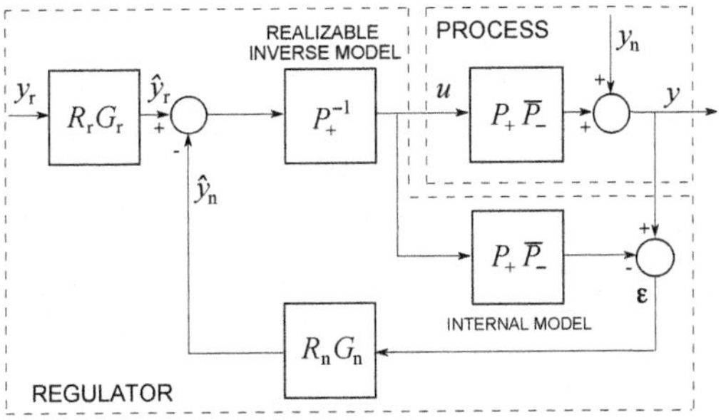

Figure 2.1.6 Generalized *internal model control* principle-based optimal control loop.

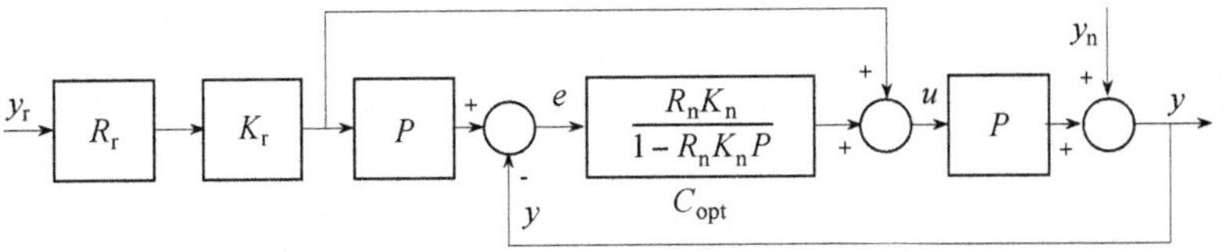

Figure 2.1.7 Generalized *internal model control* principle-based equivalent optimal control loop.

and

$$Q_r = R_r G_r P_+^{-1} = R_r K_r \quad \text{where } K_r = G_r P_+^{-1} \tag{2.1.12}$$

The obtained general control loop—owed to the *YP*—gives structurally the best regulator for stable processes. Further, optimality of the regulator can be set by the embedded transfer functions G_r and G_n. To understand this, let us consider again the most important signals of the *TDOF* closed loop in the optimal case:

$$\begin{aligned}
u_{opt} &= R_r G_r P_+^{-1} y_r - R_n G_n P_+^{-1} y_n \\
e_{opt} &= \left(1 - R_r G_r P_- e^{-sT_d}\right) y_r - \left(1 - R_n G_n P_- e^{-sT_d}\right) y_n = \left(1 - T_r^{opt}\right) y_r - S_n^{opt} y_n \\
y_{opt} &= R_r G_r P_- e^{-sT_d} y_r + \left(1 - R_n G_n P_- e^{-sT_d}\right) y_n = T_r^{opt} y_r + \left(1 - T_n^{opt}\right) y_n \\
&= T_r^{opt} y_r + S_n^{opt} y_n
\end{aligned} \tag{2.1.13}$$

where the equalities $T_r^{opt} = R_r G_r P_- e^{-sT_d}$ and $T_n^{opt} = R_n G_n P_- e^{-sT_d}$ are fulfilled.

Compare the ideal output y_{id} of (2.1.8) with the optimal output y_{opt} of (2.1.13). It can be clearly seen that the ideal and the designed transfer functions determined by the reference models R_r and R_n cannot be reached but can only be approximated. The element $P_- e^{-sT_d}$ appearing in the approximating transfer function cannot be eliminated, and therefore it is called an invariant factor. Thus the dead time e^{-sT_d} and the *IU* term P_- of the process cannot be eliminated by any regulator. In the case of CT processes, this latter term contains the unstable zeros of the nonminimum phase processes and those poles of the stable process, which could not get into the invertible P_+ of (2.1.8). In practice, only the necessary number of the slowest poles (whose number corresponds to the number of the stable zeros in *P*) of *P* is usually included in P_+, and the rest should be added to P_-. The effect of the invariant P_- can only be attenuated by the transfer functions G_r and G_n.

Formulate the deviation of the outputs of the ideal and reachable best (optimal) control loops

$$\Delta y = y_{id} - y_{opt} = R_r\left(1 - G_r P_- e^{-sT_d}\right) y_r + R_n\left(1 - G_n P_- e^{-sT_d}\right) y_n \tag{2.1.14}$$

where the error comes from the transfer function $R_x\left(1 - G_x P_- e^{-sT_d}\right)\big|_{x=r,n}$ for both the tracking and noise rejection. The minimization of this error according to different criteria (in theoretical investigations the so-called

$\mathcal{H}_2$ and $\mathcal{H}_\infty$ optimality) can be performed by the optimal choice of $G_x|_{x=r,n}$. See the optimization of G_x in Chapter 4 according to different criteria and norms.

Further equivalent forms of the best reachable optimal control loop are shown in Figure 2.1.8. The simplest realizable form must be chosen from these. Figure 2.1.8(b) gives advice on the realization of the system with dead time. The control loop shown in Figure 2.1.7 is the most general (generic) form of *TDOF* systems. The *YP* was extended for *TDOF* systems by Keviczky and Bányász. They introduced two further parameters R_r and R_n instead of Q (called *KB* parameterization). The derivation of the generic schemes and their optimization possibilities are also connected with their names.

Realizability could be discussed ad infinitum. If the design of the optimal regulator includes the optimization of G_r and G_n, then the procedure itself must also ensure the realizability of the transfer functions $G_r P_-$ and $G_n P_-$, and the realizability of the other factors (like $R_r G_r P_+^{-1}$, $R_n G_n P_+^{-1}$ and $R_n G_n P_-$) have to be also considered. As it was mentioned earlier, the theory concerning the optimality of G_r and G_n will be discussed

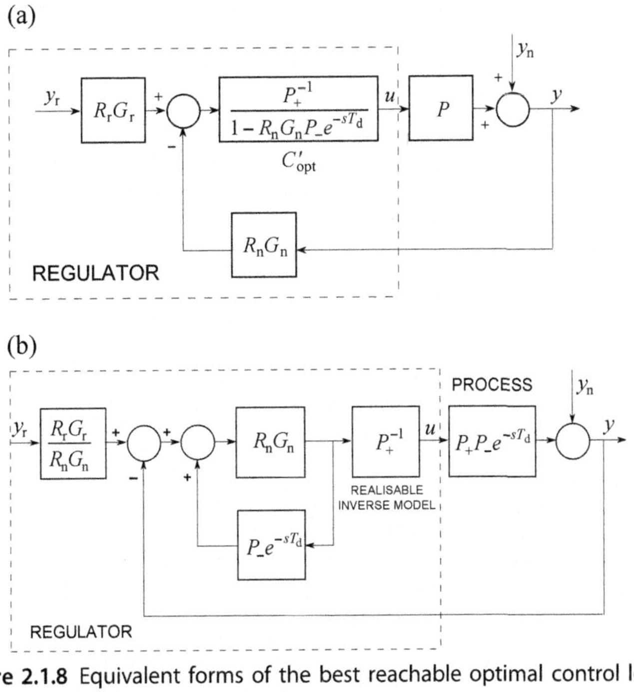

Figure 2.1.8 Equivalent forms of the best reachable optimal control loops.

in the next sections. The simplest nonoptimal choice $G_r = G_n = 1$ does not change the invariant P_-, i.e., it appears, as a consequence, unchanged in the signals of the system

$$\begin{aligned} u &= R_r P_+^{-1} y_r - R_n P_+^{-1} y_n \\ e &= \left(1 - R_r P_- e^{-sT_d}\right) y_r - \left(1 - R_n P_- e^{-sT_d}\right) y_n = (1 - T_r) y_r - S_n y_n \\ y &= R_r P_- e^{-sT_d} y_r + \left(1 - R_n P_- e^{-sT_d}\right) y_n = T_r y_r + (1 - T_n) y_n = T_r y_r + S_n y_n \end{aligned} \tag{2.1.15}$$

Furthermore, the realizability of the transfer functions $R_r P_+^{-1}$, $R_n P_+^{-1}$ and $R_n P_-$ is required. It is well seen that in this case, the realizability can be simply handled by the appropriate choice of the order and pole access of the reference models R_r and R_n, e.g., by prescribing $R_r = 1/(1 + sT_r)^j$ (and the same for R_n). The realizable, but not optimal control loop can be seen in Figure 2.1.9.

Although the regulator is theoretically realizable for CT systems, it cannot be expected in practice that an ideal dead-time element modeling the time delay of the process that can be realized in the internal positive feedback loop of the regulator and in the serial compensator. Therefore in the case of time-delay CT systems, the above discussed optimal control scheme has only theoretical importance. In some cases, the time-delay term can be approximated by higher-order Pade series. In computer-controlled cases, however, the method can be fully applied for discrete-time (DT) (sampled) controls (see later).

Example 2.1.1

Let the controlled system be a first-order time-delay lag

$$P = \frac{1}{1 + 10s} e^{-5s} \quad \text{i.e.,} \quad P_+ = \frac{1}{1 + 10s}; \quad \overline{P}_- = e^{-5s} \quad \text{and} \quad P_- = 1 \tag{2.1.16}$$

which should be speeded up by the control. Let the tracking and disturbance cancellation reference models be

$$R_r = \frac{1}{1 + 4s} \quad \text{and} \quad R_n = \frac{1}{1 + 2s} \tag{2.1.17}$$

Since $P_- = 1$, there is nothing to be optimized, i.e., $G_r = 1$ and $G_n = 1$ can be chosen. The optimal regulator is

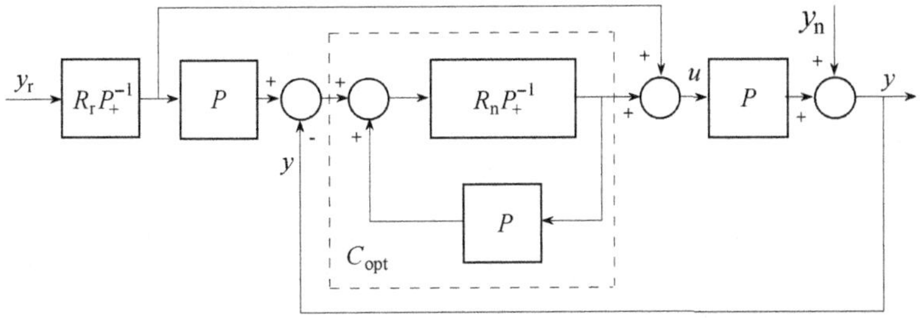

Figure 2.1.9 Realizable Youla-parameterized control loop with the selected $G_r = G_n = 1$.

$$C_{opt} = \frac{R_n G_n P_+^{-1}}{1 - R_n G_n P_- e^{-sT_d}} = \frac{1}{1 - R_n e^{-sT_d}} R_n P_+^{-1} = \frac{1}{1 - \frac{1}{1+2s} e^{-5s}} \frac{1+10s}{1+2s}$$
$$= \frac{1+10s}{1+2s-e^{-5s}} \qquad (2.1.18)$$

and the serial compensator has the form

$$R_r K_r = R_r G_r P_+^{-1} = R_r P_+^{-1} = \frac{1+10s}{1+4s} \qquad (2.1.19)$$

and thus the optimal *TDOF* control loop has the structure shown in Figure 2.1.10. Observe that $C_{opt}(s=0) = \infty$, i.e., the regulator has integrating behavior, which results from the condition $R_n(s=0) = 1$.

It can be easily checked that the output of the closed system is

$$y_{opt} = R_r e^{-sT_d} y_r + \left(1 - R_n e^{-sT_d}\right) y_n = \frac{1}{1+4s} e^{-5s} y_r + \left(1 - \frac{1}{1+2s} e^{-5s}\right) y_n \qquad (2.1.20)$$

which fully corresponds with the designed *TDOF* control loop.

Example 2.1.2

Let the controlled process be a second-order lag

$$P = \frac{(1+5s)(1+6s)}{(1+10s)(1+8s)} = P_+ \quad \text{i.e.,} \quad P_- = 1 \qquad (2.1.21)$$

Suppose that the tracking and disturbance rejection models are again of form (2.1.17). Since $P_- = 1$, there is nothing to be optimally compensated, i.e., $G_r = 1$ and $G_n = 1$ can be chosen. Now the optimal regulator is

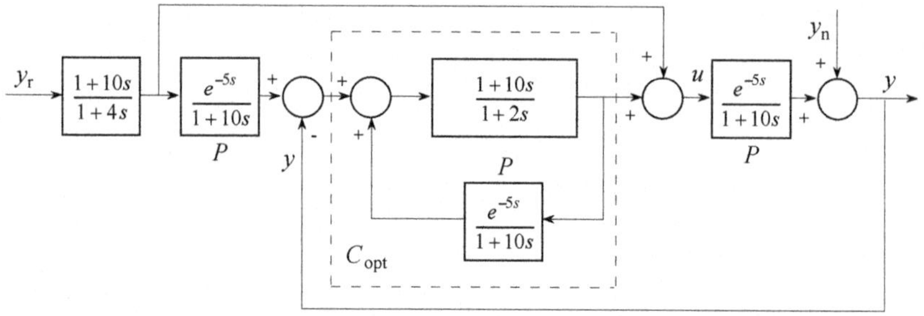

Figure 2.1.10 Optimal control loop of *Example 2.1.1*.

$$C_{opt} = \frac{R_n G_n P_+^{-1}}{1 - R_n G_n P_- e^{-sT_d}} = \frac{R_n}{1 - R_n} P^{-1} = \frac{1}{2s} \frac{(1+10s)(1+8s)}{(1+5s)(1+6s)} = C_{id} \tag{2.1.22}$$

and thus it corresponds to the ideal regulator; the serial compensator, however, has the form

$$R_r K_r = R_r G_r P_+^{-1} = R_r P^{-1} = \frac{(1+10s)(1+8s)}{(1+4s)(1+5s)(1+6s)} \tag{2.1.23}$$

It can be clearly seen that the regulator is of integrating type as also shown in Figure 2.1.11.

Note that in the ideal case the term $R_n/(1 - R_n)$ in the regulator corresponds to an integrator, whose integrating time is equal to the time constant of the first-order reference model R_n.

In the first part of this section, a general regulator parameterization and design method has been discussed. This method, based on the so-called *YP*, was suggested to the design of ODOF and TDOF control loops. The advantage of this method is that the design of the closed system is connected to two reference models, to R_r concerning the reference tracking properties, and to R_n, concerning the noise rejection properties. The design of the regulator can be given in a relatively simple closed form. The disadvantage of the method is that it can be applied only in stable processes.

Sampled Data Systems

As mentioned earlier, the transfer operators $G(\ldots)$ in Section 1.1—depending on the DT formulation—can be the functions of z or z^{-1}, as well as q or q^{-1}. Therefore, the design methods discussed here are actually polynomial (algebraic) methods, so they can be used for all of our models as

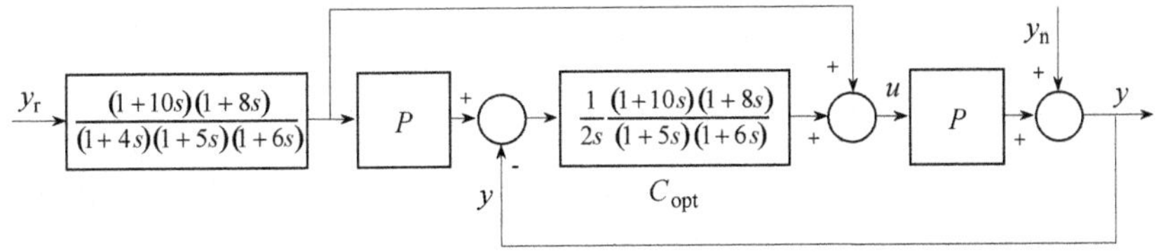

Figure 2.1.11 Optimal control loop of *Example 2.1.2.*

long as the chosen model form is applied consistently. The main advantage of the sampled-data systems is that the signal forming terms—carrying out different methods in the DT computer control systems—can be realized easier.

The presentation of the *YP*, its relationship with the *IMC* principle, the optimality, and the best reachable control were discussed very generally, so—in many cases—it is enough to replace the transfer functions with the pulse transfer functions. All the relations are valid here, too, and therefore the general statements are not repeated; instead the differences and deviances are emphasized. Thus, let the DT process now be

$$G(z^{-1}) = G_+(z^{-1})\overline{G}_-(z^{-1}) = G_+(z^{-1})G_-(z^{-1})z^{-d} \quad \text{or in short}$$

$$G = G_+\overline{G}_- = G_+G_-z^{-d} \tag{2.1.24}$$

where G_+ is stable, and its inverse is also stable and *ISR*. The inverse of $\overline{G}_-$ is *IU* and *IUNR*. In general, the inverse of the delay z^{-d} is unrealizable because it would mean an ideal predictor.

The optimal regulator obtained for the general case (see (2.1.10)) is

$$C_{opt} = \frac{R_n K_n}{1 - R_n K_n G} = \frac{Q_{opt}}{1 - Q_{opt} G} = \frac{R_n G_n G_+^{-1}}{1 - R_n G_n G_- z^{-d}} = R_n G_n C'_{opt} \tag{2.1.25}$$

where the optimal Youla parameter is

$$Q_{opt} = R_n G_n G_+^{-1} = R_n K_n \quad \text{and} \quad K_n = G_n G_+^{-1} \tag{2.1.26}$$

and

$$Q_r = R_r G_r G_+^{-1} = R_r K_r \quad \text{and} \quad K_r = G_r G_+^{-1} \tag{2.1.27}$$

For sampled systems the equivalent optimal control system corresponding to the generalized *IMC* principle is completely the same as that seen in Figure 2.1.7, whose simplified version is shown in Figure 2.1.12.

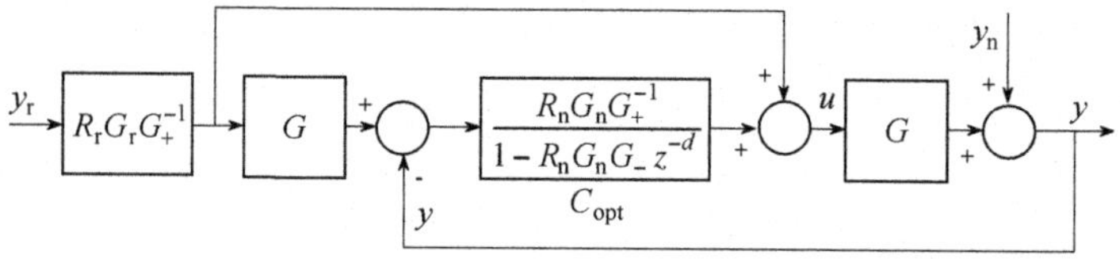

Figure 2.1.12 Optimal sampled-time control system based on the generalized *internal model control* principle.

The most important signals of the *TDOF* closed-control loop are

$$\begin{aligned} u_{opt}[k] &= R_r G_r G_+^{-1} y_r[k] - R_n G_n G_+^{-1} y_n[k] \\ e_{opt}[k] &= \left(1 - R_r G_r G_- z^{-d}\right) y_r[k] - \left(1 - R_n G_n G_- z^{-d}\right) y_n[k] \\ &= \left(1 - T_r^{opt}\right) y_r[k] - S_n^{opt} y_n[k] \\ y_{opt}[k] &= R_r G_r G_- z^{-d} y_r[k] + \left(1 - R_n G_n G_- z^{-d}\right) y_n[k] \\ &= T_r^{opt} y_r[k] + \left(1 - T_n^{opt}\right) y_n[k] = T_r^{opt} y_r[k] + S_n^{opt} y_n[k] \end{aligned} \tag{2.1.28}$$

where the equalities $T_r^{opt} = R_r G_r G_- z^{-d}$ and $T_n^{opt} = R_n G_n G_- z^{-d}$ are valid. The further equivalent forms of the best reachable optimal control systems are shown in Figure 2.1.13. (These figures are only for illustration, their realizability must be investigated in each case!)

As mentioned earlier, the theory of the optimality of G_r and G_n will be discussed later. The nonoptimal simplest choice $G_r = G_n = 1$ leaves unchanged the invariant process factor G_-, so it appears unchanged in the signals of the system.

$$\begin{aligned} u[k] &= R_r G_+^{-1} y_r[k] - R_n G_+^{-1} y_n[k] \\ e[k] &= \left(1 - R_r G_- z^{-d}\right) y_r[k] - \left(1 - R_n G_- z^{-d}\right) y_n[k] = (1 - T_r) y_r[k] - S_n y_n[k] \\ y[k] &= R_r G_- z^{-d} y_r[k] + \left(1 - R_n G_- z^{-d}\right) y_n[k] = T_r y_r[k] + (1 - T_n) y_n[k] \\ &= T_r y_r[k] + S_n y_n[k] \end{aligned} \tag{2.1.29}$$

so the realizability of the transfer functions $R_r G_+^{-1}$, $R_n G_+^{-1}$, and $R_n G_-$, respectively, must be ensured. It can be clearly seen that the realizability can be simply handled by the appropriate choice of the order and pole access of the reference models R_r and R_n. A realizable but nonoptimal control system is shown in Figure 2.1.14.

In the case of sampled-data systems, it is worth noting that with the *SRE* transformation it is always true for the delay-free part $(G_+ G_-)$ in the pulse

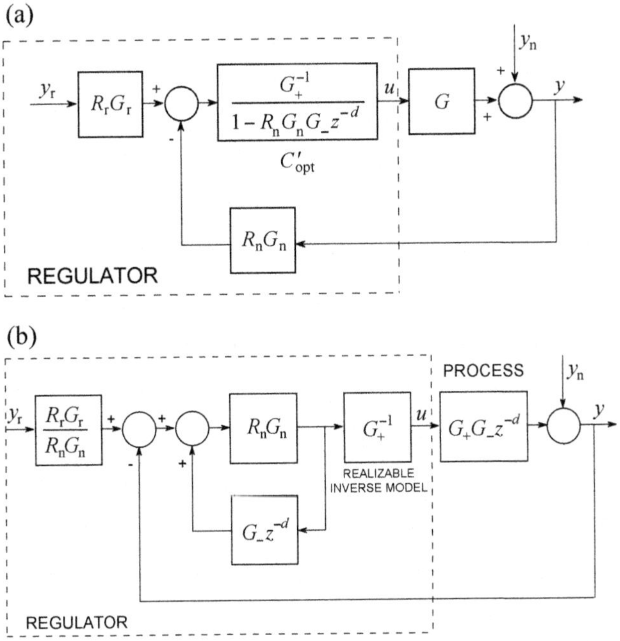

Figure 2.1.13 Equivalent forms of the best reachable optimal sampled data control loop.

transfer function of the process—independently of the pole access of the CT process—that the pole access is equal to one. Therefore, the realizability of the items $R_r G_+^{-1}$ and $R_n G_+^{-1}$ is ensured even for the first-order reference models R_r and R_n.

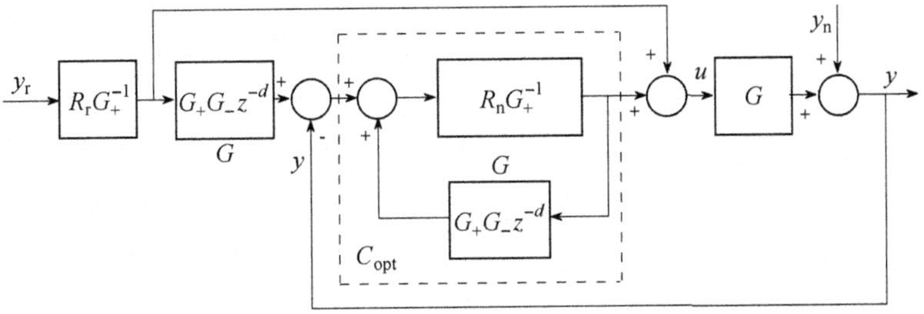

Figure 2.1.14 Youla-parameterized realizable sampled control loop with selected $G_r = G_n = 1$.

Example 2.1.3

Let the controlled system be a first-order process with delay

$$G = \frac{0.2z^{-1}}{1-0.8z^{-1}}z^{-3} = \frac{0.2z^{-4}}{1-0.8z^{-1}} \quad i.e., \quad G_+ = \frac{0.2z^{-1}}{1-0.8z^{-1}} \quad \text{and} \quad G_- = 1 \tag{2.1.30}$$

The goal of the control is to make it faster. Let the tracking and disturbance rejection reference models be

$$R_r = \frac{0.8z^{-1}}{1-0.2z^{-1}} \quad \text{and} \quad R_n = \frac{0.5z^{-1}}{1-0.5z^{-1}} \tag{2.1.31}$$

Since $G_- = 1$ there is nothing to be compensated optimally, i.e., $G_r = 1$ and $G_n = 1$ can be chosen. The optimal regulator is

$$\begin{aligned} C_{opt} &= \frac{R_n G_n G_+^{-1}}{1 - R_n G_n G_- z^{-d}} = \frac{1}{1 - R_n z^{-d}} R_n G_+^{-1} \\ &= \frac{1}{1 - \dfrac{0.5z^{-1}}{1-0.5z^{-1}}z^{-3}} \frac{0.5z^{-1}}{1-0.5z^{-1}} \frac{1-0.8z^{-1}}{0.2z^{-1}} = \frac{2.5(1-0.8z^{-1})}{1-0.5z^{-1}-0.5z^{-4}} \end{aligned} \tag{2.1.32}$$

but the serial compensation, however, has the form

$$R_r G_+^{-1} = \frac{0.8z^{-1}}{1-0.2z^{-1}} \frac{1-0.8z^{-1}}{0.2z^{-1}} = \frac{4(1-0.8z^{-1})}{1-0.2z^{-1}} \tag{2.1.33}$$

so the optimal *TDOF* control loop has the scheme shown in Figure 2.1.15. (see Figure 2.1.13(b)). Note that $C_{opt}(z=1) = \infty$, i.e., the regulator has an integrating character, which comes from the condition $R_n(z=1) = 1$.

It can be easily checked that the output of the closed loop is

$$\begin{aligned} y_{opt}[k] &= R_r z^{-d} y_r[k] + (1 - R_n z^{-d}) y_n[k] \\ &= \frac{0.8z^{-1}}{1-0.2z^{-1}}z^{-3} y_r[k] + \left(1 - \frac{0.5z^{-1}}{1-0.5z^{-1}}z^{-3}\right) y_n[k] \\ &= \frac{0.8z^{-4}}{1-0.2z^{-1}} y_r[k] + \left(1 - \frac{0.5z^{-4}}{1-0.5z^{-1}}\right) y_n[k] \end{aligned} \tag{2.1.34}$$

which fully corresponds to the designed *TDOF* control system.

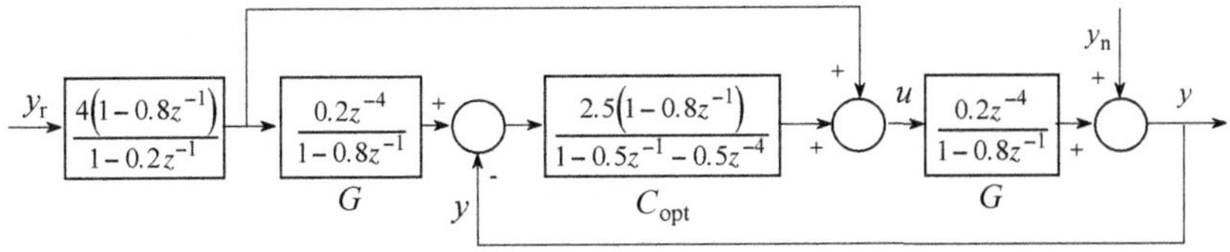

Figure 2.1.15 Optimal control loop of *Example 2.1.3.*

Substituting Unstable Zeros with Additional Time Delay

As seen in Eq. (1.1.17), all nonminimum phase transfer functions H_{nmp} can be written as a product of an all-pass term H_{ap} and a minimum-phase term H_{mp}. This phenomenon is investigated now for the unstable zero $z_1^f = q$

$$s - q = -(s+q)\frac{1 - sq^{-1}}{1 + sq^{-1}} = -(s+q)\frac{1 - sT}{1 + sT} \approx -(s+q)e^{-s\Delta T_d} \quad (2.1.35)$$

where the all-pass term $e^{-s\Delta T_d}$ is approximated by the first-order Pade approximation with delay $\Delta T_d = T/2 = 1/2q$. If the zero is given in Bode form

$$1 - sT = (1 + sT)\frac{1 - sT}{1 + sT} \approx (1 + sT)e^{-s\Delta T_d} \quad (2.1.36)$$

Thus, a CT nonminimum phase, unstable zero can be substituted with good approximation by a delay increased by ΔT_d.

Sampled Data Systems

Consider now an unstable zero $z_1^d = b$ of a DT process, where

$$z - b = (z^{-1} - b)\frac{z - b}{z^{-1} - b} = (-b)(1 - b^{-1}z^{-1})\frac{z - b}{z^{-1} - b} \quad (2.1.37)$$

Investigating the last all-pass term in the frequency domain we obtain

$$\left.\frac{z - b}{z^{-1} - b}\right|_{z=e^{j\omega T_s}} = \frac{e^{j\omega T_s} - b}{e^{-j\omega T_s} - b} \cong \frac{1 + j\omega T_s - b}{1 - j\omega T_s - b}$$
$$= \left(1 - \frac{j\omega T_s}{b-1}\right) \Big/ \left(1 + \frac{j\omega T_s}{b-1}\right) \cong e^{-j\omega \Delta T_d} \quad (2.1.38)$$

i.e., the corresponding approximating delay is $\Delta T_d = 2T_s/(b-1)$. The DT delay increment is

$$\Delta d = \frac{\Delta T_d}{T_s} = \frac{2}{b-1} \tag{2.1.39}$$

or more precisely

$$\Delta d = \text{int}\left\{\frac{2}{b-1}\right\} \tag{2.1.40}$$

It is reasonable in practice—in both CT and DT systems—to overestimate the calculated additional delays.

See further analysis on delay mismatch in Section 9.4, which evaluates robust stability and performance.

$\mathcal{V}-\mathcal{X}-\mathcal{Y}$ Form of *YP* Regulators

Using the special form of the Youla-parameterized regulator shown in Figure 2.1.13(a) and assuming $G = G_+G_- = \frac{\mathcal{B}}{\mathcal{A}} = \frac{\mathcal{B}_+}{\mathcal{A}}\mathcal{B}_-$, $G_r = G_n = 1$, and furthermore applying the reference models $R_r = \frac{\mathcal{B}_r}{\mathcal{A}_r}$ and $R_n = \frac{\mathcal{B}_n}{\mathcal{A}_n}$, we obtain the optimal regulator

$$C'_{opt} = \frac{G_+^{-1}}{1 - R_n G_- z^{-d}} = \frac{\frac{\mathcal{A}}{\mathcal{B}_+}}{1 - \frac{\mathcal{B}_n}{\mathcal{A}_n}\mathcal{B}_- z^{-d}} = \frac{\mathcal{A}_n\mathcal{A}}{\mathcal{B}_+(\mathcal{A}_n - \mathcal{B}_n\mathcal{B}_- z^{-d})} \tag{2.1.41}$$

Consequently, the output of the regulator can be computed as

$$u = C\left(\frac{\mathcal{B}_r}{\mathcal{A}_r}y_r - \frac{\mathcal{B}_n}{\mathcal{A}_n}y\right) = \frac{\mathcal{A}_n\mathcal{A}}{\mathcal{B}_+(\mathcal{A}_n - \mathcal{B}_n\mathcal{B}_- z^{-d})}\left(\frac{\mathcal{B}_r\mathcal{A}_n}{\mathcal{A}_r\mathcal{A}_n}y_r - \frac{\mathcal{B}_n\mathcal{A}_r}{\mathcal{A}_n\mathcal{A}_r}y\right) \tag{2.1.42}$$

which corresponds to the

$$\mathcal{B}_+\mathcal{A}_r\mathcal{A}_n(\mathcal{A}_n - \mathcal{B}_n\mathcal{B}_- z^{-d})u = \mathcal{A}\mathcal{A}_n^2\mathcal{B}_r y_r - \mathcal{A}\mathcal{A}_n\mathcal{B}_n\mathcal{A}_r y \tag{2.1.43}$$

polynomial equation; in short

$$\mathcal{X}u = \mathcal{V}y_r - \mathcal{Y}y \tag{2.1.44}$$

Introducing the $\mathcal{X} = x_o + \tilde{\mathcal{X}}$ notation the regulator output is

$$u = \frac{1}{x_o}\left(\mathcal{V}y_r - \mathcal{Y}y - \tilde{\mathcal{X}}u\right) \tag{2.1.45}$$

which is a very reasonable form for the direct coding of this algorithm and corresponds to (1.4.54).

2.2 OTHER CLASSICAL PARAMETERIZED REGULATORS

Before the introduction of the *YP*, there were other control design methods that—in some senses—had similar structural properties to the *YP* regulators.

Truxal–Guillemin Regulator

Truxal and Guillemin were the first to recommend a simple algebraic method for the control design of *ODOF* systems. According to the method, the required design goal needs to be formulated for the transfer function of the closed loop

$$R_n = T = \frac{CP}{1+CP} \tag{2.2.1}$$

from which a simple algebraic equation results for C

$$CP = R_n + CPR_n \tag{2.2.2}$$

Expressing the regulator we get

$$C = \frac{R_n}{1-R_n}\frac{1}{P} = C_{TG} \tag{2.2.3}$$

Note that this form is the same as the simple case of the Youla regulator C_{id} in (2.1.5). The realization of the regulator can be achieved according to Figure 2.2.1.

Thus, R_n corresponds to one of the reference models of the Youla method. For the *ODOF* case, however, $R_n = R_r$. Let the reference model be $R_n = \mathcal{B}_n/\mathcal{A}_n$, and let the process be $P = \mathcal{B}/\mathcal{A}$. So the polynomial form of the regulator is

$$C_{TG} = \frac{\mathcal{B}_n}{\mathcal{A}_n - \mathcal{B}_n}\frac{\mathcal{A}}{\mathcal{B}} \tag{2.2.4}$$

The regulator is realizable if the pole access of R_n is greater or equal to that of the process. If R_n has unity gain ($R_n(0) = 1$), then the type of the

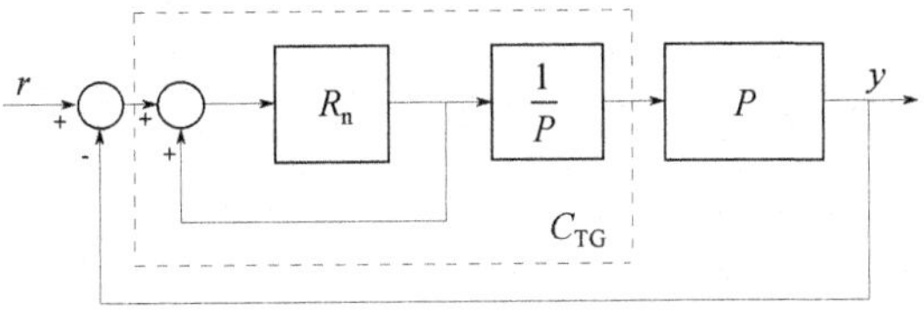

Figure 2.2.1 Realization of the Truxal–Guillemin regulator.

regulator is one. Truxal observed that the loop transfer function $L = \mathcal{N}/s^k\mathcal{D} = CP$ of type k can be established by the reference model

$$R_n = T = \frac{\mathcal{N}}{\mathcal{N} + s^k\mathcal{D}} = \frac{n_o + n_1 s + \ldots + n_{k-1}s^{k-1}}{n_o + n_1 s + \ldots + n_{k-1}s^{k-1} + s^k + \ldots d_{n_R+k}s^{n_R+k}}$$

$$= \frac{n_o + n_1 s + \ldots + n_{k-1}s^{k-1}}{n_o + n_1 s + \ldots + n_{k-1}s^{k-1} + s^k(1 + \ldots d_{n_R}s^{n_R})}; \quad n_R - k - 1 \geq n - m \tag{2.2.5}$$

where the first k terms of the denominator are equal to the numerator. This initial model can be found—with or without names—in several early publications.

The Truxal–Guillemin method can be applied for the design of regulator of *ODOF*-DT control loops. According to the method the design goal can be formulated for the transfer function of the closed system, whose delay process is

$$T = \frac{CG}{1 + CG} = \frac{CG_+ z^{-d}}{1 + CG_+ z^{-d}} = R_n z^{-d} \tag{2.2.6}$$

where it is assumed that $G_- = 1$ in the DT process (2.1.24), and hence the following simple algebraic equation is obtained for C

$$CG_+ = R_n + CG_+ z^{-d} R_n \tag{2.2.7}$$

Expressing the regulator we obtain

$$C = \frac{R_n}{1 - R_n z^{-d}}(G_+)^{-1} = \frac{R_n}{1 - R_n z^{-d}} G^{-1} = C_{TG} \tag{2.2.8}$$

Note that this form is exactly the same as the basic case of the Youla regulator (2.1.25) for ($G_n = 1$, $G_- = 1$). The realization of the regulator can be performed according to Figure 2.2.2, and realizing the regulator in computer-aided control systems is no problem at all.

Let the reference model be given in the form $R_n = \mathcal{G}_n/\mathcal{F}_n$, and the process $G = \mathcal{G}/\mathcal{F}$. The polynomial form of the regulator is obtained as

$$C_{TG} = \frac{\mathcal{G}_n}{\mathcal{F}_n - \mathcal{G}_n} \frac{\mathcal{F}}{\mathcal{G}} \tag{2.2.9}$$

The regulator can be realized if the pole access of R_n is greater or equal to that of the process. It was seen in Section 1.1. that in the DT case the pole access of the pulse transfer function of the process is unity (for the most

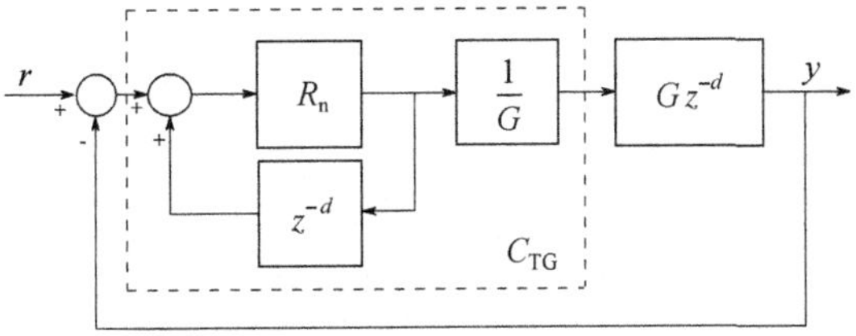

Figure 2.2.2 Realization of the Truxal-Guillemin regulator in sampled systems.

general case of the practice having zero hold, i.e., in the case of unit step equivalent transformation). Thus, in general, the regulator (2.2.8) is realizable because such R_n is usually not chosen when the pole access is less than one. If R_n has unity gain ($R_n(1) = 1$), then the type of the regulator is one.

Smith Regulator

The handling of the time delay of the processes has required particular attention on the part of designers of control loops from the beginning. First, Otto Smith [145,146] suggested a technique by means of which it was thought for a long time that the regulator can be designed without the consideration of the dead time. To understand his method, let us consider a simple dead-time process of (2.1.9)

$$P(s) = P_+(s)\overline{P}_-(s) = P_+(s)e^{-sT_d} \quad \text{or shortly} \quad P = P_+\overline{P}_- = P_+e^{-sT_d} \tag{2.2.10}$$

where P_+ is stable. Figure 2.2.3(a) shows the original idea of Smith. Since this figure is equivalent to Figure 2.2.3(b), his main goal is clearly seen, namely to change the original dead-time loop to a closed loop which does not contain the dead time and to serially connected dead time. So the regulator C_+ regulating the process P_+ can be designed by a conventional method (without considering the dead time at all).

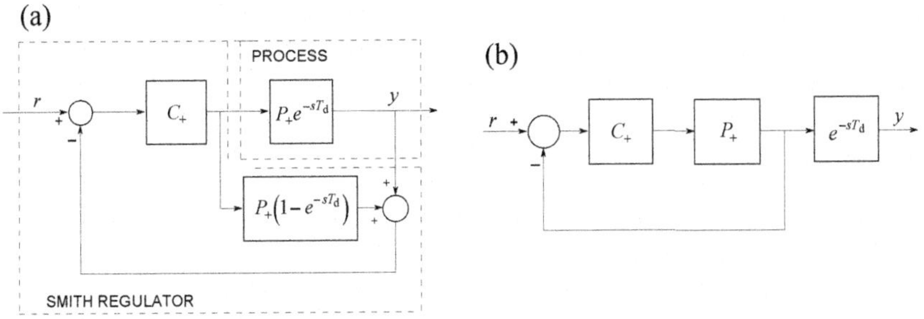

Figure 2.2.3 Block scheme of the Smith regulator.

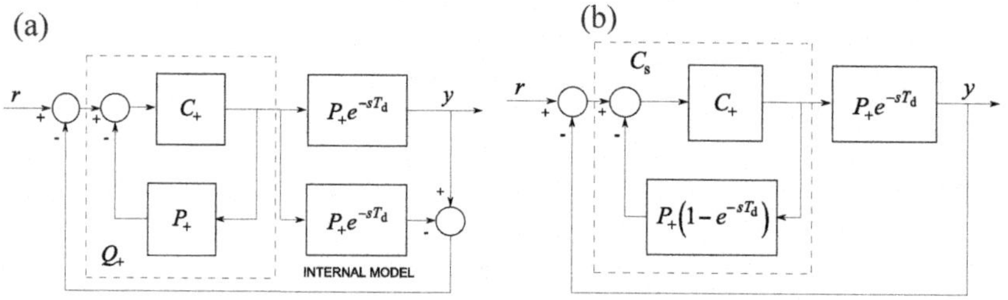

Figure 2.2.4 Equivalent Smith regulator block schemes.

By simple block manipulations, Figure 2.2.3(a) can be redrawn to the equivalent forms of Figure 2.2.4(a) and (b).

The *IMC* structure of Figure 2.2.4(a) clearly shows that the Smith regulator is a *YP* regulator with a special Youla parameter

$$Q_+ = \frac{C_+}{1 + C_+P_+} = \frac{C_+P_+}{1 + C_+P_+}P_+^{-1} = \frac{L_+}{1 + L_+}P_+^{-1} = R_+P_+^{-1} \tag{2.2.11}$$

if the regulator C_+ stabilizes the delay-free part of the process P_+. Here $L_+ = C_+P_+$ is the loop transfer function of the closed system of Figure 2.2.3(b) and furthermore the complementary sensitivity function

$$T_+ = R_+ = \frac{L_+}{1 + L_+} \tag{2.2.12}$$

is the reference model R_+. Since in the *IMC* structure the internal model predicts the output of the process, the term Smith predictor derives from this phenomenon. At the time of its introduction, the *IMC* principle and the *Y* parameterization were unknown.

Figure 2.2.4(b) shows the equivalent complete closed-loop control, where the serial (*Y*-parameterized) regulator is

$$C_s = \frac{Q_+}{1 - Q_+P_+e^{-sT_d}} = \frac{C_+}{1 + C_+P_+(1 - e^{-sT_d})} = C_+K_S \tag{2.2.13}$$

and, at the same time, also shows the inner closed loop referring to the realization. Here K_S means the serial transfer function by which the Smith regulator modifies the effect of the original regulator C_+. Thus,

$$K_S = \frac{1}{1 + C_+P_+(1 - e^{-sT_d})} = \frac{1}{1 + L_+(1 - e^{-sT_d})} \tag{2.2.14}$$

On the stability limit $L_+ = -1$, thus we get

$$K_S = \frac{1}{1 + (-1)(1 - e^{-sT_d})} = \frac{1}{1 - 1 + e^{-sT_d}} = e^{sT_d}\Big|_{\omega_c} = e^{j\omega_c T_d} \tag{2.2.15}$$

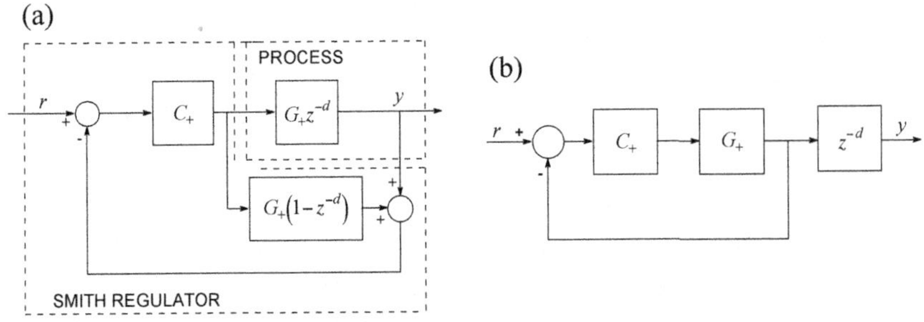

Figure 2.2.5 Scheme of the sampled data Smith regulator.

which means that the Smith regulator adds significant positive phase access to the original closed loop, and that is why it can be applied very successfully in many cases of stabilization. At the same time, it is very sensitive to any change in parameters.

With regard to the complete evaluation of the Smith regulator should also be mentioned that it can be used only for the design of the *ODOF* system, i.e., for tracking. The regulator designs the tracking of the reference signal only indirectly, as the expression $T_+ = R_+$ of (2.2.12) shows. Elaboration of the concept of *YP* shows that a simple design method is also available for *TDOF* systems in terms of both tracking and disturbance rejections via the design of the reference models.

The Smith regulator has not been used widely for CT systems. The main reason is that it cannot be expected in practice to realize an ideal CT dead time that is equivalent to the process dead time (see Example 2.1.1 in the previous section).

Let us consider a simple process with delay based on (2.1.24) in the sampled data control system

$$G(z^{-1}) = G_+(z^{-1})\overline{G}_-(z^{-1}) = G_+(z^{-1})G_-(z^{-1})z^{-d} \quad \text{or shortly}$$

$$G = G_+\overline{G}_- = G_+G_-z^{-d} \tag{2.2.16}$$

where G_+ is stable. For a DT process (2.2.16) the Smith-predictor principle is shown by the control system given in Figure 2.2.5(a). Since this control loop is equivalent to the scheme given in Figure 2.2.5(b), the goal of the control is well seen, i.e., to change the original closed loop containing the delay to a delay-free closed loop in which the delay appears to be serially connected. So the regulator C_+ of the process G_+ can also be designed by a conventional method (not taking the delay into account at all).

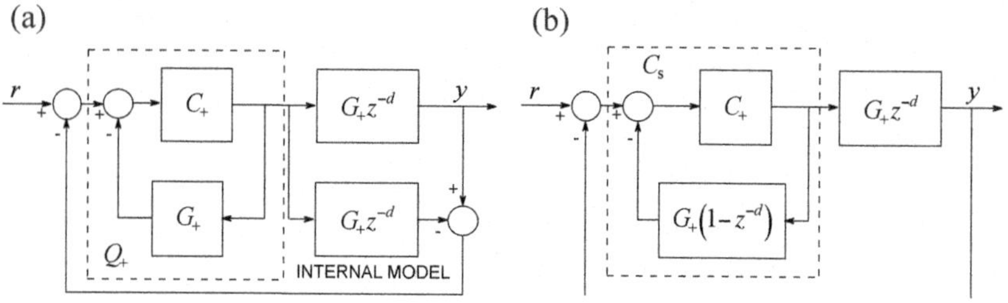

Figure 2.2.6 Schemes of equivalent sampled data Smith regulator.

Figure 2.2.5(a) can be redrawn for the equivalent forms (a) and (b) of Figure 2.2.6 by simple block manipulations.

The *IMC* structure of Figure 2.2.6(a) clearly shows that the Smith regulator is a *Y*-parameterized special regulator with Youla parameter

$$Q_+ = \frac{C_+}{1 + C_+ G_+} = \frac{C_+ G_+}{1 + C_+ G_+} G_+^{-1} = \frac{L_+}{1 + L_+} G_+^{-1} = R_+ G_+^{-1} \quad (2.2.17)$$

if the regulator C_+ stabilizes the delay-free part G_+ of the process. Here, $L_+ = C_+ G_+$ is the loop transfer function of the closed loop shown in Figure 2.2.5(b), and furthermore the complementary sensitivity function

$$T_+ = R_+ = \frac{L_+}{1 + L_+} \quad (2.2.18)$$

is the reference model R_+.

Figure 2.2.6(b) shows the equivalent complete closed loop, where the *YP* sampled data serial regulator is

$$C_s = \frac{Q_+}{1 - Q_+ G_+ z^{-d}} = \frac{C_+}{1 + C_+ G_+ (1 - z^{-d})} = C_+ K_S \quad (2.2.19)$$

which, at the same time, suggests the realization of the internal closed loop representing the mode of the realizability. Here K_S represents the serial transfer function by means of which the Smith regulator modifies the effect of the original regulator C_+. Thus,

$$K_S = \frac{1}{1 + C_+ G_+ (1 - z^{-d})} = \frac{1}{1 + L_+ (1 - z^{-d})} \quad (2.2.20)$$

In contrast to CT systems, the realization of the sampled time Smith regulator is not difficult in practice, since C_s can be easily realized partly or completely by computer-aided systems (see Appendix 3 on linear DT filters).

Dahlin Algorithm

In the early years of DT systems, the so-called Dahlin algorithm was very popular and was used to design the DT regulator of simple, first-order lag processes with dead time [40]. Let the transfer function of such a system be

$$P(s) = \frac{K}{1+sT}e^{-sT_d} \tag{2.2.21}$$

The step response equivalent DT pulse transfer function has the form

$$G(z) = \frac{g_1}{z+f_1}z^{-d} = \frac{g_1 z^{-1}}{1+f_1 z^{-1}}z^{-d} = \frac{K(1+f_1)z^{-1}}{1+f_1 z^{-1}}z^{-d} \tag{2.2.22}$$

(see (A.3.45)). Dahlin's regulator design concept was completely the same as the Truxal–Guillemin (see C_{TG} in (2.2.3)) or the *YP* (see C_{id} in (2.1.5)). The method can be interpreted more easily if it is assumed in (2.1.25) valid for DT systems that

$$G_+ = \frac{g_1 z^{-1}}{1+f_1 z^{-1}}; \quad G_- = 1; \quad G_n = 1 \tag{2.2.23}$$

and the design task is to reach the reference model

$$R_n = \frac{g_1^n z^{-1}}{1+f_1^n z^{-1}} \tag{2.2.24}$$

The optimal *Y*-parameterized regulator can be obtained in this case with a very simple calculation, i.e.,

$$C_{opt} = \frac{R_n G_+^{-1}}{1-R_n z^{-d}} = \frac{1+f_1 z^{-1}}{K(1+f_1)z^{-1}} \frac{\frac{g_1^n z^{-1}}{1+f_1^n z^{-1}}}{1-\frac{g_1^n z^{-1}}{1+f_1^n z^{-1}}z^{-d}}$$

$$= \frac{g_1^n}{K(1+f_1)} \frac{1+f_1 z^{-1}}{1+f_1^n z^{-1} - g_1^n z^{-(d+1)}} \tag{2.2.25}$$

Let the CT reference model corresponding to the DT one in (2.2.24) be

$$R_n^c = \frac{1}{1+sT_n} \tag{2.2.26}$$

On the basis of (A.3.45) (containing the *SRE* transformation of a first-order lag), it can be written that

$$g_1 = K\left(1 - e^{-x}\right); \quad f_1 = -e^{-x}; \quad x = T_s/T \tag{2.2.27}$$

and

$$g_1^n = 1 - e^{-x_n}; \quad f_1^n = -e^{-x_n}; \quad x_n = T_s/T_n \tag{2.2.28}$$

Thus, the Dahlin regulator has the form

$$C_{\text{Dahlin}} = \frac{1 - e^{-x_n}}{K(1 - e^{-x})} \frac{1 - e^{-x} z^{-1}}{1 - e^{-x_n} z^{-1} - (1 - e^{-x_n}) z^{-(d+1)}} \tag{2.2.29}$$

The complementary sensitivity function of the overall DT closed system is

$$T = \frac{g_1^n z^{-1}}{1 + f_1^n z^{-1}} z^{-d} \tag{2.2.30}$$

and the complementary sensitivity function of the overall CT closed system has the form

$$T^c = \frac{1}{1 + sT_n} e^{-sT_d} \tag{2.2.31}$$

The popularity of the Dahlin regulator in the early days can be explained by the following facts: in several industrial applications the form (2.2.21) is a good approximation for the process model, and the Dahlin regulator can be given in a fixed equation form (2.2.29) not requiring further considerations.

2.3 DEADBEAT REGULATORS

In sampled-data systems, it is possible to construct regulators which can perfectly track the unit step reference signal in finite steps, i.e., make the error signal zero in finite steps. This kind of regulator is called a *deadbeat* (*DB*) regulator [37,105].

Assume that the relative prime $G = \mathcal{G}/\mathcal{F}$ represents the DT process in the *ODOF* control loop and C_{DB} is the *DB* regulator to be designed. Supposing a unit step reference signal, the z-transform of the reference signal is $R(z) = z/(z-1) = 1/(1-z^{-1})$. The *DB* property requires that the z-transform of the error must be finite-order polynomial: $\mathcal{P}_e(z)$, i.e.,

$$E(z) = S(z)R(z) = \frac{1}{1 + C_{DB}G} R(z) = \left[1 - \frac{C_{DB}G}{1 + C_{DB}G}\right] R(z)$$
$$= \mathcal{P}_e(z) \tag{2.3.1}$$

It follows from this that both the sensitivity function and the overall pulse transfer function of the closed system must be a finite-order polynomial, i.e., the polynomials in (2.3.1)

$$S = \left(1 - z^{-1}\right)\mathcal{P}_e\left(z^{-1}\right); \quad T = \frac{C_{DB}G}{1 + C_{DB}G} = 1 - \left(1 - z^{-1}\right)\mathcal{P}_e(z)$$
$$= \mathcal{P}_y(z) \tag{2.3.2}$$

are of finite orders. Similar requirements are valid for the pulse transfer function regarding the output of the regulator, it must be also finite-order polynomial.

$$\frac{C_{DB}}{1 + C_{DB}G} = \mathcal{P}_u(z) \tag{2.3.3}$$

These kinds of transfer functions are called *finite impulse response* (*FIR*) types or moving average filters. Therefore, it can be written that

$$T = \frac{C_{DB}G}{1 + C_{DB}G} = \mathcal{P}_y(z) = \mathcal{P}_u(z)\frac{\mathcal{G}}{\mathcal{F}} \tag{2.3.4}$$

where the condition of *DB* controlling is

$$\frac{\mathcal{G}(z)}{\mathcal{F}(z)} = \frac{\mathcal{P}_y(z)}{\mathcal{P}_u(z)} = G \tag{2.3.5}$$

which in the case of relative prime process polynomials, can be fulfilled only if

$$\mathcal{P}_y(z) = \mathcal{M}(z)\mathcal{G}(z); \quad \mathcal{P}_u(z) = \mathcal{M}(z)\mathcal{F}(z) \tag{2.3.6}$$

Since in steady state the error is zero, the condition $\mathcal{P}_y(1) = 1$ must also be fulfilled. As a consequence, for the gain of the design polynomial $\mathcal{M}(z)$

$$\mathcal{M}(1) = \frac{1}{\mathcal{G}(1)} \tag{2.3.7}$$

is obtained. Finally, based on (2.3.3), (2.3.4), and (2.3.6), the form of the regulator is

$$C_{DB} = \frac{\mathcal{P}_u}{1 - \mathcal{P}_u G} = \frac{\mathcal{M}\mathcal{F}}{1 - \mathcal{M}\mathcal{G}} \tag{2.3.8}$$

Thus, the most important step of the design of the *DB* regulator is the proper choice of the design polynomial $\mathcal{M}(z)$. In the simplest cases

$\mathcal{M}(z) = z^{-j}$ is chosen, when the design of the coefficients in $\mathcal{M}(z)$ does not need any further considerations.

The termination of the input and output signals of the process happens according to (2.3.6). It is helpful to investigate the shape of the error signal by using (2.3.1) and (2.3.8)

$$\mathcal{P}_{\mathrm{e}}(z) = \frac{z(1 - \mathcal{MG})}{z - 1} = \frac{1 - \mathcal{MG}}{1 - z^{-1}} = \mathcal{N} \tag{2.3.9}$$

Here (2.3.7) is taken into account, according to which $z = 1$ is always the root of the factor $(1 - \mathcal{MG})$, since $1 - \mathcal{M}(1)\mathcal{G}(1) = 0$.

Further forms of (2.3.8) exist

$$C_{\mathrm{DB}} = \frac{\mathcal{MF}}{1 - \mathcal{MG}} = \frac{\mathcal{P}_{\mathrm{u}}}{1 - \mathcal{P}_{\mathrm{y}}} = \frac{\mathcal{P}_{\mathrm{y}}}{1 - \mathcal{P}_{\mathrm{y}}} \frac{\mathcal{F}}{\mathcal{G}} = \frac{R_{\mathrm{n}}}{1 - R_{\mathrm{n}}} \frac{\mathcal{F}}{\mathcal{G}} \tag{2.3.10}$$

where the substitution $R_{\mathrm{n}} = \mathcal{P}_{\mathrm{y}}$ is applied. Thus, again the same form is obtained as for the Truxal–Guillemin regulator in (2.2.4) or the basic Youla regulator in (2.1.5). The significant difference is that now R_{n} is a *FIR* filter, i.e., is a polynomial and (2.3.9) must be fulfilled.

To summarize the barriers in connection with the *DB* regulator design:

- it is assumed that the controlled process is stable
- it is assumed that the reference signal of the closed loop is a unit step
- the *DB* control is in operation only in the sampling points.

Note that even in those cases, when the above conditions do not exist, there is still a chance to design a *DB* regulator (e.g., polynomial or state-space technique), but other, less simple and transparent, design methods need to be applied.

Example 2.3.1

The method is illustrated for a second-order CT process with dead time. Let the transfer function of the CT process be

$$P(s) = \frac{e^{-s}}{(1 + 10s)(1 + 5s)}. \tag{2.3.11}$$

The first step is to discretize the continuous process by applying a zero-order hold. Using sampling time $T_{\mathrm{s}} = 1$ s

$$G(z) = \frac{\mathcal{G}(z)}{\mathcal{F}(z)} = \frac{0.0091(z + 0.9048)}{(z - 0.9048)(z - 0.8187)z} \tag{2.3.12}$$

is obtained for the discretized model (with *SRE* transformation). Observe the factor z in the denominator, which represents the time delay $T_d = 1$ s, since $T_s = T_d$. The time delay can be better recognized in the form

$$G(z) = \frac{0.0091(z + 0.9048)}{(z - 0.9048)(z - 0.8187)} z^{-1} \tag{2.3.13}$$

Concerning the goal of the design, the following statements can be made. Without time delay in the process, the effect of the input signal $u[0] \neq 0$, put to the zero-order hold element in the time moment $k = 0$, appears, at the earliest, in the output in the time moment $k = 1$ because of the order of the process. Consequently, with time delay of $T_d = 1$ s, the effect of $u[0] \neq 0$ appears, at the earliest, in the output in the time moment $k = 2$. It follows from this that the best tracking regulator which can be constructed will provide the discrete signal $y[k] = 1[k - 2]$ for the unit step reference signal $y_r[k] = 1[k]$. Applying a pulse transfer function, we can state the following condition (see (2.3.4)) for the transfer function of the closed system

$$T = \frac{C(z)G(z)}{1 + C(z)G(z)} = \mathcal{P}_y(z) = z^{-2},$$

from which the regulator is

$$C(z) = \frac{\mathcal{P}_y}{1 - \mathcal{P}_y} \frac{\mathcal{F}}{\mathcal{G}} = \frac{z^{-2}}{(1 - z^{-2})} \frac{1}{G(z)} = \frac{1}{G(z)(z^2 - 1)} = C_{DB}.$$

Expressing the regulator with the polynomials of the transfer function of the discretized process, we obtain

$$\begin{aligned} C(z) &= \frac{1}{G(z)(z^2 - 1)} = \frac{\mathcal{F}(z)}{\mathcal{G}(z)(z^2 - 1)} \\ &= \frac{109.9(z^3 - 1.7236z^2 + 0.7408z)}{z^3 + 0.9048z^2 - z - 0.9048} \end{aligned} \tag{2.3.14}$$

It can be clearly seen that for unit step reference signal the steady-state output is error-free, since the loop transformation function $L(z) = C(z)G(z)$ has a pole at $z = 1$, i.e., the regulator has an integrating effect. The operation of the closed loop can be followed in Figure 2.3.1. Concerning the DT instants, the output has the desired shape, but even in the sampled-data systems, the quality of the operation of the closed system is finally evaluated on the basis of the shape of the *continuous output signal*, which shows some unacceptable oscillations. Furthermore, the actuator

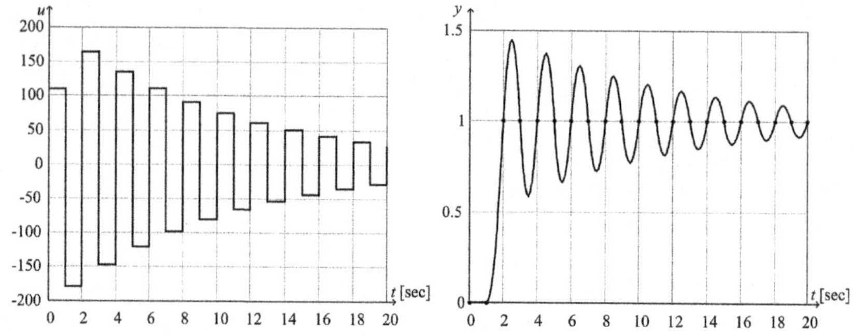

Figure 2.3.1 Deadbeat control (two steps).

signal has no *DB* character; its changes have extreme dynamics because condition (2.3.6) is not fulfilled.

Let us now investigate the so-called *hidden oscillations* between the sampling instants. These are also called *intersampling ripples* in the literature. On the basis of the time diagrams, it can be seen that the oscillations of the continuous output are caused by the oscillations of the continuous step-like input signal generated by the zero-order hold element. These values are generated by the regulator because that it has a slightly damped pole of $p_1 = -0.9048$. The relevance of this qualification can be explained in two ways. Let us first investigate exclusively the effect of the mentioned pole and look at the step response in Figure 2.3.2 of the pulse transfer function

$$G_1(z) = \frac{z+1}{z+0.9048} \tag{2.3.15}$$

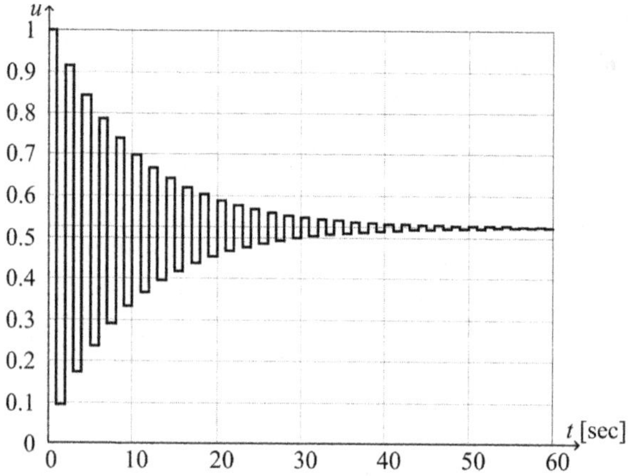

Figure 2.3.2 Unwanted dynamics in the actuating signal.

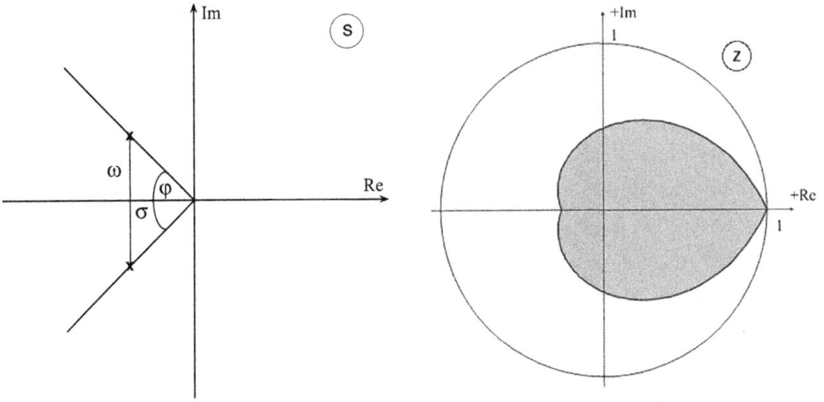

Figure 2.3.3 Region of the well-damped poles in the z-plane (see also Figure 1.3.7).

Considering only the pulse transfer function of the regulator itself, the question arises as to the location of those poles, which result in stable well-attenuated step response, i.e., for unit step input they do not generate oscillating output. In the case of CT systems, the regions of the well-damped and slightly damped poles are separated by lines belonging to constant attenuations (damping factors) in the complex frequency region s. These lines are at constant angle φ, depending on the damping, to the negative real axis: $\cos(\varphi) = \xi$. Now the mapping of these lines needs to be found in the z plane. On the s plane the points $s = \sigma + j\omega$ are on the lines of the constant damping with the condition $\omega = \sigma\sqrt{1-\xi^2}/\xi$. For a given damping, the mapping of $z = e^{sT_s}$ to the $z = e^{\sigma + j\sigma T_s\sqrt{1-\xi^2}/\xi}$ can be calculated and drawn for different σ values. For example, the corresponding form in the z plane of the constant line $\xi = 0.4$ can be seen in Figure 2.3.3. This well-damped region ($\xi > 0.4$), shown in the figure, is also called "heart form curve" in the literature. The formula $\xi = 1\Big/\sqrt{1+\pi^2/[\ln\left(|p_1|\right)]^2}$ is obtained after some long calculations showing how ξ depends on a root p_1 falling to the negative real axis.

Based on the previous investigations, it can be stated that the oscillation is generated by the regulator itself because it can be clearly seen from

$$C(z) = \frac{\mathcal{F}(z)}{\mathcal{G}(z)(z^2-1)} \tag{2.3.16}$$

that $C(z)$ has the roots of the polynomial $\mathcal{G}(z)$ as its poles (in this case $\mathcal{G}(z)$ is of first order, i.e., it has only one root). The oscillating effect of the roots depends on their positions related to the heart-type figure belonging to a given damping. In the present case, the root of $\mathcal{G}(z)$ is $z_1 = -0.9048$, which is outside the well-damped region. In order to avoid the oscillation, the direct compensation of the slightly damped roots of $\mathcal{G}(z)$, i.e., simply saying their cancellation with the corresponding poles of the regulator needs to be avoided. Separate the roots of $\mathcal{G}(z)$ in such a way that $\mathcal{G}_+(z)$ contains the well-damped roots of $\mathcal{G}(z)$ (they are inside the heart form region) and $\mathcal{G}_-(z)$ contains the slightly damped roots (outside the heart form region)

$$\mathcal{G}(z) = \mathcal{G}_+(z)\mathcal{G}_-(z) \tag{2.3.17}$$

and the condition $\mathcal{G}_-(z)\big|_{z=1} = \mathcal{G}_-(1) = 1$ must be fulfilled. For example,

$$\mathcal{G}(z) = \mathcal{G}_+(z)\mathcal{G}_-(z) = 0.01733(0.525z + 0.475), \tag{2.3.18}$$

where $\mathcal{G}_+(z) = 0.01733$ (polynomial $\mathcal{G}_+(z)$ has no pole at all) and $\mathcal{G}_-(z) = 0.525z + 0.475$ (the polynomial $\mathcal{G}_-(z)$ has one slightly damped pole). Next let m_+ and m_- be the order of the polynomials $\mathcal{G}_+(z)$ and $\mathcal{G}_-(z)$, respectively.

During the design, take the applied separation into account and modify our expectation for the pulse transfer function of the closed DT system:

$$T = \frac{C(z)G(z)}{1 + C(z)G(z)} = \mathcal{P}_y(z) = \mathcal{G}_-(z)z^{-2}z^{-m_-} = \mathcal{G}_-(z)z^{-m_- -2} \tag{2.3.19}$$

so assumption (2.3.6) is also fulfilled. Given the above condition, the regulator becomes:

$$C(z) = \frac{\mathcal{F}(z)}{\mathcal{G}_+(z)[z^3 - \mathcal{G}_-(z)]} = \frac{57.7(z^3 - 1.7236z^2 + 0.7408z)}{z^3 - 0.525z - 0.475}. \tag{2.3.20}$$

From the above, it is clear why condition $\mathcal{G}_-(z)\big|_{z=1} = \mathcal{G}_-(1) = 1$ needs to be assumed during the separation of the polynomial $\mathcal{G}(z)$: namely thanks to $\left[z^3 - \mathcal{G}_-(z)\right]_{z=1} = 0,\ \ z = 1$, is still the pole of the loop transfer function $L(z)$, i.e., it is of type one. The results obtained by the modified regulator are illustrated in Figure 2.3.4. The closed system is slowed down, the settling time increased from 2 to 3 s, in other words from two steps to three steps, but at the same time, the moderate dynamics of the actuator

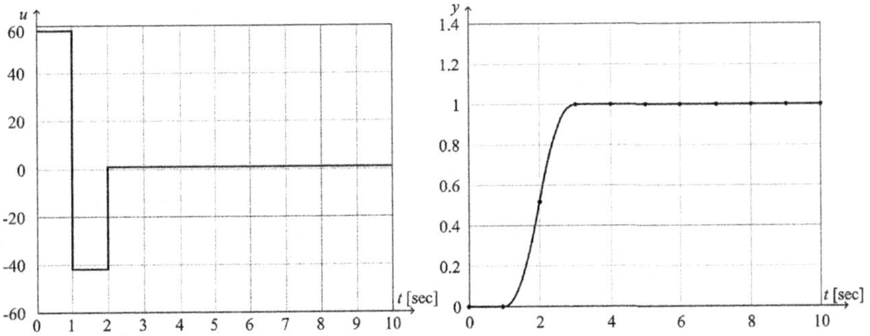

Figure 2.3.4 Deadbeat regulator (three steps).

signal can be observed. The oscillation is completely eliminated but the magnitude of the initial value of the actuator signal cannot be considered acceptable for any kind of application.

Further, slowing down can be performed in many different ways from which that solution is shown next, when the DB character is kept but the settling time is further increased depending on the order of the design polynomial. Let the slowing design polynomial be

$$T(z) = \mathcal{P}_{\mathrm{y}}(z) = 0.2z^2 + 0.3z + 0.5 \tag{2.3.21}$$

of the order $m_{\mathrm{T}} = 2$, and the condition $T(z)|_{z=1} = T(1) = 1$ is used in the specification of the closed-loop pulse transfer function according to

$$\frac{C(z)G(z)}{1 + C(z)G(z)} = \mathcal{P}_{\mathrm{y}}(z)\mathcal{G}_{-}(z)z^{-2-m_{-}-m_{\mathrm{T}}} = \mathcal{P}_{\mathrm{y}}(z)\mathcal{G}_{-}(z)z^{-5}. \tag{2.3.22}$$

The regulator becomes

$$C(z) = \frac{\mathcal{F}(z)\mathcal{P}_{\mathrm{y}}(z)}{\mathcal{G}_{+}(z)\left[z^5 - \mathcal{G}_{-}(z)\mathcal{P}_{\mathrm{y}}(z)\right]}$$
$$= \frac{11.54(z^5 - 0.2235z^4 + 0.6555z^3 - 3.198z^2 + 1.852z)}{z^5 - 0.105z^3 - 0.2525z^2 - 0.405z - 0.2375}$$

The operation of the closed loop in the time domain is seen in Figure 2.3.5.

As regards the regulator form (2.3.10), it has already been mentioned that the design equation of the *DB* regulator actually corresponds to the basic case (2.1.5) of the Youla regulator. As regards, comparison with the general case form (2.1.25) let us consider the form (2.1.24) of the DT process according to the above separation (2.3.17)

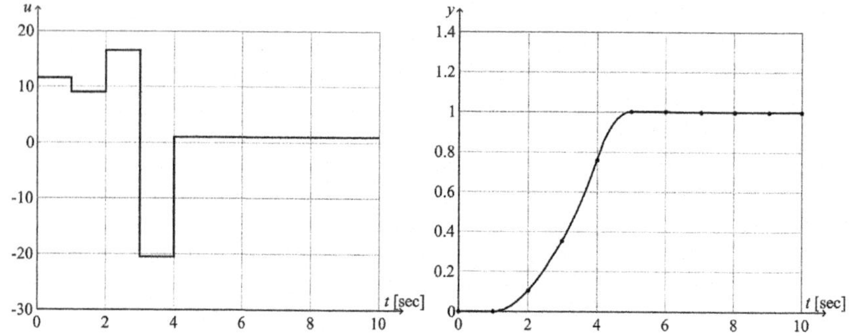

Figure 2.3.5 Deadbeat regulator (five steps).

$$G = G_+ G_- z^{-d} = \frac{\mathcal{G}_+ \mathcal{G}_-}{\mathcal{F}} z^{-2} \tag{2.3.23}$$

where

$$\mathcal{G}_+(z) = 0.01733 \quad \text{and} \quad \mathcal{G}_-(z) = 0.525z + 0.475 \tag{2.3.24}$$

Here $\mathcal{G}_+$ is the factor with acceptable inverse form and $\mathcal{G}_-$ is the factor with nonacceptable inverse form in the numerator of the pulse transfer function. Note that the inverse of $\mathcal{G}_-$ is not unstable now but is slightly damped and therefore it is also considered as unwanted factor. The form of the optimal regulator obtained for the general case according to (2.1.25) with the selected $G_r = G_n = 1$ is the following:

$$\begin{aligned} C_{\text{opt}} &= \frac{R_n G_+^{-1}}{1 - R_n G_- z^{-d}} = \frac{\mathcal{P}_y G_+^{-1}}{1 - \mathcal{P}_y G_- z^{-d}} = \frac{\mathcal{P}_y \mathcal{F}}{\mathcal{G}_+ \left(1 - \mathcal{P}_y \mathcal{B}_- z^{-d}\right)} \\ &= \frac{z^{-2} \mathcal{F}}{\mathcal{G}_+ \left(1 - z^{-2} \mathcal{G}_- z^{-1}\right)} \end{aligned} \tag{2.3.25}$$

which is completely the same as the regulator of (2.3.20)

$$C(z) = \frac{\mathcal{F}(z)}{\mathcal{G}_+(z)\left[z^3 - \mathcal{G}_-(z)\right]} = \frac{57.7(z^3 - 1.7236z^2 + 0.7408z)}{z^3 - 0.525z - 0.475}. \tag{2.3.26}$$

It is thus confirmed again that the Youla regulator is generally valid for stable processes.

2.4 PREDICTIVE REGULATORS

Assume that the pulse transfer function of the process in an *ODOF* control loop is

$$G(z^{-1}) = \frac{\mathcal{G}(z^{-1})}{\mathcal{F}(z^{-1})} z^{-d} = G_+(z^{-1}) G_-(z^{-1}); \quad G_-(z^{-1}) = z^{-d} \tag{2.4.1}$$

which corresponds to a CT process with time delay. A relationship is sought by means of which the value of the output signal at the sampling time $[k+d]$ can be estimated from the information available until the sampling time $[k]$. To achieve this, let us introduce a special polynomial equation

$$1 = \mathcal{F}\mathcal{T} + \mathcal{P}z^{-d} \tag{2.4.2}$$

whose solution is unambiguous in seeking $\mathcal{T} = 1 + t_1 z^{-1} + \ldots + t_{d-1} z^{-(d-1)}$ of order $(d-1)$ and $\mathcal{P} = p_o + p_1 z^{-1} + \ldots + t_{n-1} z^{-(n-1)}$ of order $(n-1)$, if $\mathcal{F}$ has order n. Equation (2.4.2) is a special form of the Diophantine equation discussed in Chapter 4. Using equivalent rewriting, $G(z^{-1})$ can be decomposed as

$$\begin{aligned} G &= \frac{\mathcal{G}}{\mathcal{F}} z^{-d} = \frac{\mathcal{G}\mathcal{F}\mathcal{T} + \mathcal{G}\mathcal{P}z^{-d}}{\mathcal{F}} z^{-d} = \left(\mathcal{G}\mathcal{T} + \frac{\mathcal{G}\mathcal{P}z^{-d}}{\mathcal{F}} \right) z^{-d} \\ &= \mathcal{G}\mathcal{T} z^{-d} + \mathcal{P}\left(\frac{\mathcal{G}}{\mathcal{F}} z^{-d} \right) z^{-d} = \mathcal{G}\mathcal{T} z^{-d} + \mathcal{P}G z^{-d} \end{aligned} \tag{2.4.3}$$

Apply both sides for the series of the input signal $u[k]$

$$\begin{aligned} y[k] &= \mathcal{G}\mathcal{T} z^{-d} u[k] + \mathcal{P}z^{-d} G u[k] = \mathcal{G}\mathcal{T} u[k-d] + \mathcal{P}z^{-d} y[k] \\ &= \mathcal{G}\mathcal{T} u[k-d] + \mathcal{P}y[k-d] = y[k|k-d] \end{aligned} \tag{2.4.4}$$

The equation can also be written for sampling time $(k+d)$

$$y[k+d] = \mathcal{G}\mathcal{T} u[k] + \mathcal{P}y[k] = y[k+d|k] \tag{2.4.5}$$

where $y[k+d|k]$ is the estimate or prediction of the series $y[k]$ in the sampling time $[k+d]$. Note that the prediction is error-free and it uses only the information available until the time instant $[k]$ concerning both the input and the output. Both $\mathcal{G}\mathcal{T}$ and $\mathcal{P}$ are polynomials of z^{-1} and in (2.4.5) their coefficients weight only the current and the previous values of signals u and y. Based on the d-step predictor, a special so-called predictive regulator can be constructed. If the goal is to track the output of a reference model R_r, then the equation of the regulator is

$$R_r y_r[k] = y[k+d|k] = \mathcal{G}\mathcal{T} u[k] + \mathcal{P}y[k] \tag{2.4.6}$$

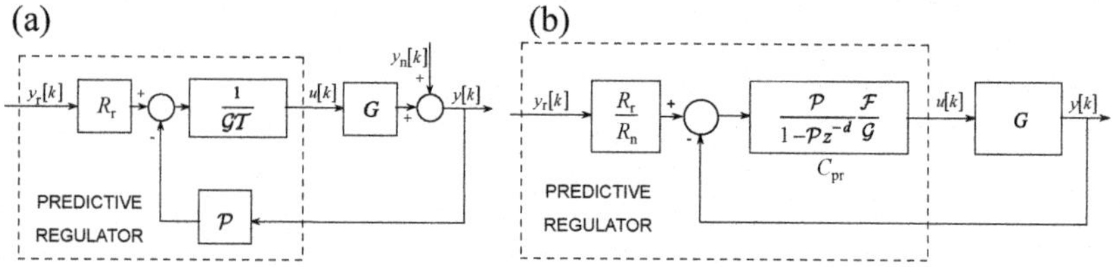

Figure 2.4.1 Forms of the predictive regulator.

whence the input signal is

$$u[k] = \frac{R_r y_r[k] - \mathcal{P}y[k]}{\mathcal{GT}} \tag{2.4.7}$$

The direct mapping of (2.4.7) can be seen in Figure 2.4.1(a); the equivalent block scheme of Figure 2.4.1(b), however, shows the conventional closed-loop control.

Thus, the predictive regulator has the form

$$C_{pr} = \frac{\mathcal{P}}{1 - \mathcal{P}z^{-d}}\frac{\mathcal{F}}{\mathcal{G}} = \frac{\mathcal{P}}{\mathcal{GT}} \tag{2.4.8}$$

i.e., the reference model for the disturbance rejection is now

$$R_n = \mathcal{P} \tag{2.4.9}$$

which does not depend on the designer but comes from (2.4.2). The predictive regulator is formally equal to a Youla regulator where $G_- = z^{-d}$. The transfer characteristic of the complete control loop is

$$y = R_r z^{-d} y_r + \left(1 - R_n z^{-d}\right) y_n = R_r z^{-d} y_r + \left(1 - \mathcal{P}z^{-d}\right) y_n \tag{2.4.10}$$

according to which the predictive regulator completely solves the problem for reference signal tracking, but R_n cannot be designed. The d-step predictor of (2.4.5) is linear in terms of parameters and therefore it is easy to apply parameter estimation (identification) techniques to determine the parameters of the regulator. From the above control design principle, a new, widely applied computer-controlled method has been developed, *model predictive control (MPC)*.

For the noise rejection behavior of the closed system, a method is introduced which penalizes the change or variance of the input. So the dynamics of the closed loop, albeit in a restricted way, are acceptable subject to the proper choice of the penalty weights.

CHAPTER 3

Feedback Regulators

All of the *Youla parameterization* (*YP*) regulators, discussed up to now, apply pole cancellation techniques. This is one reason why they can be used only for stable processes. If the process has an unstable pole, another kind of approach is required. These regulators are called, in general, *pole placement regulators,* because they provide the desired assignment of the poles of the closed system for any process poles. These methods reach the control goal via the direct design of the feedback. Before going into details, however, we need to overview the basic design principles concerning the desired pole assignment of the closed loop.

POLE PLACEMENT WITH POLE CANCELLATION

Consider the closed control system shown in Figure 3.1, where the regulator $C = \mathcal{A}/\mathcal{X}$ is used to place the poles of the closed control system according to the characteristic equation $\mathcal{R} = 0$ ($\mathcal{R}$ is the design polynomial) by the cancellation of the process poles. To do this, $\mathcal{X}$ needs to be expressed by the equation $\mathcal{R} = \mathcal{X} + \mathcal{B}$. The complementary sensitivity function of the closed loop is

$$T = \frac{\frac{\mathcal{A}}{\mathcal{X}}\frac{\mathcal{B}}{\mathcal{A}}}{1 + \frac{\mathcal{A}}{\mathcal{X}}\frac{\mathcal{B}}{\mathcal{A}}} = \frac{\mathcal{A}\mathcal{B}}{\mathcal{A}\mathcal{X} + \mathcal{A}\mathcal{B}} = \frac{\mathcal{B}}{\mathcal{X} + \mathcal{B}} = \frac{\mathcal{B}}{\mathcal{R}} \tag{3.1}$$

The regulator is

$$C = \frac{\mathcal{A}}{\mathcal{X}} = \frac{\mathcal{A}}{\mathcal{R} - \mathcal{B}} = \frac{\frac{\mathcal{B}}{\mathcal{R}}}{1 - \frac{\mathcal{B}}{\mathcal{R}}}\frac{\mathcal{A}}{\mathcal{B}} = \frac{R_{\mathrm{r}}}{1 - R_{\mathrm{r}}}P^{-1} \tag{3.2}$$

and actually the ideal Youla regulator with reference model $R_{\mathrm{r}} = R_{\mathrm{n}} = \mathcal{B}/\mathcal{R}$. This regulator places the poles in $\mathcal{R}$ and leaves the zeros in $\mathcal{B}$ untouched.

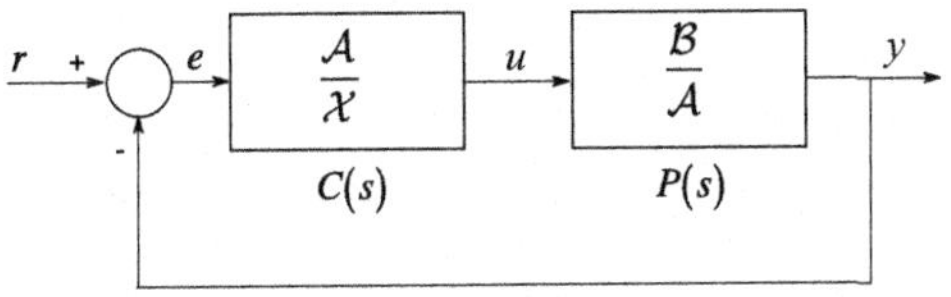

Figure 3.1 Pole-canceling regulator.

In the case of a $\hat{P} = \hat{\mathcal{B}}/\hat{\mathcal{A}}$ model-based control design

$$\hat{C} = \frac{\hat{\mathcal{A}}}{\hat{\mathcal{X}}} = \frac{\hat{\mathcal{A}}}{\mathcal{R} - \hat{\mathcal{B}}} \tag{3.3}$$

so the model-based complementary sensitivity function in the real closed loop is

$$\hat{T} = \frac{\hat{C}P}{1 + \hat{C}P} = \frac{\frac{\hat{\mathcal{A}}}{\mathcal{R}-\hat{\mathcal{B}}}\frac{\mathcal{B}}{\mathcal{A}}}{1 + \frac{\hat{\mathcal{A}}}{\mathcal{R}-\hat{\mathcal{B}}}\frac{\mathcal{B}}{\mathcal{A}}} = \frac{\hat{\mathcal{A}}\mathcal{B}}{\mathcal{A}(\mathcal{R} - \hat{\mathcal{B}}) + \hat{\mathcal{A}}\mathcal{B}}$$
$$= \frac{\hat{\mathcal{A}}\mathcal{B}}{\hat{\mathcal{A}}\mathcal{B} - \mathcal{A}\hat{\mathcal{B}} + \mathcal{A}\mathcal{R}}\bigg|_{\hat{P} \to P} = \frac{\mathcal{A}\mathcal{B}}{\mathcal{A}\mathcal{R}} \tag{3.4}$$

Thus, the characteristic equation of the closed system has the form $\mathcal{A}\mathcal{R} = 0$, so the method cannot be applied for unstable processes.

POLE PLACEMENT WITH FEEDBACK REGULATOR

An another solution when the regulator is put in the feedback is shown in Figure 3.2.

Now the task is again to place the poles of the closed system according to the characteristic equation $\mathcal{R} = 0$ ($\mathcal{R}$ is the design polynomial). To do this, $\mathcal{K}$ needs to be determined from the equation $\mathcal{R} = \mathcal{K} + \mathcal{A}$. The complementary sensitivity function of the closed system is

$$T = \frac{\frac{\mathcal{B}}{\mathcal{A}}}{1 + \frac{\mathcal{K}}{\mathcal{B}}\frac{\mathcal{B}}{\mathcal{A}}} = \frac{\mathcal{B}}{\mathcal{A} + \mathcal{K}} = \frac{\mathcal{B}}{\mathcal{R}} \tag{3.5}$$

and thus this regulator places the poles in $\mathcal{R}$ and leaves the zeros in $\mathcal{B}$ untouched.

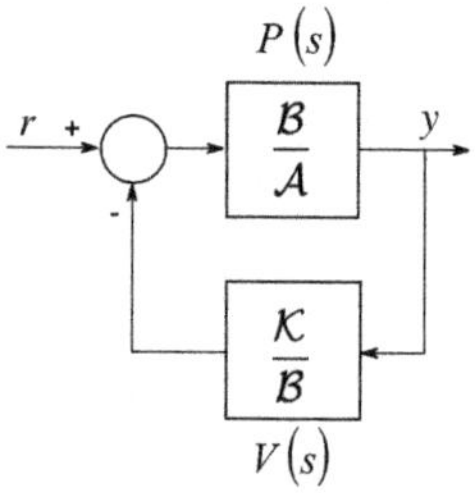

Figure 3.2 Regulator in the feedback.

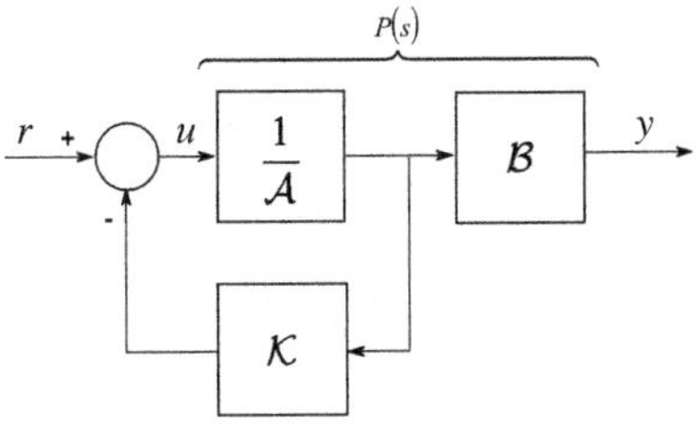

Figure 3.3 The regulator feeds back the internal signal of the process.

In the case of model-based design, the feedback regulator is

$$\hat{\mathcal{V}} = \frac{\hat{\mathcal{K}}}{\hat{\mathcal{B}}} = \frac{\mathcal{R} - \hat{\mathcal{A}}}{\hat{\mathcal{B}}} \tag{3.6}$$

so the model-based complementary sensitivity function is

$$\hat{T} = \frac{\frac{\mathcal{B}}{\mathcal{A}}}{1 + \frac{\hat{\mathcal{K}}}{\hat{\mathcal{B}}}\frac{\mathcal{B}}{\mathcal{A}}} = \frac{\mathcal{B}\hat{\mathcal{B}}}{\mathcal{A}\hat{\mathcal{B}} + \hat{\mathcal{K}}\mathcal{B}} = \frac{\mathcal{B}\hat{\mathcal{B}}}{\mathcal{A}\hat{\mathcal{B}} + \left(\mathcal{R} - \hat{\mathcal{A}}\right)\mathcal{B}} = \frac{\hat{\mathcal{B}}}{\frac{\mathcal{A}\hat{\mathcal{B}} - \mathcal{B}\hat{\mathcal{A}}}{\mathcal{B}} + \mathcal{R}}\Bigg|_{\hat{P} \to P} = \frac{\mathcal{B}}{\mathcal{R}} \tag{3.7}$$

The characteristic equation of the closed system has the form $\mathcal{R} = 0$ and it does not depend on the unstable property of the process.

The block diagram in Figure 3.2 can be redrawn as Figure 3.3. (The state feedback methods are discussed in detail in Section 3.1, and the same control principle is represented in Figure 3.1.5(c) among the schemes showing the equivalent transfer function representations for state feedback.)

POLE PLACEMENT WITH CHARACTERISTIC POLYNOMIAL DESIGN

The characteristic polynomial $\mathcal{R}$ of the closed-loop control can be directly designed by algebraic methods. In Figure 3.4, the regulator $C = \mathcal{Y}/\mathcal{X}$ is the quotient of two polynomials. It will be shown in Section 3.3 that under certain conditions, the *Diophantine equation* (*DE*) $\mathcal{A}\mathcal{X} + \mathcal{B}\mathcal{Y} = \mathcal{R}$ can be

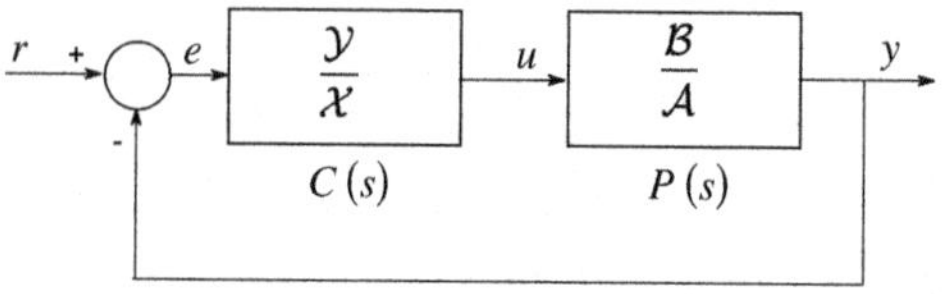

Figure 3.4 Direct control design on the basis of the characteristic polynomial.

solved for $\mathcal{X}$ and $\mathcal{Y}$. Thus from the characteristic equation $\mathcal{R} = 0$ the regulator can be directly determined.

The complementary sensitivity function of the closed system is

$$T = \frac{\frac{\mathcal{Y}}{\mathcal{X}}\frac{\mathcal{B}}{\mathcal{A}}}{1 + \frac{\mathcal{Y}}{\mathcal{X}}\frac{\mathcal{B}}{\mathcal{A}}} = \frac{\mathcal{B}\mathcal{Y}}{\mathcal{A}\mathcal{X} + \mathcal{B}\mathcal{Y}} = \frac{\mathcal{B}\mathcal{Y}}{\mathcal{R}} \tag{3.8}$$

and thus this regulator also places the poles in $\mathcal{R}$ and leaves the zeros in $\mathcal{B}$ untouched, but in the nominator $\mathcal{Y}$ appears, which depends on the desired properties and also on *DE*.

In model-based control design the regulator becomes $\hat{C} = \hat{\mathcal{Y}}/\hat{\mathcal{X}}$, where $\hat{\mathcal{X}}$ and $\hat{\mathcal{Y}}$ are obtained from the *DE* $\hat{\mathcal{A}}\hat{\mathcal{X}} + \hat{\mathcal{B}}\hat{\mathcal{Y}} = \mathcal{R}$ using the polynomials $\hat{\mathcal{A}}$ and $\hat{\mathcal{B}}$ of the model. The model-based complementary sensitivity function is now

$$\hat{T} = \frac{\frac{\hat{\mathcal{Y}}}{\hat{\mathcal{X}}}\frac{\mathcal{B}}{\mathcal{A}}}{1 + \frac{\hat{\mathcal{Y}}}{\hat{\mathcal{X}}}\frac{\mathcal{B}}{\mathcal{A}}} = \frac{\mathcal{B}\hat{\mathcal{Y}}}{\mathcal{A}\hat{\mathcal{X}} + \mathcal{B}\hat{\mathcal{Y}}} \tag{3.9}$$

Substitute the computed polynomial

$$\hat{\mathcal{Y}} = \frac{\mathcal{R} - \hat{\mathcal{A}}\hat{\mathcal{X}}}{\hat{\mathcal{B}}} \tag{3.10}$$

into the numerator of $\hat{T}$, where the result is

$$\hat{T} = \frac{\mathcal{B}\frac{\mathcal{R} - \hat{\mathcal{A}}\hat{\mathcal{X}}}{\hat{\mathcal{B}}}}{\mathcal{A}\hat{\mathcal{X}} + \mathcal{B}\frac{\mathcal{R} - \hat{\mathcal{A}}\hat{\mathcal{X}}}{\hat{\mathcal{B}}}} = \left.\frac{\left(\mathcal{R} - \hat{\mathcal{A}}\hat{\mathcal{X}}\right)}{\frac{\mathcal{A}\hat{\mathcal{B}} - \hat{\mathcal{A}}\mathcal{B}}{\mathcal{B}}\hat{\mathcal{X}} + \mathcal{R}}\right|_{\hat{P} \to P} = \frac{\mathcal{B}\mathcal{Y}}{\mathcal{R}} \tag{3.11}$$

Thus, the characteristic equation of the closed system has the form $\mathcal{R} = 0$ and it does not depend on the unstable character of the process. The general polynomial methods of the regulators directly designing the characteristic equation are discussed in detail in Section 3.3.

3.1 CONTROL LOOPS WITH STATE FEEDBACK

It was shown in Section 1.1 how processes are represented in state space. In many cases this kind of description is only available and the transfer function of the controlled system is unavailable. This partly explains why control design methodology directly based on state-space description has been evolved. Let us consider the state-space representation of a *linear time-invariant* (*LTI*) process to be controlled such as

$$\frac{d\boldsymbol{x}}{dt} = \dot{\boldsymbol{x}} = \boldsymbol{A}\boldsymbol{x} + \boldsymbol{b}u \qquad y = \boldsymbol{c}^{\mathrm{T}}\boldsymbol{x} \tag{3.1.1}$$

which corresponds to (1.1.10) for the case of $d = 0$. This, as mentioned earlier, does not violate the generality, because it is very rare for the model to contain a proportional channel directly affecting the output. The block scheme of (3.1.1) is shown in Figure 3.1.1, where the thick lines present vector variables.

Here u and y are the input and output signals of the process, respectively, and $\boldsymbol{x}$ is the state vector. According to the equivalent transfer function (1.1.11), we obtain

$$P(s) = \boldsymbol{c}^{\mathrm{T}}(s\boldsymbol{I} - \boldsymbol{A})^{-1}\boldsymbol{b} = \frac{\mathcal{B}(s)}{\det(s\boldsymbol{I} - \boldsymbol{A})} = \frac{\mathcal{B}(s)}{\mathcal{A}(s)} = \frac{b_1 s^{n-1} + \ldots + b_{n-1}s + b_n}{s^n + a_1 s^{n-1} + \ldots + a_{n-1}s + a_n} \tag{3.1.2}$$

Figure 3.1.2 shows the so-called classical closed control system directly fitting the state equation description, where r denotes the reference signal. In the closed loop the state vector is fed back with the linear proportional vector $\boldsymbol{k}^{\mathrm{T}}$ according to the expression below

$$u = k_r r - \boldsymbol{k}^{\mathrm{T}}\boldsymbol{x} \tag{3.1.3}$$

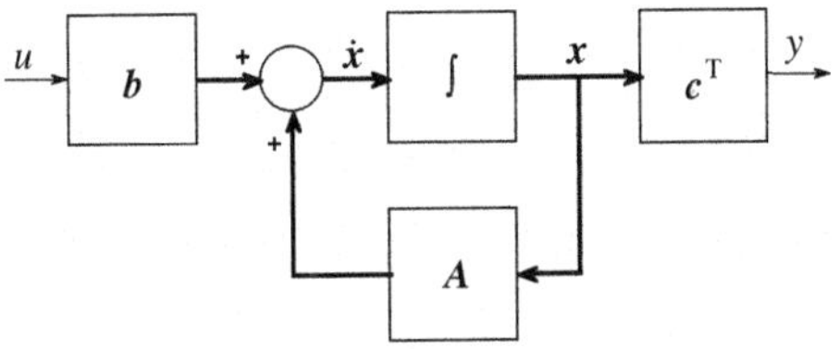

Figure 3.1.1 Block diagram of the state-space equation of the *linear time-invariant* system.

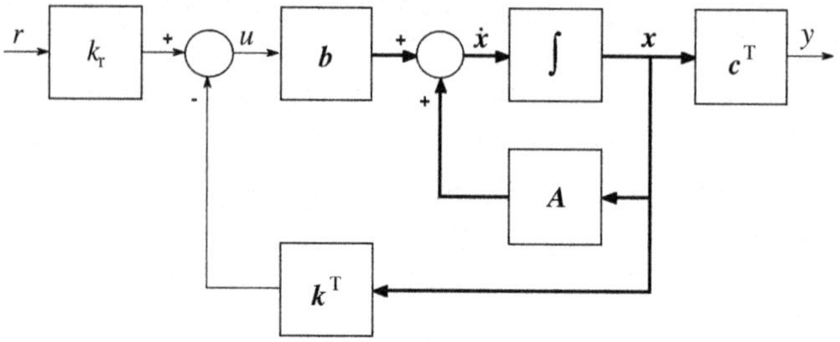

Figure 3.1.2 Linear regulator with state feedback.

Based on Figure 3.1.2 the state equation of the complete closed system can be easily written as

$$\frac{d\boldsymbol{x}}{dt} = \left(\boldsymbol{A} - \boldsymbol{b}\boldsymbol{k}^{T}\right)x + k_r\boldsymbol{b}r$$
$$y = \boldsymbol{c}^{T}\boldsymbol{x} \tag{3.1.4}$$

i.e., with the state feedback the dynamics represented by the original system matrix $\boldsymbol{A}$ is modified by the dyadic product $\boldsymbol{b}\boldsymbol{k}^T$ to $(\boldsymbol{A}-\boldsymbol{b}\boldsymbol{k}^T)$.

The transfer function of the closed-loop control is [65]

$$T_{ry}(s) = \frac{Y(s)}{R(s)} = \boldsymbol{c}^{T}\left(s\boldsymbol{I} - \boldsymbol{A} + \boldsymbol{b}\boldsymbol{k}^{T}\right)^{-1}\boldsymbol{b}k_r = \frac{\boldsymbol{c}^{T}(s\boldsymbol{I} - \boldsymbol{A})^{-1}\boldsymbol{b}k_r}{1 + \boldsymbol{k}^{T}(s\boldsymbol{I} - \boldsymbol{A})^{-1}\boldsymbol{b}}$$
$$= \frac{k_r}{1 + \boldsymbol{k}^{T}(s\boldsymbol{I} - \boldsymbol{A})^{-1}\boldsymbol{b}}P(s) = \frac{k_r\mathcal{B}(s)}{\mathcal{A}(s) + \boldsymbol{k}^{T}\boldsymbol{\Psi}(s)\boldsymbol{b}} \tag{3.1.5}$$

which derives from the comparison of equations valid for the Laplace transforms $\boldsymbol{X}(s) = (s\boldsymbol{I}-\boldsymbol{A})^{-1}\boldsymbol{b}U(s)$ (see (A.2.2)), $U(s) = k_r\boldsymbol{R}(s)-\boldsymbol{k}^T\boldsymbol{X}(s)$ (see (3.1.3)), and $Y(s) = \boldsymbol{c}^T\boldsymbol{X}(s)$ (see (3.1.1)) using the matrix inversion lemma (see details in Section A.2.5.4 of Appendix 2). Note that the state feedback leaves the zeros of the process untouched and only the poles of the closed-loop system can be designed by $\boldsymbol{k}^T$.

The so-called calibration factor k_r [12] is introduced in order to make the gain of T_{ry} equal to unity ($T_{ry}(0) = 1$). The open loop is obviously not of type 1, so it cannot provide zero error and unity static transfer gain. It can be ensured only if the condition

$$k_r = \frac{-1}{\boldsymbol{c}^{T}\left(\boldsymbol{A} - \boldsymbol{b}\boldsymbol{k}^{T}\right)^{-1}\boldsymbol{b}} = \frac{\boldsymbol{k}^{T}\boldsymbol{A}^{-1}\boldsymbol{b} - 1}{\boldsymbol{c}^{T}\boldsymbol{A}^{-1}\boldsymbol{b}} \tag{3.1.6}$$

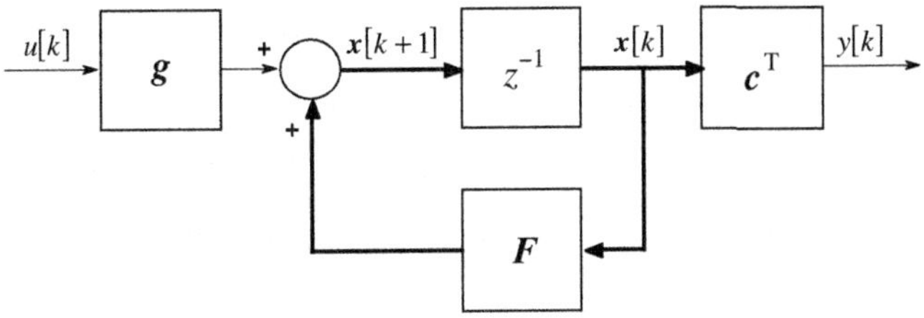

Figure 3.1.3 Block diagram of the state equation of the *linear time-invariant* sampled system.

is fulfilled (see Section A.2.5.5 in Appendix 2). The above special control loop is called state feedback.

Sampled Data Systems

Let us investigate the state feedback for sampled systems. For this purpose consider the state equation of an *LTI* sampled data process to be controlled using the results of Section A.3.2 for the case $d_c = 0$

$$\begin{aligned} \boldsymbol{x}[k+1] &= \boldsymbol{F}\boldsymbol{x}[k] + \boldsymbol{g}u[k] \\ y[k] &= \boldsymbol{c}^{\mathrm{T}}\boldsymbol{x}[k] \end{aligned} \tag{3.1.7}$$

The block diagram presented by the above equations is seen in Figure 3.1.3.

Here $u[k]$ and $y[k]$ are the process input and output, respectively, and $\boldsymbol{x}[k]$ denotes the state vector. The equivalent pulse transfer function is now

$$\begin{aligned} G(z) &= \boldsymbol{c}^{\mathrm{T}}(z\boldsymbol{I} - \boldsymbol{F})^{-1}\boldsymbol{g} = \frac{\mathcal{G}(z)}{\det(z\boldsymbol{I} - \boldsymbol{F})} = \frac{\mathcal{G}(z)}{\mathcal{F}(z)} \\ &= \frac{g_1 z^{n-1} + \ldots + g_{n-1}z + g_n}{z^n + f_1 z^{n-1} + \ldots + f_{n-1}z + f_n} \end{aligned} \tag{3.1.8}$$

A classical closed-loop control directly fitting the state equation description is shown in Figure 3.1.4, where the reference signal is denoted

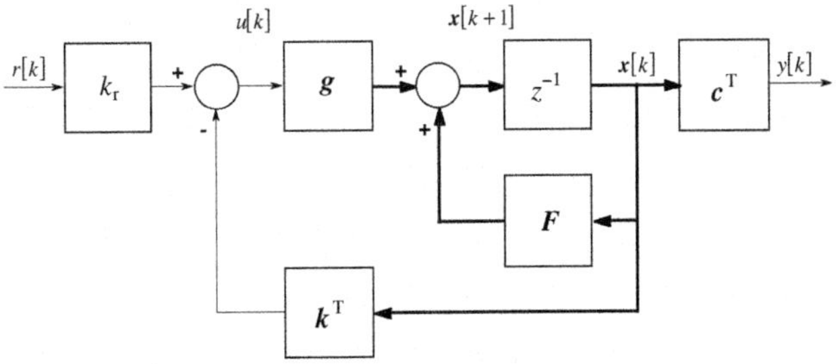

Figure 3.1.4 Linear discrete-time control with state feedback.

by $r[k]$. The closed loop is formed by the feedback from the state vector via the linear proportional feedback vector $\boldsymbol{k}^{\mathrm{T}}$ in the form

$$u[k] = k_{\mathrm{r}}r[k] - \boldsymbol{k}^{\mathrm{T}}\boldsymbol{x}[k] \tag{3.1.9}$$

Based on Figure 3.1.4, the state equation of the complete closed-loop system can be written as

$$\begin{aligned} \boldsymbol{x}[k+1] &= \left(\boldsymbol{F} - \boldsymbol{g}\boldsymbol{k}^{\mathrm{T}}\right)\boldsymbol{x}[k] + k_{\mathrm{r}}\boldsymbol{g}r[k] \\ y[k] &= \boldsymbol{c}^{\mathrm{T}}\boldsymbol{x}[k] \end{aligned} \tag{3.1.10}$$

i.e., the dynamics concerning the original system matrix $\boldsymbol{F}$ is modified by the dyadic product $\boldsymbol{g}\boldsymbol{k}^{\mathrm{T}}$ to $(\boldsymbol{F} - \boldsymbol{g}\boldsymbol{k}^{\mathrm{T}})$.

The pulse transfer function of the closed-loop control is

$$\begin{aligned} T_{\mathrm{ry}}(z) &= \frac{Y(z)}{R(z)} = \boldsymbol{c}^{\mathrm{T}}\left(z\boldsymbol{I} - \boldsymbol{F} + \boldsymbol{g}\boldsymbol{k}^{\mathrm{T}}\right)^{-1}\boldsymbol{g}k_{\mathrm{r}} = \frac{\boldsymbol{c}^{\mathrm{T}}(z\boldsymbol{I} - \boldsymbol{F})^{-1}\boldsymbol{g}k_{\mathrm{r}}}{1 + \boldsymbol{k}^{\mathrm{T}}(z\boldsymbol{I} - \boldsymbol{F})^{-1}\boldsymbol{g}} \\ &= \frac{k_{\mathrm{r}}}{1 + \boldsymbol{k}^{\mathrm{T}}(z\boldsymbol{I} - \boldsymbol{F})^{-1}\boldsymbol{g}}G(z) = \frac{k_{\mathrm{r}}\mathcal{G}(z)}{\mathcal{F}(z) + \boldsymbol{k}^{\mathrm{T}}\boldsymbol{\Psi}(z)\boldsymbol{g}} \end{aligned} \tag{3.1.11}$$

which derives from the comparison of the Laplace transform forms $\boldsymbol{X}(z) = (z\boldsymbol{I} - \boldsymbol{F})^{-1}\boldsymbol{g}U(z)$, $U(z) = k_{\mathrm{r}}R(z) - \boldsymbol{k}^{\mathrm{T}}\boldsymbol{X}(z)$, and $Y(z) = \boldsymbol{c}^{\mathrm{T}}\boldsymbol{X}(z)$, respectively, using the matrix inversion lemma (the proof for continuous-time (CT) systems is in Appendix A.2.5.4). Note that the state feedback leaves the zeros of the process untouched, and only the poles of the closed system can be designed by $\boldsymbol{k}^{\mathrm{T}}$.

Introduce the so-called calibration factor k_{r}, by means of which the gain of the $\boldsymbol{T}_{\mathrm{ry}}$ can be set to unity, i.e., $T_{\mathrm{ry}}(1) = 1$. Obviously the open loop is not of an integrating type, so it cannot provide zero error and unity static gain. In order to reach this, the parameters of the process have to be known and the condition

$$k_{\mathrm{r}} = \frac{-1}{\boldsymbol{c}^{\mathrm{T}}\left(\boldsymbol{F} - \boldsymbol{g}\boldsymbol{k}^{\mathrm{T}}\right)^{-1}\boldsymbol{g}} = \frac{\boldsymbol{k}^{\mathrm{T}}\boldsymbol{F}^{-1}\boldsymbol{g} - 1}{\boldsymbol{c}^{\mathrm{T}}\boldsymbol{F}^{-1}\boldsymbol{g}} \tag{3.1.12}$$

must be fulfilled (see A.2.5.5 for CT systems). The above closed-loop control is called state feedback.

Pole Placement by State Feedback

The most natural design method of state feedback is the so-called pole placement. In this case, the feedback vector $\boldsymbol{k}^{\mathrm{T}}$ needs to be chosen to make the characteristic equation of the closed loop equal to the prescribed, so-called design polynomial $\mathcal{R}(s)$ [12,64], i.e.,

$$\begin{aligned}\mathcal{R}(s) &= s^n + r_1 s^{n-1} + \ldots + r_{n-1}s + r_n = \det\left(s\boldsymbol{I} - \boldsymbol{A} + \boldsymbol{b}\boldsymbol{k}^{\mathrm{T}}\right) \\ &= \mathcal{A}(s) + \boldsymbol{k}^{\mathrm{T}}\boldsymbol{\Psi}(s)\boldsymbol{b}\end{aligned} \tag{3.1.13}$$

The solution always exists if the process is controllable. (It is reasonable if the order of $\mathcal{R}$ is equal to that of $\mathcal{A}$.) In the exceptional case when the transfer function of the controlled system is known, the canonical state equations can be directly written. Based on the controllable canonical form (A.2.37), the system matrices are

$$\boldsymbol{A}_{\mathrm{c}} = \begin{bmatrix} -a_1 & -a_2 & \ldots & -a_{n-1} & -a_n \\ 1 & 0 & \ldots & 0 & 0 \\ 0 & 1 & & 0 & 0 \\ \vdots & \vdots & \ddots & \vdots & \vdots \\ 0 & 0 & 0 & 1 & 0 \end{bmatrix}; \quad \boldsymbol{c}_{\mathrm{c}}^{\mathrm{T}} = \left[b_1, b_2, \ldots, b_n\right];$$

$$\boldsymbol{b}_{\mathrm{c}} = \left[1, 0, \ldots, 0\right]^{\mathrm{T}} \tag{3.1.14}$$

Considering the special forms of $\boldsymbol{A}_{\mathrm{c}}$ and $\boldsymbol{b}_{\mathrm{c}}$, it can be seen that the design equation

$$\begin{aligned}\boldsymbol{A}_{\mathrm{c}} - \boldsymbol{b}_{\mathrm{c}}\boldsymbol{k}_{\mathrm{c}}^{\mathrm{T}} &= \begin{bmatrix} -a_1 & -a_2 & \ldots & -a_{n-1} & -a_n \\ 1 & 0 & \ldots & 0 & 0 \\ 0 & 1 & \ldots & 0 & 0 \\ \vdots & \vdots & \ddots & \vdots & \vdots \\ 0 & 0 & 0 & 1 & 0 \end{bmatrix} - \begin{bmatrix} 1 \\ 0 \\ 0 \\ \vdots \\ 0 \end{bmatrix} \boldsymbol{k}_{\mathrm{c}}^{\mathrm{T}} \\ &= \begin{bmatrix} -r_1 & -r_2 & \ldots & -r_{n-1} & -r_n \\ 1 & 0 & \ldots & 0 & 0 \\ 0 & 1 & \ldots & 0 & 0 \\ \vdots & \vdots & \ddots & \vdots & \vdots \\ 0 & 0 & 0 & 1 & 0 \end{bmatrix}\end{aligned} \tag{3.1.15}$$

and the selected

$$\boldsymbol{k}^{\mathrm{T}} = \boldsymbol{k}_{\mathrm{c}}^{\mathrm{T}} = [r_1 - a_1, r_2 - a_2, \ldots, r_n - a_n] \tag{3.1.16}$$

ensure the characteristic equation $(\mathcal{R}(s) = 0)$ of (3.1.13), i.e., the prescribed poles. The choice of the calibration factor can be determined by simple calculation

$$k_{\mathrm{r}} = \frac{a_n + (r_n - a_n)}{b_n} = \frac{r_n}{b_n} \tag{3.1.17}$$

Based on Equations (3.1.4) and (3.1.6), it can be seen that in the case of state feedback pole placement the transfer function results in

$$T_{\mathrm{ry}}(s) = \frac{k_{\mathrm{r}}\mathcal{B}(s)}{\mathcal{R}(s)} \tag{3.1.18}$$

as shown in (3.1.5).

The most common case of state feedback is when not the transfer function but the state-space form of the control system is given. In relation to (A.2.57) it has been observed that all controllable systems can be described in a controllable canonical form by using the transformation matrix $\boldsymbol{T}_{\mathrm{c}} = \boldsymbol{M}_{\mathrm{c}}^{\mathrm{c}}(\boldsymbol{M}_{\mathrm{c}})^{-1}$. This linear transformation also refers to the feedback vector

$$\begin{aligned} \boldsymbol{k}^{\mathrm{T}} &= \boldsymbol{k}_{\mathrm{c}}^{\mathrm{T}}\boldsymbol{T}_{\mathrm{c}} = \boldsymbol{k}_{\mathrm{c}}^{\mathrm{T}}\boldsymbol{M}_{\mathrm{c}}^{\mathrm{c}}\boldsymbol{M}_{\mathrm{c}}^{-1} \\ \boldsymbol{k}^{\mathrm{T}} &= \boldsymbol{b}_{\mathrm{c}}^{\mathrm{T}}\boldsymbol{M}_{\mathrm{c}}^{-1}\mathcal{R}(\boldsymbol{A}) = [0, 0, \ldots, 1]\boldsymbol{M}_{\mathrm{c}}^{-1}\mathcal{R}(\boldsymbol{A}) \end{aligned} \tag{3.1.19}$$

The design relating to the controllable canonical form (3.1.16), together with the linear transformation relationship corresponding to the first row of the noncontrollable form (3.1.19), is known as the Bass–Gura algorithm [21]. The algorithm in the second row of (3.1.19) is called Ackermann method [1,2] after its elaborator (see A.2.5.6. of Appendix A.2).

In the Bass–Gura algorithm, the inverse of the controllability matrix $\boldsymbol{M}_{\mathrm{c}}$ needs to be determined by the general system matrices $\boldsymbol{A}$ and $\boldsymbol{b}$ on the one hand and the controllability matrix $\boldsymbol{M}_{\mathrm{c}}^{\mathrm{c}}$ of the controllable canonical form (see (A.2.51)) on the other. Since this latter term depends only on the coefficients a_i in the denominator of the process transfer function, the denominator needs to be calculated: $\mathcal{A}(s) = \det(s\boldsymbol{I} - \boldsymbol{A})$. Since $[0, 0, \ldots, 1]\boldsymbol{M}_{\mathrm{c}}^{-1}$ is the last row of the inverse of the controllability matrix, and $\mathcal{R}(\boldsymbol{A})$ also needs to be calculated; with the Ackermann method it is not necessary to calculate $\mathcal{A}(s)$.

It is worth mentioning that the state feedback formally corresponds to a conventional *proportional-derivative* (*PD*) control and therefore overactuating peaks are expected at the input of the process because the pole placement tries to make the process faster. In practice, however, the actuator usually limits the amplitude of the peaks, which needs to be taken into account during the design of the poles of the characteristic polynomial $\mathcal{R}(s)$.

It can be clearly seen that state feedback formally corresponds to a serial compensation $\boldsymbol{R}_s = \mathcal{A}(s)/\mathcal{R}(s)$ (Figure 3.1.5(a)). The real operation and effect of the state feedback can be easily understood by the equivalent block schemes using the transfer functions shown in Figure 3.1.5. The "regulator" $\boldsymbol{R}_f(s)$ of the closed loop is in the feedback line (Figure 3.1.5(b)). The transfer function of the closed loop (3.1.18) is

$$T_{ry}(s) = \frac{k_r\mathcal{B}(s)}{\mathcal{R}(s)} = \frac{k_r\mathcal{B}(s)}{\mathcal{A}(s)+\mathcal{B}(s)} = \frac{k_r P(s)}{1+K_k(s)P(s)} = \frac{k_r\mathcal{A}(s)}{\mathcal{R}(s)}\frac{\mathcal{B}(s)}{\mathcal{A}(s)}$$
$$= k_r R_s(s)P(s) \tag{3.1.20}$$

where

$$R_f = K_k(s) = \frac{\mathcal{K}(s)}{\mathcal{B}(s)} = \frac{\mathcal{R}(s)-\mathcal{A}(s)}{\mathcal{B}(s)} = \frac{\boldsymbol{k}^T(s\boldsymbol{I}-\boldsymbol{A})^{-1}\boldsymbol{b}}{\boldsymbol{c}^T(s\boldsymbol{I}-\boldsymbol{A})^{-1}\boldsymbol{b}} \tag{3.1.21}$$

and the calibration factor is

$$k_r = \frac{\boldsymbol{k}^T\boldsymbol{A}^{-1}\boldsymbol{b}-1}{\boldsymbol{c}^T\boldsymbol{A}^{-1}\boldsymbol{b}} = \frac{1+K_k(0)P(0)}{P(0)} \tag{3.1.22}$$

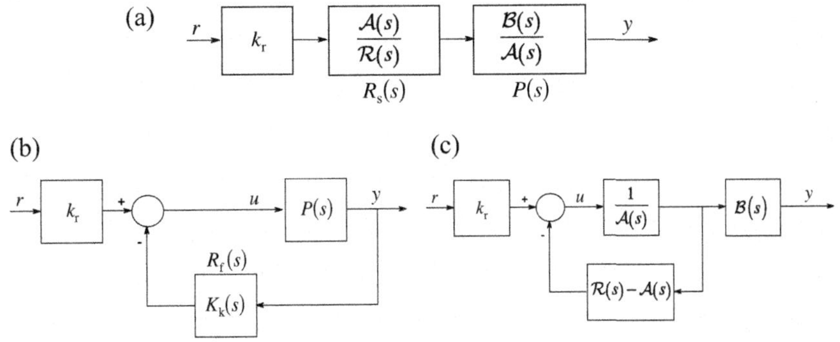

Figure 3.1.5 Equivalent schemes to the state feedback design using transfer functions and polynomials.

Given the block schemes of Figure 3.1.5 it can be stated that state feedback also stabilizes the unstable terms, since because of the effect of the polynomial $\mathcal{K}(s) = \mathcal{R}(s) - \mathcal{A}(s)$ there is a pole placement for any process, so with the stable $\mathcal{R}(s)$ the stabilization is fulfilled. The feedback polynomial $\mathcal{K}(s)$ corresponds formally to $\boldsymbol{k}^{\mathrm{T}}$. The fact that the numerator $\mathcal{B}(s)$ of the process is present in the denominator of $K_{\mathrm{k}}(s)$ needs special consideration. The regulator can be applied only for minimum-phase (inverse stable) processes, where the roots of $\mathcal{B}(s)$ are stable. As a consequence of this special character of the state feedback, however, here $\mathcal{B}(s)$ is not substituted by its model $\hat{\mathcal{B}}(s)$, but the method itself realizes the exact $1/\mathcal{B}(s)$.

Further methods have been developed for the calculation of the pole placement state feedback vector $\boldsymbol{k}^{\mathrm{T}}$. Of these, the so-called Mayne–Murdoch method [12,130] is briefly shown here, on the basis of which useful statements can be made. In the Bass–Gura [21] and Ackermann methods [1,2] the controllable canonical form has a special role. A similarly important canonical form is the diagonal form. Let the diagonal form $\boldsymbol{A}_{\mathrm{d}} = \mathbf{diag}[\lambda_1,\ldots,\lambda_n]$ have the eigenvalues λ_i (these are the roots of $\mathcal{A}(s)$) and let the design polynomial $\mathcal{R}(s)$ has prescribed root values of $\{\mu_1,\ldots,\mu_n\}$. Assuming single eigenvalues, the Mayne–Murdoch method [12,130] gives the following closed expression for the product $k_i^{\mathrm{d}} b_i^{\mathrm{d}}$

$$k_i^{\mathrm{d}} b_i^{\mathrm{d}} = \frac{\Pi_{j=1}^{n}\left(\lambda_i - \mu_j\right)}{\Pi_{\substack{j=1\\ i\neq j}}^{n}\left(\lambda_i - \lambda_j\right)} \qquad i = 1, \ldots, n \tag{3.1.23}$$

whence k_i^{d} can be easily determined (see the proof in A.2.5.7 of Appendix A2). Here the coefficient b_i^{d} is an element of the parameter vector $\boldsymbol{b}^{\mathrm{d}} = [b_1^d, \ldots, b_n^d]^{\mathrm{T}} = [\beta_1, \ldots, \beta_n]^{\mathrm{T}}$ of the diagonal form (see also (A.2.28)). The most interesting consequence of (3.1.23) is that it clearly shows that the absolute value of the feedback gain k_i^{d}, required by the pole placement, increases in directly proportion to the so-called "moving" distance $\left|\lambda_i - \mu_j\right|$, i.e., to the distance between the poles of the open and closed loop.

Sampled Data Systems

For sampled systems the feedback vector $\boldsymbol{k}^{\mathrm{T}}$ needs to be chosen to provide the desired polynomial $\mathcal{R}(z)$ for the characteristic equation of the closed system, i.e., in discrete-time (DT) case

$$\mathcal{R}(z) = z^n + r_1 z^{n-1} + \ldots + r_{n-1} z + r_n = \prod_{i=1}^{n} (z - z_i)$$
$$= \det\left(z\boldsymbol{I} - \boldsymbol{F} + \boldsymbol{g}\boldsymbol{k}^{\mathrm{T}}\right) = \mathcal{F}(z) + \boldsymbol{k}^{\mathrm{T}}\Psi(z)\boldsymbol{g} \tag{3.1.24}$$

The solution always exists, if the process is controllable. If the pulse transfer function of the system to be controlled is known, then it is an exceptional case, because the canonical state equations can be directly written. Based on the controllable canonical forms (A.2.37) and (A.3.59), the system matrices can be obtained as

$$\boldsymbol{F}_{\mathrm{c}} = \begin{bmatrix} -f_1 & -f_2 & \ldots & -f_{n-1} & -f_n \\ 1 & 0 & \ldots & 0 & 0 \\ 0 & 1 & \ldots & 0 & 0 \\ \vdots & \vdots & \ddots & \vdots & \vdots \\ 0 & 0 & 0 & 1 & 0 \end{bmatrix}; \quad \boldsymbol{c}_{\mathrm{c}}^{\mathrm{T}} = \left[g_1, g_2, \ldots, g_n\right];$$
$$\boldsymbol{g}_{\mathrm{c}} = [1, 0, \ldots, 0]^{\mathrm{T}} \tag{3.1.25}$$

Taking the special forms of $\boldsymbol{F}_{\mathrm{c}}$ and $\boldsymbol{g}_{\mathrm{c}}$, it can be clearly seen that according to the design equation

$$\boldsymbol{F}_{\mathrm{c}} - \boldsymbol{g}_{\mathrm{c}}\boldsymbol{k}_{\mathrm{c}}^{\mathrm{T}} = \begin{bmatrix} -f_1 & -f_2 & \ldots & -f_{n-1} & -f_n \\ 1 & 0 & \ldots & 0 & 0 \\ 0 & 1 & \ldots & 0 & 0 \\ \vdots & \vdots & \ddots & \vdots & \vdots \\ 0 & 0 & 0 & 1 & 0 \end{bmatrix} - \begin{bmatrix} 1 \\ 0 \\ 0 \\ \vdots \\ 0 \end{bmatrix} \boldsymbol{k}_{\mathrm{c}}^{\mathrm{T}}$$
$$= \begin{bmatrix} -r_1 & -r_2 & \ldots & -r_{n-1} & -r_n \\ 1 & 0 & \ldots & 0 & 0 \\ 0 & 1 & \ldots & 0 & 0 \\ \vdots & \vdots & \ddots & \vdots & \vdots \\ 0 & 0 & 0 & 1 & 0 \end{bmatrix} \tag{3.1.26}$$

the choice of

$$\boldsymbol{k}^{\mathrm{T}} = \boldsymbol{k}_{\mathrm{c}}^{\mathrm{T}} = [r_1 - f_1, r_2 - f_2, \ldots, r_n - f_n] \tag{3.1.27}$$

ensures the characteristic equation (3.1.24), i.e., the desired poles. The value of the calibration factor can be given by a simple computation

$$k_{\mathrm{r}} = \frac{f_n + (r_n - f_n)}{g_n} = \frac{r_n}{g_n} \tag{3.1.28}$$

It can be clearly seen from equations (3.1.21) and (3.1.23) that the overall pulse transfer function of the closed-loop system is

$$T_{\mathrm{ry}}(z) = \frac{k_{\mathrm{r}}\mathcal{G}(z)}{\mathcal{R}(z)} \tag{3.1.29}$$

as it was already mentioned in connection with (3.1.20).

The most frequently appearing case of the state feedback is when instead of the pulse transfer function, the state-space form of the controlled system is given. In connection with (A.2.57), it has already been noted that all controllable systems can be written in controllable canonical form by using the transformation matrix $\boldsymbol{T}_{\mathrm{c}} = \boldsymbol{M}_{\mathrm{c}}^{\mathrm{c}}(\boldsymbol{M}_{\mathrm{c}})^{-1}$. This similarity transformation has an effect on the feedback vector, too

$$\boldsymbol{k}^{\mathrm{T}} = \boldsymbol{k}_{\mathrm{c}}^{\mathrm{T}}\boldsymbol{T}_{\mathrm{c}} = \boldsymbol{k}_{\mathrm{c}}^{\mathrm{T}}\boldsymbol{M}_{\mathrm{c}}^{\mathrm{c}}\boldsymbol{M}_{\mathrm{c}}^{-1} = \boldsymbol{g}_{\mathrm{c}}^{\mathrm{T}}\boldsymbol{M}_{\mathrm{c}}^{-1}\mathcal{R}(\boldsymbol{F}) = [0, 0, \ldots, 1]\boldsymbol{M}_{\mathrm{c}}^{-1}\mathcal{R}(\boldsymbol{F}) \tag{3.1.30}$$

To compute (3.1.30), the inverse of the controllability matrix needs to be constructed by the system matrices $\boldsymbol{F}$ and $\boldsymbol{g}$, on the one hand. On the other hand, the controllability matrix $\boldsymbol{M}_{\mathrm{c}}^{\mathrm{c}}$ of the controllable canonical form needs to be generated (see (A.2.51)). Since this latter depends only on the coefficients a_i in the denominator of the process pulse transfer function, the denominator needs to be computed: $\mathcal{F}(z) = \det(z\boldsymbol{I} - \boldsymbol{F})$. The same is valid for the computation of $\mathcal{R}(\boldsymbol{F})$ in the second form. The computation method of the pole placement state feedback vector shown above is the Ackermann method [1,2].

Note that the transformation properties, the canonical forms, and the controllability/observability concepts of the CT and DT state equations are completely the same. Therefore, the state feedback techniques for DT systems are also very similar to the methods for CT systems.

Observer-Based State Feedback

The method of state feedback shown in the previous section requires the direct measurement of the state vector of the state equation describing the process. This, however, can be fulfilled very rarely, generally only in the case of the low-order dynamics (e.g., in mechanical systems measuring the values of distance, velocity, and acceleration). Thus, the usefulness of the method depends on the possible measurement or estimation of the state vector. To determine the state vector, the so-called observer principle has been elaborated. This method requires knowledge of the system matrices $\boldsymbol{A}$, $\boldsymbol{b}$, and $\boldsymbol{c}^{\mathrm{T}}$; by means of them the exact model of the process is realized. Using the same excitation as that applied in the original process, this model (*observer*) provides the estimated values $\hat{\boldsymbol{x}}$ and $\hat{y}$ of the variables $\boldsymbol{x}$ and y. The state feedback is realized by using $\hat{\boldsymbol{x}}$. The principle is shown in Figure 3.1.6.

Strictly speaking, the estimated values $\hat{\boldsymbol{A}}$, $\hat{\boldsymbol{b}}$, and $\hat{\boldsymbol{c}}^{\mathrm{T}}$ in the observer should have been used instead of $\boldsymbol{A}$, $\boldsymbol{b}$, and $\boldsymbol{c}^{\mathrm{T}}$. The value of the observer, however, is that it applies not only a parallel model but calculates an error $\varepsilon = y - \hat{y}$ from the deviation of the original and estimated output values of the process, and has a feedback via a proportional feedback vector $\boldsymbol{l}$ to the

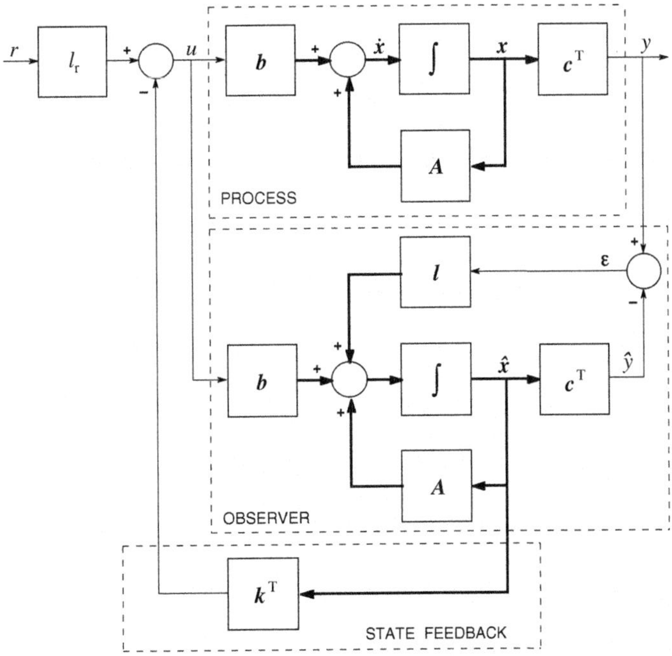

Figure 3.1.6 Observer-based state feedback.

input of the integrator of the observer. This feedback is in operation until the error exists, i.e., until the outputs of the process and the observer become equal. This operation can tolerate considerable error in the knowledge of the system matrices (see the last section of this chapter).

It can be seen in the figure that now the state feedback is

$$u = k_r r - \boldsymbol{k}^T \hat{\boldsymbol{x}} \tag{3.1.31}$$

so only $\hat{\boldsymbol{x}}$ is used instead of $\boldsymbol{x}$. After a very complex derivation, the details of which are not discussed here, we obtain the overall closed-loop transfer function in the form

$$T_{ry}(s) = \frac{k_r P(s)}{1 + \boldsymbol{k}^T (s\boldsymbol{I} - \boldsymbol{A})^{-1} \boldsymbol{b}} = \frac{k_r \mathcal{B}(s)}{\mathcal{R}(s)} \tag{3.1.32}$$

which, perhaps surprisingly, is exactly the same as (3.1.18), i.e., the case of state feedback without observer. (The detailed proof can be seen in A.2.5.8 of Appendix A.2.) This means that the tracking property of the closed loop does not depend on the choice of the vector $\boldsymbol{l}$. (The theoretical explanation for these phenomena is that the observer is the noncontrollable part of the whole closed loop.) The feedback "regulator" introduced in Figure 3.1.5 can also be determined now as

$$R_f = \boldsymbol{k}^T \left(s\boldsymbol{I} - \boldsymbol{A} + \boldsymbol{b}\boldsymbol{k}^T + \boldsymbol{l}\boldsymbol{c}^T\right)^{-1} \boldsymbol{l} = \frac{\boldsymbol{k}^T \left(s\boldsymbol{I} - \boldsymbol{A} + \boldsymbol{b}\boldsymbol{k}^T\right)^{-1} \boldsymbol{l}}{1 + \boldsymbol{c}^T \left(s\boldsymbol{I} - \boldsymbol{A} + \boldsymbol{b}\boldsymbol{k}^T\right)^{-1} \boldsymbol{l}} \tag{3.1.33}$$

a more complex form than the (3.1.21).

To investigate the operation of the observer, let us determine a new state vector error as

$$\tilde{\boldsymbol{x}} = \boldsymbol{x} - \hat{\boldsymbol{x}} \tag{3.1.34}$$

which can also be written as

$$\frac{d\tilde{\boldsymbol{x}}}{dt} = \left(\boldsymbol{A} - \boldsymbol{l}\boldsymbol{c}^T\right) \hat{\boldsymbol{x}} \tag{3.1.35}$$

and is very similar to (3.1.4) without excitation. For the design of the observers a method very similar to what was used in the case of the state feedback is applied, whereby with the selection of $\boldsymbol{l}$ our goal is to ensure the dynamics of (3.1.21) with the characteristic polynomial

$$\det\left(s\boldsymbol{I} - \boldsymbol{A} + \boldsymbol{l}\boldsymbol{c}^T\right) = \mathcal{Q}(s) = s^n + q_1 s^{n-1} + \ldots + q_{n-1} s + q_n \tag{3.1.36}$$

The solution always exists if the process is observable. (It is reasonable to assume that the order of $\mathcal{Q}$ is equal to that of $\mathcal{A}$.) It is an exceptional case when the transfer function of the process to be controlled is known and the canonical state equations can be directly written. Based on the observable canonical form of (A.2.44), the system matrices are [12]

$$\boldsymbol{A}_\text{o} = \begin{bmatrix} -a_1 & 1 & 0 & \dots & 0 \\ -a_2 & 0 & 1 & \dots & 0 \\ \vdots & \vdots & \vdots & \ddots & \vdots \\ -a_{n-1} & 0 & 0 & \dots & 1 \\ -a_n & 0 & 0 & \dots & 0 \end{bmatrix}; \quad \boldsymbol{c}_\text{o}^\text{T} = [1, 0, \dots, 0]; \tag{3.1.37}$$

$$\boldsymbol{b}_\text{o} = [b_1, b_2, \dots, b_n]^\text{T}$$

Considering the special form of $\boldsymbol{A}_\text{o}$ and $\boldsymbol{c}_\text{o}^\text{T}$, it can be clearly seen that according to the design equation

$$\boldsymbol{A}_\text{o} - \boldsymbol{l}_\text{o}\boldsymbol{c}_\text{o}^\text{T} = \begin{bmatrix} -a_1 & 1 & 0 & \dots & 0 \\ -a_2 & 0 & 1 & \dots & 0 \\ \vdots & \vdots & \vdots & \ddots & \vdots \\ -a_{n-1} & 0 & 0 & \dots & 1 \\ -a_n & 0 & 0 & \dots & 0 \end{bmatrix} - \boldsymbol{l}_\text{o}[1, 0, \dots, 0]$$

$$= \begin{bmatrix} -q_1 & 1 & 0 & \dots & 0 \\ -q_2 & 0 & 1 & \dots & 0 \\ \vdots & \vdots & \vdots & \ddots & \vdots \\ -q_{n-1} & 0 & 0 & \dots & 1 \\ -q_n & 0 & 0 & \dots & 0 \end{bmatrix} \tag{3.1.38}$$

the selection of

$$\boldsymbol{l} = \boldsymbol{l}_\text{o} = [q_1 - a_1, q_2 - a_2, \dots, q_n - a_n]^\text{T} \tag{3.1.39}$$

ensures the characteristic equation of (3.1.36), i.e., the desired poles.

The general case is when the state-space form of the process to be controlled is given instead of its transfer function. Referring to equation (A.2.67) it has been noted that all observable systems can be written in observable canonical form by using the transformation matrix $\boldsymbol{T}_\text{o} = (\boldsymbol{M}_\text{o}^\text{o})^{-1}\boldsymbol{M}_\text{o}$. This similarity transformation also has effect on the feedback vector

$$\boldsymbol{l} = (\boldsymbol{T}_\text{o})^{-1}\boldsymbol{l}_\text{o} = \boldsymbol{M}_\text{o}^{-1}\boldsymbol{M}_\text{o}^\text{o}\boldsymbol{l}_\text{o} \tag{3.1.40}$$

To calculate (3.1.40), the inverse of the observability matrix $\boldsymbol{M}_{\rm o}$ is required together with the general system matrices $\boldsymbol{A}$ and $\boldsymbol{c}^{\rm T}$. Similarly, the observability matrix $\boldsymbol{M}_{\rm o}^{\rm o}$ of the observable canonical form needs to be formed (see (A.2.61)). Since this latter depends only on the coefficients a_i in the denominator of the transfer function of the process, so the denominator needs to be calculated: $\mathcal{A}(s) = \det(s\boldsymbol{I} - \boldsymbol{A})$. This calculation method of the observer vector is called the Ackermann method [1,2].

There is an interesting similarity in the design methods of the dynamics of the observer and the state feedback, often called duality, i.e., they correspond to each other under the condition: $\boldsymbol{A} \leftrightarrow \boldsymbol{A}^{\rm T}$, $\boldsymbol{b} \leftrightarrow \boldsymbol{c}^{\rm T}$, $\boldsymbol{k} \leftrightarrow \boldsymbol{l}^{\rm T}$, $\boldsymbol{M}_{\rm c}^{\rm c} \leftrightarrow (\boldsymbol{M}_{\rm o}^{\rm o})^{\rm T}$.

Based on the equations of the error (3.1.34) and the process (3.1.1), the joint equation of the state feedback and the observer is

$$\frac{\rm d}{{\rm d}t}\begin{bmatrix}\boldsymbol{x}\\ \tilde{\boldsymbol{x}}\end{bmatrix} = \begin{bmatrix}\boldsymbol{A}-\boldsymbol{b}\boldsymbol{k}^{\rm T} & \boldsymbol{b}\boldsymbol{k}^{\rm T}\\ \boldsymbol{0} & \boldsymbol{A}-\boldsymbol{l}\boldsymbol{c}^{\rm T}\end{bmatrix}\begin{bmatrix}\boldsymbol{x}\\ \tilde{\boldsymbol{x}}\end{bmatrix} + \begin{bmatrix}k_{\rm r}\boldsymbol{b}\\ \boldsymbol{0}\end{bmatrix} r \tag{3.1.41}$$

$$e = y - \hat{y} = \boldsymbol{c}^{\rm T}\tilde{\boldsymbol{x}}$$

Since the system matrix of the right-hand side is block diagonal, the characteristic equation of the closed loop is

$$\det\left(s\boldsymbol{I} - \boldsymbol{A} + \boldsymbol{b}\boldsymbol{k}^{\rm T}\right)\det\left(s\boldsymbol{I} - \boldsymbol{A} + \boldsymbol{l}\boldsymbol{c}^{\rm T}\right) = \mathcal{R}(s)\mathcal{Q}(s) \tag{3.1.42}$$

Thus the polynomial is the product of two terms: the first term relates to the state feedback, the other one to the observer. It is important to note that $\mathcal{Q}(s)$, in spite of (3.1.42), does not appear in the transfer function $T_{\rm ry}(s)$ of the closed loop of (3.1.5). This can be explained by the redefinition of the whole system given in the block diagram of Figure 3.1.6, which applies appropriate transfer functions.

Equation (3.1.42) of the observer-based state feedback, according to which the state feedback and the characteristic equation of the observer are independent, is also called the separation principle.

It is advisable to choose poles in $\mathcal{Q}(s)$ which are three to five times faster than the poles of the process $\mathcal{A}(s)$ or those in the characteristic systems of the overall closed system $\mathcal{R}(s)$. This solution used to ensure that the primary pole placement state feedback could have quite accurate estimation on the state variables. It is worth noting that the observer almost always appears in the form of computer simulation, and therefore the overexcitation required by the observer-based pole placement means high peaks in the actuator signal only inside the computer in the state-space model of the process. This means that realizability does not seem to be as critical as in the state feedback regulator itself.

Sampled Data Systems

Similarly to the case of CT systems, in DT systems observer-based state feedback methods require knowledge of the system matrices $\boldsymbol{F}$, $\boldsymbol{g}$, and $\boldsymbol{c}^{\mathrm{T}}$, by means of which an accurate copy of the process is constructed, and using the same excitation signal this model (observer) provides the estimated values $\hat{\boldsymbol{x}}[k]$ and $\hat{y}[k]$ of the variables $\boldsymbol{x}[k]$ and $y[k]$, respectively. The state feedback is performed by $\hat{\boldsymbol{x}}[k]$. The realization can be followed in Figure 3.1.7. Strictly speaking, $\hat{\boldsymbol{F}}$, $\hat{\boldsymbol{g}}$, and $\hat{\boldsymbol{c}}^{\mathrm{T}}$ need to be applied instead of $\boldsymbol{F}$, $\boldsymbol{g}$, and $\boldsymbol{c}^{\mathrm{T}}$ in the observer. But the value of the observer is that as well as providing a parallel model it also constructs an error $\varepsilon[k] = y[k] - \hat{y}[k]$ from the deviation of the original and estimated output of the process and feeds it back to the input of the observer delay via a proportional feedback vector $\boldsymbol{l}$. This feedback operates until the error exists, i.e., until the outputs of the

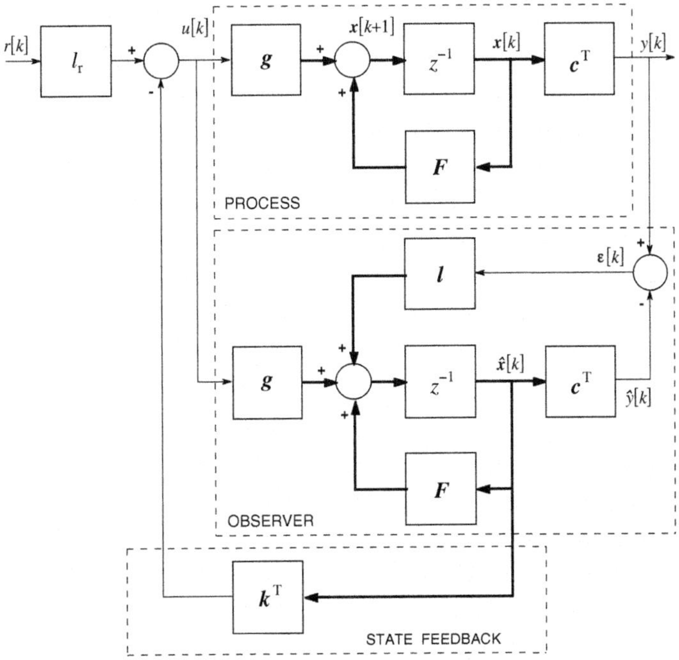

Figure 3.1.7 Observer-based state feedback (*SFO* scheme).

process and the observer become the same. With the knowledge of the system matrices this operating mode can compensate for relatively large errors (see the last section of this chapter). It is also clear from the figure that now the state feedback has the form

$$u[k] = k_r r[k] - \boldsymbol{k}^T \hat{\boldsymbol{x}}[k] \tag{3.1.43}$$

and thus $\hat{\boldsymbol{x}}[k]$ appears instead of $\boldsymbol{x}[k]$.

After a long and complex derivation, the details of which are not discussed here, the transfer function of the complete closed system is obtained as

$$\begin{aligned} T_{ry}(z) &= \frac{\left[\boldsymbol{c}^T(z\boldsymbol{I}-\boldsymbol{F})^{-1}\boldsymbol{g}\right]\left[1-\boldsymbol{k}^T\left(z\boldsymbol{I}-\boldsymbol{F}+\boldsymbol{g}\boldsymbol{k}^T+\boldsymbol{l}\boldsymbol{c}^T\right)^{-1}\boldsymbol{b}\right]k_r}{1+\left[\boldsymbol{l}^T\left(z\boldsymbol{I}-\boldsymbol{F}+\boldsymbol{g}\boldsymbol{l}^T+\boldsymbol{l}\boldsymbol{c}^T\right)^{-1}\boldsymbol{g}\right]\left[\boldsymbol{c}^T(z\boldsymbol{I}-\boldsymbol{F})^{-1}\boldsymbol{g}\right]} \\ &= \boldsymbol{c}^T\left(z\boldsymbol{I}-\boldsymbol{F}+\boldsymbol{g}\boldsymbol{k}^T\right)^{-1}\boldsymbol{g}k_r = \frac{\boldsymbol{c}^T(z\boldsymbol{I}-\boldsymbol{F})^{-1}\boldsymbol{g}k_r}{1+\boldsymbol{k}^T(z\boldsymbol{I}-\boldsymbol{F})^{-1}\boldsymbol{g}} \\ &= \frac{k_r G(z)}{1+\boldsymbol{k}^T(z\boldsymbol{I}-\boldsymbol{F})^{-1}\boldsymbol{g}} = \frac{k_r \mathcal{G}(z)}{\mathcal{R}(z)} \end{aligned} \tag{3.1.44}$$

which is, perhaps surprisingly, precisely the same as (3.1.29), i.e., the case of state feedback without observer. This means that the tracking behavior of the closed system does not depend on the choice of vector $\boldsymbol{l}$. To examine the operation of the observer let us construct the vector of the state error

$$\tilde{\boldsymbol{x}}[k] = \boldsymbol{x}[k] - \hat{\boldsymbol{x}}[k] \tag{3.1.45}$$

and also

$$\tilde{\boldsymbol{x}}[k] = \left(\boldsymbol{F} - \boldsymbol{l}\boldsymbol{c}^T\right)\hat{\boldsymbol{x}}[k] \tag{3.1.46}$$

which is very similar to (3.1.10) without excitation. Very similar methods to those for state feedback can be used for the design of the observers, where with the selection of $\boldsymbol{l}$ the goal is to ensure the system dynamics (3.1.46) by the characteristic polynomial

$$\det\left(z\boldsymbol{I} - \boldsymbol{F} + \boldsymbol{l}\boldsymbol{c}^T\right) = \mathcal{Q}(z) = z^n + q_1 z^{n-1} + \ldots + q_{n-1}z + q_n \tag{3.1.47}$$

The solution always exists if the process is observable. (It is reasonable if the order of $\mathcal{Q}$ is equal to that of $\mathcal{F}$.) If the transfer function of the process to be controlled is known, then it is an exceptional case, because then the canonical forms can be directly written. In this case, the system matrices are

$$\boldsymbol{F}_{\mathrm{o}} = \begin{bmatrix} -f_1 & 1 & 0 & \dots & 0 \\ -f_2 & 0 & 1 & \dots & 0 \\ \vdots & \vdots & \vdots & \ddots & \vdots \\ -f_{n-1} & 0 & 0 & \dots & 1 \\ -f_n & 0 & 0 & \dots & 0 \end{bmatrix}; \quad \boldsymbol{c}_{\mathrm{o}}^{\mathrm{T}} = [1, 0, \dots, 0]; \qquad (3.1.48)$$

$$\boldsymbol{g}_{\mathrm{o}} = [g_1, g_2, \dots, g_n]^{\mathrm{T}}$$

based on the observable canonical forms (A.2.44), (A.3.60).

Taking the special forms of $\boldsymbol{F}_{\mathrm{o}}$ and $\boldsymbol{c}_{\mathrm{o}}^{\mathrm{T}}$ into account, it is clear that according to the design equation

$$\boldsymbol{F}_{\mathrm{o}} - \boldsymbol{l}_{\mathrm{o}}\boldsymbol{c}_{\mathrm{o}}^{\mathrm{T}} = \begin{bmatrix} -f_1 & 1 & 0 & \dots & 0 \\ -f_2 & 0 & 1 & \dots & 0 \\ \vdots & \vdots & \vdots & \ddots & \vdots \\ -f_{n-1} & 0 & 0 & \dots & 1 \\ -f_n & 0 & 0 & \dots & 0 \end{bmatrix} - \boldsymbol{l}_{\mathrm{o}}[1, 0, \dots, 0] = \begin{bmatrix} -q_1 & 1 & 0 & \dots & 0 \\ -q_2 & 0 & 1 & \dots & 0 \\ \vdots & \vdots & \vdots & \ddots & \vdots \\ -q_{n-1} & 0 & 0 & \dots & 1 \\ -q_n & 0 & 0 & \dots & 0 \end{bmatrix} \qquad (3.1.49)$$

the selection of

$$\boldsymbol{l} = \boldsymbol{l}_{\mathrm{o}} = [q_1 - f_1, q_2 - f_2, \dots, q_n - f_n]^{\mathrm{T}} \qquad (3.1.50)$$

ensures the characteristic equation (3.1.47), i.e., the desired poles.

The general case is when the state-space equation of the process is given instead of its pulse transfer function, as now. It has already been noted with regard to (A.2.67) that all observable systems can be written in observable canonical form by the use of the transformation matrix $\boldsymbol{T}_{\mathrm{o}} = (\boldsymbol{M}_{\mathrm{o}}^{\mathrm{o}})^{-1}\boldsymbol{M}_{\mathrm{o}}$. This similarity transformation has an effect on the feedback vector, too,

$$l = (T_o)^{-1} l_o = M_o^{-1} M_o^o l_o \tag{3.1.51}$$

To compute (3.1.51), the inverse of the observability matrix M_o needs to be constructed by the general system matrices F and c^T. On the other hand, the observability matrix M_o^o of the observable canonical form must be also given (see (A.2.61)). Since this latter depends only on the coefficients f_i in the denominator of the transfer function of the process, the determination of the denominator needs to be computed: $\mathcal{F}(z) = \det(zI - F)$. The computation method of the observer vector shown above is named the Ackermann method [1,2].

There is an interesting similarity between the design methods of the dynamics of the state feedback and of the observer. It is the so-called duality, i.e., they are corresponding to each other under the condition $F \leftrightarrow F^T$, $g \leftrightarrow c^T$, $k \leftrightarrow l^T$, $M_c^c \leftrightarrow (M_o^o)^T$.

Based on the state error (3.1.45) and the equations of the process (3.1.7), the joint equation of the state feedback and the observer is

$$\begin{bmatrix} x[k+1] \\ \tilde{x}[k+1] \end{bmatrix} = \begin{bmatrix} F - gk^T & gk^T \\ 0 & F - lc^T \end{bmatrix} \begin{bmatrix} x[k] \\ \tilde{x}[k] \end{bmatrix} + \begin{bmatrix} k_r g \\ 0 \end{bmatrix} r[k] \tag{3.1.52}$$
$$e[k] = y[k] - \hat{y}[k] = c^T \tilde{x}[k]$$

Since the right-hand side system matrix is a block diagonal, the characteristic equation of the closed system is

$$\det(zI - F + gk^T)\det(zI - F + lc^T) = \mathcal{R}(z)\mathcal{Q}(z) \tag{3.1.53}$$

Thus the polynomial is the product of two factors: one is connected to the state feedback, the other relates to the observer. It is important to remark that, contrary to (3.1.53), $\mathcal{Q}(z)$ does not appear in the transfer function $T_{ry}(z)$ (see (3.1.29) and (3.1.44)).

Equation (3.1.53) representing observer-based state feedback, and according to which the characteristic equations of the state feedback and observer are independent, is also named the separation principle.

Observer-Based State Feedback Using Equivalent Transfer Functions

The block diagram containing transfer functions has already been applied in Figure 3.1.5. A further generalized form of the approach used there can also be applied, which is shown in Figure 3.1.8.

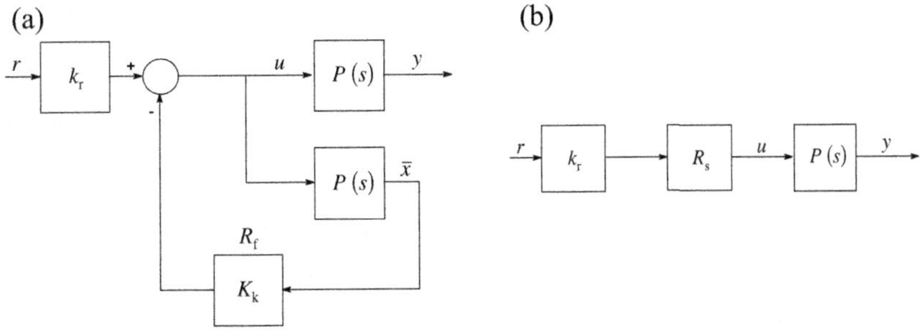

Figure 3.1.8 Further equivalent schemes for state feedback with transfer functions.

It follows from Figure 3.1.8 that the resulting equivalent serial compensator is now again

$$R_s = \frac{1}{1 + R_f P} = \frac{1}{1 + K_k P} = \frac{\mathcal{A}(s)}{\mathcal{A}(s) + \mathcal{K}(s)} = \frac{\mathcal{A}(s)}{\mathcal{R}(s)} \tag{3.1.54}$$

It needs to be stated that R_s is a virtual term used only for demonstrating the final signal formation, i.e., $k_r R_s P$ ensures the same T_{ry} as (3.1.20). If the pole cancellation represented by R_s is to be performed by a serial compensator, then it cannot be applied for unstable processes, since the unstable zeros and poles cannot be eliminated by cancellation. The signal $\bar{x}$ (which is not the same as $\boldsymbol{x}$) introduced in Figure 3.1.8 shows that finally both the state feedback and the observer are *single-input single-output* subsystems and can be performed by transfer functions, i.e., it is always possible to find equivalent representations for the input and output. Applying this approach and Figure 3.1.6, the block diagram using transfer functions can be drawn as shown in Figure 3.1.9.

After a long transformation procedure and block manipulations, the block diagram of Figure 3.1.9 can be traced back to a very simple unity feedback closed loop, shown in Figure 3.1.10. Here the relationship (3.1.21) defining K_k is also used, and K_l is introduced in a similar way

$$K_k(s) = \frac{\mathcal{K}(s)}{\mathcal{B}(s)}; \quad K_l(s) = \frac{\mathcal{L}(s)}{\mathcal{B}(s)} \tag{3.1.55}$$

where the design polynomial equations

$$\mathcal{K}(s) = \mathcal{R}(s) - \mathcal{A}(s) \text{ and } \mathcal{L}(s) = \mathcal{Q}(s) - \mathcal{A}(s) \tag{3.1.56}$$

result from the conditions of the two kinds of pole placements.

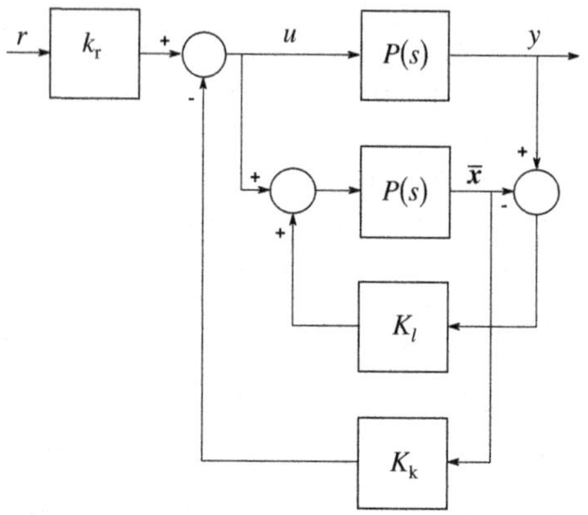

Figure 3.1.9 State feedback and observer using transfer functions.

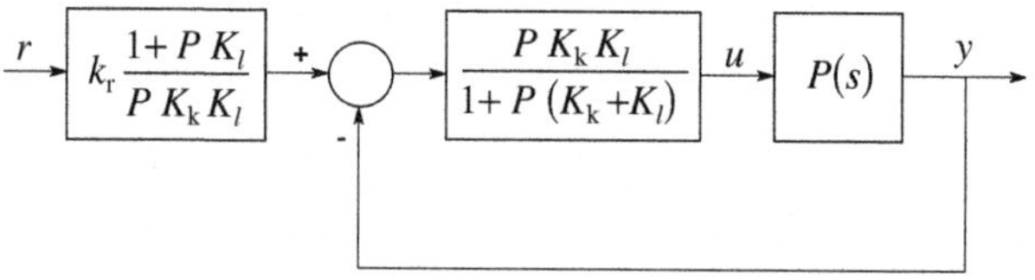

Figure 3.1.10 Reduced block diagram of the state feedback and observer.

In Figure 3.1.10 the overall transfer function of the inner closed loop

$$\frac{P^2 K_k K_l}{1+P(K_k+K_l)+P^2 K_k K_l} = \frac{PK_k}{1+PK_k}\,\frac{PK_l}{1+PK_l} = \frac{\mathcal{K}}{\mathcal{A}+\mathcal{K}}\,\frac{\mathcal{L}}{\mathcal{A}+\mathcal{L}} = \frac{\mathcal{K}}{\mathcal{R}}\,\frac{\mathcal{L}}{\mathcal{Q}} \tag{3.1.57}$$

has a special form, but its denominator completely corresponds to the characteristic equation (3.1.42), i.e., represents two serially connected independent closed loops (see Figure 3.1.11). This character is called the separation principle of the state feedback and the observer. To ensure stability both

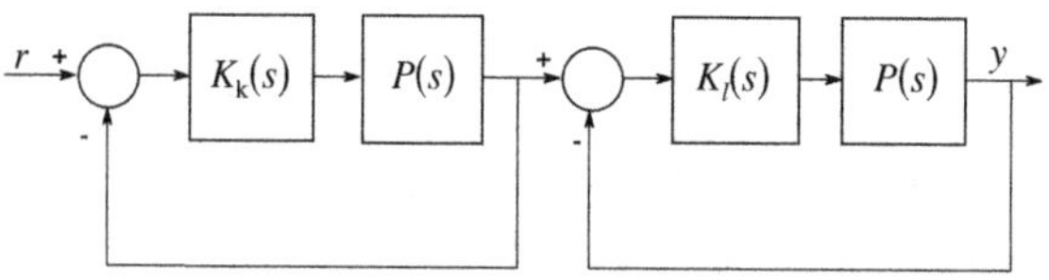

Figure 3.1.11 Equivalent observer block diagrams of the inner system.

loops must be stable. This can be achieved by proper pole placement design.

At the same time, the transfer function of the whole system is

$$\begin{aligned} T_{ry}(s) &= k_r \frac{1+PK_k}{PK_l K_k} \frac{PK_l}{1+PK_l} \frac{PK_k}{1+PK_k} = \frac{k_r P}{1+PK_l} \\ &= \frac{k_r \frac{\mathcal{B}}{\mathcal{A}}}{1+\frac{\mathcal{B}}{\mathcal{A}}\frac{\mathcal{K}}{\mathcal{B}}} = \frac{k_r \mathcal{B}}{\mathcal{A}+\mathcal{K}} = \frac{k_r \mathcal{B}(s)}{\mathcal{R}(s)} \end{aligned} \tag{3.1.58}$$

which is completely the same as (3.1.32). As expected the poles of the observer do not appear in T_{ry}. The inner character of the whole system can be better seen from the final block diagram shown in Figure 3.1.12 for the tracking properties.

This simple structure is not valid for the disturbance rejection capabilities of the closed loop. This can be simply seen if the sensitivity function of the closed loop is constructed as

$$\frac{1}{1+\frac{P^2 K_k K_l}{1+P(K_k+K_l)}} = \frac{1+P(K_k+K_l)}{1+P(K_k+K_l)+P^2 K_k K_l} = \left(1+\frac{\mathcal{L}}{\mathcal{R}}\right)\left(1-\frac{\mathcal{L}}{\mathcal{Q}}\right) \tag{3.1.59}$$

which shows that both $\mathcal{R}$ and $\mathcal{Q}$ appear in the transfer function of the disturbance rejection according to (3.1.42). Equation (3.1.59) has a special form, since formally it is the product of the output noise rejection transfer functions of two serially connected closed loops, while it is known that the tracking properties are indeed resulting a product of the transfer functions, but this phenomenon is not valid for the sensitivity functions. Note that the resulting noise rejection properties are not independent of the tracking ones, and therefore the joint application of the state feedback and the observer is not appropriate for realizing an actual *two-degree-of-freedom (TDOF)* control loop.

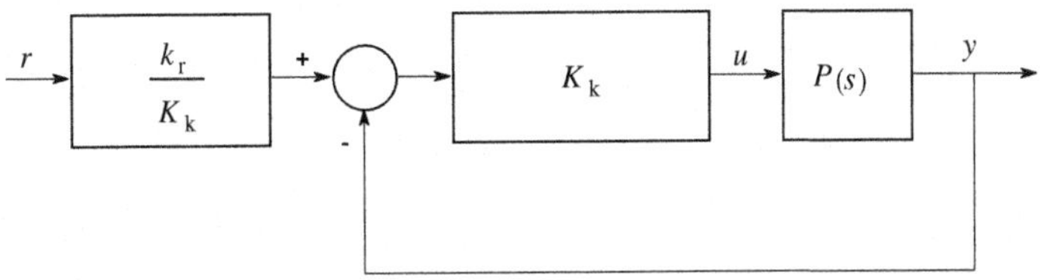

Figure 3.1.12 Reduced block diagram of the state feedback and the observer for the tracking properties.

Two-Step Design Methods with State Feedback

It was seen in the discussion of the state feedback-based control that the most advantageous features of the method are that:

- applicability does not depend on whether the process is stable or not
- the tracking property does not depend on the applied observer and thus it can be directly designed
- the method is not very sensitive to exact knowledge of the parameter matrices of the state equation

(This latter feature is usually demonstrated by experimental and simulation examples, but it can be proved that the error with an observer can be reduced by the $[1 + K_l(s)P(s)]$ part of the original, compared with the modeling error obtained by the simple parallel model of the state equation of the process, as would be obtained via a closed loop $1/[1 + K_l(s)P(s)]$. So it can be reduced by the feedback $K_l(s)$ of the observer in a given frequency region (see the last section of this chapter). If the model of the process is applied, which is quite conventional practice, then both loops of Figure 3.1.11 must be robust stable.)

The unfavorable (unwanted) features are that:

- the state feedback is basically a zero-type control and therefore the remaining error can be eliminated by the calibration factor, which, in the case of using a process model, never provides accurate results
- the state feedback cannot change the zeros of the process
- the disturbance rejection property cannot be designed directly

Because of the latter features, further steps are usually required in the state feedback-based control systems. The necessity for the calibration factor can best be avoided by using the cascade integrating regulator shown in Figure 3.1.13.

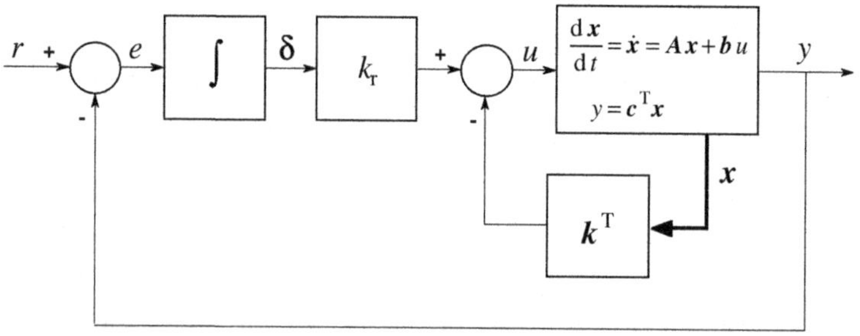

Figure 3.1.13 Joint state feedback and integrating regulator.

Instead of (3.1.4), the joint state equation of the closed loop can be written as

$$\dot{\boldsymbol{x}}^*(t) = \begin{bmatrix} \dot{\boldsymbol{x}}(t) \\ \dot{\delta}(t) \end{bmatrix} = \begin{bmatrix} \boldsymbol{A} & \boldsymbol{0} \\ \boldsymbol{c}^{\mathrm{T}} & 0 \end{bmatrix} \begin{bmatrix} \boldsymbol{x}(t) \\ \delta(t) \end{bmatrix} + \begin{bmatrix} \boldsymbol{b} \\ 0 \end{bmatrix} u(t) + \begin{bmatrix} \boldsymbol{0} \\ -1 \end{bmatrix} r(t)$$
$$= \left(\boldsymbol{A}^* - \boldsymbol{b}^* \boldsymbol{k}_*^{\mathrm{T}}\right) \boldsymbol{x}^*(t) + \boldsymbol{v}^* r(t) \tag{3.1.60}$$

by introducing the new state variable $\delta(t)$, which is the integral of the error $e(t) = r(t) - y(t)$ in the outer loop. In this extended state equation, the notations

$$\boldsymbol{A}^* = \begin{bmatrix} \boldsymbol{A} & \boldsymbol{0} \\ \boldsymbol{c}^{\mathrm{T}} & 0 \end{bmatrix}; \quad \boldsymbol{b}^* = \begin{bmatrix} \boldsymbol{b} \\ 0 \end{bmatrix}; \quad \boldsymbol{v}^* = \begin{bmatrix} \boldsymbol{0} \\ -1 \end{bmatrix} \tag{3.1.61}$$

and the new extended feedback equation

$$u(t) = -[\boldsymbol{k}^T \quad k_{\mathrm{r}}] \begin{bmatrix} \boldsymbol{x}(t) \\ \delta(t) \end{bmatrix} = -\boldsymbol{k}_*^{\mathrm{T}} \boldsymbol{x}^*(t) = k_{\mathrm{r}} \int_0^t e(\tau) \mathrm{d}\tau - \boldsymbol{k}^{\mathrm{T}} \boldsymbol{x}(t) \tag{3.1.62}$$

are considered. Equation (3.1.62) clearly shows the integrating effect. The term $\boldsymbol{k}^{\mathrm{T}}\boldsymbol{x}(t)$, however, can be considered as the generalization of the differentiating effect.

Thus, the closed-loop control including integrator can be formulated by a state equation of an order greater by one, where as well as the feedback vector $\boldsymbol{k}^{\mathrm{T}}$ and the coefficient k_{r} also needs to be determined. To design the extended system, the characteristic polynomial $\mathcal{R}^*(s)$ of the order $(n + 1)$ is required, and then the design equation (3.1.16) of the Ackermann method can be directly applied here too. If the process is not presented in the transfer function form, then first the general state equation needs to be transformed into the controllable canonical form, as shown in (3.1.19).

Note that the extended task cannot be solved sequentially, i.e., in such a way that the $\boldsymbol{k}^{\mathrm{T}}$ relating to $\mathcal{R}(s)$ is determined, followed by k_{r} based on $\mathcal{R}^*(s) = \mathcal{R}(s)(s - s_{n+1})$. The task must be solved in one step for $\boldsymbol{k}_*^{\mathrm{T}}$ by $\mathcal{R}^*(s)$.

Integrating effects can be also included in the design of the state feedback for a modified process $P^*(s) = P(s)/s$ instead of the transfer function $P(s)$. Note that the state feedback vectors obtained for the previous case and for this approach are not equal!

Obviously, besides the *I*-regulator, other—higher-order—regulators can be applied, but the pole placement is not always automatically given by the Ackermann method and can result in complex nonlinear equation systems.

In the case of observer-based state feedback, at the feedback of the observer error, not only zero-type but one-type or higher-type regulators can also be applied by the above methods.

The untouched zeros of the process can be modified by a serial compensator

$$K_s(s) = G_s(s)\frac{\mathcal{N}(s)}{\mathcal{B}_+(s)} \qquad (3.1.63)$$

where the nominator of the process $\mathcal{B}(s) = \mathcal{B}_+(s)\mathcal{B}_-(s)$ is assumed according to the method applied in Section 2.1. Here $\mathcal{B}_+$ has stable zeros, but $\mathcal{B}_-$ contains unstable zeros. To the realizability $\mathcal{N}(s)/\mathcal{B}_+(s)$ must be proper, so only as many zeros as the number of the stable zeros in the process can be placed in the transfer function of the closed loop. Finally the resulting transfer function has the form

$$T_{ry}(s) = \frac{\mathcal{N}(s)}{\mathcal{R}(s)} k_r G_s(s)\mathcal{B}_-(s) \qquad (3.1.64)$$

where the effect of the invariant $\mathcal{B}_-(s)$ can be optimally attenuated by the filter $G_s(s)$. In many cases, however, the simple, but not optimal, $G_s(s) = 1$ is used.

An acceptable design of the disturbance rejection feature can be reached by the application of the Youla-parameterized regulator in the outer cascade loop. This can be done because, with state feedback, any process, even an unstable one, can be stabilized. The qualitative control of the unstable processes has two stages in general. In the first stage, the process is stabilized by the regulator and then the required qualitative goals can be reached by a second outer control loop or even *TDOF* structures.

The state feedback-based stabilizing regulator can only be applied for processes without time delay. If the process has a considerable time delay, then one possibility is to employ rational fraction functions (1.1.22), (1.1.23). Another is to use computer-based sampled-data control (see Appendix 3).

Sampled Data Systems

In DT systems, the necessity for the calibration factor can be eliminated in the simplest way by applying a cascade integrating regulator as seen in Figure 3.1.14.

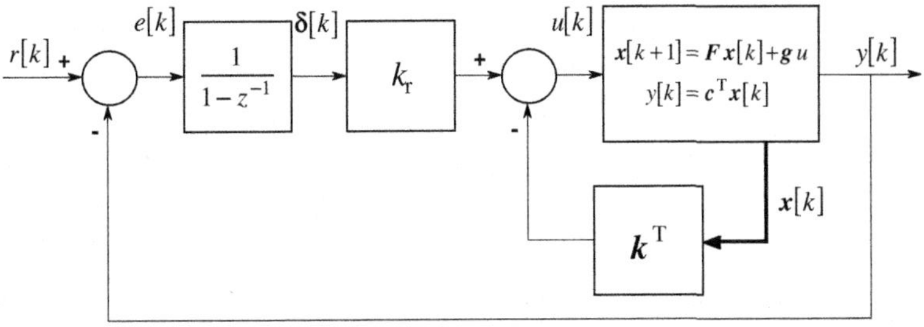

Figure 3.1.14 Joint use of the state feedback and the integrating regulator.

The joint state equation of the closed system, which substitutes equation (3.1.10), can be written as

$$\begin{aligned}\dot{\boldsymbol{x}}^*[k+1] &= \begin{bmatrix} \dot{\boldsymbol{x}}[k+1] \\ \dot{\delta}[k+1] \end{bmatrix} = \begin{bmatrix} \boldsymbol{F} & \boldsymbol{0} \\ \boldsymbol{c}^{\mathrm{T}} & 0 \end{bmatrix} \begin{bmatrix} \boldsymbol{x}[k] \\ \delta[k] \end{bmatrix} + \begin{bmatrix} \boldsymbol{g} \\ 0 \end{bmatrix} u[k] + \begin{bmatrix} \boldsymbol{0} \\ -1 \end{bmatrix} r[k] \\ &= \left(\boldsymbol{F}^* - \boldsymbol{g}^* \boldsymbol{k}_*^{\mathrm{T}} \right) \boldsymbol{x}^*[k] + \boldsymbol{v}^* r[k] \end{aligned} \tag{3.1.65}$$

by introducing a new state variable $\delta[k]$, which is the integral of the error $e[k] = r[k] - y[k]$ of the outer loop, where the notations

$$\boldsymbol{F}^* = \begin{bmatrix} \boldsymbol{F} & \boldsymbol{0} \\ \boldsymbol{c}^{\mathrm{T}} & 0 \end{bmatrix}; \quad \boldsymbol{g}^* = \begin{bmatrix} \boldsymbol{g} \\ 0 \end{bmatrix}; \quad \boldsymbol{v}^* = \begin{bmatrix} \boldsymbol{0} \\ -1 \end{bmatrix} \tag{3.1.66}$$

and the new extended feedback equation

$$u[k] = -[\, \boldsymbol{k}^{\mathrm{T}} \quad k_{\mathrm{r}} \,] \begin{bmatrix} \boldsymbol{x}[k] \\ \delta[k] \end{bmatrix} = -\boldsymbol{k}_*^{\mathrm{T}} \boldsymbol{x}^*[k] = \frac{k_{\mathrm{r}}}{1 - z^{-1}} e[k] - \boldsymbol{k}^{\mathrm{T}} \boldsymbol{x}[k] \tag{3.1.67}$$

are taken into account. Equation (3.1.67) clearly shows the integrating effect. The item $\boldsymbol{k}^{\mathrm{T}}\boldsymbol{x}[k]$, however, can be considered as the generalization of the derivative effect.

Thus, the closed-loop control with an integrator can be described by a state equation which has a dimension higher by one than the earlier one, and now k_{r} must be determined as well as $\boldsymbol{k}^{\mathrm{T}}$. For the design of the extended system the characteristic polynomial $\mathcal{Q}^*(z)$ with an order greater by one needs to be applied, so the design equation (3.1.30) of the Ackermann method can be directly applied here. If the process is not given in the transfer function form, then the general state equation needs to be rewritten first into a controllable canonical form, as shown in (3.1.30).

Note that the extended task cannot be solved sequentially, i.e., by determining first $\boldsymbol{k}^{\mathrm{T}}$ belonging to $\mathcal{Q}(z)$, and then k_{r} based on $\mathcal{Q}^*(z) = \mathcal{Q}(z)(z - z_{n+1})$. The task needs to be solved in one step for $\boldsymbol{k}_*^{\mathrm{T}}$ on the basis of $\mathcal{Q}^*(z)$.

Integrating effects can also be included by designing the state feedback for a modified process $G^*(z) = zG(z)/(z-1)$ instead of the transfer function $G(z)$. Note that the feedback vectors obtained for the earlier case and for this latter approach are not the same!

Obviously, besides the *I*-regulator, higher-order regulators can also be applied. The solution of the pole placement, however, cannot be obtained automatically by the Ackermann method [1,2], and may lead to a complicated nonlinear equation system.

State feedback, observer, *I* or higher-order regulator can also be applied instead of the type 0 regulator in the error feedback of the observer by using the methods shown above.

The untouched zeros of the process can be compensated by a serial compensator

$$K_{\mathrm{s}}(z) = G_{\mathrm{s}}(z)\frac{\mathcal{N}(z)}{\mathcal{G}_+(z)} \tag{3.1.68}$$

where it is assumed, according to the method applied in Section 2.1, that the numerator of the process is $\mathcal{G}(z) = \mathcal{G}_+(z)\mathcal{G}_-(z)$. Here $\mathcal{G}_+$ contains stable zeros and $\mathcal{G}_-$ unstable zeros, respectively. $\mathcal{N}(z)/\mathcal{G}_+(z)$ needs to be proper, so only as many zeros can be placed in the transfer function of the closed system as the number of the stable zeros in the process.

Finally the loop transfer function has the form

$$T_{\mathrm{ry}}(z) = \frac{\mathcal{N}(z)}{\mathcal{R}(z)}k_{\mathrm{r}}G_{\mathrm{s}}(z)\mathcal{G}_-(z) \tag{3.1.69}$$

where the effect of the invariant $\mathcal{G}_-(z)$ can be attenuated optimally by the filter $G_{\mathrm{s}}(z)$. In many cases simple, but not optimal, $G_{\mathrm{s}}(z) = 1$ is chosen.

A favorable design of the disturbance rejection feature can be obtained by applying a *YP* regulator in the outer cascade loop. This can be done since the state feedback is capable of stabilizing unstable processes, too. In general, the control of unstable processes has two steps. In the first step the process is stabilized; then in the second step, via a second outer loop, the required quality goals can be ensured even by a *TDOF* structure.

The stabilizing regulator using state feedback can be applied only for delay-free processes. If the process has significant delay, then the only possibility is to switch to a sampled-data control using a general polynomial method (see Section 3.3).

Examples of Stabilization

YP, unlike the state feedback regulators, is the simplest and most accurate method for the control of stable processes. Its advantage is that the design of the Keviczky–Bányász-parameterized *TDOF* regulators is also very simple. It was mentioned above that for unstable systems the two-step design method is recommended, whereby the first step (the inner loop) is always a stabilizing state feedback. Simple CT and DT examples of the stabilization of unstable processes are shown below.

Example 3.1.1

Let the transfer function of the unstable CT process be

$$P(s) = \frac{-8}{(s+2)(s-\mathbf{4})} = \frac{1}{(1+0.5s)(1-\mathbf{0.25}s)} \tag{3.1.70}$$

where the pole written in bold type is unstable. The equivalent design form is

$$P(s) = \frac{-8}{s^2 - 2s - 8} = \frac{-8}{\mathcal{A}(s)}; \qquad \mathcal{A}(s) = (s+2)(s-\mathbf{4}) = s^2 - 2s - 8 = s^2 + a_1 s + a_2 \tag{3.1.71}$$

For stabilization, let us mirror the unstable pole $p_2^c = 4$ to the left-hand side plane of the imaginary axis, and by this stable pole $p_2^c = -4$, let us seek the poles of the overall closed system, i.e., let the design polynomial be

$$\mathcal{R}(s) = (s+2)(s+4) = s^2 + 6s + 8 = s^2 + r_1 s + r_2 \tag{3.1.72}$$

The necessary state feedback vector can easily be calculated as

$$\boldsymbol{k}^{\mathrm{T}} = [\, r_1 - a_1 \quad r_2 - a_2 \,] = [\, 6 - (-2) \quad 8 - (-8) \,] = [\, 8 \quad 16 \,] \tag{3.1.73}$$

Example 3.1.2

Let the transfer function of the DT unstable process be

$$G(z) = \frac{-0.2z}{(z-0.8)(\boldsymbol{z}-\mathbf{2})} = \frac{-0.2z^{-1}}{(1-0.8z^{-1})(\mathbf{1}-\mathbf{2}\boldsymbol{z}^{-1})} \tag{3.1.74}$$

where the pole written in bold type is unstable. The equivalent design form is

$$G(z) = \frac{-0.2z}{z^2 - 2.8z + 1.6} = \frac{-0.2z}{\mathcal{F}(z)}$$
$$\mathcal{F}(z) = (z - 0.8)(z - 2) = z^2 - 2.8z + 1.6 = z^2 + f_1 z + f_2 \tag{3.1.75}$$

For the stabilization, let us mirror the unstable pole $p_2^{\mathrm{d}} = 2$ to the inner part of the unit circle, and by this stable pole $p_2^{\mathrm{d}} = 0.5$, let us seek the poles of the overall closed system, i.e., let the design polynomial be

$$\mathcal{Q}(z) = (z - 0.8)(z - 0.5) = z^2 - 1.3z + 0.4 = z^2 + q_1 z + q_2 \tag{3.1.76}$$

The state feedback vector can easily be calculated as

$$\begin{aligned} \boldsymbol{k}^{\mathrm{T}} &= \left[q_1 - f_1 \quad q_2 - f_2 \right] = \left[-1.3 - (-2.8) \quad 0.4 - (1.6) \right] \\ &= \left[1.5 \quad -1.2 \right] \end{aligned} \tag{3.1.77}$$

Effect of Modeling Error on the State Feedback Regulators

As mentioned earlier, in practice, calculations can always be based simply on the estimated parameters $\hat{\boldsymbol{A}}, \hat{\boldsymbol{b}}, \hat{\boldsymbol{c}}^{\mathrm{T}}$ of model $\hat{P}$ instead of the theoretical parameter matrices $\boldsymbol{A}$, $\boldsymbol{b}$, $\boldsymbol{c}^{\mathrm{T}}$ belonging to the real process P. Using the equivalent transfer function representations, the state feedback can be realized according to Figure 3.1.15 instead of Figure 3.1.8.

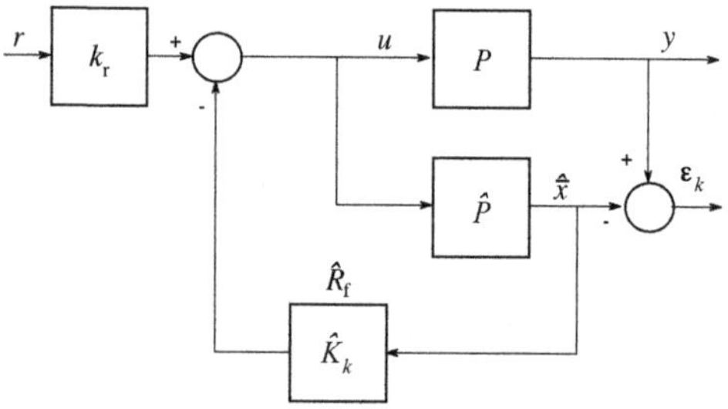

Figure 3.1.15 Model-based state feedback.

Here $\hat{R}_f = \hat{K}_k$ is calculated on the basis of the model, i.e.,

$$\hat{K}_k = \frac{\mathcal{K}(s)}{\hat{\mathcal{B}}(s)}; \quad \mathcal{K}(s) = \mathcal{R}(s) - \hat{\mathcal{A}}(s) \tag{3.1.78}$$

The model-based overall transfer function of the closed system shown in Figure 3.1.15 is

$$\hat{T}_{ry} = \frac{k_r P}{1 + \hat{K}_k \hat{P}} = \frac{k_r \mathcal{B}}{\mathcal{R}} \frac{\hat{\mathcal{A}}}{\mathcal{A}} = T_{ry} \frac{\hat{\mathcal{A}}}{\mathcal{A}} \tag{3.1.79}$$

where the process model is

$$\hat{P} = \frac{\hat{\mathcal{B}}}{\hat{\mathcal{A}}} = \frac{\hat{b}_1 s^{n-1} + \ldots + \hat{b}_{n-1} s + \hat{b}_n}{s^n + \hat{a}_1 s^{n-1} + \ldots + \hat{a}_{n-1} s + \hat{a}_n} \tag{3.1.80}$$

For the characterization of the model accuracy use the additive Δ and relative ℓ errors (see (1.1.41) and (1.1.42)).

Similarly, the relative uncertainty of the complementary sensitivity function calculated from the model can be defined as

$$\ell_T = \frac{T_{ry} - \hat{T}_{ry}}{\hat{T}_{ry}} = \frac{\mathcal{A} - \hat{\mathcal{A}}}{\hat{\mathcal{A}}} = \ell_{\mathcal{A}} \tag{3.1.81}$$

which is equal to the relative error $\ell_{\mathcal{A}}$ of the denominator of the process model $\hat{\mathcal{A}}$.

The error $\boldsymbol{\varepsilon}_k$ shown in the figure is the error in the model output

$$\boldsymbol{\varepsilon}_k = \left(P - \hat{P}\right) u = \frac{k_r \hat{\mathcal{B}}}{\mathcal{R}} \ell r = T_{ry} \frac{\hat{\mathcal{B}}}{\mathcal{B}} \ell r = \hat{P} \ell u \tag{3.1.82}$$

The block diagram of model-based state feedback with equivalent transfer functions is shown in Figure 3.1.16.

Here $\hat{K}_l$ is calculated from the model

$$\hat{K}_l = \frac{\mathcal{L}(s)}{\hat{\mathcal{B}}(s)}; \quad \mathcal{L}(s) = \mathcal{Q}(s) - \hat{\mathcal{A}}(s) \tag{3.1.83}$$

After long, but simple, calculations the error $\boldsymbol{\varepsilon}_l$ in the observer can be obtained as

$$\boldsymbol{\varepsilon}_l = \frac{\hat{P}}{1 + \hat{K}_l \hat{P}} \ell u = \frac{\hat{\mathcal{B}}}{\mathcal{Q}} \ell u = \frac{1}{1 + \hat{K}_l \hat{P}} \boldsymbol{\varepsilon}_k \tag{3.1.84}$$

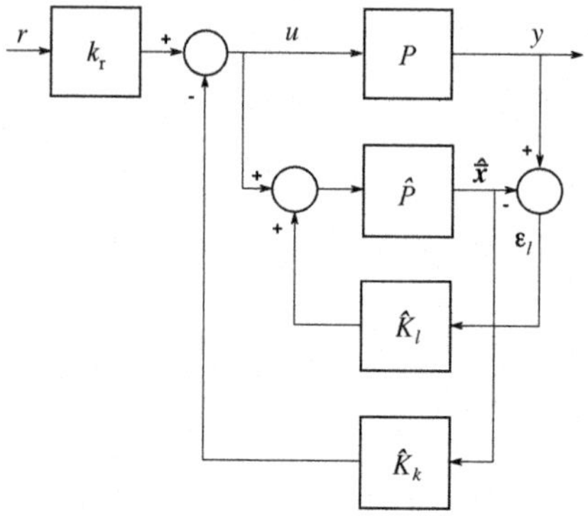

Figure 3.1.16 Model-based state feedback with observer.

which has a significant role in explaining the incidence of the observer-based state feedback methods. Thus the observer decreases the original model error ε_k by $1 + \hat{K}_l\hat{P}$. With fast observer poles in $\mathcal{Q}$, very little error ε_l can be reached in the important frequency regions. In fact, in the model-based design the design equations (3.1.16) and (3.1.39) are only approximations, and specific factorizability property (3.1.57) does not exist. Equation (3.1.57) is substituted by the expression

$$\left.\frac{\hat{P}^2 K_k K_l(1+\ell)}{1+\hat{P}(K_k+K_l)+\hat{P}^2 K_k K_l(1+\ell)}\right|_{\ell\to 0} = \frac{PK_k}{1+PK_k}\,\frac{PK_l}{1+PK_l} = \frac{\mathcal{K}}{\mathcal{R}}\,\frac{\mathcal{L}}{\mathcal{Q}} \tag{3.1.85}$$

which depends on the relative error, but it returns (3.1.57) in the case of $\ell\to 0$. There is no other way to use the model-based design expressions

$$\hat{\boldsymbol{k}}^{\mathrm{T}} = \hat{\boldsymbol{k}}_{\mathrm{c}}^{\mathrm{T}} = [r_1 - \hat{a}_1, r_2 - \hat{a}_2, \ldots, r_n - \hat{a}_n] \tag{3.1.86}$$

$$\hat{\boldsymbol{l}} = \hat{\boldsymbol{l}}_{\mathrm{o}} = [q_1 - \hat{a}_1, q_2 - \hat{a}_2, \ldots, q_n - \hat{a}_n]^{\mathrm{T}} \tag{3.1.87}$$

It should be noted that the concept of the observer is largely owed to Luenberger [124–126], who elaborated the first versions of realizable methods. Do not forget that the application of the observers has ensured the wide-ranging popularity of the state feedback methods. (In any case the excellent capabilities of Luenberger are proved by the fact that when he left the engineering for the mathematical finance he continued to achieve remarkable results.)

Observer-Based Youla Regulator

For open-loop stable processes, the *all realizable stabilizing* (*ARS*) model-based regulator $\hat{C}$ is the Youla-parametrized one:

$$\hat{C}(\hat{P}) = \frac{Q}{1 - Q\hat{P}}\bigg|_{\hat{P}\to P} = \frac{Q}{1 - QP} = C(P) \tag{3.1.88}$$

where the "parameter" Q ranges over all proper ($Q(\omega = \infty)$ is finite), stable transfer functions [94,127] (see Figure 3.1.17(a)).

It is important to know that the Youla parameterized closed loop with the ARS regulator is equivalent to the well-known form of the so-called *internal model control* (*IMC*) principle [127] based structure (see Section 2.1) shown in Figure 3.1.17(b).

Q is anyway the transfer function from r to u and the closed-loop transfer function (i.e., CSF) for $\hat{P} = P$, when $\ell \to 0$

$$\hat{T}_{ry} = \frac{\hat{C}P}{1 + \hat{C}P} = QP\frac{1+\ell}{1+(1-QP)\ell}\bigg|_{\ell\to 0} = QP = T_{ry} \tag{3.1.89}$$

is linear (and hence convex) in Q. (Linear functions are both convex and/or concave.)

It is interesting to compute the relative error ℓ_T of $\hat{T}_{ry}$

$$\ell_T = \frac{T_{ry} - \hat{T}_{ry}}{\hat{T}_{ry}} = \frac{T_{ry}}{\hat{T}_{ry}} - 1 = \frac{QP}{\frac{QP}{1+Q(P-\hat{P})}} - 1 = Q(P - \hat{P})$$
$$= QP\frac{\ell}{1+\ell} = T_{ry}\frac{\ell}{1+\ell} \tag{3.1.90}$$

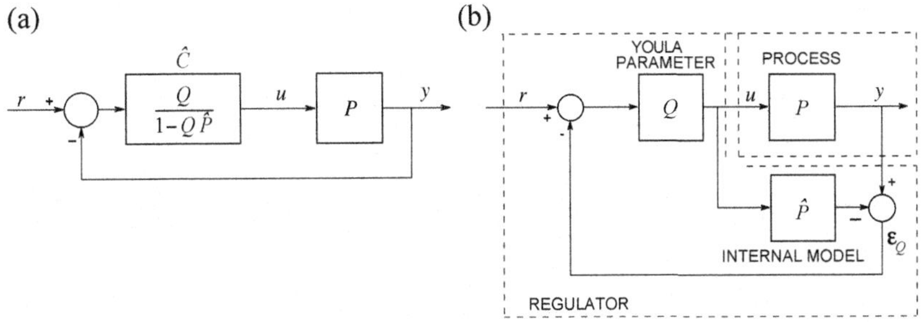

Figure 3.1.17 The equivalent *internal model control* structure of an *all realizable stabilizing* regulator.

The equivalent *IMC* structure performs the feedback from the model error ε_Q. Similarly to the *SFO* scheme it is possible to construct an internal closed loop, which virtually reduces the model error to

$$\varepsilon_l = \frac{1}{1+\hat{K}_l\hat{P}}\left(y-\hat{P}u\right) = \frac{1}{1+\hat{K}_l\hat{P}}\varepsilon_Q = \frac{1}{1+\hat{L}_l}\varepsilon_Q = \hat{H}\varepsilon_Q; \quad \hat{L}_l = \hat{K}_l\hat{P}$$

and performs the feedback from ε_l (see Figure 3.1.18), where $\hat{L}_l$ is the internal loop transfer function. In this case, the resulting closed loop will change to the scheme shown in Figure 3.1.19.

This means that the introduction of the observer feedback changes the Youla-parametrized regulator to

$$\hat{C}'\left(\hat{P}'\right) = \frac{Q}{1-Q\frac{\hat{P}}{1+\hat{K}_l\hat{P}}} = \frac{Q\left(1+\hat{K}_l\hat{P}\right)}{1+\hat{K}_l\hat{P}-Q\hat{P}} \tag{3.1.91}$$

The form of $\hat{C}'$ shows that the regulator virtually controls a fictitious plant $\hat{P}'$, which is also demonstrated in Figure 3.1.19. Here the fictitious plant is

$$\hat{P}' = \frac{\hat{P}}{1+\hat{K}_l\hat{P}} = \frac{\hat{P}}{1+\hat{L}_l} \tag{3.1.92}$$

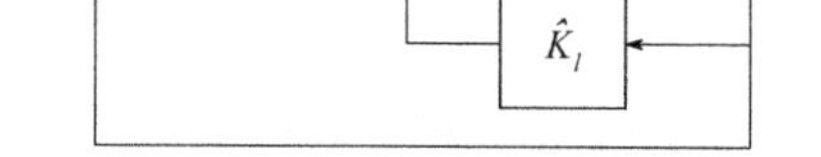

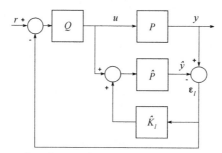

Figure 3.1.18 The observer-based *internal model control* structure.

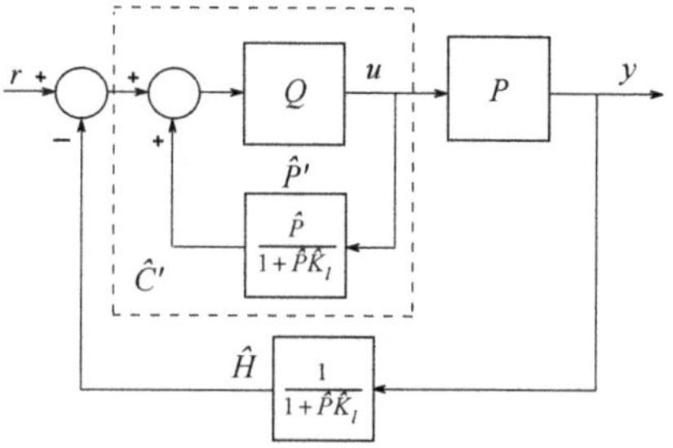

Figure 3.1.19 Equivalent closed loop for the observer-based *internal model control* structure.

The closed-loop transfer function is now

$$\hat{T}'_{\mathrm{ry}} = \frac{\hat{C}'P}{1+\hat{C}'P} = \frac{QP\left(1+\hat{K}_l\hat{P}\right)}{1+\hat{K}_l\hat{P}-Q\hat{P}+QP} = QP\left.\frac{1}{1+QP\frac{1}{1+\hat{K}_l\hat{P}}\frac{\ell}{1+\ell}}\right|_{\ell\to 0}$$

$$= QP = T_{\mathrm{ry}} \tag{3.1.93}$$

The relative error ℓ'_{T} of $\hat{T}'_{\mathrm{ry}}$ becomes

$$\ell'_{\mathrm{T}} = \frac{T_{\mathrm{ry}}-\hat{T}'_{\mathrm{ry}}}{\hat{T}'_{\mathrm{ry}}} = \frac{T_{\mathrm{ry}}}{\hat{T}'_{\mathrm{ry}}} - 1 = \frac{QP}{\frac{QP\left(1+\hat{K}_l\hat{P}\right)}{1+Q\left(P-\hat{P}\right)+\hat{K}_l\hat{P}}} - 1$$

$$= QP\,\frac{\ell}{1+\ell}\,\frac{1}{\left(1+\hat{K}_l\hat{P}\right)} = \ell_{\mathrm{T}}\frac{1}{1+\hat{L}_l} \tag{3.1.94}$$

which is smaller than ℓ_{T}. The reduction is by $\hat{H} = 1/(1+\hat{L}_l)$.

3.2 STATE FEEDBACK *LINEAR QUADRATIC (LQ)* REGULATORS

The method shown previously can perform arbitrary (stabilizing) pole placement by means of the so-called state feedback from the state vector of the process. Further optimization tasks can also be solved. The goal is to optimally control the *LTI* process (3.1.1) through the minimization of a complex optimality criterion

$$I = \frac{1}{2}\int_0^{\infty}\left[\boldsymbol{x}^{\mathrm{T}}(t)\boldsymbol{W}_{\mathrm{x}}\boldsymbol{x}(t) + W_{\mathrm{u}}u^2(t)\right]\mathrm{d}t \tag{3.2.1}$$

Here $\boldsymbol{W}_{\mathrm{x}}$ is a real symmetrical positive semidefinite matrix weighting the state vector and W_{u} is a positive constant weighting the excitation. The solution minimizing the criterion is a state feedback

$$u(t) = -\boldsymbol{k}_{\mathrm{LQ}}^{\mathrm{T}}\boldsymbol{x}(t) \tag{3.2.2}$$

(see (3.1.3)), where the feedback vector $\boldsymbol{k}_{\mathrm{LQ}}^{\mathrm{T}}$ takes the form

$$\boldsymbol{k}_{\mathrm{LQ}}^{\mathrm{T}} = \frac{1}{W_{\mathrm{u}}}\boldsymbol{b}^{\mathrm{T}}\boldsymbol{P} \tag{3.2.3}$$

Here the symmetrical positive semidefinite matrix $\boldsymbol{P}$ comes from the solution of the algebraic Riccati matrix equation [38,39]

$$\boldsymbol{P}\boldsymbol{A} + \boldsymbol{A}^{\mathrm{T}}\boldsymbol{P} - \frac{1}{W_{\mathrm{u}}}\boldsymbol{P}\boldsymbol{b}\boldsymbol{b}^{\mathrm{T}}\boldsymbol{P} = -\boldsymbol{W}_{\mathrm{x}} \tag{3.2.4}$$

Since this Riccati equation is nonlinear in $\boldsymbol{P}$, it has no explicit algebraic solution. The *CAD* (*computer-aided design*) systems frequently used in control techniques, however, generally provide several numerical algorithms for the solution of this equation. This regulator is called the LQ (*linear quadratic: linear regulator–quadratic criterion*) regulator.

The state equation of the *LQ* regulator-based closed loop is

$$\frac{\mathrm{d}\boldsymbol{x}}{\mathrm{d}t} = \left(\boldsymbol{A} - \boldsymbol{b}\boldsymbol{k}_{\mathrm{LQ}}^{\mathrm{T}}\right)\boldsymbol{x} = \overline{\boldsymbol{A}}\boldsymbol{x}; \quad \overline{\boldsymbol{A}} = \boldsymbol{A} - \boldsymbol{b}\boldsymbol{k}_{\mathrm{LQ}}^{\mathrm{T}} \tag{3.2.5}$$

The details of the *LQ*-based method are given A.2.5.9 of Appendix A.2. (The above regulator is very simple, but its derivation is quite time consuming.)

If the transfer function of the process is known, then the controllable canonical form can be easily given. For a special $\boldsymbol{A}_{\mathrm{c}}$ and $\boldsymbol{b}_{\mathrm{c}}$, equation (3.1.16) gives the classical state feedback design algorithm. With the *LQ* method the feedback vector $\boldsymbol{k}_{\mathrm{LQ}}^{\mathrm{T}}$ is obtained from the design (from the solution of the Riccati equation). So turning back the derivation of (3.1.16) the characteristic polynomial $\mathcal{R}(s)$ of the resulting closed-loop system can be given by its coefficients as

$$\left[r_1, r_2, \ldots, r_n\right]^{\mathrm{T}} = \boldsymbol{k}_{\mathrm{LQ}}^{\mathrm{T}} + \left[a_1, a_2, \ldots, a_n\right]^{\mathrm{T}} \tag{3.2.6}$$

It is also possible to apply an observer for constructing the state vector in *LQ* control.

In engineering practice, it is simpler to solve the stabilizing task by pole placement state feedback, since there the prescribed poles are directly known. It is evident, however, that in this case the quality of the transient processes is less known. The *LQ* regulator, as well as stabilization, makes it possible to design even the quality of the transient processes, but it needs long-term practice to determine the proper weighting matrix $\boldsymbol{W}_{\mathrm{x}}$ and weighting factor W_{u}, usually via a trial-and-error method.

The simpler version of the *LQ* regulator is when, instead of the states, only the square of the output is weighted, similarly to the input, i.e., instead of (3.2.1), the criterion

$$I = \frac{1}{2}\int_{0}^{\infty}\left[W_y y^2(t) + W_u u^2(t)\right]\mathrm{d}t \tag{3.2.7}$$

is used. This task (in the case of $d = 0$), after some identical manipulations, can be traced back to the original *LQ* regulator

$$W_y y^2 = y W_y y = \boldsymbol{x}^{\mathrm{T}}\boldsymbol{c} W_y \boldsymbol{c}^{\mathrm{T}}\boldsymbol{x} = \boldsymbol{x}^{\mathrm{T}}\left(\boldsymbol{c} W_y \boldsymbol{c}^{\mathrm{T}}\right)\boldsymbol{x} = \boldsymbol{x}^{\mathrm{T}}\left(W_y \boldsymbol{c}\boldsymbol{c}^{\mathrm{T}}\right)\boldsymbol{x} \tag{3.2.8}$$

with selection of a weighting matrix like

$$\boldsymbol{W}_{\mathrm{x}} = W_y \boldsymbol{c}\boldsymbol{c}^{\mathrm{T}} \tag{3.2.9}$$

Observe that the state feedback $\boldsymbol{k}_{\mathrm{LQ}}^{\mathrm{T}}$ leaves the process zeros untouched.

Sampled Data Systems

The design method of the sampled state feedback *LQ* regulator is formally completely the same as for the CT processes shown above. The goal is to control optimally the DT *LTI* process (A.3.11)–(A.3.12) through the minimization of a complex optimality criterion

$$I = \frac{1}{2}\sum_{k=0}^{\infty}\left\{\boldsymbol{x}^{\mathrm{T}}[k]\boldsymbol{W}_{\mathrm{x}}\boldsymbol{x}[k] + W_u u^2[k]\right\} \tag{3.2.10}$$

Here $\boldsymbol{W}_{\mathrm{x}}$ is a real symmetric positive semidefinite matrix weighting the DT state vector and W_u is a positive scalar weighting the DT actuator signal. The solution optimizing the criterion is a state feedback in the form

$$u[k] = -\boldsymbol{k}_{\mathrm{LQ}}^{\mathrm{T}}\boldsymbol{x}[k] \tag{3.2.11}$$

(see (3.2.3)), where $\boldsymbol{k}_{\mathrm{LQ}}^{\mathrm{T}}$ is the feedback vector

$$\boldsymbol{k}_{\mathrm{LQ}}^{\mathrm{T}} = \frac{1}{W_u}\boldsymbol{g}^{\mathrm{T}}\boldsymbol{P} \tag{3.2.12}$$

Here the symmetric positive semidefinite matrix $\boldsymbol{P}$ is the solution of the algebraic Riccati equation

$$\boldsymbol{P}\boldsymbol{F} + \boldsymbol{F}^{\mathrm{T}}\boldsymbol{P} - \frac{1}{W_u}\boldsymbol{P}\boldsymbol{g}\boldsymbol{g}^{\mathrm{T}}\boldsymbol{P} = -\boldsymbol{W}_{\mathrm{x}} \tag{3.2.13}$$

The (algebraic) Riccati equation is nonlinear in $\boldsymbol{P}$, and therefore it does not have an explicit algebraic solution. The *CAD* systems used in control engineering, however, have several numerical algorithms for the solution of the above equation.

The state equation of the closed system obtained by the *LQ* regulator takes the form

$$\mathbf{x}[k+1] = \left(\mathbf{F} - \mathbf{g}\mathbf{k}_{\mathrm{LQ}}^{\mathrm{T}}\right)\mathbf{x}[k] = \overline{\mathbf{F}}\,\mathbf{x}[k]; \quad \overline{\mathbf{F}} = \mathbf{F} - \mathbf{g}\mathbf{k}_{\mathrm{LQ}}^{\mathrm{T}} \qquad (3.2.14)$$

(The derivation of the *LQ* regulator for CT systems can be found in A.2.5.9 of Appendix A.2; the derivation of the DT regulator can be found similarly.)

If the transfer function of the process is known, then the controllable canonical form can be easily written in analogy with the CT equation (3.1.16) for the special $\mathbf{F}_{\mathrm{c}}$ and $\mathbf{g}_{\mathrm{c}}$ formed by (3.1.26), according to the design algorithm (3.1.27) of the classical DT state feedback. The feedback vector $\mathbf{k}_{\mathrm{LQ}}^{\mathrm{T}}$ comes from the *LQ* control design (from the solution of the Riccati equation). So by turning back the derivation of (3.1.27), the coefficients of the characteristic polynomial $\mathcal{Q}(z)$ of the closed-loop system are given by

$$\left[q_1, q_2, \ldots, q_n\right]^{\mathrm{T}} = \mathbf{k}_{\mathrm{LQ}}^{\mathrm{T}} + \left[f_1, f_2, \ldots, f_n\right]^{\mathrm{T}} \qquad (3.2.15)$$

In the case of *LQ* control, it is also possible to apply an observer for the determination of the state vector.

Design of the *LQ* Regulator in Frequency Domain

Design methods based on the *LQ* regulator are very popular in the literature despite the fact that the weights $\mathbf{W}_{\mathrm{x}}$ and W_{u} are usually chosen by trial-and-error methods. Perhaps this explains why attention has recently been focused on pole placement techniques: these methods are relatively simple and precise.

The *LQ* state feedback vector $\mathbf{k}_{\mathrm{LQ}}^{\mathrm{T}}$, minimizing the cost function (3.2.1), can be computed not only by solving the Riccati equation but by other methods. Of these, the method solving the design reference differential equation in the frequency domain is most appropriate. This method is discussed below.

The predecessor of the general criterion (3.2.1) was the square integral of the control error

$$I_2 = \int_0^{\infty} e^2(t)\mathrm{d}t = \frac{1}{2\pi j}\int_{-\infty}^{\infty} E(-s)E(s)\mathrm{d}s = \frac{1}{\pi}\int_0^{\infty} \left|E(j\omega)\right|^2 \mathrm{d}\omega \qquad (3.2.16)$$

where $e(t)$ is the error signal and $E(s)$ is the Laplace transform of the error. If $E(s)$ is strictly proper then this expression is called the Parseval theory and the expression (3.2.16) can be appraised by different methods. This criterion was very popular among experts in control engineering, because such evaluation methods are analytic and the minimum can be given in closed form (Wiener problem [159]). Unfortunately, the obtained optimal regulator could not often be applied in practice because it usually produced an unacceptably high (20–25%) overstep in the case of unit step disturbances. Therefore, instead of the square integral of the error, another criterion should be sought which provides optimal performance in practice.

As an obvious generalization of the square integral of the error the criterion

$$I_{2(n)} = \int_0^\infty \left[x^2 + \tau_1^2 \dot{x}^2 + \ldots + \tau_n^2 \left(\overset{(n)}{x} \right)^2 \right] \mathrm{d}t \tag{3.2.17}$$

is obtained whose equivalent form is [30]

$$I_{2(n)} = \int_0^\infty \left[x + \alpha \dot{x} + \cdots + \alpha_n \overset{(n)}{x} \right]^2 \mathrm{d}t + c_o \tag{3.2.18}$$

where $x(\infty) = \dot{x}(\infty) = \ldots = \overset{(n-1)}{x}(\infty) = 0$ and $c_o = \alpha_1 x_o^2$, $x_o = x(0)$. The connection between the two kinds of forms is given by the so-called Rekasius–Feldbaum equations [45,140]

$$\begin{aligned} \tau_1^2 &= \alpha_1^2 - 2\alpha_o\alpha_2 \\ \tau_2^2 &= \alpha_2^2 - 2\alpha_1\alpha_3 + 2\alpha_o\alpha_4 \\ \tau_3^2 &= \alpha_3^2 - 2\alpha_2\alpha_4 + 2\alpha_1\alpha_3 - 2\alpha_o\alpha_4 \\ &\vdots \\ \tau_n^2 &= \alpha_n^2 \end{aligned} \tag{3.2.19}$$

Note that (3.2.18) has its minimum where $x(t)$ fulfills the reference differential equation

$$\alpha_n \overset{(n)}{x} + a_{n-1} \overset{(n-1)}{x} + \cdots + \alpha_1 x + \alpha_o = 0; \quad \alpha_o = 1 \tag{3.2.20}$$

Here $x(t)$ is more general than the simple error signal, since arbitrary state variables can appear in a dynamic system.

Introduce the following orthogonal deconvolution, i.e., factorization of the weighting matrix $\boldsymbol{W}_\mathrm{x}$

$$W_x = G^T G \tag{3.2.21}$$

Here this weighting can also be considered as the virtual state vector $\boldsymbol{v}$ obtained via the transformation which takes over the roles of the original state vector $\boldsymbol{x}$

$$\boldsymbol{v} = G\boldsymbol{x} \tag{3.2.22}$$

Analyzing the LQ problem, it can be proved that the closed feedback system obtained by $\boldsymbol{k}_{LQ}^T$ is stable if the process in the plane of the state variables $\boldsymbol{v}$ is observable, i.e., the observability matrix

$$\boldsymbol{M}_o = \begin{bmatrix} G^T \\ G^T\boldsymbol{A} \\ \vdots \\ G^T\boldsymbol{A}^{n-1} \end{bmatrix} \tag{3.2.23}$$

is regular.

The simplest case of the deconvolution (3.2.21) is the dyadic deconvolution, when

$$\boldsymbol{W}_x = \boldsymbol{h}\boldsymbol{h}^T; \quad \boldsymbol{h} = [h_1, \ldots, h_{n-1}, h_n]^T \tag{3.2.24}$$

Here the coefficients in $\boldsymbol{h}$ correspond to the coefficients of the reference differential equation (3.2.20).

The solution of the LQ problem (3.2.1) in the frequency domain was derived by Kalman [65,38], according to whom the condition for the minimum is

$$W_u\left|1 + \boldsymbol{k}_{LQ}^T\boldsymbol{\Phi}(s)\boldsymbol{b}\right|^2 = W_u + \left|\boldsymbol{h}^T\boldsymbol{\Phi}(s)\boldsymbol{b}\right|^2 \tag{3.2.25}$$

See the definition of $\boldsymbol{\Phi}(s) = (s\boldsymbol{I} - \boldsymbol{A})^{-1}$ in (A.2.3). $\boldsymbol{A}$ and $\boldsymbol{b}$ are the parameters of the state equation. Without violating generality it can always be assumed that $W_u \equiv 1$, which actually means the normalization (scaling) by W_u. In this case, the Kalman equation [38] can be further simplified as

$$\left|1 + \boldsymbol{k}_{LQ}^T\boldsymbol{\Phi}(s)\boldsymbol{b}\right|^2 = 1 + \left\|\boldsymbol{h}^T\boldsymbol{\Phi}(s)\boldsymbol{b}\right\|_2^2 = 1 + \left|\boldsymbol{h}^T\boldsymbol{\Phi}(s)\boldsymbol{b}\right|^2 \tag{3.2.26}$$

or the equivalent form

$$\left|1 + \boldsymbol{k}_{LQ}^T\frac{\boldsymbol{\Psi}(s)}{\mathcal{A}(s)}\boldsymbol{b}\right|^2 = 1 + \left\|\boldsymbol{h}^T\frac{\boldsymbol{\Psi}(s)}{\mathcal{A}(s)}\boldsymbol{b}\right\|_2^2 = 1 + \left|\boldsymbol{h}^T\frac{\boldsymbol{\Psi}(s)}{\mathcal{A}(s)}\boldsymbol{b}\right|^2 \tag{3.2.27}$$

where $\boldsymbol{\Psi}(s) = \mathbf{adj}(s\boldsymbol{I} - \boldsymbol{A})$ (see (A.2.3)). Introduce the polynomials of $(n-1)$ order

$$\begin{aligned} \boldsymbol{k}_{\mathrm{LQ}}^{\mathrm{T}}\boldsymbol{\Phi}(s)\boldsymbol{b} &= k_1 s^{n-1} + \ldots + k_{n-1}s + k_n = \mathcal{K}_{\mathrm{LQ}}(s) \\ \boldsymbol{h}^{\mathrm{T}}\boldsymbol{\Phi}(s)\boldsymbol{b} &= h_1 s^{n-1} + \ldots + h_{n-1}s + h_n = \mathcal{H}(s) \end{aligned} \tag{3.2.28}$$

and use the identity valid for complex functions

$$\{Z(s)\}^2 = |Z(s)|^2 = Z(s)Z(-s) \tag{3.2.29}$$

so the new form of the Kalman equation (3.2.26) is obtained as

$$\begin{aligned} &\underbrace{\left[\mathcal{A}(s) + \boldsymbol{k}_{\mathrm{LQ}}^{\mathrm{T}}\boldsymbol{\Psi}(s)\boldsymbol{b}\right]}_{\mathcal{R}(s)}\underbrace{\left[\mathcal{A}(-s) + \boldsymbol{k}_{\mathrm{LQ}}^{\mathrm{T}}\boldsymbol{\Psi}(-s)\boldsymbol{b}\right]}_{\mathcal{R}(-s)} \\ &= \mathcal{A}(s)\mathcal{A}(-s) + \underbrace{\left[\boldsymbol{h}^{\mathrm{T}}\boldsymbol{\Psi}(s)\boldsymbol{b}\right]}_{\mathcal{H}(s)}\underbrace{\left[\boldsymbol{h}^{\mathrm{T}}\boldsymbol{\Psi}(-s)\boldsymbol{b}\right]}_{\mathcal{H}(-s)} \end{aligned} \tag{3.2.30}$$

which is a square polynomial equation. Thus, the final square polynomial equation is

$$\begin{aligned} &\mathcal{R}(s)\mathcal{R}(-s) = \mathcal{A}(s)\mathcal{A}(-s) + \mathcal{H}(s)\mathcal{H}(-s) \quad \text{or} \\ &|\mathcal{R}(s)|^2 = |\mathcal{A}(s)|^2 + |\mathcal{H}(s)|^2 \end{aligned} \tag{3.2.31}$$

where $\mathcal{R}(s)$ is the design characteristic polynomial, $\mathcal{A}(s)$ is the denominator of the process, and $\mathcal{H}(s)$ contains the coefficients of the reference differential equation.

It is worth noting that the polynomial square equations can be given by spectrum factor decomposition. Since

$$\mathcal{R}(s)\mathcal{R}(-s) = [\mathcal{A}(s)\mathcal{A}(-s) + \mathcal{H}(s)\mathcal{H}(-s)]^+[\mathcal{A}(s)\mathcal{A}(-s) + \mathcal{H}(s)\mathcal{H}(-s)]^- \tag{3.2.32}$$

the solution can only be positive factor (all poles and zeros of $[\ldots]^+$ have negative real values, so all poles and zeros of $[\ldots]^-$ have positive real values)

$$\mathcal{R}(s) = [\mathcal{A}(s)\mathcal{A}(-s) + \mathcal{H}(s)\mathcal{H}(-s)]^+ \tag{3.2.33}$$

The solution of the square polynomial equation is neither more difficult nor simpler than the solution of the Riccati equation. Because of its nonlinear character, only an iterative numerical solution can be performed. Since there are several techniques for the numerical solution of the Riccati equation in the available programs and packages, the solution of the Kalman equation is not supplied. Do not forget, however, that only the solution in the frequency domain shown above gives the connection between the coefficients of the reference differential equation and the eligible weights of the LQ problem.

Observe that, if $W_u \to \infty$, then the solution goes to the equality $\mathcal{R}(s) = \mathcal{A}(s)$, and if $W_u \to 0$, then the condition $\boldsymbol{h}^T\boldsymbol{x} = 0$ should be fulfilled.

Example 3.2.1

Consider a simple first-order process, where

$$\mathcal{A}(s) = s + a_1; \quad \mathcal{R}(s) = s + r_1 \quad \text{and} \quad \mathcal{H}(s) = h_1 \tag{3.2.34}$$

Based on (3.2.31), the square polynomial equation is

$$-s^2 - r_1 s + r_1 s + r_1^2 = -s^2 - a_1 s + a_1 s + a_1^2 + h_1^2 \tag{3.2.35}$$

whose solution is

$$r_1^2 = a_1^2 + h_1^2 > 0 \tag{3.2.36}$$

or in another form

$$k_1 = r_1 - a_1 = \sqrt{a_1^2 + h_1^2} - a_1 > 0 \tag{3.2.37}$$

Using the decomposition to spectrum factors, it can be written as

$$\begin{aligned} s + r_1 &= \left[(s + a_1)(-s + a_1) + h_1^2\right]^+ = \left[a_1^2 - s^2 + h_1^2\right]^+ = \left[h_1^2 + a_1^2 - s^2\right]^+ \\ &= \left\{\left[\sqrt{h_1^2 + a_1^2} - s\right]\left[\sqrt{h_1^2 + a_1^2} + s\right]\right\}^+ = \sqrt{h_1^2 + a_1^2} + s \end{aligned} \tag{3.2.38}$$

After comparing the coefficients, we obtain $r_1 = \sqrt{a_1^2 + h_1^2}$.

The solution can also be verified by solving the Riccati equation. In this task

$$\boldsymbol{A} = -a_1; \quad \boldsymbol{b} = 1; \quad W_u = 1; \quad \boldsymbol{W}_x = h_1^2; \quad \boldsymbol{P} = p_{11} \tag{3.2.39}$$

Equation (3.2.4) now becomes

$$p_{11}(-a_1) + (-a_1)p_{11} - \frac{1}{1}p_{11}(1)(1)p_{11} = -h_1^2 \tag{3.2.40}$$

which means an equation of the second order in p_{11}

$$p_{11}^2 + 2a_1 p_{11} - h_1^2 = 0 \tag{3.2.41}$$

whose roots are

$$p_{11} = \frac{-2a_1 \pm \sqrt{4a_1^2 + 4h_1^2}}{2} = -a_1 \pm \sqrt{a_1^2 + h_1^2} \tag{3.2.42}$$

Since $\boldsymbol{P}$ is positive and semidefinite, the inequality $p_{11} \geq 0$ must be fulfilled, and thus a possible solution is

$$p_{11} = -a_1 + \sqrt{a_1^2 + h_1^2} \tag{3.2.43}$$

Thus, the state feedback vector is

$$\boldsymbol{k}_{\mathrm{LQ}}^{\mathrm{T}} = \frac{1}{W_{\mathrm{u}}}\boldsymbol{b}^{\mathrm{T}}\boldsymbol{P} = \frac{1}{1}(1)\left(p_{11}\right) = p_{11} = k_1 = -a_1 + \sqrt{a_1^2 + h_1^2} > 0 \tag{3.2.44}$$

In the case of state feedback, the controllable canonical form is $k_1 = r_1 - a_1$, and finally we retrieve the previous result $r_1 = \sqrt{a_1^2 + h_1^2}$.

Example 3.2.2

Let the process be second order, where

$$\mathcal{A}(s) = s^2 + a_1 s + a_2; \quad \mathcal{R}(s) = s^2 + r_1 s + r_2; \quad \mathcal{H}(s) = h_1 s + h_2 \tag{3.2.45}$$

Passing over the detailed calculations, we obtain the solution of (3.2.30) as

$$\begin{aligned} r_1^2 &= 2(r_2 - a_2) + \left(a_1^2 + h_1^2\right) \\ r_2^2 &= a_2^2 + h_2^2 \end{aligned} \tag{3.2.46}$$

and

$$\begin{aligned} h_1 &= \sqrt{r_1^2 - a_1^2 - 2(r_2 - a_2)} = \sqrt{\left(r_1^2 - 2r_2\right) - \left(a_1^2 - 2a_2\right)} \\ h_2 &= \sqrt{r_2^2 - a_2^2} \end{aligned} \tag{3.2.47}$$

On a Special Anomaly of *LQ* Regulators

Some recent findings of the authors proved that the (3.2.1) form of the quadratic criterion does not allow poles to be placed arbitrarily [113,114].

For example, it is not possible to slow down poles in the closed state feedback loop. A new criterion

$$I = \frac{1}{2}\int_0^\infty \left[\boldsymbol{x}^{\mathrm{T}}(t)\boldsymbol{W}_{\mathrm{x}}\boldsymbol{x}(t) + \boldsymbol{x}^{\mathrm{T}}(t)\boldsymbol{w}_{\mathrm{xu}}u(t) + W_{\mathrm{u}}u^2(t)\right]dt \tag{3.2.48}$$

was proposed in [115,116], which solves the problem and offers a possible way to overcome this anomaly of the standard method.

3.3 GENERAL POLYNOMIAL METHOD FOR REGULATOR DESIGN

It was shown in Section 2.1 for stable processes that *YP* can be applied for the design of optimal regulators. The only disadvantage of this general method is that it cannot be applied for unstable processes, largely because all methods are actually based on pole cancellation, which is not permitted for unstable processes. The methods discussed in Section 3 try to reach the control goal via the direct design of the feedback. Since in this case there is no direct pole cancellation, these methods are general and can also be applied for unstable processes. The methods described in Section 3, using state feedback or the *LQ* criterion, allocate the poles of the closed system directly or indirectly. Since the characteristic equation of the closed system is obtained via polynomial operations, it is possible to introduce a new method operating on the polynomials of the numerator and denominator of the transfer functions.

Find the regulator $C(s)$ in the form of regular rational fractional function

$$C(s) = \frac{\mathcal{Y}(s)}{\mathcal{X}(s)} = \frac{\mathcal{Y}}{\mathcal{X}} \tag{3.3.1}$$

Let the prescribed stable characteristic polynomial of the closed loop be $\mathcal{R}(s)$, i.e., the characteristic equation $\mathcal{R}(s) = 0$. Similarly to the state feedback, the stabilization and the design of the quality attributes are also performed via pole placement. Let the regular transfer function of the delay-free process be

$$P(s) = \frac{\mathcal{B}(s)}{\mathcal{A}(s)} = \frac{\mathcal{B}}{\mathcal{A}} \tag{3.3.2}$$

The characteristic equation representing the design goal is

$$\mathcal{A}(s)\mathcal{X}(s) + \mathcal{B}(s)\mathcal{Y}(s) = \mathcal{A}\mathcal{X} + \mathcal{B}\mathcal{Y} = \mathcal{R} = \mathcal{R}(s) \tag{3.3.3}$$

where $\mathcal{A}$, $\mathcal{B}$, and $\mathcal{R}$ are known, and the unknown parameters to be determined are in the polynomials $\mathcal{X}$ and $\mathcal{Y}$. Equation (3.3.3) is known as the *DE*. Since the process is not assumed to be stable, the resulting regulator is often called a stabilizing regulator.

The *DE* has a solution [100], if and only if all common divisors of $\mathcal{A}$ and $\mathcal{B}$ are, at the same time, also the common divisor of $\mathcal{R}$. If $\mathcal{A}$ and $\mathcal{B}$ are relative primes (they do not have common divisor), then *DE* always has a solution for any $\mathcal{R}$: the number of solutions is infinite. If a pair $\mathcal{X}_o$, $\mathcal{Y}_o$ fulfills the equation then the pair

$$\mathcal{X} = \mathcal{X}_o + \mathcal{D}\mathcal{B}; \quad \mathcal{Y} = \mathcal{Y}_o - \mathcal{D}\mathcal{A} \tag{3.3.4}$$

is also the solution of *DE,* where $\mathcal{D}$ is an arbitrary polynomial. If the process polynomials are relative primes (if $\mathcal{A} \neq 0$) then *DE* always has the solution $\mathcal{X}_o$, $\mathcal{Y}_o$, where either $\mathcal{Y}_o = 0$ or $\deg\{\mathcal{Y}_o\} < \deg\{\mathcal{A}\}$. The latter solution $\mathcal{X}_o$, $\mathcal{Y}_o$ is called the minimum solution, because there is no other solution $\mathcal{X}$, $\mathcal{Y}$ where the order of the polynomials is less than the order of $\mathcal{Y}_o$.

In the infinite number of solutions of *DE,* there exists a special solution where the assumption

$$\deg\{\mathcal{X}\} < \deg\{\mathcal{B}\} \tag{3.3.5}$$

is fulfilled for the orders. Similarly there exists a solution for which the inequality

$$\deg\{\mathcal{Y}\} < \deg\{\mathcal{A}\} \tag{3.3.6}$$

is valid. Both conditions are simultaneously fulfilled, if

$$\deg\{\mathcal{A}\} + \deg\{\mathcal{B}\} \geq \deg\{\mathcal{R}\} \tag{3.3.7}$$

If (3.3.6) exists then *DE* has special solution of minimum order where

$$\deg\{\mathcal{X}\} = \deg\{\mathcal{Y}\} \tag{3.3.8}$$

If (3.3.6) is not valid then a solution can be obtained where $\mathcal{X}$ or $\mathcal{Y}$ is minimum.

In practice two basic cases can be distinguished:

- Let $\mathcal{R}(s)$ be an arbitrary polynomial of order $\deg\{\mathcal{R}\} = 2\deg\{\mathcal{A}\} - 1$. Then the solution of *DE* can be sought by the control polynomials of order $\deg\{\mathcal{X}\} = \deg\{\mathcal{A}\} - 1$ and $\deg\{\mathcal{Y}\} = \deg\{\mathcal{A}\} - 1$, i.e., the resulting regulator becomes proper.
- Let $\mathcal{R}(s)$ be an arbitrary polynomial of order $\deg\{\mathcal{R}\} = 2\deg\{\mathcal{A}\}$. Then the solution of *DE* can be sought by the control polynomials of order $\deg\{\mathcal{X}\} = \deg\{\mathcal{A}\}$ and $\deg\{\mathcal{Y}\} = \deg\{\mathcal{A}\} - 1$, i.e., the resulting regulator becomes strictly proper.

So if the process has order n then a stabilizing regulator of order $(n-1)$ is searched for, since in this case *DE* always has a solution. It is clear from (3.3.4) that $\mathcal{Y}$ can have less order than $\mathcal{A}$. In principle $\mathcal{X}$ can have less order than $\mathcal{B}$, but the obtained regulator cannot be realized. Therefore, the stabilizing regulator is sought as a regulator of order $(n-1)$.

It seems to be a reasonable choice if the order of $\mathcal{R}$ is equal to the order of $\mathcal{A}$. In a good case it is possible to find a suitable stabilizing regulator of corresponding order and pole access to a process with pole access greater than one, but the procedure cannot be systematized and the resulting solution (3.3.7) is not minimum.

Equation (3.3.4) is also valid for the rational fractional function $D = \mathcal{G}/\mathcal{D}$. Here, besides $\mathcal{R}$, $\mathcal{D}$ also appears in the denominator of the overall transfer function of the closed system. Equation (3.3.4) parameterizes all stabilizing regulators by D. Parameter D is called the Youla–Kučera parameter.

It can be seen clearly that the transfer function of the *one-degree-of-freedom* (*ODOF*) stabilized closed control loop shown in Figure 3.3.1 is

$$T = \frac{\mathcal{B}\mathcal{Y}}{\mathcal{A}\mathcal{X} + \mathcal{B}\mathcal{Y}} = \frac{\mathcal{Y}}{\mathcal{R}}\mathcal{B} = R'_{\mathrm{n}}\mathcal{B} \tag{3.3.9}$$

It is clear from (3.3.9) that stabilization is performed, but the numerator of the process and the polynomial $\mathcal{Y}$, deriving from the solution of the *DE*, appear in the numerator of the overall transfer function. Note that none of these can be directly influenced, so the numerator of the transfer function of the closed systems cannot be designed. (Observe the similarities to the results obtained by state feedback.)

A *TDOF* control loop can be constructed where the reference signal tracking, at least, can be designed. This loop is shown in Figure 3.3.2(a). Figure 3.3.2(b) presents an equivalent block scheme which can be directly compared to the generic *TDOF* control loop shown in Figure 2.1.8(a), obtained by a *YP* regulator for a stable process. Here the regulator, however, is obviously different.

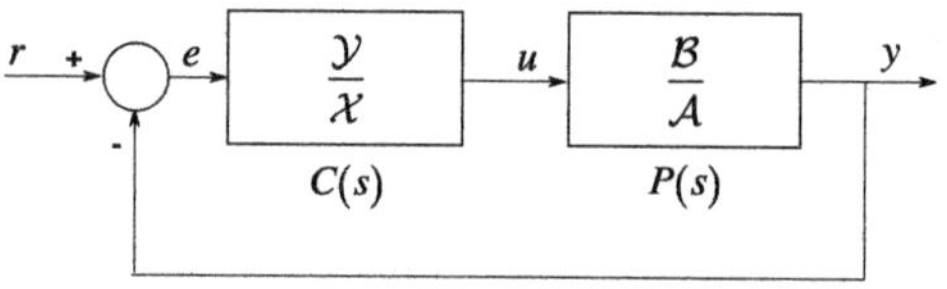

Figure 3.3.1 *One-degree-of-freedom* stabilized closed-loop control.

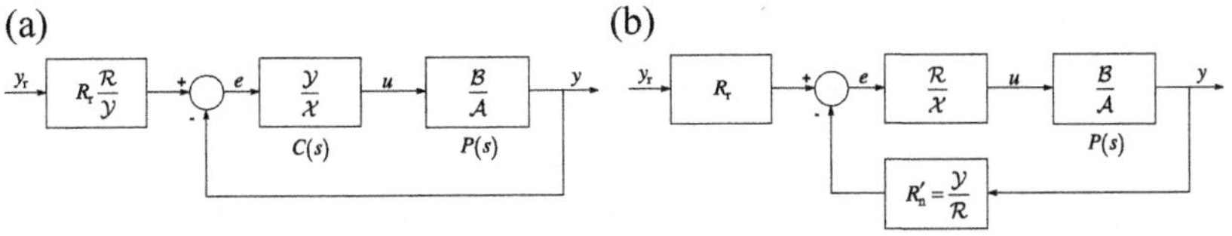

Figure 3.3.2 *Two-degree-of-freedom* stabilized closed control loop.

The transfer function of the control loops shown in Figure 3.3.2 is

$$T_r = R_r\mathcal{B} \tag{3.3.10}$$

Here only the numerator of the process appears, $\mathcal{Y}$ disappears, and R_r is independent of R_n, and thus it is really a *TDOF* control system. The disturbance rejection property can be calculated from T as

$$S = 1 - T = 1 - R'_n\mathcal{B} = 1 - \frac{\mathcal{Y}}{\mathcal{R}}\mathcal{B} \tag{3.3.11}$$

It was seen in the discussion of the *YP* regulator that in the numerator of the transfer function of the process only the stable zeros can be canceled. This method can be extended for the stable poles of the denominator in the design method using *DE*. Assume that the transfer function of the process is

$$P(s) = P_+(s)P_-(s) \quad \text{or shortly} \quad P = P_+P_- \tag{3.3.12}$$

where P_+ is stable and its inverse is also stable (*stable inverse stable* (*SIS*)). P_- is unstable and its inverse is also unstable (*unstable inverse unstable* (*UIU*)). Thus a practical factorization is

$$P = \frac{\mathcal{B}}{\mathcal{A}} = \frac{\mathcal{B}_+\mathcal{B}_-}{\mathcal{A}_+\mathcal{A}_-} = \left(\frac{\mathcal{B}_+}{\mathcal{A}_+}\right)\left(\frac{\mathcal{B}_-}{\mathcal{A}_-}\right) = P_+P_- \tag{3.3.13}$$

Here $\mathcal{A}_+$ contains the stable poles and $\mathcal{A}_-$ the unstable poles of the process. Similarly $\mathcal{B}_+$ contains the stable zeros and $\mathcal{B}_-$ the unstable zeros. The *DE* should be constructed to make possible the cancellation of the stable roots $\mathcal{B}_+$ and $\mathcal{A}_+$. In order to define the design procedure in a completely general way, predefined polynomials $\mathcal{Y}_d$ and $\mathcal{X}_d$ are introduced in the numerator and denominator of the regulator. The following design *DE* can be written for this most general case

$$\begin{aligned} (\mathcal{A}_+\mathcal{A}_-)(\mathcal{B}_+\mathcal{X}_d\mathcal{X}) + (\mathcal{B}_+\mathcal{B}_-)(\mathcal{A}_+\mathcal{Y}_d\mathcal{Y}) &= \mathcal{R} = \mathcal{A}_+\mathcal{B}_+\mathcal{R} \\ \mathcal{A} \qquad \mathcal{X} \qquad + \qquad \mathcal{B} \qquad \mathcal{Y} \qquad &= \mathcal{R}' \end{aligned} \tag{3.3.14}$$

A lower-order DE can be reached by simplifying with the common factors

$$\begin{aligned}(\mathcal{A}_-\mathcal{X}_\mathrm{d})\mathcal{X}' + (\mathcal{B}_-\mathcal{Y}_\mathrm{d})\mathcal{Y}' &= \mathcal{R} \\ \mathcal{A}'\mathcal{X}' + \mathcal{B}'\mathcal{Y}' &= \mathcal{R}\end{aligned} \tag{3.3.15}$$

where $\mathcal{A}' = \mathcal{A}_-\mathcal{X}_\mathrm{d}$ and $\mathcal{B}' = \mathcal{B}_-\mathcal{Y}_\mathrm{d}$ are known, and the regulator is obtained as

$$C = \frac{\mathcal{Y}}{\mathcal{X}} = \frac{\mathcal{A}_+\mathcal{Y}_\mathrm{d}\mathcal{Y}'}{\mathcal{B}_+\mathcal{X}_\mathrm{d}\mathcal{X}'} \tag{3.3.16}$$

Clearly, the stabilizing regulator cancels only the stable zeros and poles, and introduces the desired polynomials $\mathcal{Y}_\mathrm{d}$ and $\mathcal{X}_\mathrm{d}$ into the numerator and denominator. The Youla regulator is an integrating one, if the unity gain concerning the reference model is guaranteed: $R_\mathrm{n}(\omega = 0) = R_\mathrm{n}(s = 0) = 1$. This cannot be automatically guaranteed for the stabilizing regulator resulting from DE. It can be guaranteed only if $\mathcal{X}_\mathrm{d}$ brings a pole $s = 0$ into the denominator.

Since $\mathcal{Y}_\mathrm{d}$ can now be considered as the numerator of the reference model, and $\mathcal{R}$, however, as the denominator, in the general case, the corrected reference model

$$R'_\mathrm{n} = \frac{\mathcal{Y}_\mathrm{d}}{\mathcal{R}} \tag{3.3.17}$$

is dependent only on us, so it can be designed in full.

Equivalent block diagrams of the general stabilized control loop are shown in Figure 3.3.3.

It can be easily checked that the transfer function of the whole loop is

$$T_\mathrm{r} = P_\mathrm{r}G_\mathrm{r}\mathcal{B}_- \tag{3.3.18}$$

and the sensitivity function of the closed loop is

$$S = 1 - P'_\mathrm{w}\mathcal{Y}'\mathcal{B}_- \tag{3.3.19}$$

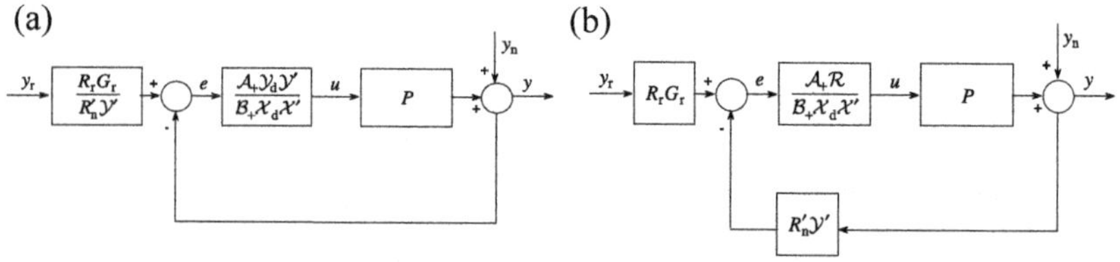

Figure 3.3.3 *Two-degree-of-freedom* general stabilized closed control loop.

So the transfer characteristic of the whole closed control loop is

$$y = T_r y_r + S y_n = R_r G_r \mathcal{B}_- y_r + \left(1 - R_n' \mathcal{Y}' \mathcal{B}_-\right) y_n \tag{3.3.20}$$

It follows that the filter G_r can be freely chosen and can be optimized to attenuate the effect of $\mathcal{B}_-$. Unfortunately the same does not apply to the optimal design respecting disturbance rejection, because here $\mathcal{Y}'$ results from the modified *DE* (3.3.15), so it cannot be freely chosen and therefore the attenuation of the effect of $\mathcal{Y}'$ cannot be easily solved, as seen in the *YP* of the tracking problem (3.3.20).

The form of the resulting stabilizing regulator shown in (3.3.16) can be further simplified as

$$C = \frac{\mathcal{A}_+ \mathcal{Y}_d \mathcal{Y}'}{\mathcal{B}_+ \mathcal{X}_d \mathcal{X}'} = \frac{\left(\frac{\mathcal{Y}_d}{\mathcal{R}}\right) \mathcal{Y}' \mathcal{A}}{\mathcal{B}_+ \left(1 - \frac{\mathcal{Y}_d}{\mathcal{R}} \mathcal{Y}' \mathcal{B}_-\right)} = \frac{P_w' \mathcal{Y}'}{1 - P_w' \mathcal{Y}' \mathcal{B}_-} \frac{\mathcal{A}}{\mathcal{B}_+} \tag{3.3.21}$$

which is very similar to the form of the optimal Youla regulator (2.1.10). Observe that although only the stable factors $\mathcal{A}_+$ and $\mathcal{B}_+$ are canceled, formally the regulator cancels the whole denominator of the process.

If the feature obtained for the disturbance rejection in (3.3.20) cannot be accepted, an outer cascade control loop should be applied, which has already been designed by the *YP* since the system has already been stabilized by the internal loop. This two-step method was discussed in detail in Section 3.1.

Example 3.3.1

Let the controlled system be a first-order ($n = 1$) unstable process

$$P = \frac{\mathcal{B}}{\mathcal{A}} = \frac{0.5}{1 - 0.5s} = \frac{-1}{s - 2} \tag{3.3.22}$$

whose pole $p = 2$ is on the right-hand side of the complex plane. Find the regulator $C = \mathcal{Y}/\mathcal{X}$, which stabilizes the process by prescribing the characteristic polynomial $\mathcal{R}(s) = s + 2 = 0$. The regulator is sought in the form of order $n - 1 = 0$, which can be ensured by the structure

$$C = \frac{\mathcal{Y}}{\mathcal{X}} = \frac{K}{1} = K \tag{3.3.23}$$

i.e., by a proportional regulator. Based on (3.3.3), it can be written that

$$\begin{aligned} &\mathcal{A}\mathcal{X} + \mathcal{B}\mathcal{Y} = \mathcal{R} \\ &(s - 2) - K = s + 2 \end{aligned} \tag{3.3.24}$$

Table 3.3.1 Different stabilizing regulators

$D(s)=\mathcal{G}/\mathcal{D}$	$C(s)$	$T(s)$
0	-4	$\frac{4}{s+2}$
1	$\frac{s-6}{2}$	$-\frac{s-6}{s+2}$
$1+s$	$\frac{s^2-s-6}{s+2}$	$-\frac{s^2-s-6}{s+2}$
$\frac{s+2}{s+1}$	$\frac{s^2-4s-8}{2s+3}$	$-\frac{s^2-4s-8}{(s+1)(s+2)}$

where $C=K=-4$ is obtained for the regulator. Thus, to stabilize an unstable pole, a proportional regulator is required with a gain equal to the double of the unstable pole and a positive feedback.

It can be checked by simple computation that the transfer function of the closed loop is

$$T=\frac{4}{s+2}=\frac{2}{1+0.5s} \tag{3.3.25}$$

and thus the unstable pole could be mirrored on the imaginary axis and the system consequently stabilized. The static gain of the closed-loop system is not unity, because the regulator is proportional and not integrating. For better quality, in terms of performance, it is reasonable to use an outer cascade control loop, as seen with the state feedback regulators.

Based on the expression (3.3.4), the resulting stabilizing regulators $C(s)$ and $T(s)$ are given for different parameters $D(s)=\mathcal{G}/\mathcal{D}$ in Table 3.3.1. The first row of the Table 3.3.1 contains the first solution obtained in (3.3.23) and (3.3.25).

It is clear that only the first regulator can be realized, so the other solutions have only theoretical importance. For higher-order processes the expressions are more complicated, but even for these cases it is reasonable to summarize the different order solutions in tables and choose the lowest order realizable regulator. Similarly, it may also be reasonable to give the solutions as lower order than the $(n-1)$-order regulator.

Example 3.3.2

Let the controlled system be a first-order ($n=1$) stable process

$$P=\frac{\mathcal{B}}{\mathcal{A}}=\frac{1}{1+10s}=\frac{0.1}{s+0.1} \tag{3.3.26}$$

which we would like to speed up. Assuming an *ODOF* system, the design goal is expressed in the reference models

$$R_r = R_n = \frac{1}{1+2s} = \frac{0.5}{s+0.5} \tag{3.3.27}$$

The Youla regulator is now

$$C_{opt} = C_{id} = \frac{R_n P_+^{-1}}{1-R_n} = \left(1 - \frac{1}{1+2s}\right)^{-1} \frac{1+10s}{1+2s} = \frac{1+10s}{2s} \tag{3.3.28}$$

an integrating one, so the transfer function of the closed loop is

$$T(s) = \frac{1}{1+2s} \tag{3.3.29}$$

For the *DE* design, based on (3.3.27), the characteristic equation is $\mathcal{R}(s) = s + 0.5 = 0$. As in the previous example, the regulator is sought again in the form of $n - 1 = 0$ degree, and thus the proportional regulator (3.3.23) is applied. Equation (3.3.3) is now

$$\begin{gathered} \mathcal{A}\mathcal{X} + \mathcal{B}\mathcal{Y} = \mathcal{R} \\ (s+0.1) + 0.1K = s + 0.5 \end{gathered} \tag{3.3.30}$$

where $C = K = 4$ is obtained for the regulator. It can be easily checked that the transfer function of the closed system is

$$T = \frac{0.4}{s+0.5} = \frac{0.8}{1+2s} \tag{3.3.31}$$

The prescribed pole -0.5 is successfully placed, but the control loop is zero type, so for the gain of T the value 0.8 is obtained. The above two examples show how the *YP* can be reasonably applied for stable processes, whereas for stabilizing unstable processes the application of *DE* or the state feedback discussed in Section 3.1 can provide solutions.

The stabilizing regulator obtained by *DE* can be applied only for delay-free processes. If the process has significant dead-time, then there is a possibility of approximating the delay by means of rational fractional function (Pade approximation). The other possibility is to use a sampled-data control system.

Sampled Data Systems

Assume that the pulse transfer function of the process is

$$\begin{gathered} \mathcal{G}(z^{-1}) = \mathcal{G}_+(z^{-1})\overline{\mathcal{G}}_-(z^{-1}) = \mathcal{G}_+(z^{-1})\mathcal{G}_-(z^{-1})z^{-d} \text{ in brief} \\ \mathcal{G} = \mathcal{G}_+\overline{\mathcal{G}}_- = \mathcal{G}_+\mathcal{G}_-z^{-d} \end{gathered} \tag{3.3.32}$$

where G_+ is *SIS*, $\overline{G}_-$ is *UIU*, and G_- is also *UIU*. Here, in general, the inverse of the time delay part cannot be realized, because it would be an ideal predictor. Thus, a reasonable factorization of the process is

$$G=\frac{\mathcal{G}}{\mathcal{F}}z^{-d}=\frac{\mathcal{G}_+\mathcal{G}_-}{\mathcal{F}_+\mathcal{F}_-}z^{-d}=\left(\frac{\mathcal{G}_+}{\mathcal{F}_+}\right)\left(\frac{\mathcal{G}_-}{\mathcal{F}_-}\right)z^{-d}=G_+G_-z^{-d} \tag{3.3.33}$$

Here $\mathcal{F}_+$ contains the stable poles, $\mathcal{F}_-$ the unstable ones. Similarly, $\mathcal{G}_+$ includes the stable zeros, $\mathcal{G}_-$ the unstable ones. The general design *DE* for discrete systems is simply obtained from (3.3.14) by formally exchanging $\mathcal{G}_-$ for $\mathcal{G}_-z^{-d}$. The new form of (3.3.14) becomes

$$\begin{aligned}(\mathcal{F}_+\mathcal{F}_-)(\mathcal{G}_+\mathcal{X}_d\mathcal{X}')+(\mathcal{G}_+\mathcal{G}_-z^{-d})(\mathcal{F}_+\mathcal{Y}_d\mathcal{Y}')&=\mathcal{R}'=\mathcal{F}_+\mathcal{G}_+\mathcal{R}\\ \mathcal{F}\qquad\mathcal{X}\qquad+\qquad\mathcal{G}\qquad\mathcal{Y}\qquad&=\mathcal{R}'\end{aligned} \tag{3.3.34}$$

The modified *DE* is

$$\begin{aligned}(\mathcal{F}_-\mathcal{X}_d)\mathcal{X}'+(\mathcal{G}_-z^{-d}\mathcal{Y}_d)\mathcal{Y}'&=\mathcal{R}'\\ \mathcal{F}'\ \ \mathcal{X}'+\qquad\mathcal{G}'\qquad\mathcal{Y}'&=\mathcal{R}'\end{aligned} \tag{3.3.35}$$

where $\mathcal{F}'=\mathcal{F}_-\mathcal{X}_d$ and $\mathcal{G}'=\mathcal{G}_-z^{-d}\mathcal{Y}_d$ are known and the regulator is obtained again as

$$C=\frac{\mathcal{Y}}{\mathcal{X}}=\frac{\mathcal{F}_+\mathcal{Y}_d\mathcal{Y}'}{\mathcal{G}_+\mathcal{X}_d\mathcal{X}'}=\frac{\left(\frac{\mathcal{Y}_d}{\mathcal{R}}\right)\mathcal{Y}'\mathcal{F}}{\mathcal{G}_+\left(1-\frac{\mathcal{Y}_d}{\mathcal{R}}\mathcal{Y}'\mathcal{G}_-z^{-d}\right)}=\frac{P'_w\mathcal{Y}'}{1-P'_w\mathcal{Y}'\mathcal{G}_-z^{-d}}\,\frac{\mathcal{F}}{\mathcal{G}_+} \tag{3.3.36}$$

The Youla regulator is integrating, if the unity gain is ensured for the reference model: $R_n(\omega=0)=R_n(z=1)=1$. This cannot be automatically guaranteed for the stabilizing regulator coming from *DE*. This solution is guaranteed if $\mathcal{X}_d$ brings the pole $z=1$ into the denominator. To solve the *DE* equation (3.3.4), the equation should be formed by the powers of z.

The transfer characteristic of the whole control loop is

$$y=T_r y_r+S y_n=R_r G_r \mathcal{G}_- z^{-d} y_r+\left(1-R'_n\mathcal{Y}'\mathcal{G}_- z^{-d}\right)y_n \tag{3.3.37}$$

It is clear that the filter G_r can be chosen arbitrarily and can be optimized to attenuate the effect of $\mathcal{G}_-$. Unfortunately the same cannot be said about the optimization of the disturbance rejection. Here $\mathcal{Y}'$ comes from the modified *DE* (3.3.4), so it cannot be chosen arbitrarily, therefore the attenuation of the effect of $\mathcal{G}_-$ cannot be solved as easily as it was seen at the *YP* and tracking properties (3.3.6).

Example 3.3.3

Let the controlled system be a first-order ($n = 1$), unstable DT process

$$G(z^{-1}) = \frac{\mathcal{G}(z^{-1})}{\mathcal{F}(z^{-1})} = \frac{-0.2z^{-1}}{1 - 1.2z^{-1}} = \frac{-0.2}{z - 1.2} \tag{3.3.38}$$

whose pole $p = 1.2$ is outside the unit circle. Determine the regulator $C = \mathcal{Y}/\mathcal{X}$ which stabilizes the process by prescribing the characteristic polynomial $\mathcal{R}(z) = z - 0.2 = 0$. The regulator is sought in the form of an $n - 1 = 0$ order, which can be achieved by the structure

$$C = \frac{\mathcal{Y}}{\mathcal{X}} = \frac{K}{1} = K \tag{3.3.39}$$

i.e., by a proportional regulator. Given (3.3.35), it can be written that

$$\begin{aligned} &\mathcal{F}\mathcal{X} + \mathcal{G}\mathcal{Y} = \mathcal{R} \\ &(z - 1.2) - 0.2K = z - 0.2 \end{aligned} \tag{3.3.40}$$

whence $C = K = -5$ is obtained for the regulator. It can be checked by simple computation that the pulse transfer function of the closed system is

$$T = \frac{1}{z - 0.2} = \frac{z^{-1}}{1 - 0.2z^{-1}} \tag{3.3.41}$$

and thus the unstable pole is successfully allocated to the prescribed place inside the unit circle, by means of which the system is stabilized. The static transfer of the closed loop is not unity, because the regulator is proportional and not integrating. To obtain a better quality control, it is reasonable to apply a further outer cascade loop, as seen for the state feedback control.

Example 3.3.4

Let the controlled system be a first-order ($n = 1$), stable DT process

$$G(z^{-1}) = \frac{\mathcal{G}(z^{-1})}{\mathcal{F}(z^{-1})} = \frac{0.2z^{-1}}{1 - 0.8z^{-1}} = \frac{0.2}{z - 0.8} \tag{3.3.42}$$

and let the goal be to make it faster. Assuming an *ODOF* system, our design goal is expressed by the reference models

$$R_r = R_n = \frac{0.8z^{-1}}{1 - 0.2z^{-1}} = \frac{0.8}{z - 0.2} \tag{3.3.43}$$

Now the Youla regulator is of integrating type, i.e.,

$$C_{\text{opt}} = C_{\text{id}} = \frac{R_n G_+^{-1}}{1 - R_n} = \frac{1}{1 - \dfrac{0.8z^{-1}}{1 - 0.2z^{-1}}} \frac{0.8z^{-1}}{1 - 0.2z^{-1}} \frac{1 - 0.8z^{-1}}{0.2z^{-1}}$$
$$= 4\frac{1 - 0.8z^{-1}}{1 - z^{-1}} \tag{3.3.44}$$

(because the root of the denominator is at $z = 1$), and the transfer function of the closed system is

$$T = \frac{0.8z^{-1}}{1 - 0.2z^{-1}} = \frac{0.8}{z - 0.2} \tag{3.3.45}$$

whose static transfer is unity corresponding to the control of type 1.

Based on (3.3.12), the characteristic equation for the design by *DE* is $\mathcal{R}(z) = z - 0.2 = 0$. Now the regulator is also sought in the form of order $n - 1 = 0$, and thus in line with (3.3.8) a proportional regulator is applied. Equation (3.3.9) becomes

$$\begin{aligned} \mathcal{F}\mathcal{X} + \mathcal{G}\mathcal{Y} &= \mathcal{R} \\ (z - 0.8) + 0.2K &= z - 0.2 \end{aligned} \tag{3.3.46}$$

where the regulator is $C = K = 3$. It can be checked by simple computation that the overall transfer function of the closed-loop system is now

$$T = \frac{0.6}{z - 0.2} = \frac{0.6z^{-1}}{1 - 0.2z^{-1}} \tag{3.3.47}$$

The prescribed pole 0.2 is successfully allocated, but the loop is of type 0 and therefore the gain of T is 0.75. The above two examples are representing well the practice, i.e., for stable systems the *YP* should be applied, whereas for stabilization of unstable systems, the application of *DE*, or the state feedback discussed in Section 3.1 can provide the solution.

Example 3.3.5

Let the controlled system be a first-order ($n = 1$), unstable, time-delay DT process

$$P(z^{-1}) = \frac{\mathcal{G}(z^{-1})}{\mathcal{F}(z^{-1})} = \frac{-0.2z^{-1}}{1 - 1.2z^{-1}}z^{-1} = \frac{-0.2z^{-2}}{1 - 1.2z^{-1}} = \frac{-0.2}{z(z - 1.2)} \tag{3.3.48}$$

whose pole $p = 1.2$ is outside the unit circle. Find the regulator $C = \mathcal{Y}/\mathcal{X}$ which stabilizes the process by prescribing the characteristic polynomial $\mathcal{R}(z) = (z - 0.2)^2 = z^2 - 0.4z + 0.04$. The process is formally second order, which is why $\mathcal{R}$ is chosen as a second-order polynomial. The regulator, however, is sought in first-order form, i.e., $n - 1 = 1$. To investigate *DE*, it seems reasonable to choose a first-order structure for the regulator

$$C = \frac{\mathcal{Y}}{\mathcal{X}} = \frac{y_o}{1 + x_1 z^{-1}} = \frac{y_o z}{z + x_1} \tag{3.3.49}$$

The number of parameters to be determined is two, the *DE* is

$$\begin{gathered} \mathcal{F}\mathcal{X} + \mathcal{G}\mathcal{Y} = \mathcal{R} \\ (z^2 - 1.2z)(z + x_1) - 0.2y_o z = z^2 - 0.4z + 0.04 \end{gathered} \tag{3.3.50}$$

and solving the equation for y_o, y_1, and x_1, we obtain

$$C = \frac{-5z}{z + 0.8} = \frac{-5}{1 + 0.8z^{-1}} \tag{3.3.51}$$

Simple computations can be checked that the transfer function of the closed loop is

$$T = \frac{1}{(z - 0.2)^2} = \frac{z^{-2}}{(1 - 0.2z^{-1})^2} \tag{3.3.52}$$

The prescribed double poles on the place 0.2 could have been allocated but the control loop is of zero type, therefore the gain of T is not unity and its value is 1.5625. In this way quite complex problems can be solved with a regulator of relatively simple structure and an unstable process with delay time stabilized.

CHAPTER 4

Concept of the Best Achievable Control

In the design of a regulator, the best achievable control depends on independent circumstances. The most important is that the design task should be solved under constraints $u \in \mathbb{U}$ regarding the output of the regulator. Here $\mathbb{U}$ means the admissible region of u, with the usual amplitude constraint $\mathbb{U}: |u| \leq 1$. This limit depends on the equipment used.

There are other constraints concerning the controlled process. The most important is the time delay of the process which is invariant for any control methods, i.e., its effect cannot be eliminated. A regulator with time delay cannot be realized in continuous-time (CT) systems (see the Smith regulator, or any general Youla regulator).

Furthermore, such invariant factors are the unstable zeros of the process; they cannot be cancelled or eliminated by any method. The unpleasant effect of the invariant zeros on the transient processes, however, can be compensated or attenuated to some extent. (This compensation can be performed optimally with the embedded filters G_r and G_n; see Section 2.1.)

Thus, both the time delay and the unstable zeros can be considered independent properties of the process which cannot be changed by any control design methods but only by the redesign of the process.

During the recent discussion of the control design methods, the transfer function P of the process was assumed to be known. In reality the exact transfer function of the process is not known and only its model $\hat{P}$ is available. This distinction has been used so far only in discussion of the concept of robust stability in Section 1.3 and the model-based observer in Section 3.1. In terms of closed-loop control design, it should be noted that the complementary sensitivity function

$$\hat{T} = \frac{C\hat{P}}{1 + C\hat{P}} \tag{4.1}$$

resulting from the *one-degree-of-freedom* (*ODOF*) model-based design is not equal to the real case

$$T = \frac{CP}{1+CP} = \hat{T}\frac{1+\ell}{1+\hat{T}\ell} \tag{4.2}$$

Here ℓ means the relative uncertainty of the process model of (1.3.15).

DECOMPOSITION OF SENSITIVITY FUNCTION

The sensitivity function of the real closed loop can be written in the following decomposed form:

$$\begin{aligned} S &= \underbrace{(1-R_n)}_{S_{des}} + \overbrace{\underbrace{(R_n-\hat{T})}_{S_{real}} - \underbrace{(T-\hat{T})}_{S_{mod}}}^{S_{perf}} = S_{des} + S_{real} + S_{mod} \\ &= \underbrace{(1-R_n)}_{S_{des}} + \underbrace{(R_n-T)}_{S_{perf}} = S_{des} + S_{perf} = \underbrace{(1-\hat{T})}_{S_{contr}} + S_{mod} \\ &= \underbrace{(1-\hat{T})}_{S_{des}+S_{real}} + S_{mod} = S_{contr} + S_{mod} \end{aligned} \tag{4.3}$$

Here $S_{des} = (1-R_n)$ is the design loss, $S_{real} = (R_n - \hat{T})$ is the realizability loss, $S_{mod} = -(T-\hat{T}) = \hat{T} - T$ refers to the modeling loss in the sensitivity function. In its other form $S_{contr} = (1-\hat{T})$ means the decomposed term referring to the control loss, $S_{perf} = (R_n - T)$ to the performance loss. Each term can be interpreted and explained very easily. The meaning of the reference model R_n has been discussed in the previous sections. It is obvious that the trivial equalities $S = 1 - T$ and $\hat{S} = 1 - \hat{T}$ exist, where

$$\hat{S} = \frac{1}{1+C\hat{P}} \quad \text{and} \quad S = \frac{1}{1+CP} = \hat{S}\frac{1}{1+\hat{T}\ell} = \hat{S} + S_{mod} \tag{4.4}$$

The term S_{mod} can be further simplified by means of linear approximation for the case $\ell \to 0$

$$S_{mod} = S - \hat{S} = \hat{T} - T = -\frac{\hat{T}\hat{S}\ell}{1+\hat{T}\ell} = -\hat{T}S\ell\Big|_{\ell\to 0} \approx -\hat{T}\hat{S}\ell \tag{4.5}$$

It is evident that $\left|\hat{T}\hat{S}\right|$ has its maximum at the cutoff frequency ω_c, so the model must be most accurate in the vicinity of this frequency.

For the *two-degree-of-freedom* (*TDOF*) control loop, the complete transfer function corresponding to the concept of the complementary sensitivity function is generally obtained by adding an extension such as $T_r = FT$ (Figure 1.2.2.). For the model-based control

$$T_r = \hat{T}_r \frac{1+\ell}{1+\hat{T}\ell} \tag{4.6}$$

as in (4.2).

Obviously, the trivial expression $S_r = 1 - T_r$ and the triple decomposition introduced in (4.3)

$$S_r = (1 - R_r) + \left(R_r - \hat{T}_r\right) - \left(T_r - \hat{T}_r\right) = S^r_{des} + S^r_{real} + S^r_{mod} \tag{4.7}$$

also exist.

The meaning of the reference model R_r has also been discussed in the above sections. The term S^r_{mod} can be further simplified

$$S^r_{mod} = \hat{T}_r - T_r = -\frac{\hat{T}_r \hat{S}\ell}{1+\hat{T}\ell} = -\hat{T}_r S\ell\big|_{\ell\to 0} \approx -\hat{T}_r \hat{S}\ell \tag{4.8}$$

The ideal control loop has to follow the signals prescribed by R_r and R_n (more precisely $1 - R_n$), and thus the ideal output of the closed loop is according to (2.1.8)

$$y_{id} = y^o = R_r y_r - (1 - R_n) y_n = y^o_r + y^o_n \tag{4.9}$$

Theoretically, instead of (4.9) only

$$y = T_r y_r - S y_n = T_r y_r - (1 - T) y_n \tag{4.10}$$

can be obtained, and even this should be modified according to the model-based design

$$\hat{y} = \hat{T}_r y_r - \hat{S} y_n = \hat{T}_r y_r - \left(1 - \hat{T}\right) y_n \tag{4.11}$$

The deviation between the ideal (y^o) and the theoretically achievable output is

$$\Delta y = y^o - y = (R_r - T_r) y_r - (R_n - T) y_n = S^r_{perf}\, y_r - S_{perf}\, y_n \tag{4.12}$$

where S^r_{perf} refers to the performance loss concerning the tracking, $S^n_{perf} = S_{perf}$ the performance loss concerning the disturbance rejection. Similar expressions can be obtained for the deviation between the ideal output y^o and $\hat{y}$ obtained by the model-based design

$$\Delta\hat{y} = y^o - \hat{y} = \left(R_r - \hat{T}_r\right) y_r - \left(R_n - \hat{T}\right) y_n = S^r_{real} y_r - S_{real} y_n \tag{4.13}$$

Note that in the above expressions the terms S_{real} and $S^{\mathrm{r}}_{\mathrm{real}}$ can be made zero only in the case of inverse stable systems, whereas they can never be made zero for *inverse unstable* (*IU*) systems. So the realizability losses appear in $\Delta\hat{y}$, the performance losses in Δy. The relationship between the two errors is

$$\Delta y = \Delta\hat{y} - \left(S^{\mathrm{r}}_{\mathrm{mod}} y_{\mathrm{r}} - S_{\mathrm{mod}} y_{\mathrm{n}}\right) \tag{4.14}$$

The above triple decomposition of the sensitivity functions gives a good insight into the limit-optimality (limits of the optimality) of closed-loop control systems, i.e., the characterization of the best control achievable. As regards this distinction optimality criteria need to be created for each term, i.e.,

$$\begin{aligned} J_{\mathrm{dist}} &\leq J^{\mathrm{n}}_{\mathrm{des}} + J^{\mathrm{n}}_{\mathrm{real}} + J^{\mathrm{n}}_{\mathrm{mod}} = \|S_{\mathrm{des}}\| + \|S_{\mathrm{real}}\| + \|S_{\mathrm{mod}}\| \\ J_{\mathrm{track}} &\leq J^{\mathrm{r}}_{\mathrm{des}} + J^{\mathrm{r}}_{\mathrm{real}} + J^{\mathrm{r}}_{\mathrm{mod}} = \left\|S^{\mathrm{r}}_{\mathrm{des}}\right\| + \left\|S^{\mathrm{r}}_{\mathrm{real}}\right\| + \left\|S^{\mathrm{r}}_{\mathrm{mod}}\right\| \end{aligned} \tag{4.15}$$

both for the tracking and disturbance rejection behaviors. Here the notation $\|\ldots\|$ is used to express the optimality criterion. In strict mathematical analysis this notation is used to refer to the chosen norm of the transfer function. It is not an easy task to optimize all three terms simultaneously. In practice, iterative techniques are used, whereby the optimization problem is solved step-by-step. The optimization of all three criteria is discussed in detail below.

DECOMPOSITION OF SENSITIVITY FUNCTION FOR *YP* REGULATORS

The expressions for the real and model-based *YP* regulators are

$$C = \frac{Q}{1 - QP}; \quad \hat{C} = \frac{Q}{1 - Q\hat{P}} \tag{4.16}$$

according to which (4.1) and (4.2) change to

$$T = \frac{\hat{C}P}{1 + \hat{C}P} = \frac{Q\hat{P}(1+\ell)}{1 + Q\hat{P}\ell}; \quad \hat{T} = \frac{C\hat{P}}{1 + C\hat{P}} = Q\hat{P} \tag{4.17}$$

The best achievable complementary sensitivity function (which has sense only in the *YP* case) is

$$T_* = \frac{CP}{1 + CP} = QP = Q\hat{P}(1+\ell) = \hat{T}(1+\ell) \tag{4.18}$$

The sensitivity functions of the model-based and the real closed systems are

$$\hat{S} = \frac{1}{1+\hat{C}\hat{P}} = 1 - Q\hat{P} \tag{4.19}$$

and

$$S = \frac{1}{1+\hat{C}P} = \frac{1-Q\hat{P}}{1+Q\hat{P}\ell} = \frac{\hat{S}}{1+\hat{T}\ell} \tag{4.20}$$

The best sensitivity function which is realizable and corresponds to T_* is

$$S_* = \frac{1}{1+CP} = 1 - QP = 1 - Q\hat{P}(1+\ell) = \hat{S} - \hat{T}\ell \tag{4.21}$$

The decomposition of the sensitivity function of the YP closed control loop according to (4.3) is now

$$\begin{aligned} S &= (1-R_n) + \left(R_n - \hat{T}\right) - \left(T - \hat{T}\right) = S_{des} + S_{real} + S_{mod} \\ &= (1-R_n) + \left(R_n - Q\hat{P}\right) - \frac{Q\hat{P}\left(1-Q\hat{P}\right)}{1+Q\hat{P}\ell}\ell \end{aligned} \tag{4.22}$$

where the modeling loss in the case of linear approximation for $\ell \to 0$ is

$$S_{mod} = -\left.\frac{Q\hat{P}\left(1-Q\hat{P}\right)}{1+Q\hat{P}\ell}\ell\right|_{\ell\to 0} \approx -Q\hat{P}\left(1-Q\hat{P}\right)\ell \tag{4.23}$$

Based on $S_* = 1 - T_*$ the following expression is obtained for the realizable best sensitivity function

$$\begin{aligned} S_* &= 1 - QP = S_{des} + S_{real} + S^*_{mod} = (1-R_n) + \left(R_n - Q\hat{P}\right) - QM\ell \\ &= S_{contr} + S^*_{min} = (1-R_n) + (R_n - QP) = \end{aligned} \tag{4.24}$$

where

$$S^*_{mod} \approx -Q\hat{P}\ell = \frac{1}{S}S_{mod} = -\hat{T}\ell \tag{4.25}$$

This latter expression significantly differs from (4.5), because in the case of modeling loss when $\hat{P} = P$, the YP closed loop virtually opens; therefore here the weighting $\hat{S}$ is missing. The maximum of $\left|\hat{T}\right|$ is usually not at the cutoff frequency ω_c and it is close to unity in a relatively wide low-frequency region.

The decomposition of the tracking sensitivity function can be expressed similarly to (4.22)

$$\begin{aligned} S_r &= 1 - T_r = (1 - R_r) + \left(R_r - Q_r\hat{P}\right) - \left(T_r - \hat{T}_r\right) \\ &= S^{r}_{des} + S^{r}_{real} + S^{r}_{mod} \end{aligned} \tag{4.26}$$

where

$$S^{r}_{mod} = -\left.\frac{Q_r\hat{P}\left(1 - Q\hat{P}\right)}{1 + Q\hat{P}\ell}\ell\right|_{\ell \to 0} \approx -Q_r\hat{P}\left(1 - Q\hat{P}\right)\ell \tag{4.27}$$

DIRECT OPTIMIZATION OF SENSITIVITY FUNCTION

To optimize a closed loop, it is not necessary to follow the decomposition of the sensitivity function according to (4.3) in spite of the fact that the decomposition introduced here gives a clear explanation of the most important inner components determining the quality of the control. Methods published so far apply the direct optimization of the sensitivity function. The sensitivity function of the *YP* closed loop is

$$S = 1 - R_n G_n P_- = S_n \tag{4.28}$$

and the tracking sensitivity function is

$$S_r = 1 - R_r G_r P_- \tag{4.29}$$

The general energy norm introduced in (A.5.22) for optimization is used in the form

$$J^{x}_{\mathcal{E}_2} = \|Y_x(1 - R_x G_x P_-)\|_{\mathcal{E}_2} = \left\|\frac{1}{s^j}(1 - R_x G_x P_-)\right\|_{\mathcal{E}_2} \tag{4.30}$$

For $j = 0$, this norm gives the $\mathcal{H}_2$ operator, whereas for $j \geq 1$, it gives the $\mathcal{L}_2$ norm

$$J^{x}_{\mathcal{E}_2} = \begin{cases} \mathcal{H}_2 \text{ norm}; & j = 0 \\ \mathcal{L}_2 \text{ norm}; & j \geq 1 \end{cases} \tag{4.31}$$

In general, the $\mathcal{L}_2$ norm is used for real functions, but its original definition allows us to use it also for complex functions, as here. The optimality

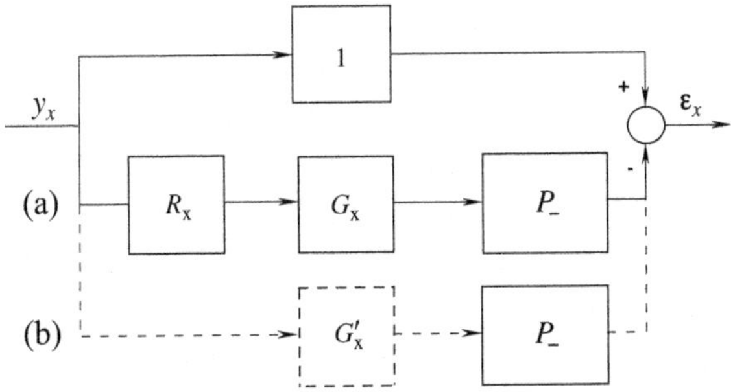

Figure 4.1 The error signal optimized by norm $\mathcal{E}_2$.

criterion (4.30) measures the energy of the error signal resulting from the sensitivity functions S or S_r under the excitation of the general form $Y_x = \mathcal{L}(y_x) = s^{-j}$. Thus, the loss functions of the YP closed loop which are to be optimized take the form $\boldsymbol{\varepsilon}_x = Y_x(1 - R_x G_x P_-)$ (see Figure 4.1), where the index x refers to the noise rejection or tracking tasks. The goal of the optimization is to find the minimum

$$
\begin{aligned}
G_x &= \arg\left\{\min_{G_x}\left[J_{\mathcal{E}_2}^{x}\right]\right\} = \arg\left\{\min_{G_x}\left[\|Y_x S_x\|_{\mathcal{E}_2}\right]\right\} \\
&= \arg\left\{\min_{G_x}\left[\|Y_x(1 - R_x G_x P_-)\|_{\mathcal{E}_2}\right]\right\}
\end{aligned}
\tag{4.32}
$$

To calculate the norm $\mathcal{E}_2$ the integral of (A.5.22) must be finite. Y_x is a low-pass filter but $(1 - R_x G_x P_-)$ is a high-pass filter (if the type number of the control is higher than zero), and therefore this condition is fulfilled for a stable closed system. Note that the $\mathcal{H}_2$ case of the optimization is equal to the so-called *mean square error* or *minimum variance* control paradigm if the excitation (y_n or y_r) is white noise, i.e., $j = 0$ and $Y_x = 1$. This task is known as the Wiener paradigm in the classical control engineering literature.

Assume that the inverse stable reference models take the form

$$
R_x = \frac{\mathcal{B}_x}{\mathcal{A}_x}
\tag{4.33}
$$

where $\mathcal{B}_x$ and $\mathcal{A}_x$ are relative primes. Let the IU invariant factor of the process be $P_- = \mathcal{B}_-$, where $\mathcal{B}_-(0) = 1$ and $\mathcal{B}_-(s) = \prod_{i=1}^{m_-}(1 - s/z_i)$.

Here z_i are the zeros of $\mathcal{B}_-$ and m_- is the number of *IU* zeros. It can be written for the norm $\mathcal{E}_2$ in line with (4.30) that

$$J^{x}_{\mathcal{E}_2} = \left\| \frac{1}{s^j}\left(1 - \frac{\mathcal{B}_x}{\mathcal{A}_x} G_x \mathcal{B}_-\right) \right\|_{\mathcal{E}_2} = \left\| \frac{\mathcal{B}_-}{\mathcal{B}_-^-} \right\|_{\mathcal{E}_2} \left\| \frac{\mathcal{B}_-^-}{\mathcal{B}_- s^j} - \frac{\mathcal{B}_x G_x \mathcal{B}_-^-}{\mathcal{A}_x s^k} \right\|_{\mathcal{E}_2} \tag{4.34}$$

$\mathcal{B}_-^-$ is a polynomial created by mirroring the zeros of $\mathcal{B}_-$ on the imaginary axis; thus for the new zeros the following conditions must be fulfilled: $\mathrm{Re}\{z_i^*\} = -\mathrm{Re}\{z_i\}$ and $\mathrm{Im}\{z_i^*\} = \mathrm{Im}\{z_i\}$. So, $\mathcal{B}_-^-$ becomes a stable polynomial and the factor $\mathcal{B}_-/\mathcal{B}_-^-$ is called the norm-keeping inner function (in the context of signal-forming it is called the all-pass function (1.1.16)), and thus $\left\|\mathcal{B}_-/\mathcal{B}_-^-\right\|_{\mathcal{E}_2} = 1$.

To minimize (4.34), the classical method of the Wiener paradigm [159] needs to be applied, i.e., the argument of (4.34) must be decomposed to causal and noncausal factors, so the first term should be decomposed as

$$\frac{\mathcal{B}_-^-}{\mathcal{B}_- s^j} = \frac{\mathcal{R}}{\mathcal{B}_-} + \frac{\mathcal{K}}{s^j} = \frac{\mathcal{R} s^j + \mathcal{K}\mathcal{B}_-}{\mathcal{B}_- s^j} \tag{4.35}$$

Here $\mathcal{R}/\mathcal{B}_-$ is the noncausal factor and $\mathcal{K}/s^j$ is the causal factor (obviously the polynomials $\mathcal{R}$ and $\mathcal{K}$ introduced here have no relation to the notations applied in the Section 3.1). The new form of (4.34) is now

$$J^{x}_{\mathcal{E}_2} = \left\| \frac{\mathcal{R}}{\mathcal{B}_-} + \frac{\mathcal{K}}{s^j} - \frac{\mathcal{B}_x G_x \mathcal{B}_-^-}{\mathcal{A}_x s^j} \right\|_{\mathcal{E}_2} \tag{4.36}$$

The minimum can be reached, according to the orthogonality principle of the Wiener method, by making the causal term equal to zero, i.e., the condition of the minimum is

$$\frac{\mathcal{K}}{s^j} - \frac{\mathcal{B}_x G_x \mathcal{B}_-^-}{\mathcal{A}_x s^j} = 0 \tag{4.37}$$

The optimality of the $\mathcal{E}_2$ norm, i.e., the error signal $\boldsymbol{\varepsilon}_x$ of minimum energy can be ensured by the embedded filter

$$G_x = \frac{\mathcal{A}_x \mathcal{K}}{\mathcal{B}_x \mathcal{B}_-^-} \tag{4.38}$$

$\mathcal{R}$ and $\mathcal{K}$ can be calculated from the solution of a special *Diophantine equation* (*DE*)

$$\mathcal{B}_-^- = \mathcal{R} s^j + \mathcal{K}\mathcal{B}_- \tag{4.39}$$

The expression (4.39) results from the comparison of the two sides of (4.35). Observe that substituting G_x, optimum according to (4.38), to $\|Y_x(1 - R_x G_x P_-)\|_{\mathcal{E}_2}$, the simpler form $\|Y_x(1 - G'_x P_-)\|_{\mathcal{E}_2}$ is obtained (see Figure 4.1(b)), where

$$G'_x = \mathcal{K}/\mathcal{B}_-^- \tag{4.40}$$

Thus, the optimization cancels all stable factors. The block diagram of Figure 4.1(b) corresponds to the so-called classical Nehari problem [132], where the dynamics of P_- should be cancelled in such a way ($G_x P_- \stackrel{?}{=} 1$) that its inverse cannot be created. In this case, the optimization tries to find the closest realizable inverse. For the minimum of $J^{x}_{\mathcal{E}_2}$

$$J^{x}_{\mathcal{E}_2} = \left\| \frac{\mathcal{R}}{\mathcal{B}_-} \right\|_{\mathcal{E}_2} \tag{4.41}$$

is obtained. Finally, the $\mathcal{E}_2$ optimal *YP*-regulator is obtained as (based on (2.1.10) and (4.38) for $T_d = 0$)

$$C_* = \frac{\mathcal{K}\mathcal{A}}{\mathcal{B}_+\left(\mathcal{B}_-^- - \mathcal{K}P_-\right)} \tag{4.42}$$

Looking at the error signal $\varepsilon_x = Y_x(1 - R_x G_x P_-)$ and (2.1.10), we see that an integrating regulator can be obtained if and only if the reference models R_x have unity gain and the condition $G_x P_-(\omega = 0) = 1$ can be ensured during the optimization. Look at

$$G_x\mathcal{B}_- = \frac{\mathcal{A}_x\mathcal{K}\mathcal{B}_-}{\mathcal{B}_x\mathcal{B}_-^-} = \frac{\mathcal{A}_x}{\mathcal{B}_x}\left(1 - \frac{\mathcal{K}}{\mathcal{B}_-^-}s^j\right) \tag{4.43}$$

where it can be seen that the condition $G_x\mathcal{B}_-|_{s\to 0} = 1$ for an integrating regulator can be fulfilled if and only if $j \geq 1$, but it is not fulfilled by the optimization of the $\mathcal{H}_2$ norm according to the equivalence relationship (4.31). This is partly why the generalized $\mathcal{E}_2$ norm and the so-called general excitation in the form $Y_x(s) = s^{-j}$ are introduced.

(The interesting result obtained for the integrating regulator is not surprising, because if the Parseval theorem [37] is applied for the computation of the $\mathcal{H}_2$ norm, then the pole access of the error transfer function must be one, at least, since the sensitivity function is usually a high-pass filter. This condition is fulfilled in the above solution, if $j \geq 1$.)

As regards the optimal solution, it should be noted that G'_x does not depend on the reference model R_x but only on $\mathcal{B}_-$ and j. Therefore, further criteria need to be found whose optimization depends on the design goal. This condition is fulfilled in Section 4.2, if the realizability loss is minimized.

During the optimization the real process and its model are not distinguished. In this case, only one thing can be known for certain: if the assumed model is equal to the real process, then this procedure optimizes the sensitivity function of the real system. In the case of a general model, it is very difficult to judge how far the obtained optimum is from the ideal case.

Example 4.1

Let the reference model be $R_x = 1/(1+sT_w)$ and the *IU* factor $\mathcal{B}_- = 1 - sT$ according to the definition $\mathcal{B}_-^- = 1 + sT$. The solution of the *DE* (4.39) for $j = 0$ is $\mathcal{R} = 2$ and $\mathcal{K} = -1$. The $\mathcal{E}_2$ optimal G_x filter (now also $\mathcal{H}_2$ optimal) is

$$G_x = -\frac{1+sT_w}{1+sT} \tag{4.44}$$

and the optimal *YP*-regulator

$$C_* = \frac{\mathcal{A}(-1)}{\mathcal{B}_+[(1+sT)-(-1)(1-sT)]} = \frac{\mathcal{A}}{2\mathcal{B}_+} \tag{4.45}$$

is an integrating regulator. Solve the task for $j = 1$, then the solution $\mathcal{R} = 2T$ and $\mathcal{K} = 1$ is obtained. Finally, the $\mathcal{E}_2$ optimal regulator takes the form

$$C_* = \frac{\mathcal{A}(+1)}{\mathcal{B}_+[(1+sT)-(+1)(1-sT)]} = \frac{\mathcal{A}}{2\mathcal{B}_+ sT} \tag{4.46}$$

and it is a non integrating one.

Sampled Data System

Obviously the above results can be repeated for DT control loops, if the transfer functions are substituted by pulse transfer functions. In sampled-data systems the time delay must not be assumed to be zero, because the form z^{-d} provides a rational fractional function. When an integrator is assumed, then the root $z = 1$ must appear in the denominator. To construct $\mathcal{B}_-^-$, the unstable zeros must be mirrored on the unity circle instead of the imaginary axis; thus the simple transformation $z_i^* = 1/z_i$ needs to be applied.

SPECIAL METHODS

It was seen in discussion of the *YP* regulators that the effect of the invariant factors of the process can be optimized by the proper choice of the embedded filters G_x (these methods have been used above and will be further discussed in Section 4.2). A very interesting approach is shown below where the optimization is performed in a different way from the minimization of the classical scalar-vector function. The complementary sensitivity function of the model-based closed system (4.1) is determined by the loop transfer function $\hat{L} = C\hat{P}$ of the system

$$\hat{L} = C\hat{P} = \frac{R_n G_n \hat{P}_+^{-1}}{1 - R_n G_n \hat{\bar{P}}_-} \hat{P} = \frac{R_n G_n \hat{\bar{P}}}{1 - R_n G_n \hat{\bar{P}}_-} \tag{4.47}$$

The cutoff frequency can be obtained from the condition $|\hat{L}| = |C\hat{P}| = 1$, i.e., from the equation

$$\left|R_n G_n \hat{\bar{P}}\right| = \left|1 - R_n G_n \hat{\bar{P}}_-\right| \tag{4.48}$$

For the nonrealizable, ideal regulator this equation leads to the condition

$$|R_n| = |1 - R_n| \quad \text{i.e.,} \quad |R_n(\omega_c)| = |1 - R_n(\omega_c)| \tag{4.49}$$

This latter equation constitutes a very simple geometrical tool for the determination of the cutoff frequency ω_c. Namely, ω_c can be found on the Nyquist diagram of the frequency function $R_n(j\omega)$ where this curve has an intersection with the vertical line at point $(0.5 + 0j)$. At this point, the distance from $(1 + 0j)$, i.e., the right side of (4.49) is equal to its left side. The geometry can be followed in Figure 4.2. It can also be seen in the

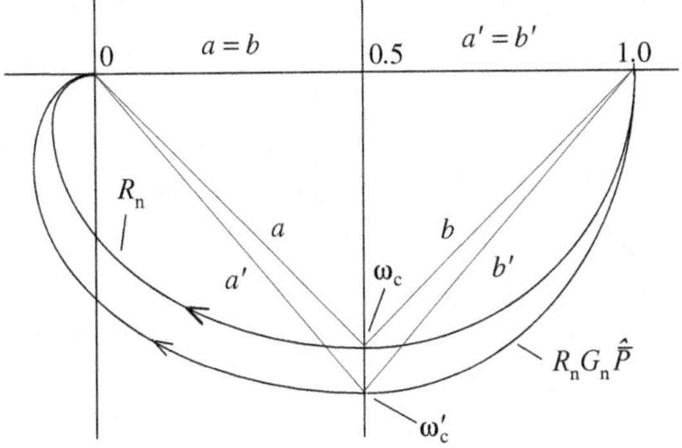

Figure 4.2 Change of the cutoff frequency owed to the invariant factor $\hat{\bar{P}}$ and its compensation.

figure that the geometrical rule is exactly the same for the invariant factors and their compensation, i.e., for Equation (4.48). The cutoff frequency ω_c will change to the value ω'_c even if G_n is an all-pass filter, as shown in Section 4.2.

According to the figure the embedded filter G_n is optimal and provides the least loss of quality. Finding such a filter is no easy task. Filters with very positive phase access in the vicinity of ω'_c are generally appropriate, however.

EMPIRICAL RELATIONSHIPS

Besides the optimization of the sensitivity function components, several empirical relationships can be found in control engineering practice and it is recommended that these are taken into account before application.

It was seen in the investigation of the best achievable control systems that the basic constraint derives from the saturation of the actuator signal and the process dynamics itself. One of the most important dynamical limits is the time delay of the process, which cannot be eliminated, since the system cannot respond in a shorter time than the time delay. The first-order Pade approximation of a dead-time lag has already been discussed, according to which

$$e^{-sT_d} \approx \frac{1 - sT_d/2}{1 + sT_d/2} = \frac{s - 2/T_d}{s + 2/T_d} = \frac{s - z_j}{s + z_j} \quad (4.50)$$

i.e., the right-hand zero z_j approximation corresponds to a dead-time lag $T_d = 2/z_j$. This also means, at the same time, that the right-hand zeros of the process correspond to, in any way, certain limits. Unstable zeros of small value mean considerable delay time.

It can be assumed that the unstable poles of the process can also result in constraints. It can be expected that a fast regulator is required to stabilize an unstable process.

Summarizing, the constraints derive from the dead time and the unstable, right-hand zeros (z_j) and poles (p_j) of the process dynamics. Based on fundamental theoretical considerations and practice, the constraints are as follows [12]:

- the *right-hand unstable* (*RU*) zero z_j has the following limit for the cutoff-frequency

$$\omega_c < 0.5 z_j \quad (4.51)$$

(A slow *RU* zero has an especially bad effect.)

- the dead-time has also limit for the cutoff frequency, based on the practice

$$\omega_c < 0.5\frac{1}{T_d} \tag{4.52}$$

- an RU pole requires high cutoff frequency

$$\omega_c > 2p_j \tag{4.53}$$

- the systems, which have both RU zeros and poles simultaneously cannot be controlled, in general, unless the poles and zeros are far enough from each other.

$$p_j > 6z_j \tag{4.54}$$

- unstable dead-time systems cannot be controlled, unless the separation (distance) condition

$$p_j < 0.16\frac{1}{T_d} \tag{4.55}$$

is fulfilled.

Sampled Data System

For discrete systems all the above constraints exist. In addition, whereas in CT systems the nonminimum phase processes are quite rare, in DT systems, however, it has been counted on the pulse transfer functions of all processes, where the pole access is, at least, two. (The effect of the fractional time delay has the same result.)

The sampling time itself does not constitute a constraint as it is only a design parameter, though it can become a constraint in the case of slow computer solutions.

Among the constraints in discrete systems the problem of measurement noise should be mentioned [12]. This noise can derive from the operation of the sensor, from its inner electronics, or from the A/D transformer. The noises usually appear in the high frequency region; therefore the uncertainty caused by the noise limits the high-frequency gain A_∞ of the regulator. For simplicity, assume that the A/D and D/A transformers generally use 12-bit arithmetic. The 12 bits correspond to 4096 levels; thus a conversion error or measurement noise of 1 bit, gained by 4096, can reach the level of the whole signal region.

In practice, there should be no uncertainty caused by measurement noise higher than 5%. This means that the high frequency gain of the regulator must be less than 200, i.e., $A_\infty < 200$.

4.1 OPTIMIZATION OF DESIGN LOSS

The regulators always invoke zeros to accelerate the process. To demonstrate this, let us investigate the case shown in Figure 4.1.1, where a phase-lead element is connected in serial before the first-order lag element. In the first moment, a signal value of 10 appears at the output of the phase-lead element and at the input of the first-order lag element for the effect of a unit step signal. The first-order lag starts with a high gradient in order to reach this value as soon as possible with its time constant and by the time the input signal is settling down the output has reached its steady state value. The cost of the acceleration is the so-called overexcitation, i.e., the ratio of the initial and final signal values at the input of the lag. The acceleration can be reached only by overexcitation greater than one. In many cases, it is worth applying pole cancellation (see Sections 2 and 3), when zeros are invoked to cancel the undesirable poles which cause slow operation, and instead a pole ensuring more favorable behavior is inserted.

Thus it is obvious that the overexcitation means the control equipments will have an initial peak at their output as a response to unit step commands or disturbances. The problem is that the output of the regulators in the closed-loop control, or the output of the actuators gaining the signal for the proper level, is always amplitude restricted.

$$|u(t)| \le U_{\max} \tag{4.1.1}$$

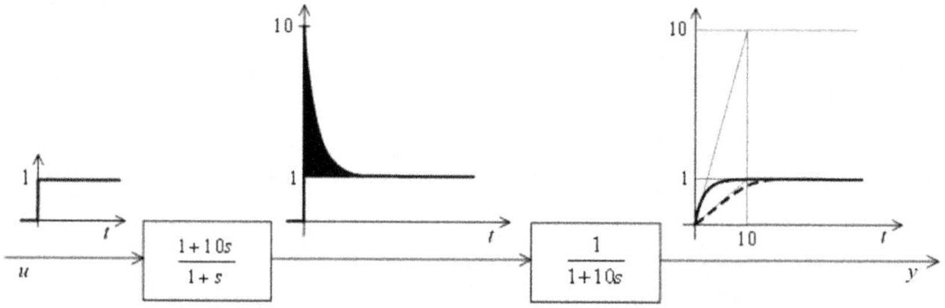

Figure 4.1.1 Insertion of a zero may accelerate the system at the cost of overexcitation.

This means that the arbitrary jump, i.e., significant change between the initial and final values, and thus arbitrary overexcitation, cannot be performed. It was seen in Section 2.4 that the poles can be cancelled by the insertion of zeros, which results in the acceleration of the process. This acceleration always requires overexcitation. The optimal control methods discussed earlier apply, without exception, a certain kind of pole cancellation, i.e., overexcitation (see Figure 4.1.2). The above-mentioned amplitude restriction means that the calculated optimal regulator parameters cannot be realized. Thus the reachable acceleration depends, to a large extent, on the value of the overexcitation applicable.

The design of the regulator signal range requires careful work. The working point is usually set in the middle of the signal domain, and the possible signal changes are sized without reaching the limitations (Figure 4.1.3).

The actuator signal $u(t)$ has an important role in control systems, because it is the input of the process and the physical constraints appear here. As an effect of the change of the reference signal $r(t)$ transient phenomena appear according to the system dynamics. In general, during the transient settlement the actuator signal reaches a much higher value than its static value, and this phenomenon is described by dynamic overexcitation. Its definition is

$$u_t = \frac{U_{max}}{u(t \to \infty)} \cong \frac{u(0)}{u_\infty} \tag{4.1.2}$$

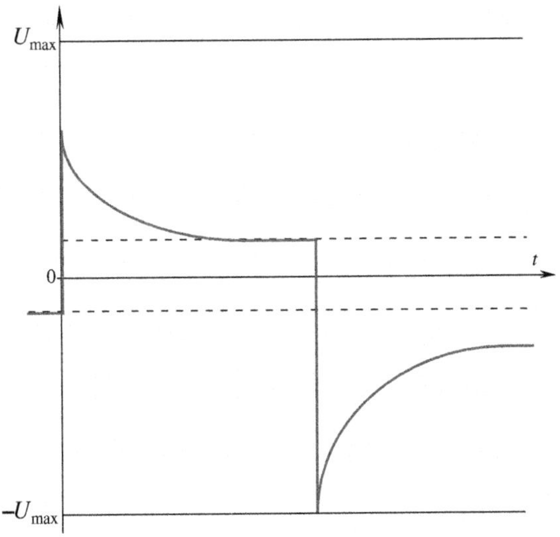

Figure 4.1.2 Typical regulator output (actuator signal) in case of overexcitation.

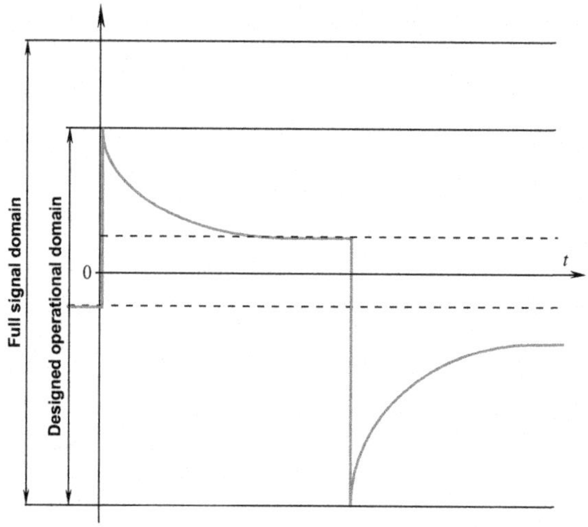

Figure 4.1.3 Design of the signal region of the regulator output.

In general, the maximum value is the initial value; therefore, for simplicity it is approximated by this value. The dynamic overexcitation can be interpreted in this way only if u_∞ is not equal to zero. This can happen if the process contains an integrator, since the system can reach its steady state only if the input of the integrator becomes zero. In this case, the dynamic overexcitation is substituted by U_{max}.

In many cases, not only the amplitude of the regulator output has limits but its changing speed also has practical constraints. Consider the control valves of pipeline networks where it takes some time until the motor can move the valve from one position to another. Such so-called speed-type limitations can be handled only with difficulty by analytical methods; the solution may be simulation or practical experimentation. Here the method is the same as that recommended earlier: the design requirements must be reduced.

Summing up, we can say that the fastest reachable control primarily depends on the constraints concerning the regulator output. These constraints depend on the type of the device applied in the technology. Thus improvements can be made only by choosing another control device or by the redesign of the whole process.

The optimization of the first term in the decomposition of the sensitivity function (4.3) means the determination of the best (fastest) reachable reference models $R_r = R_r^{opt}$ and $R_n = R_n^{opt}$, i.e., the solution of the task under the constraints

$$R_r^{opt} = \arg\left\{ \min_{R_r} \left(J_{des}^r \right) \Big|_{u \in U} \right\} = \arg\left\{ \min_{R_r} \|1 - R_r\| \Big|_{u \in U} \right\}$$
$$R_n^{opt} = \arg\left\{ \min_{R_n} \left(J_{des}^n \right) \Big|_{u \in U} \right\} = \arg\left\{ \min_{R_n} \|1 - R_n\| \Big|_{u \in U} \right\} \tag{4.1.3}$$

where the chosen criteria $J_{des}^r = \|1 - R_r\|$ and $J_{des}^n = \|1 - R_n\|$ state that each reference model should approach the ideal unity. This task must be solved under the constraints for the regulator output $u \in \mathbb{U}$. Here $\mathbb{U}$ means the allowable region for u, i.e., $\mathbb{U}$: $|u| \leq 1$ (see (4.1.1) or (4.1.2)).

The optimization task (4.1.3) is very difficult because the solution is always on the border of the limited region. There is no analytical solution except for some low-order simple cases. The optimal reference models are usually determined by simulation (CAD tools). Note that under the given constraints, faster reference models cannot be used to solve the task (4.1.3). Vice versa, if under the constraints and design goal there is no solution for the reference models then the only option is to prescribe a less demanding (usually slower) model. Thus, the best (fastest) reachable response of the closed system basically depends on the constraints of the control output. Of course, Equation (4.1.3) contains the applied regulator and also the process in a complex way; thus it is a closed loop. Therefore, the optimality of the regulator depends on the process, the model, and the invariant factors.

Redesign of Reference Model

Even in the case of a very careful design, it can happen that the overactuated output of the obtained regulator is beyond the acceptable signal domain. Then the original design goals need to be reduced. The advantage of the *Keviczky–Bányász* (*KB*) parameterization of generic *TDOF* control loops is that only the problematic (overdemanding) reference models R_r and R_n need to be changed to accommodate less demanding design conditions. Usually this can be performed only step by step via an iterative procedure. The steps can contain model simulation and also experiments on the real process. Therefore, the optimization is often termed the redesign of the reference model. In the case of low-order reference models the time constant of the model (i.e., the bandwidth) can be determined by explicit design expression if the process model and the amplitude constraint $U_{\max}$ are known.

As an example, let us consider the *ODOF YP* control loop, where $R_r = R_n = R_x$ and choose a first-order reference model in the form

$$R_r = R_n = \frac{1}{1 + sT_x} \tag{4.1.4}$$

Investigate the ideal regulator C_{id} of (2.1.5), in this case the output of the regulator

$$u_{id} = R_x P^{-1} y_x \tag{4.1.5}$$

The dynamic overexcitation caused by the regulator is

$$u_t = \frac{R_x P^{-1}\big|_{s\to\infty}}{R_x P^{-1}\big|_{s\to 0}} \tag{4.1.6}$$

where the numerator is

$$R_x P^{-1}\big|_{s\to\infty} = \frac{1}{1 + sT_x} \frac{\mathcal{A}(s)}{\mathcal{B}(s)}\bigg|_{s\to\infty} = \frac{a_n}{T_x b_{n-1}} \tag{4.1.7}$$

and the denominator is

$$R_x P^{-1}\big|_{s\to 0} = \frac{1}{1 + sT_x} \frac{\mathcal{A}(s)}{\mathcal{B}(s)}\bigg|_{s\to 0} = \frac{\mathcal{A}(0)}{\mathcal{B}(0)} \tag{4.1.8}$$

Finally, the expression

$$u_t = \frac{\dfrac{\alpha_n}{T_x \beta_{n-1}}}{\dfrac{\mathcal{A}(0)}{\mathcal{B}(0)}} = \frac{\mathcal{B}(0) a_n}{\mathcal{A}(0) T_x b_{n-1}} \tag{4.1.9}$$

is obtained for (4.1.6) where $R_x P^{-1}$ is assumed to be regular. From the limit for the initial step, the possible minimum time constant of the reference model is obtained as

$$T_x \geq \frac{a_n}{b_{n-1} U_{\max}} \tag{4.1.10}$$

Thus, only a transient process, slower than the above, is required during the design.

Sampled Data Systems

The decomposition of the control error shown in (4.3) is also valid for DT systems and all expressions can be applied without any change.

Note that in the DT case, the fastest reachable first-order reference model can be determined from the amplitude constraints given for the output of the regulator

$$|u[k]| = U_{\max} \tag{4.1.11}$$

if a *YP* regulator is applied. Let the first-order reference model of unity gain be

$$R_x = \frac{(1+f_x)z^{-1}}{1+f_x z^{-1}} = \frac{(1+f_x)}{z+f_x} \tag{4.1.12}$$

Investigate the ideal DT regulator corresponding to (2.1.5). In this case, the output of the regulator is

$$u_{id} = R_x G^{-1} y_x \tag{4.1.13}$$

The dynamic overexcitation is now

$$u_t = \frac{R_x G^{-1}|_{z\to\infty}}{R_x G^{-1}|_{z\to 1}} \tag{4.1.14}$$

where the numerator is

$$\left.\frac{(1+f_x)z^{-1}}{1+f_x z^{-1}} \frac{\mathcal{F}(z)}{\mathcal{G}(z)}\right|_{z\to\infty} = \frac{1+f_x}{g_1} \tag{4.1.15}$$

but the denominator is

$$\left.\frac{(1+f_x)z^{-1}}{1+f_x z^{-1}} \frac{\mathcal{F}(z)}{\mathcal{G}(z)}\right|_{z\to 1} = \frac{\mathcal{F}(1)}{\mathcal{G}(1)} \tag{4.1.16}$$

Finally, the expression

$$u_t = \frac{\mathcal{G}(1)(1+f_x)}{\mathcal{F}(1)g_1} \tag{4.1.17}$$

is obtained for (4.1.14). From the limit for the initial step, the possible greatest value for reference model parameter is obtained as

$$f_x \le g_1 U_{\max} - 1 \tag{4.1.18}$$

The parameter f_x must, of course, also satisfy the inequality condition derived from the stability and realizability

$$-1 \leq f_x \leq 0 \tag{4.1.19}$$

which is fulfilled only if $g_1 U_{max} \leq 1$. To satisfy this the sampling time might have to be changed.

The possible value domain of the actuating signal can be checked, in simple cases, by simulation of the closed-loop control. The simulation, however, requires knowledge of the process model, and obviously in this case the whole closed loop should be a model-based one. But the violation of the limits for the signals can be tested only by experiments with real closed loops. In practice, if the model is known, then first the redesign of the reference model is performed by the model-based closed loop, and if the design goal is satisfied then the investigation continues with experiments on the real system. The redesign technique can be performed by the iterative joint identification and control algorithm discussed later in Section 11.2. In each iterative step, the reference model can easily be redesigned on the basis of the identified model and the computed optimal control. The eventual violation of the amplitude limit can be checked before the application of the actuating signal and if the signal is too high it can be modified to the allowable value. If the iteration is convergent then the reference model converges to the fastest permissible model.

Note that if the applied regulator is not the ideal *YP* one, then the quotient $\mathcal{F}(z)/\mathcal{G}(z)$ in (4.1.15) changes according to $\mathcal{F}(z)/\mathcal{G}_+(z)$ or $\mathcal{F}_+(z)/\mathcal{G}_+(z)$, which derive from the realizability of the regulator.

4.2 OPTIMIZATION OF REALIZABILITY LOSS

Based on (4.15), the goal of the task is to optimize the realizability loss terms J^{r}_{real} and J^{n}_{real} according to the following expressions:

$$\begin{aligned} G_r^{opt} &= \arg\left\{ \min_{G_r} \left(J^{r}_{real} \right) \right\} = \arg\left\{ \min_{G_r} \left\| R_r - \hat{T}_r \right\| \right\} \\ G_n^{opt} &= \arg\left\{ \min_{G_n} \left(J^{n}_{real} \right) \right\} = \arg\left\{ \min_{G_n} \left\| R_n - \hat{T} \right\| \right\} \end{aligned} \tag{4.2.1}$$

This optimization task is basically discussed for *YP* control loops, where the optimality can be ensured by the selection of $G_r = G_r^{opt}$ and $G_n = G_n^{opt}$ (see Section 2.1). As mentioned earlier, the condition $R_r = \hat{T}_r$ and $R_n = \hat{T}$ can theoretically be reached for the *inverse stable and realisable (ISR)* case which provides the trivial solution $G_r = G_n = 1$. The optimal transfer functions need to be determined for the more general *IU* case.

Optimization Method Using the $\mathcal{E}_2$ Norm

Instead of the criterion (4.32) defined directly for the sensitivity function, let us use now the model-based loss function

$$G_x = \arg\left\{\min_{G_x}\left[\left\| Y_x R_x\left(1 - G_x \hat{P}_-\right)\right\|_{\mathcal{E}_2}\right]\right\}$$
$$= \arg\left\{\min_{G_x}\left[\left\|\frac{R_x}{s^j}\left(1 - G_x \hat{P}_-\right)\right\|_{\mathcal{E}_2}\right]\right\} \tag{4.2.2}$$

The scope of the task can be seen in Figure 4.2.1, where $Y_x = \mathcal{L}(y_x) = s^{-j}$ and the error function takes the form $\hat{\varepsilon}_x = R_x(1 - G_x\hat{P}_-)$. Note that this is a real optimization task, because all initial terms (R_x and $\hat{P}_-$) depend only on us.

To solve the problem, let us apply here the Wiener method [158,159] used in Chapter 4.

Determine the $\mathcal{E}_2$ norm for the error signal $\hat{\varepsilon}_x$

$$J_{\text{real}}^{x} = \left\| Y_x R_x\left(1 - G_x\hat{\mathcal{B}}_-\right)\right\|_{\mathcal{E}_2} = \left\|\frac{R_x}{s^j}\left(1 - G_x\hat{\mathcal{B}}_-\right)\right\|_{\mathcal{E}_2}$$
$$= \left\|\frac{\hat{\mathcal{B}}_-}{\hat{\mathcal{B}}_-^-}\right\|_{\mathcal{E}_2}\left\|\frac{\hat{\mathcal{B}}_-^-\mathcal{B}_x}{\hat{\mathcal{B}}_-\mathcal{A}_x s^j} - \frac{\mathcal{B}_x G_x \hat{\mathcal{B}}_-^-}{\mathcal{A}_x s^j}\right\|_{\mathcal{E}_2} \tag{4.2.3}$$

where $\hat{\mathcal{B}}_-^-$ is model-based but its meaning is the same as in Chapter 4. Thus, it is stable and the factor $\left\|\hat{\mathcal{B}}_-/\hat{\mathcal{B}}_-^-\right\|_{\mathcal{E}_2}$ preserves the norm. Apply the decomposition (4.35) for this task

$$\frac{\hat{\mathcal{B}}_-^-\mathcal{B}_x}{\hat{\mathcal{B}}_-\mathcal{A}_x s^j} = \frac{\mathcal{R}}{\hat{\mathcal{B}}_-} + \frac{\mathcal{K}}{\mathcal{A}_x s^j} = \frac{\mathcal{R}\mathcal{A}_x s^j + \mathcal{K}\hat{\mathcal{B}}_-}{\hat{\mathcal{B}}_-\mathcal{A}_x s^j} \tag{4.2.4}$$

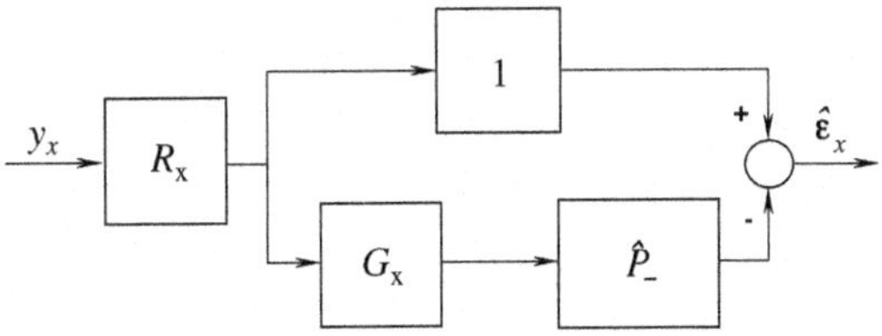

Figure 4.2.1 Scheme for the optimization of realizability loss.

where $\mathcal{R}/\hat{\mathcal{B}}_-$ is the noncausal and $\mathcal{K}/\mathcal{A}_x s^j$ the causal term. The new form of (4.2.4) is

$$J^{x}_{\mathcal{E}_2} = \left\| \frac{\mathcal{R}}{\hat{\mathcal{B}}_-} + \frac{\mathcal{K}}{\mathcal{A}_x s^j} - \frac{\mathcal{B}_x G_x \hat{\mathcal{B}}_-^-}{\mathcal{A}_x s^j} \right\|_{\mathcal{E}_2} \tag{4.2.5}$$

According to the orthogonality principle of the Wiener method [158], the minimum can be obtained by making the causal term equal to zero, i.e., the condition of the minimum is

$$\frac{\mathcal{K}}{\mathcal{A}_x s^j} - \frac{\mathcal{B}_x G_x \hat{\mathcal{B}}_-^-}{\mathcal{A}_x s^j} = 0 \tag{4.2.6}$$

The optimality of the $\mathcal{E}_2$ norm, i.e., the error signal $\hat{\varepsilon}_x$ of minimum energy, can be guaranteed by the inner filter

$$G_x = \frac{\mathcal{K}}{\mathcal{B}_x \hat{\mathcal{B}}_-^-} \tag{4.2.7}$$

The value of $\mathcal{R}$ and $\mathcal{K}$ can be obtained by the solution of the *DE*

$$\hat{\mathcal{B}}_-^- \mathcal{B}_x = \mathcal{R}\mathcal{A}_x s^j + \mathcal{K}\hat{\mathcal{B}}_- \tag{4.2.8}$$

This equation is obtained through comparison of the two sides of Equation (4.2.4). The minimum of (4.2.5) is formally the same as (4.41), but now $\mathcal{R}$ derives from (4.2.8).

The $\mathcal{E}_2$ optimal *YP* regulator can be obtained by using the expressions (2.1.10) and (4.2.7)

$$C_* = \frac{\mathcal{K}\hat{\mathcal{A}}}{\hat{\mathcal{B}}_+\left(\mathcal{A}_n \hat{\mathcal{B}}_-^- - \mathcal{K}\hat{\mathcal{B}}_-\right)} \tag{4.2.9}$$

Look again at the product

$$G_x \hat{\mathcal{B}}_- = \frac{\mathcal{K}\hat{\mathcal{B}}_-}{\mathcal{B}_x \hat{\mathcal{B}}_-^-} = 1 - \frac{\mathcal{A}_x \mathcal{K}}{\mathcal{B}_x \hat{\mathcal{B}}_-^-} s^j \tag{4.2.10}$$

where it is clear that the condition $G_x \hat{\mathcal{B}}_-\big|_{s\to 0} = 1$ of the integrating regulator is fulfilled if and only if $j \geq 1$. Thus, it cannot be performed by the optimization of the $\mathcal{H}_2$ norm based on the equivalence expression (4.31).

It is proved again that it is necessary to introduce the generalized $\mathcal{E}_2$ norm, and the logically connected general excitation $Y_x(s) = s^{-j}$.

Example 4.2.1

Let the reference model be $R_x = 1/(1 + sT_w)$ and the IU term $\mathcal{B}_- = 1 - sT$. So, according to the definition, $\mathcal{B}_-^- = 1 + sT$. The solution of the DE for $j = 0$ is

$$\mathcal{R} = \frac{2T}{T_w + T} = \frac{2}{1+a} = b; \quad \mathcal{K} = \frac{T_w - T}{T_w + T} = \frac{a-1}{a+1}; \quad a = \frac{T_w}{T} \tag{4.2.11}$$

The $\mathcal{E}_2$ optimal (also the $\mathcal{H}_2$ optimal) filter G_x is

$$G_x = \frac{T_w - T}{T_w + T}\frac{1}{1+sT} = \frac{a-1}{a+1}\frac{1}{1+sT} \tag{4.2.12}$$

It can be easily checked that the gain of $G_x\hat{\mathcal{B}}_-$ is now not equal to unity and the integrating regulator cannot be obtained. Solve the task for $j = 1$. The solution of the DE is

$$\mathcal{R} = \frac{2T^2}{T_w + T} = \frac{2T}{1+a} = bT; \quad b = \frac{2T}{T_w + T} = \frac{2}{1+a} \tag{4.2.13}$$

$$\mathcal{K} = 1 + T_g s = 1 + bT_w s; \qquad T_g = bT_w$$

The optimal filter obtained is

$$G_x = \frac{1 + T_g s}{1 + sT} \tag{4.2.14}$$

It can be easily verified that the gain of $G_x\hat{\mathcal{B}}_-$ is unity and therefore an integrating YP regulator is obtained.

Optimization Method Using the $\mathcal{E}_\infty$ Norm

The general supreme norm introduced in (A.5.23) is used for optimization in the following form:

$$J^x_{\mathcal{E}_\infty} = \left\| Y_x\left(R_x - G'_x P_-\right)\right\|_{\mathcal{E}_\infty} = \left\| \frac{1}{s^j}\left(R_x - G'_x P_-\right)\right\|_{\mathcal{E}_\infty} \tag{4.2.15}$$

This norm is equal to the $\mathcal{H}_2$ norm for $j = 0$

$$J^{\mathrm{x}}_{\mathcal{E}_\infty} == \begin{cases} \mathcal{H}_\infty \text{ norm}; & j = 0 \\ \mathcal{E}_\infty \text{ norm}; & j \geq 1 \end{cases} \tag{4.2.16}$$

The optimality criterion (4.2.15) measures the maximum absolute value of the frequency characteristic of the error signal $\hat{\varepsilon}_{\mathrm{x}} = Y_{\mathrm{x}}(R_{\mathrm{x}} - G'_{\mathrm{x}}P_-)$ obtained for the general excitation $Y_{\mathrm{x}} = \mathcal{L}(y_{\mathrm{x}}) = s^{-j}$ of the *YP* model-based closed control loop (see Figure 4.2.2), where the subscript x refers to the disturbance rejection or tracking task. Thus, the task of the optimization is the solution of the minimum seeking as

$$G_{\mathrm{x}} = \arg\left\{\min_{G_{\mathrm{x}}}\left[J^{\mathrm{x}}_{\mathcal{E}_\infty}\right]\right\} = \arg\left\{\min_{G_{\mathrm{x}}}\left[\|Y_{\mathrm{x}}S_{\mathrm{x}}\|_{\mathcal{E}_\infty}\right]\right\}$$
$$= \arg\left\{\min_{G_{\mathrm{x}}}\left[\left\|Y_{\mathrm{x}}\left(R_{\mathrm{x}} - G'_{\mathrm{x}}P_-\right)\right\|_{\mathcal{E}_\infty}\right]\right\} \tag{4.2.17}$$

Here the reduced embedded filter G' introduced according to Figure 4.1(b), and expression (4.40) is assumed.

The minimization task is solved in the framework of the optimal interpolation theory, which is known as the Nevanlinna–Pick problem [134,135] in the literature. If $\hat{P}_-$ has m_- zeros then the supreme minimum W^{o} of the general $\|W\|_{\mathcal{E}_\infty}$ can be sought in the form

$$W^{\mathrm{o}} = \begin{cases} \mu\dfrac{\mathcal{H}^{\#}}{\mathcal{H}}, & \text{if } m_- \geq 1 \\ 0, & \text{if } m_- = 0 \end{cases} \tag{4.2.18}$$

Here $\mathcal{H}$ is an $(m_- - 1)$ order Hurwitz polynomial [158]. The polynomial $\mathcal{H}^{\#}$ can be obtained by mirroring the zeros of $\mathcal{H}$ on the imaginary axis

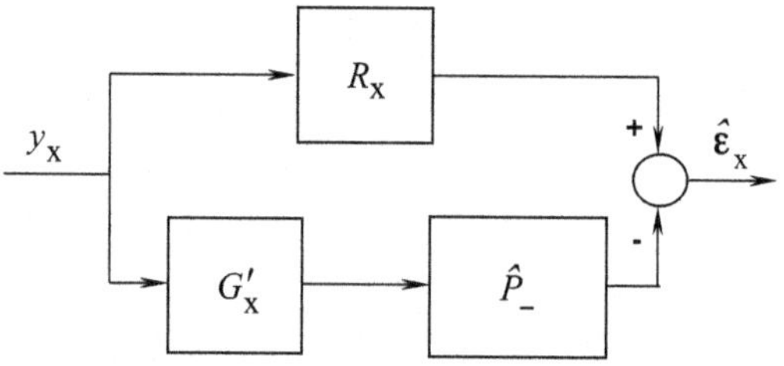

Figure 4.2.2 Scheme used for the optimization of the realizability loss $\mathcal{E}_\infty$.

(see the derivation of $\mathcal{B}_-^-$). The constant μ and the coefficients of $\mathcal{H}$ can be clearly determined by the so-called interpolation constraints of m_- pieces

$$W(z_i) = \mu \frac{\mathcal{H}^{\#}(z_i)}{\mathcal{H}(z_i)} = \frac{1}{s^j}\left(R_x - G'_x \hat{P}_-\right)\Big|_{z_i} = \frac{R_x(z_i)}{z_i^j} = r_i; \quad i = 1, 2, \ldots, m_- \tag{4.2.19}$$

where the disjunct zeros of $\hat{P}_-$ are denoted by $z_1, z_2, \ldots, z_{m_-}$. (The multiplicity of the zeros can be simply handled by taking further constraints into consideration.) The $(m_- - 1)$ coefficients of $\mathcal{H}$ and μ, together, mean m_- unknown parameters, which can be clearly determined from the m_- equations of (4.2.19). Note that this means a nonlinear equation system. By rearranging (4.2.19), we see that $\hat{P}_-$ is the divider of the expression $R_x - s^j W$, so

$$R_x - s^j W = \frac{\mathcal{B}_x}{\mathcal{A}_x} - \mu \frac{s^j \mathcal{H}^{\#}}{\mathcal{H}} = \frac{\mathcal{N}}{\mathcal{D}} \hat{P}_- = \frac{\mathcal{N}}{\mathcal{D}} \hat{\mathcal{B}}_- \tag{4.2.20}$$

can be written, where $R_x = \mathcal{B}_x / \mathcal{A}_x$ means the reference model, $\hat{P}_- = \hat{\mathcal{B}}_-$ the invariant (*IU*) factor. The embedded filter is obtained as

$$G'_x = \frac{\mathcal{N}}{\mathcal{D}} \tag{4.2.21}$$

By comparing the two sides of (4.2.20), we obtain the expressions for calculating the denominator and numerator of G'_x

$$\mathcal{D} = \mathcal{A}_x \mathcal{H} \tag{4.2.22}$$

and

$$\mathcal{N} = \frac{\mathcal{B}_x \mathcal{H} - \mathcal{A}_x \mu s^j \mathcal{H}^{\#}}{\hat{\mathcal{B}}_-} \tag{4.2.23}$$

Because of features mentioned earlier, the division must not have residue. To simplify the numerical solution of Equation (4.2.19), the equations can be arranged into a *quasi-linear* form as

$$\mu \mathcal{H}^{\#}(z_i) = r_i \mathcal{H}(z_i); \quad i = 1, 2, \ldots, m_- \tag{4.2.24}$$

which can be easily written in matrix form as

$$\boldsymbol{r} = \boldsymbol{F}(\mu, \boldsymbol{r}, z)\boldsymbol{h}(\mu, \mathcal{H}) \tag{4.2.25}$$

where the following notations are introduced

$$\boldsymbol{F}(\boldsymbol{r}, \mu, \boldsymbol{z}) = \begin{bmatrix} 1 & f_{11} & \dots & f_{1m_1} \\ 1 & f_{22} & \dots & f_{2m_1} \\ \vdots & \vdots & \ddots & \vdots \\ 1 & f_{m_-1} & \dots & f_{m_-m_1} \end{bmatrix} \tag{4.2.26}$$

$$\boldsymbol{r} = \begin{bmatrix} r_1 & \dots & r_{m_-} \end{bmatrix}^{\mathrm{T}}; \quad \boldsymbol{z} = \begin{bmatrix} z_1 & \dots & z_{m_-} \end{bmatrix}^{\mathrm{T}} \tag{4.2.27}$$

$$\boldsymbol{h}(\mu, h) = [\mu, h_1, \dots, h_{m1}]^{\mathrm{T}}; \quad m_1 = m_- - 1 \tag{4.2.28}$$

The coefficients of the matrix $\boldsymbol{F}(\mu, \boldsymbol{r}, \boldsymbol{z})$ are

$$f_{ik} = \sum_{k=1}^{m_1} \left[\mu \left(z_i^* \right)^k - r_i (z_i)^k \right]; \quad z_i^* = \operatorname{sign}(h_i) z_i \tag{4.2.29}$$

The *quasi-linear* designation is used because the elements f_{ik} also depend on μ. Several iterative methods can be used to solve the nonlinear equations system (4.2.25). Here, the so-called relaxation method is considered as it ensures fairly fast convergence in practice. The form of the iterative algorithm in the l-th step is

$$\boldsymbol{h}(\mu_{l+1}, h_{l+1}) = [\boldsymbol{F}(\mu_l, \boldsymbol{r}, \boldsymbol{z})]^{-1} \boldsymbol{r} \tag{4.2.30}$$

The iterative method provides μ and the polynomial $\mathcal{H}$ in $\boldsymbol{h}$. $\mathcal{D}$ and $\mathcal{N}$ are computed by (4.2.22) and (4.2.23). In the case of a few (one or two) *IU* zeros, Eqn (4.2.24) can be solved by the manual method.

The minimum of the loss function is simply obtained from (4.2.17), since

$$\min\left\{ \|W(j\omega)\|_{\mathcal{E}_\infty} \right\} = \|W^{\mathrm{o}}(j\omega)\|_{\mathcal{E}_\infty} = \mu \tag{4.2.31}$$

Investigate the product $G'_{\mathrm{x}} \hat{\mathcal{B}}_-$, key element providing the integrating effect

$$\begin{aligned} G'_{\mathrm{x}} \hat{\mathcal{B}}_- &= \frac{\mathcal{B}_{\mathrm{x}} \mathcal{H} - \mathcal{A}_{\mathrm{x}} \mu s^j \mathcal{H}^\#}{\mathcal{A}_{\mathrm{x}} \mathcal{H}} = \frac{\mathcal{B}_{\mathrm{x}}}{\mathcal{A}_{\mathrm{x}}} - \frac{\mu \mathcal{H}^\#}{\mathcal{H}} s^j \quad \text{or} \\ G_{\mathrm{x}} \hat{\mathcal{B}}_- &= \frac{\mathcal{A}_{\mathrm{x}}}{\mathcal{B}_{\mathrm{x}}} \frac{\mathcal{B}_{\mathrm{x}} \mathcal{H} - \mathcal{A}_{\mathrm{x}} \mu s^j \mathcal{H}^\#}{\mathcal{A}_{\mathrm{x}} \mathcal{H}} = 1 - \frac{\mathcal{A}_{\mathrm{x}} \mu \mathcal{H}^\#}{\mathcal{B}_{\mathrm{x}}} s^j \end{aligned} \tag{4.2.32}$$

which clearly shows that $G'_x\hat{\mathcal{B}}_-\big|_{s\to 0} = G_x\hat{\mathcal{B}}_-\big|_{s\to 0} = 1$ is fulfilled only if $j \geq 1$. Thus, the minimization of the $\mathcal{H}_\infty$ norm according to the classical criterion $\left\| S^x_{\text{real}} \right\|_\infty$ cannot provide integrating control.

Example 4.2.2

Consider again the reference model $R_x = 1/(1 + sT_w)$ and the *IU* term $\mathcal{B}_- = 1 - sT$. Since $m_- = 1$, the interpolation polynomial $\mathcal{H}$ is of zero-order, i.e., $\mathcal{H} = \mathcal{H}^\# = 1$ is constant. μ is simply obtained from the interpolation constraint (4.2.24)

$$\mu = R_x\left(z_1 = \frac{1}{T}\right) = \frac{T}{T_w + T} \tag{4.2.33}$$

and from (4.2.21) $\mathcal{D} = 1 + sT_w$. Consider first the case of $j = 0$. As regards the orders of the right-hand side polynomials of (4.2.23), only the scalar solution $\mathcal{N} = k$ can be accepted. Therefore,

$$1 - k(1 - sT) = (1 + sT_w)\mu \tag{4.2.34}$$

which gives the μ above, and

$$\mathcal{N} = k = \frac{T_w}{T_w + T} \tag{4.2.35}$$

Finally, the embedded filter is obtained as a result of the optimization

$$G'_x = \frac{\mathcal{N}}{\mathcal{D}} = \frac{\mathcal{N}}{\mathcal{A}_x\mathcal{H}} = \frac{T_w}{T_w + T}\frac{1}{1 + sT_w} \tag{4.2.36}$$

which is also $\mathcal{H}_\infty$ optimal. It is clear that $G'_x\hat{\mathcal{B}}_-\big|_{s\to 0} \neq 1$, and thus it is not integrating.

Example 4.2.3

Consider now the case $j = 1$. The computation of $\mathcal{D}$ remains and only (4.2.24) changes as

$$1 - \mathcal{N}(1 - sT) = (1 + sT_w)\mu s^j \tag{4.2.37}$$

and the first order solution $\mathcal{N} = k(1 + s\tau)$ is sought in the form

$$1 - k(1 + s\tau)(1 - sT) = (1 + sT_w)\mu s^j \tag{4.2.38}$$

By comparing the coefficients, we obtain

$$k = 1; \quad \mu = \frac{T}{T_w + T} T; \quad \tau = \frac{T_w}{T_w + T} T \tag{4.2.39}$$

The resulting optimal inner filter is

$$G_x = \frac{\mathcal{N}}{\mathcal{D}} = \frac{\mathcal{N}}{\mathcal{A}_x \mathcal{H}} = \frac{1 + s\tau}{1 + sT_w} \tag{4.2.40}$$

where $G'_x \hat{\mathcal{B}}_-\big|_{s\to 0} = 1$ and thus the optimal regulator is integrating.

Example 4.2.4

Consider now a second-order *IU* polynomial $\hat{\mathcal{B}}_- = (1 - sT_1)(1 - sT_2)$, whose time constants are $T_1 = 1$ and $T_2 = 2$. Choose again a first-order reference model with time constant $T_w = 0.5$. In this case $\mathcal{H}(s) = 1 + h_1 s$ is of first order, since $m_- = 2$ and $m_1 = 1$. This is already a real nonlinear case that cannot be solved by elementary methods, although an analytical solution can be assumed. Equation (4.2.24) can be rearranged into a second-order nonlinear equation system as

$$\begin{aligned} &\mu^2 + \mu \frac{T_1 + T_2}{T_1 - T_2}(r_2 - r_1) - r_1 r_2 = 0 \\ &h_1^2 + h_1 \frac{r_2 + r_1}{r_2 - r_1}(T_2 - T_1) - T_1 T_2 = 0 \end{aligned} \tag{4.2.41}$$

As regards the roots (both μ and h_1 must be positive) the following solution is obtained for the case $j = 0$

$$\mu = 0.9572; \quad h_1 = 0.1789 \tag{4.2.42}$$

$\mathcal{N}$ is obtained by a division according to (4.2.24), which is very simple now, and $\mathcal{N} = 0.0428$. The $\mathcal{E}_\infty$ optimal embedded filter is

$$G'_x = \frac{\mathcal{N}}{\mathcal{D}} = \frac{\mathcal{N}}{\mathcal{A}_x \mathcal{H}} = \frac{0.0428}{(1 + 0.5s)(1 + 0.1789s)}; \quad G_x = \frac{0.0428}{1 + 0.1789s} \tag{4.2.43}$$

The iterative numerical method (4.2.30) provides the following parameters for the case of $j = 1$:

$$\mu = 3.1397; \quad h_1 = 0.6498 \tag{4.2.44}$$

$\mathcal{N}$ is obtained according to (4.2.24)

$$\mathcal{N} = 1 + 0.5s \tag{4.2.45}$$

The $\mathcal{E}_\infty$ optimal inner filter is

$$G'_x = \frac{\mathcal{N}}{\mathcal{A}_x h} = \frac{1+0.5s}{(1+0.5s)(1+0.6498s)} = \frac{1}{1+0.6498s};$$
$$G_x = \frac{1+0.5s}{1+0.6498s} \tag{4.2.46}$$

Sampled Data Systems

The above optimization problems are solved for CT systems. The process is assumed to be delay free. Next, the optimization of the criterion $\mathcal{E}_2$ for processes with time delay is solved for DT systems, whose pulse transfer function, according to (1.1.32), takes the form

$$G(z^{-1}) = \frac{g_1 + g_2 z^{-1} + \ldots + g_n z^{-(n-1)}}{1 + f_1 z^{-1} + f_2 z^{-2} + \ldots + f_n z^{-n}} z^{-(d+1)}$$
$$= \frac{\mathcal{G}_+\mathcal{G}_-}{\mathcal{F}} z^{-d} \text{ and } \hat{G} = \frac{\hat{\mathcal{G}}_+\hat{\mathcal{G}}_-}{\hat{\mathcal{F}}} z^{-d_m} \tag{4.2.47}$$

where $\hat{G}$ is the model. To understand DT systems, consider Figure 4.2.3, where the expressions of the d-step ahead predictions are also presented.

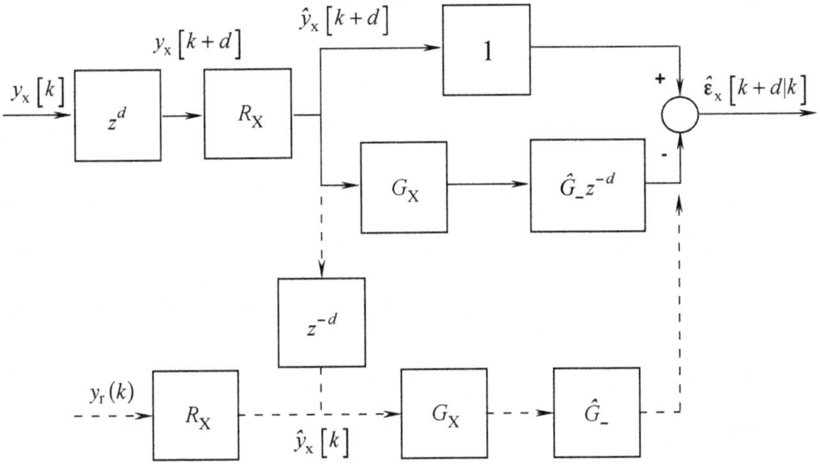

Figure 4.2.3 Explanation of a d-step ahead prediction to optimize the realizability loss $\mathcal{E}_2$.

The task, analogous with the optimization in (4.2.2), is formulated now as

$$\begin{aligned} G_x &= \arg\left\{\min_{G_x}\left[\left\|R_x\left(1 - G_x\hat{\mathcal{G}}_- z^{-d}\right)\right\|_{\mathcal{H}_2}\right]\right\} \\ &= \arg\left\{\min_{G_x}\left[\left\|R_x\left(1 - G_x\hat{\mathcal{G}}_- z^{-d}\right)\right\|_{\mathcal{H}_2}\right]\right\} \end{aligned} \tag{4.2.48}$$

where the simplest cases $Y_x = 1$ and $d_m = d$ are considered and the subscript x may refer either to the reference signal (r) or the output noise (n). In this task, the norms $\mathcal{E}_2$ and $\mathcal{H}_2$ are equal. Formulate the $\mathcal{H}_2$ norm for the error signal $\hat{\boldsymbol{\varepsilon}}_x$

$$\begin{aligned} J^x_{\text{real}} &= \left\|R_x\left(1 - G_x\hat{\mathcal{G}}_- z^{-d}\right)\right\|_{\mathcal{H}_2} = \left\|\frac{\hat{\mathcal{G}}_- z^{-d}}{\hat{\mathcal{G}}_-^-}\right\|_{\mathcal{H}_2}\left\|\frac{\mathcal{B}_x\hat{\mathcal{G}}_-^-}{\mathcal{A}_x\hat{\mathcal{G}}_- z^{-d}} - G_x\frac{\mathcal{B}_x\hat{\mathcal{G}}_-^-}{\mathcal{A}_x}\right\|_{\mathcal{H}_2} \\ &= \left\|\frac{\mathcal{B}_x\hat{\mathcal{G}}_-^-}{\mathcal{A}_x\hat{\mathcal{G}}_- z^{-d}} - G_x\frac{\mathcal{B}_x\hat{\mathcal{G}}_-^-}{\mathcal{A}_x}\right\|_{\mathcal{H}_2} \end{aligned} \tag{4.2.49}$$

Contrary to the CT systems, $\hat{\mathcal{G}}_-^-$ is now a polynomial and is derived by mirroring the zeros of $\hat{\mathcal{G}}_-$ on the unit circle, thus the condition $z_i^* = 1/z_i$ must hold for the new zeros z_i^*. In (4.2.8), the expressions $\left|z^{\pm jd\omega}\right| = 1$ and $\left\|\hat{\mathcal{G}}_-/\hat{\mathcal{G}}_-^-\right\|_{\mathcal{H}_2} = 1$ are taken into consideration, which means that they are norm-keeping factors (this fact can easily be checked by simple examples). The minimization of (4.2.47) is repeated by the Wiener method, i.e., the argument is decomposed for causal and noncausal parts

$$\frac{\mathcal{B}_x\hat{\mathcal{G}}_-^-}{\mathcal{A}_x\hat{\mathcal{G}}_- z^{-d}} = \frac{\mathcal{R}}{\hat{\mathcal{G}}_- z^{-d}} + \frac{\mathcal{K}}{\mathcal{A}_x} = \frac{\mathcal{A}_x\mathcal{R} + \mathcal{K}\hat{\mathcal{G}}_- z^{-d}}{\mathcal{A}_x\hat{\mathcal{G}}_- z^{-d}} \tag{4.2.50}$$

where $\mathcal{R}/\hat{\mathcal{G}}_- z^{-d}$ is a noncausal term, $\mathcal{K}/\mathcal{A}_x$ a causal term, however. Then it can be written that

$$J^x_{\text{real}} = \left\|\frac{\mathcal{R}}{\hat{\mathcal{G}}_- z^{-d}}\right\|_{\mathcal{H}_2} + \left\|\frac{\mathcal{K}}{\mathcal{A}_x} - G_x\frac{\mathcal{B}_x\hat{\mathcal{G}}_-^-}{\mathcal{A}_x}\right\|_{\mathcal{H}_2} \geq \left\|\frac{\mathcal{R}}{\hat{\mathcal{G}}_- z^{-d}}\right\|_{\mathcal{H}_2} \tag{4.2.51}$$

The loss function has its minimum, when the second term is zero, and thus

$$\frac{\mathcal{K}}{\mathcal{A}_x} - G_x\frac{\mathcal{B}_x\hat{\mathcal{G}}_-^-}{\mathcal{A}_x} = 0 \tag{4.2.52}$$

The optimality of the $\mathcal{H}_2$ norm can be guaranteed by the following embedded filter:

$$G_x = \frac{\mathcal{K}}{\mathcal{B}_x \hat{\mathcal{G}}_-^-} \tag{4.2.53}$$

$\mathcal{R}$ and $\mathcal{K}$ can be obtained by solving a special *DE*

$$\mathcal{B}_x \hat{\mathcal{G}}_-^- = \mathcal{A}_x \mathcal{R} + \mathcal{K} \hat{\mathcal{G}}_- z^{-d} \tag{4.2.54}$$

This equation can be derived by comparing the two sides of (4.2.50).

The $\mathcal{H}_2$ optimal *YP* regulator can be obtained by using the expressions (2.1.25) and (4.2.53)

$$C_* = \frac{\mathcal{K}\hat{\mathcal{F}}}{\hat{\mathcal{G}}_+ \left(\mathcal{A}_n \hat{\mathcal{G}}_-^- - \mathcal{K}\hat{\mathcal{G}}_- z^{-d}\right)} \tag{4.2.55}$$

The minimum of (4.2.49) is formally the same as (4.41), but now $\mathcal{R}$ comes from (4.2.54). Thus, the minimum of the loss function $J_{\mathrm{real}}^{\mathrm{x}}$ is

$$\begin{aligned} \min_{G_x} \left\{ J_{\mathrm{real}}^{\mathrm{x}} \right\} &= \left\| \frac{\mathcal{R}}{\hat{\mathcal{G}}_- z^{-d}} \right\|_{\mathcal{H}_2} = \left\| \frac{\hat{\mathcal{G}}_- z^{-d}}{\hat{\mathcal{G}}_-^-} \right\|_{\mathcal{H}_2} \left\| \frac{\mathcal{R}}{\hat{\mathcal{G}}_- z^{-d}} \right\|_{\mathcal{H}_2} \\ &= \left\| \frac{\mathcal{R}}{\hat{\mathcal{G}}_-^-} \right\|_{\mathcal{H}_2} = \left\| R_x \left(1 - G_x \hat{\mathcal{G}}_- z^{-d}\right) \right\|_{\mathcal{H}_2} \end{aligned} \tag{4.2.56}$$

where the form $\mathcal{R}/\hat{\mathcal{G}}_-^-$ is rearranged as

$$\frac{\mathcal{R}}{\hat{\mathcal{G}}_-^-} = \frac{\mathcal{B}_x \hat{\mathcal{G}}_-^- - \mathcal{K}\hat{\mathcal{G}}_- z^{-d}}{\mathcal{A}_x \hat{\mathcal{G}}_-^-} = R_x \left(1 - G_x \hat{\mathcal{G}}_- z^{-d}\right) \tag{4.2.57}$$

and the *DE* in (4.2.54) is used. Thus, the task (4.2.49) actually minimizes the $\mathcal{H}_2$ norm of the error $\hat{\varepsilon}[k+d|k]$ obtained from the d-step ahead prediction under the excitation y_x

$$\hat{\varepsilon}[k+d|k] = R_x \left(1 - G_x \hat{\mathcal{G}}_- z^{-d}\right) y_x[k] \tag{4.2.58}$$

The condition for the integrating regulator can now be investigated in the form $G_x \hat{\mathcal{G}}_-|_{z \to 1} = 1$ valid for DT systems and assuming the excitation $Y_x(z) = (z-1)^{-j}$. Of course, in these cases the $\mathcal{E}_2$ norm is always applied, as it was made at the optimization of the CT systems.

The product $\mathcal{B}_x \hat{\mathcal{G}}_-^-$ is at the left side of the DE (4.2.54). If the term $\hat{\mathcal{G}}_-^-$, derived by mirroring the zeros, is arranged according to the powers of z^{-1}, similarly to other polynomials, then the form $\hat{\mathcal{G}}_-^- = z^{m_-}\hat{\hat{\mathcal{G}}}_-^-$ is obtained where m_- is the number of the IU zeros. Therefore, it is reasonable to assume the form $\mathcal{B}_x = \overline{\mathcal{B}}_x z^{-m_-}$ in the numerator of the reference model in order to ensure the solvability of the DE. (In exceptional cases, solutions can be obtained for other forms of $\mathcal{B}_x$, but the generality cannot be guaranteed.) The term z^{-m_-} introduced in the reference model means that the time delay is virtually increased to the value $d' = d + m_-$.

4.3 OPTIMIZATION OF MODELING LOSS

It was shown in Chapter 4 that the contribution of modeling loss to the sensitivity function (4.5) of the closed loop is

$$S_{\text{mod}} = S - \hat{S} = \hat{T} - T = -\frac{\hat{T}\hat{S}\ell}{1 + \hat{T}\ell} = -\hat{T}S\ell\big|_{\ell\to 0} \approx -\hat{T}\hat{S}\ell \tag{4.3.1}$$

and its contribution to the sensitivity function (4.8) valid for the tracking properties is

$$S_{\text{mod}}^{\text{r}} = \hat{T}_{\text{r}} - T_{\text{r}} = -\frac{\hat{T}_{\text{r}}\hat{S}\ell}{1 + \hat{T}\ell} = -\hat{T}_{\text{r}}S\ell\big|_{\ell\to 0} \approx -\hat{T}_{\text{r}}\hat{S}\ell = H_{\text{KB}}\ell \tag{4.3.2}$$

In these forms, the weighting filters $|\hat{T}\hat{S}|$ and $|\hat{T}_{\text{r}}\hat{S}|$ generally have their minima in the medium frequency region, since $\hat{T}$ and $\hat{T}_{\text{r}}$ are low-pass filters. $\hat{S}$ is, however, a high-pass filter. $|\hat{T}\hat{S}|$ has its maximum at the cut-off frequency ω_c, and therefore the model should be most accurate in the region of these frequencies.

It is an interesting phenomenon that in the case of the Youla-parameterized regulators this weighting changes to the form

$$S_{\text{mod}}^{*} \approx -Q\hat{P}\ell = \frac{1}{S}S_{\text{mod}} = -\hat{T}\ell \tag{4.3.3}$$

Here the maximum of $|\hat{T}|$ is usually not at the cutoff frequency ω_c and its value is close to unity in a relatively wide range of the low-frequency region. With an iterative method, it is reasonable to use the H_{KB} in (4.3.2) first and then, through gradual correction of the model, change to $\hat{T} = R_n G_n \hat{P}_-$.

It is important to note, for all three forms above, that the minimization techniques of the norms applied in the previous two sections cannot be used here to determine the best model $\hat{P}$, since the values of all three sensitivity terms are zero in the ideal case $\hat{P} = P$ $(\ell = 0)$. A criterion should be found which fits the practical modeling tasks. One such method is a so-called min-max task, according to which the optimization of the modeling loss $J^{\mathrm{r}}_{\mathrm{mod}}$ means the application of a special, optimal excitation $y_{\mathrm{r}} = y^{\mathrm{opt}}_{\mathrm{r}}$ as reference signal, followed by the determination of the optimal process model $\hat{P} = \hat{P}^{\mathrm{opt}}$ obtained as the solution of the following min-max problem

$$\hat{P}^{\mathrm{opt}} = \arg\left\{\min_{\hat{P}}\left[\max_{y_{\mathrm{r}}}\left(J^{\mathrm{r}}_{\mathrm{mod}}\right)\right]\right\} = \arg\left\{\min_{\hat{P}}\left[\max_{y_{\mathrm{r}}}\left\|S^{\mathrm{r}}_{\mathrm{mod}}\right\|\right]\right\} \tag{4.3.4}$$

This task has two steps: the optimal reference signal (depending on the criterion) usually ensures maximum output variances under amplitude-restricted y_{r}. Measuring the resulting output of the closed system, the process model should be determined by a certain modeling (identification) method, which provides the minimum of the chosen optimality criterion $\left\|S^{\mathrm{r}}_{\mathrm{mod}}\right\|$ of the modeling error. The task (4.3.4) is called the *worst case* identification task.

Consider the model-based version of Figure 2.1.7 where the regulator $\hat{C}$ and the inner filters $\hat{K}_{\mathrm{r}}$ and $\hat{K}_{\mathrm{n}}$ are computed from the model of the process $\hat{P}$.

The model-based Youla regulator is obtained from (2.1.10) and (4.16) as

$$\hat{C} = \frac{\hat{Q}}{1 - \hat{Q}\hat{P}} = \frac{R_{\mathrm{n}}\hat{K}_{\mathrm{n}}}{1 - R_{\mathrm{n}}\hat{K}_{\mathrm{n}}\hat{P}} \tag{4.3.5}$$

where

$$\hat{Q} = R_{\mathrm{n}}G_{\mathrm{n}}\hat{P}_{+}^{-1} = R_{\mathrm{n}}\hat{K}_{\mathrm{n}}; \quad \hat{K}_{\mathrm{n}} = G_{\mathrm{n}}\hat{P}_{+}^{-1} \tag{4.3.6}$$

$$\hat{Q}_{\mathrm{r}} = R_{\mathrm{r}}G_{\mathrm{r}}\hat{P}_{+}^{-1} = R_{\mathrm{r}}\hat{K}_{\mathrm{r}}; \quad \hat{K}_{\mathrm{r}} = G_{\mathrm{r}}\hat{P}_{+}^{-1} \tag{4.3.7}$$

In many practical cases the identification of the process model should be performed without opening the closed loop. Several methods exist (see later in Chapter 10), but here a method is discussed which uses the advantages and special features of *KB* parameterization (it virtually opens the closed loop). According to this method, the identification is performed between the internal virtual input $\hat{u}$ and the output y of the closed loop shown in Figure 4.3.1. Besides the favorable prediction features discussed in Chapter 10, the problem of the *circulating noise* can be avoided by this method. The strange thing about this method is that the value (absolute value) of the control error e is equal to the identification error $\boldsymbol{\varepsilon}_{\mathrm{ID}} = \boldsymbol{\varepsilon}_{\mathrm{KB}}$. For DT system

$$\begin{aligned}\boldsymbol{\varepsilon}_{\mathrm{ID}} = \boldsymbol{\varepsilon}_{\mathrm{KB}} &= -e = y - y_{\mathrm{m}} = y - \hat{P}\hat{u}\\ &= \left.\frac{\left(R_{\mathrm{r}} G_{\mathrm{r}} \hat{P}_{-} z^{-d_{\mathrm{m}}}\right)\left(1 - R_{\mathrm{n}} G_{\mathrm{n}} \hat{P}_{-} z^{-d_{\mathrm{m}}}\right)}{1 + R_{\mathrm{n}} G_{\mathrm{n}} \hat{P}_{-} z^{-d_{\mathrm{m}}} \ell}\right|_{\ell \to 0} \ell y_{\mathrm{r}}\\ &\approx \left(R_{\mathrm{r}} G_{\mathrm{r}} \hat{P}_{-} z^{-d_{\mathrm{m}}}\right)\left(1 - R_{\mathrm{n}} G_{\mathrm{n}} \hat{P}_{-} z^{-d_{\mathrm{m}}}\right)\ell y_{\mathrm{r}} = H_{\mathrm{KB}} \ell y_{\mathrm{r}}\end{aligned} \tag{4.3.8}$$

where $d_{\mathrm{m}} \neq d$ is the time delay of the model. A further fact is that in the case of $\ell \to 0$ the weighting filters in both (4.3.2) and (4.3.8) are the same and denoted by H_{KB}. Thus, the model output

$$y_{\mathrm{m}} = \hat{P}\hat{u} \neq \hat{P}u = \hat{y} \tag{4.3.9}$$

is not equal to the model output $\hat{y}$ seen in the previous sections.

The above method can be summarized as the determination of the optimal model $\hat{P}^{\mathrm{opt}}$ belonging to the model class $\mathbb{M}$, and can be written as

$$\hat{P}^{\mathrm{opt}} = \arg\left\{ \min_{\hat{P} \in \mathbb{M}} \|\boldsymbol{\varepsilon}_{\mathrm{ID}}\|_{y_{\mathrm{r}} = \arg\left\{ \sup_{y_{\mathrm{r}}} \|\boldsymbol{\varepsilon}_{\mathrm{ID}}\| \right\}} \right\} \tag{4.3.10}$$

where the norm to be applied has not yet been noted. Thus, the optimal reference signal $y_{\mathrm{r}}^{\circledast}$ is obtained via the solution of the supremum problem

$$y_{\mathrm{r}}^{\circledast} = \arg\left\{ \sup_{y_{\mathrm{r}} \in \mathbb{U}} \left\| H_{\mathrm{KB}} y_{\mathrm{r}} \right\| \right\} \tag{4.3.11}$$

Here $\mathbb{U}$ is the region of the usually amplitude-restricted input signals ($\mathbb{U}$: $|y_r| \leq 1$).

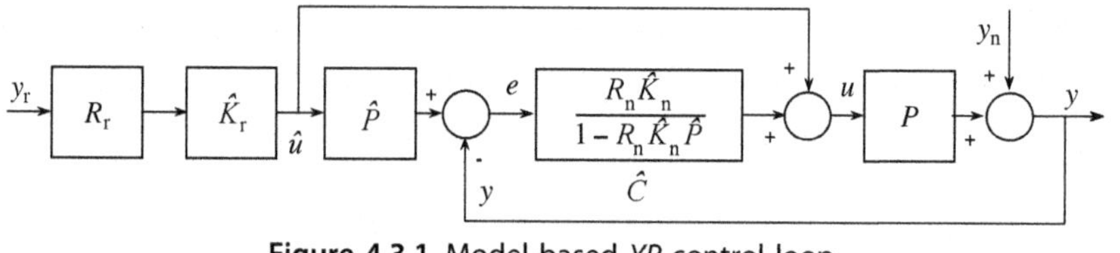

Figure 4.3.1 Model-based *YP* control loop.

If the process is *IS*, then we can select $G_n\hat{P}_- = 1$ and $G_r\hat{P}_- = 1$, and thus

$$H_{KB} = R_r z^{-d_m}\left(1 - R_n z^{-d_m}\right) \tag{4.3.12}$$

depends only on the design requirements, and so on the selection of R_n and R_r. Therefore, in this case the optimal exciting reference signal can be determined before the identification, and it can be applied in each step.

If the process is *IU*,

$$H_{KB} = \left(R_r G_r \hat{P}_- z^{-d_m}\right)\left(1 - R_n G_n \hat{P}_- z^{-d_m}\right) \tag{4.3.13}$$

it depends on the invariant component $\hat{P}_-$ of the process. Obviously in this case only a learning adaptive procedure can be applied, when in each step the lately obtained $\hat{P}_-$ is used to compute H_{KB} and the optimal reference signal is generated advisedly.

Consider the solution of the supreme problem. Given the expressions (A.5.17), (A.5.19), and (A.5.20) in the appendix it can be stated that the $\mathcal{H}_\infty$ norm is closely connected to the gain referring to the norms $\mathcal{L}_2$, $\mathcal{L}_\infty$ of the input and output signals of a linear filter

$$\|H(j\omega)\|_\infty \geq \frac{\|y(t)\|_2}{\|u(t)\|_2} \tag{4.3.14}$$

or (see Appendix 5)

$$\|H(j\omega)\|_\infty \geq \frac{\text{pow}[y(t)]}{\text{pow}[u(t)]}; \quad \|H(j\omega)\|_\infty \geq \frac{\text{pow}[y(t)]}{\|u(t)\|_\infty} \tag{4.3.15}$$

In the case of amplitude constrained input signals, these latter expressions indicate that the biggest lower limit for the $\|H(j\omega)\|_\infty$ norm can be obtained by maximizing the variance of the filter output $y(t)$.

The signal of maximum variance is generated in DT systems where the input of the filter $H_{KB}(z)$ is $y_r[k]$, and its output is denoted by $v[k]$. Thus,

$$\begin{aligned} v[k] &= H_{KB} y_r[k] = \left(1 - R_n G_n \hat{P}_- z^{-d_m}\right) R_r G_r \hat{P}_- z^{-d_m} y_r[k] \\ &= \frac{g_0 + \tilde{\mathcal{G}}(z^{-1})}{1 + \tilde{\mathcal{D}}(z^{-1})} z^{-d_m} y_r[k] \end{aligned} \tag{4.3.16}$$

It can be rewritten into a difference equation linear in parameters as

$$\begin{aligned} v[k + d_m] &= g_0 y_r(k) + \tilde{\mathcal{G}}(z^{-1}) y_r(k-1) + \tilde{\mathcal{D}}(z^{-1}) v[k + d_m - 1] \\ &= g_0 y_r[k] + \boldsymbol{g}^T[k-1]\boldsymbol{q} = g_0 y_r[k] + b \end{aligned} \tag{4.3.17}$$

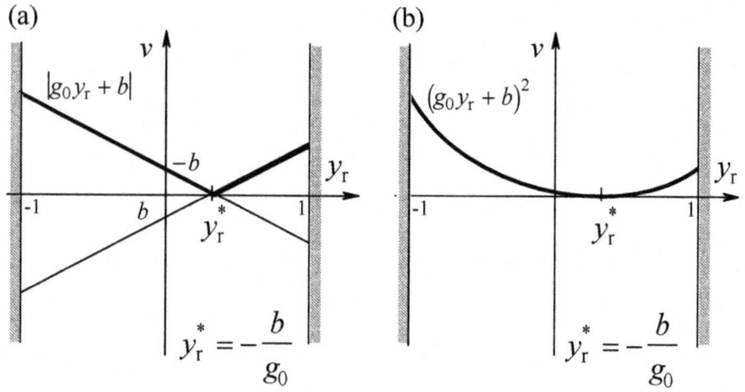

Figure 4.3.2 Derivation of the algorithm providing optimal reference signal.

Here

$$\begin{aligned} \boldsymbol{q} &= [g_0, g_1, \ldots; d_1, d_2, \ldots]^{\mathrm{T}} \\ \boldsymbol{g}[k-1] &= [y_r[k-1], \ldots; -v[k+d_m-1], \ldots]^{\mathrm{T}} \end{aligned} \tag{4.3.18}$$

The length of the polynomials $\mathcal{G}$ and $\mathcal{D}$ depends on the pulse transfer functions R_r, R_n, G_r, G_n, $\hat{P}_-$, and d_m. To generate the so-called locally optimal reference signal series $y_r^{\circledast}[k]$, the following task should be solved

$$\begin{aligned} y_r^{\circledast}[k] &= \arg \sup_{\|y_r\|_{\mathcal{L}_2}=1} \|v[k+d_m]\|_{\mathcal{L}_2} \\ &= \arg \sup_{|y_r|^2 \le 1} |v[k+d_m]|^2; \quad k = 1, 2, \ldots, n \end{aligned} \tag{4.3.19}$$

Since $y_r[k]$ and $v[k+d_m]$ are scalar, the task can be reduced for the problem

$$y_r^{\circledast}[k] = \arg \sup_{|y_r| \le 1} |v[k+d_m]|^2; \quad k = 1, 2, \ldots, n \tag{4.3.20}$$

The solution results in a very simple algorithm

$$y_r^{\circledast}[k] = -\mathrm{sign}[y_r^*] = -\mathrm{sign}\left[-\frac{\boldsymbol{g}^{\mathrm{T}}[k-1]\boldsymbol{q}}{g_0}\right]; \quad y_r^* = -\frac{\boldsymbol{g}^{\mathrm{T}}[k-1]\boldsymbol{q}}{g_0} \tag{4.3.21}$$

Its verification can be followed in Figure 4.3.2 (section (a) shows the case of absolute values, section (b) stands for the square values).

CHAPTER 5

Conventional *PID* Regulator

Proportional-integral-derivative (*PID*) regulators are configured to react proportionally to the current error value, taking into consideration the past error signal history with the integral of the error signal, and counting the future error signal trend by means of the differential quotient of the error. Figure 5.1 demonstrates [12] that the *PID* regulator calculates the actuating variable (the control signal) with the *P* effect proportional to the actual value of the error, the *I* effect, i.e., the integral of the error reflecting its past values, and the *D* effect, i.e., the gradient of the error signal (tendency of future behavior).

PID regulators are used worldwide; more than 90% of industrial control systems work with this type of regulator. In industrial process control the most frequently applied regulators are *PID* regulators, as they meet quality specifications, they have an easily realizable simple structure, and the effect of parameter changes can easily be evaluated. The design of *PID* regulators can be discussed in the frequency domain as a design method using a special pole-zero cancellation technique. The reason why they are discussed in this book is that their pole cancellation tuning method is very similar to the design principle of Youla regulators.

The transfer function of the ideal *PID* regulator can be given in the following two forms:

$$C_{\mathrm{PID}} = A_{\mathrm{P}}\left(1 + \frac{1}{sT_{\mathrm{I}}} + sT_{\mathrm{D}}\right) = A_{\mathrm{P}} + k_{\mathrm{I}}\frac{1}{s} + k_{\mathrm{D}}s = \frac{A_{\mathrm{P}}}{T_{\mathrm{I}}}\frac{1 + sT_{\mathrm{I}} + s^2T_{\mathrm{I}}T_{\mathrm{D}}}{s} \tag{5.1}$$

Here the regulator parameters are the proportional transfer gain A_{P}, the integrating time constant T_{I}, and the differentiating time constant T_{D}. The unit step response of the regulator is expressed as

$$v(t) = \mathcal{L}^{-1}\left[\frac{1}{s}C_{\mathrm{PID}}(s)\right] = A_{\mathrm{P}} + \frac{A_{\mathrm{P}}}{T_{\mathrm{I}}}t + A_{\mathrm{P}}T_{\mathrm{D}}\delta(t) \quad t \geq 0 \tag{5.2}$$

and is shown in Figure 5.2.

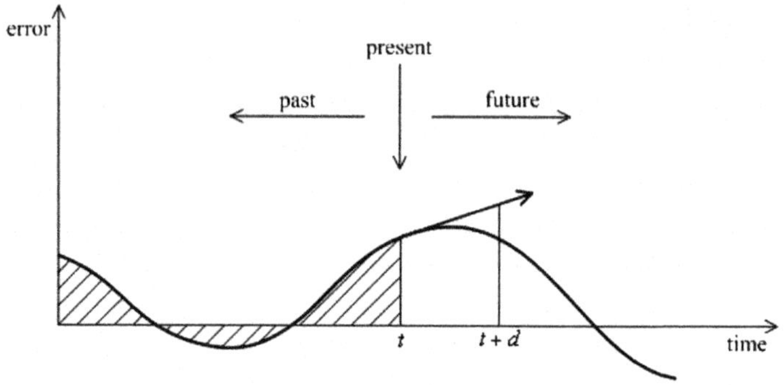

Figure 5.1 The *PID* regulator calculates the actuating variable from the current, past, and future trend (the slope) of the error signal [12].

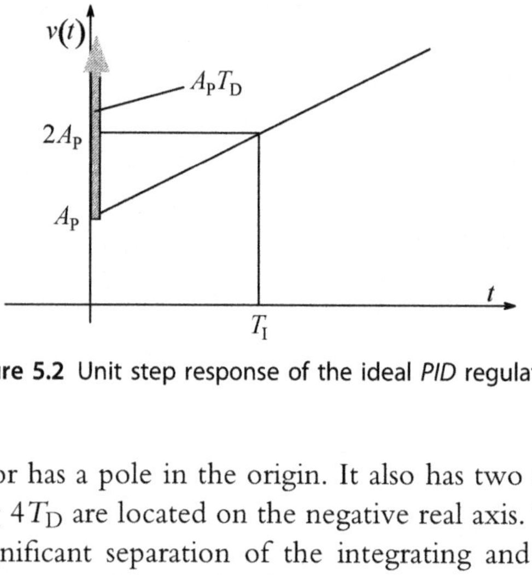

Figure 5.2 Unit step response of the ideal *PID* regulator.

The regulator has a pole in the origin. It also has two zeros which in the case of $T_I \geq 4T_D$ are located on the negative real axis. This condition requires the significant separation of the integrating and differentiating time constants, i.e., at least a fourfold distance of the corresponding breakpoints in the amplitude–frequency diagram. The ideal *PID* regulator cannot be realized and is used only for theoretical investigations and explanations.

An approximating but realizable *PID* regulator can be obtained by ensuring the realizability of the *D* effect (combining it with a serially connected lag element with a small time constant):

$$\widehat{C}_{PID} = A_P\left(1 + \frac{1}{sT_I} + \frac{sT_D}{1 + sT}\right) = \frac{A_P}{T_I}\frac{1 + s(T_I + T) + s^2T_I(T_D + T)}{s(1 + sT)} \tag{5.3}$$

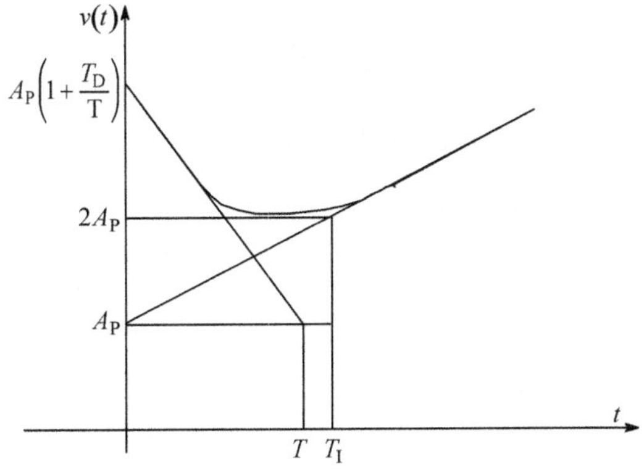

Figure 5.3 Unit step response of the approximate *PID* regulator.

The unit step response of this regulator is

$$\widehat{v}(t) = \mathcal{L}^{-1}\left[\frac{1}{s}\widehat{C}_{\mathrm{PID}}(s)\right] = A_{\mathrm{P}} + \frac{A_{\mathrm{P}}}{T_{\mathrm{I}}}t + \frac{A_{\mathrm{P}}T_{\mathrm{D}}}{T}e^{-t/T} \quad t \geq 0 \tag{5.4}$$

and is shown in Figure 5.3.

Now the regulator has one pole again in the origin but another pole appears at point $-1/T$, the result of the approximation of the differential effect ensuring realizability. The two zeros are still located on the negative real axis if condition $T_{\mathrm{D}} \leq (T_{\mathrm{I}} - T)^2/4T_{\mathrm{I}}$ is fulfilled. This condition again requires a significant separation of the integrating and the differentiating time constants, i.e., a significant distance between the corresponding breakpoint frequencies.

In the case of the Bode diagram-based design method, it might be useful to use the *PID* regulator in the following form:

$$\widehat{C}_{\mathrm{PID}} = A_{\mathrm{P}}\frac{(1+sT_{\mathrm{I}})(1+sT_{\mathrm{D}})}{sT_{\mathrm{I}}(1+sT)} \tag{5.5}$$

The advantage of this form is that it locates the breakpoint frequencies exactly at points $1/T_{\mathrm{I}}$, $1/T_{\mathrm{D}}$, and $1/T$, which belong to the given integrating, differentiating time constants, and to the time constant of the lag element. Therefore, this form is more suitable for designing the behavior of the regulator in the frequency region. (Note that the time constants of the differentiating term in (5.3) and (5.5) are slightly different!)

Figure 5.4 shows the unit step response of the *PID* regulator (5.3) and its Bode plot for the case of $A_{\mathrm{P}} = 1$. If the gain changes, the Bode amplitude

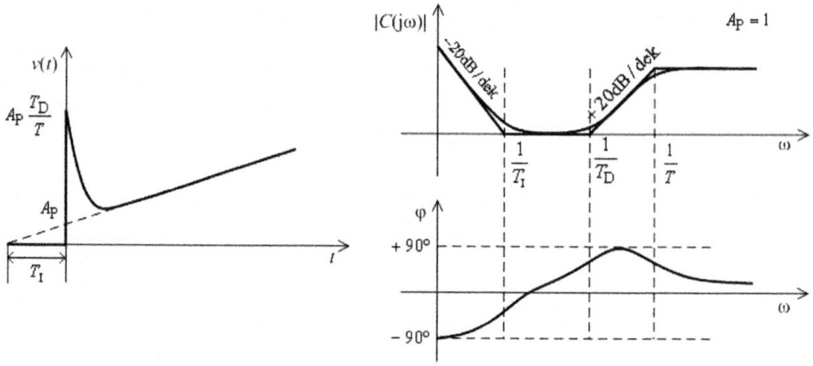

Figure 5.4 The unit step response and the approximating Bode diagram of the *PID* regulator.

diagram is shifted in parallel. This form entails the combination of two classical regulators (*PI* and *PD*) connected in series.

The classical design method of the *PID* regulators is very simple [105]. The integrating time is set by the choice $T_I = \max\{T_i\} = T_1$, whereas $T_D = T_2$ is chosen for the differentiating time. Consequently the straight-line section of slope -20 dB/decade on the Bode diagram of the loop transfer function is lengthened to the extent allowed by the structure of the regulator. In the control loop, after determination of the maximal integrating gain $K_{max}(\varphi_{to})$ belonging to the prescribed phase margin φ_{to}, the integrating gain factor of the regulator is obtained according to $K_I = A_P/T_I = K_{max}/P(0)$. Regarding the choice of the time constant T of the approximating differentiating effect, the considerations discussed in the design of *PD* regulators in [105] are also valid here, but the overexcitation in the control loop containing the integrating effect is calculated in a different way. As the output of an integrator changes until its input reaches the zero value, the steady state of the control system is reached if the error signal settles at zero. In the case of unit step reference signal, the initial jump of the *PID* regulator is $A_P T_D/T$. The steady-state value of the process input is $1/P(0)$. The calculated overexcitation is now $\eta = A_P P(0) T_D/T = K T_D/T$. Thus, the overexcitation is the product of the loop gain and the pole placement ratio.

It is clear that the design method cancels the two biggest time constants of the process. It is worth using the following simplified process model to formulate the algorithm of (5.5) for the ideal (but nonrealizable) case $T = 0$

$$P = P_+ = \frac{1}{\mathcal{A}_2} \tag{5.6}$$

where

$$\mathcal{A}_2 = (1 + sT_1)(1 + sT_2) \tag{5.7}$$

According to (2.1.5) the ideal regulator is

$$C = \frac{R_n P^{-1}}{1 - R_n} = \frac{R_n(1 + sT_1)(1 + sT_2)}{1 - R_n} \tag{5.8}$$

Let the reference model R_n be of first order

$$R_n = \frac{1}{1 + sT_n} \tag{5.9}$$

which means that the first term of the regulator is an integrator

$$\frac{R_n}{1 - R_n} = \frac{\frac{1}{1+sT_n}}{1 - \frac{1}{1+sT_n}} = \frac{1}{1 + sT_n - 1} = \frac{1}{sT_n} \tag{5.10}$$

whose integrating time is equal to the time constant of the reference model. Thus, the resulting regulator corresponds to the design principle, i.e., it is an ideal *PID* regulator

$$C = \frac{(1 + sT_1)(1 + sT_2)}{sT_n} = \frac{T_1}{T_n}\frac{(1 + sT_1)(1 + sT_2)}{sT_1} \tag{5.11}$$

with $T_I = T_1$, $T_D = T_2$, and $A_P = T_1/T_n$. The obtained disturbance rejection behavior can be written as Sy_n:

$$(1 - R_n)y_n = \left(1 - \frac{1}{1 + sT_n}\right)y_n = \frac{sT_n}{1 + sT_n}y_n \tag{5.12}$$

For $T \neq 0$, the algorithm of (5.5) is more complex. Assume the transfer function

$$P = P_+P_- = \frac{\mathcal{B}}{\mathcal{A}_2\tilde{\mathcal{A}}}; \qquad P_+ = \frac{1 + sT}{\mathcal{A}_2} = \frac{1 + sT}{(1 + sT_1)(1 + sT_2)};$$
$$P_- = \frac{\mathcal{B}}{(1 + sT)\tilde{\mathcal{A}}} \tag{5.13}$$

The regulator is obtained by the combination of the terms (2.1.5) and (2.1.10) as

$$\begin{aligned} C &= \left.\frac{R_n P_+^{-1}}{1 - R_n P_-}\right|_{P_- \approx 1} = \frac{R_n P_+^{-1}}{1 - R_n} = \frac{R_n}{1 - R_n}\frac{(1 + sT_1)(1 + sT_2)}{1 + sT} \\ &= \frac{(1 + sT_1)(1 + sT_2)}{sT_n(1 + sT)} = \frac{T_1}{T_n}\frac{(1 + sT_1)(1 + sT_2)}{sT_1(1 + sT)} \end{aligned} \tag{5.14}$$

The resulting approximating *PID* regulator is realizable and takes the form of (5.5), whose parameters are the same as those of (5.11), i.e., $T_{\mathrm{I}} = T_1$, $T_{\mathrm{D}} = T_2$, and $A_{\mathrm{P}} = T_1/T_{\mathrm{n}}$. Here the reference model is equal to (5.9). The disturbance rejection behavior is now Sy_{n}

$$\left(1 - R_{\mathrm{n}}\frac{\mathcal{B}}{(1+sT)\tilde{\mathcal{A}}}\right)y_{\mathrm{n}} \tag{5.15}$$

which more precisely fulfills $P_{-} \approx 1$ in (5.14), i.e., $\mathcal{B} \approx (1+sT)\tilde{\mathcal{A}}$, and better approximates the ideal expression (5.12). This approximation can be influenced by the choice of T: with a relatively small value, the low-frequency characteristics of the two sides can be equal in a wide region. Unfortunately in practice the small value of T is limited by the amplitude restrictions for the initial overexcitation $\eta = A_{\mathrm{P}}P(0)T_{\mathrm{D}}/T = KT_{\mathrm{D}}/T$ of the *PID* regulator (see Section 4.1).

The effect of the dead-time can be considered relatively simply in the case of series compensation, as

$$H_{\mathrm{H}}(s) = e^{-sT_{\mathrm{d}}} \Rightarrow H_{\mathrm{H}}(j\omega) = e^{-j\omega T_{\mathrm{d}}} = e^{-j\varphi_{\mathrm{d}}} \tag{5.16}$$

That means that $L(j\omega)$, the frequency characteristic of the loop transfer function is modified by an element of unity amplitude and phase angle $\varphi_{\mathrm{d}} = -\omega T_{\mathrm{d}}$, so only the phase characteristic is changed. This can be taken into account by prescribing a required phase margin of $\varphi_{\mathrm{to}}^{\mathrm{d}} = \varphi_{\mathrm{to}} + \omega T_{\mathrm{d}}$ instead of the original φ_{to}.

Sampled Data Systems

Creating the sampled version of the ideal *PID* regulator according to expression (5.1), let us approximate the integrating element with the right-hand rectangle rule and the differentiating effect by backward difference. Then the pulse transfer function is:

$$\begin{aligned} C_{\mathrm{PID}}(z) &= A_{\mathrm{P}}\left(1 + \frac{1}{T_{\mathrm{I}}}\frac{T_s z}{z-1} + T_{\mathrm{D}}\frac{z-1}{T_s z}\right) \\ &= A_{\mathrm{P}} + \frac{A_{\mathrm{P}}}{T_{\mathrm{I}}}\frac{T_s z}{z-1} + A_{\mathrm{P}}T_{\mathrm{D}}\frac{z-1}{T_s z} \end{aligned} \tag{5.17}$$

Transforming the right side of the expression to obtain a common denominator, we obtain a second-order discrete-time (DT) filter

$$C_{\mathrm{PID}}(z) = \frac{q_o + q_1 z^{-1} + q_2 z^{-2}}{1 - z^{-1}} \tag{5.18}$$

where

$$q_o = A_P\left(1+\frac{T_s}{T_I}+\frac{T_D}{T_s}\right);\quad q_1 = -A_P\left(1+\frac{2T_D}{T_s}\right);\quad q_2 = A_P\frac{T_D}{T_s} \tag{5.19}$$

For the realization of the DT version of the approximate *PID* regulator given by (5.3), the following *SRE* form can be used:

$$\widehat{C}_{PID}(z) = A_P + \frac{A_P T_s}{T_I}\frac{z^{-1}}{1-z^{-1}} + \frac{A_P T_D}{T_s}\frac{1-z^{-1}}{1-e^{-T_s/T}z^{-1}} \tag{5.20}$$

After some conversions this regulator also shows the form of a second-order DT filter:

$$\widehat{C}_{PID}(z) = \frac{q_o + q_1 z^{-1} + q_2 z^{-2}}{(1-z^{-1})(1-e^{-T_s/T}z^{-1})} = \frac{q_o + q_1 z^{-1} + q_2 z^{-2}}{1 + r_1 z^{-1} + r_2 z^{-2}} \tag{5.21}$$

where

$$q_o = A_P\left(1+\frac{T_D}{T}\right);\quad q_1 = -A_P\left(1+e^{-T_s/T} - \frac{T_s}{T_I} + \frac{2T_D}{T}\right) \tag{5.22}$$

$$q_2 = A_P\left\{e^{-T_s/T}\left(1-\frac{T_s}{T_I}\right)+\frac{T_D}{T}\right\} \tag{5.23}$$

$$r_1 = -\left(1+e^{-T_s/T}\right);\quad r_2 = e^{-T_s/T} \tag{5.24}$$

The DT form of the continuous-time (CT) approximate *PID* regulator according to (5.5) is given by the following DT pulse transfer function:

$$\widehat{C}_{PID}(z) = K_C\frac{\left(z-z_1^{cd}\right)\left(z-z_2^{cd}\right)}{(z-1)\left(z-e^{-T_s/T}\right)} = K_C\frac{\left(1-z_1^{cd}z^{-1}\right)\left(1-z_2^{cd}z^{-1}\right)}{(1-z^{-1})(1-e^{-T_s/T}z^{-1})}$$
$$= \tilde{C}_{PID}\left(z^{-1}\right) \tag{5.25}$$

(The upper index *cd* refers to the DT form of the regulator.) The parameterization of the regulator is formally the same as for CT *PID* regulators, but now $z_1^{cd} = p_1^{pd} = e^{-T_s/T_1}$ and $z_2^{cd} = p_2^{pd} = e^{-T_s/T_2}$ (the upper index *pd* refers to the DT form of the process). The regulator cancels the pole corresponding to the two biggest time constants of the process, since $T_1 = \max\{T_i\}$. In the order of the time constants, T_2 is the second one.

The gain K_C of the regulator should be set on the basis of the frequency function of the CT process $G(j\omega) \approx e^{-j\omega T_s/2}P(j\omega)$ modified by sampling.

This means that the absolute frequency function of the process remains unchanged at $a_d(j\omega) \approx a(j\omega)$, but the phase frequency function changes unfavorably to $\varphi_d(j\omega) = \varphi(j\omega) - \omega T_s/2$. This effect can be taken into consideration simply by changing the phase access φ_{to} prescribed for CT systems to a more severe one such as $\varphi_{to}^d = \varphi_{to} + \omega T_s/2$.

The initial and final values of the unit step response of the *PID* regulator given by (5.25) are

$$t = 0; \quad \lim_{z\to\infty} \frac{z}{z-1} K_C \frac{\left(z - z_1^{cd}\right)\left(z - z_2^{cd}\right)}{(z-1)\left(z - e^{-T_s/T}\right)} = K_C \tag{5.26}$$

$$t = \infty; \quad \frac{1}{P(0)} \tag{5.27}$$

Thus, the value of the overexcitation ensuring acceleration is

$$\eta = \frac{K_C}{P(0)} \tag{5.28}$$

It seems that K_C is the gain of the DT *PID* regulator, although this is not actually the case. K_C is a multiplier factor. If the gain of $\widehat{C}_{PID}$ is normalized to one, then in the form

$$\widehat{C}_{PID}(z) = k_C \frac{\left(1 - e^{-T_s/T}\right)}{\left(1 - z_1^{cd}\right)\left(1 - z_2^{cd}\right)} \frac{\left(z - z_1^{cd}\right)\left(z - z_2^{cd}\right)}{(z-1)\left(z - e^{-T_s/T}\right)} \tag{5.29}$$

k_C is the actual gain of the regulator. From comparison of the two forms, we obtain

$$K_C = k_C \frac{\left(1 - e^{-T_s/T}\right)}{\left(1 - z_1^{cd}\right)\left(1 - z_2^{cd}\right)} \tag{5.30}$$

and so

$$\eta = \frac{k_C}{P(0)} \frac{\left(1 - e^{-T_s/T}\right)}{\left(1 - z_1^{cd}\right)\left(1 - z_2^{cd}\right)} = \frac{k_C}{P(0)} \frac{\left(1 - e^{-T_s/T}\right)}{\left(1 - e^{-T_s/T_1}\right)\left(1 - e^{-T_s/T_2}\right)} \tag{5.31}$$

p_1^{cd}, and respectively T, can be determined from the permitted largest over-excitation η_{max} value

$$p_1^{cd} \geq 1 - \frac{\eta_{max} P(0)}{k_C} \left(1 - e^{-T_s/T_1}\right)\left(1 - e^{-T_s/T_2}\right) \tag{5.32}$$

For stable, correctly sampled processes $0 \le p^{\mathrm{cd}} \le 1$ and $0 \le z_2^{\mathrm{cd}} = p_2^{\mathrm{pd}} \le 1$. The time constant T corresponding to the borderline case $p_1^{\mathrm{cd}} = e^{-T_s/T}$ is given by

$$T = \frac{-T_s}{\ln\left[1 - \frac{\eta_{\max} P(0)}{k_C}\left(1 - e^{-T_s/T_1}\right)\left(1 - e^{-T_s/T_2}\right)\right]} \tag{5.33}$$

Again, the determination of the gain of the regulator for given phase access can be done according to the classical method, i.e., changing the phase access φ_{to} prescribed for CT systems to $\varphi_{\mathrm{to}}^{\mathrm{s}} = \varphi_{\mathrm{to}} + \omega_{\mathrm{c}} T_{\mathrm{s}}/2$.

SECOND-ORDER CT PROCESS WITH DEAD-TIME

Consider the CT process given by the transfer function

$$P(s) = \frac{e^{-sT_{\mathrm{d}}}}{(1 + sT_1)(1 + sT_2)} \tag{5.34}$$

where the dead-time T_{d} is assumed to be a multiple integer of the sampling time T_{s}.

$$T_{\mathrm{d}} = dT_{\mathrm{s}} \quad (d = 0, 1, 2, \ldots) \tag{5.35}$$

This model approximates the dynamics of several industrial processes. Then the joint DT model of the CT process and the zero-order hold element (*SRE* transformed) is

$$G(z) = z^{-d} \frac{K_{\mathrm{P}}(z - z_1)}{(z - p_1)(z - p_2)} = \frac{K_{\mathrm{P}}(z - z_1)}{z^d (z - p_1)(z - p_2)} \tag{5.36}$$

which can be rewritten as

$$G'\left(z^{-1}\right) = \frac{g_1' z^{-1} + g_2' z^{-2}}{1 + f_1' z^{-1} + f_2' z^{-2}} z^{-d} = \frac{g_{\mathrm{o}}'\left(1 + \gamma z^{-1}\right)}{1 + f_1' z^{-1} + f_2' z^{-2}} z^{-d'} \tag{5.37}$$

Note that in the case of δ-transformation (see A.2.52), we obtain

$$G''\left(q^{-1}\right) = \frac{b_{\mathrm{o}}'' q^{-2}}{1 + d_1'' q^{-1} + d_2'' q^{-2}} q^{-d} = \frac{b_{\mathrm{o}}''}{1 + d_1'' q^{-1} + d_2'' q^{-2}} q^{-d''} \tag{5.38}$$

where $d'' = d + 2$. (Observe that this form has certain numerical advantages for small sampling times.) Both expressions can be given by the general pulse transfer function $G = \mathcal{G} z^{-k}/\mathcal{F}$ but they differ in the values of $\mathcal{G}$ and k.

The classical pole canceling design method of the DT conventional *PID* regulator can be written in the regulator form

$$C(z) = \frac{q_o + q_1 z^{-1} + q_2 z^{-2}}{1 - z^{-1}} G_F(z) = \frac{q_o \mathcal{F}(z)}{1 - z^{-1}} G_F(z) = \frac{K_I \mathcal{F}(z)}{g_o(1 - z^{-1})} G_F(z), \tag{5.39}$$

where $G_F(z)$ is a serial filter inside the regulator. The remaining loop transfer function obtained by this compensation is

$$L(z) = \frac{q_o \mathcal{G}}{1 - z^{-1}} G_F(z) z^{-k} \Big|_{G_F=1} = \frac{q_o g_o (1 + \gamma z^{-1})}{1 - z^{-1}} z^{-k} = \frac{K_I (1 + \gamma z^{-1})}{1 - z^{-1}} z^{-k}. \tag{5.40}$$

In this closed loop, nice (1–5%) overshoot belongs to the phase access $\varphi_t = 60°$. To compute the integrated loop gain K_I, a nonlinear equation should be solved

$$x = \frac{1 - 2\mathrm{arctg}\frac{\gamma \sin x}{1+\gamma \cos x}}{2d - 1} = g(x). \tag{5.41}$$

Using this solution, we obtain

$$K_I = \frac{\sqrt{2(1 - \cos x)}}{\sqrt{1 + 2\gamma \cos x + \gamma^2}} \tag{5.42}$$

The curve series obtained from the latter two equations for the optimal values of K_I, as a function of d and γ, are presented in Figure 5.5. For positive γ values, using the Taylor series expansion of the above two functions, an approximating explicit form is obtained for K_I:

$$K_I = \frac{1}{2k(1 + \gamma) - (1 - \gamma)}; \quad (\gamma > 0) \tag{5.43}$$

For the frequent case of $\gamma = 0$ and using the δ-transformation, we obtain (see Bányász–Keviczky method [13])

$$K_I^o = \frac{1}{2k - 1} = \frac{1}{2k - 1}\Big|_{k=d''} = \frac{1}{2(d + 2) - 1} = \frac{1}{2d + 3}; \quad (\gamma \equiv 0) \tag{5.44}$$

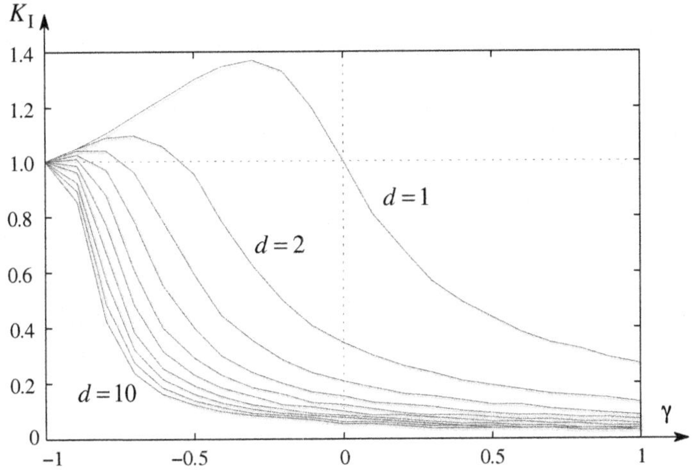

Figure 5.5 Optimal K_I as a function of d and γ.

It is simple to check that

$$\frac{1}{T_I^o} = \lim_{T_s \to 0} \frac{K_I^o}{T_s} \approx \lim_{T_s \to 0} \frac{1}{2T_d + 3T_s} = \frac{1}{2T_d} \tag{5.45}$$

which corresponds to $\varphi_{to} = 60°$ for CT systems.

Note that, for the region $\gamma < 0$, a simple serial filter

$$G_F(z) = \frac{1}{1 + \gamma z^{-1}} \tag{5.46}$$

can be applied. This case corresponds to the (5.21) form of the DT-approximating *PID* regulator. This, which is actually a zero-canceling technique, can be usefully applied in the case when the zero $z_1 = -\gamma$ is stable. If the zero is unstable, the unwanted overshoot of the closed-loop transfer function can be successfully decreased by the pole $p_1 = 1/z_1$. This pole can be assigned by the following serial filter

$$G_F(z) = \frac{1+\gamma}{\gamma} \frac{1}{1 + \frac{1}{\gamma} z^{-1} - K_I^o \frac{1-\gamma}{\gamma} z^{-d}} \tag{5.47}$$

(see the Hetthéssy–Keviczky–Bányász method [59]).

OBSERVER-BASED *PID* REGULATOR

The ideal form of a Youla regulator based on reference model design [94] is

$$C_{id} = \frac{(R_n P^{-1})}{1 - (R_n P^{-1}) P} = \frac{Q}{1 - QP} = \frac{R_n}{1 - R_n} P^{-1} \tag{5.48}$$

when the inverse of the process is realizable and stable. Here the operation of R_n can be considered a reference model (desired system dynamics). It is generally required that the reference model has to be strictly proper with unit static gain, i.e., $R_n(\omega = 0) = 1$.

For a simple, but robust *PID* regulator design method assume that the process can be well approximated by its two major time constants, i.e.,

$$P \cong \frac{\mathcal{A}}{\mathcal{A}_2} \tag{5.49}$$

where

$$\mathcal{A}_2 = (1 + sT_1)(1 + sT_2) \tag{5.50}$$

According to (2.1.5), the ideal Youla regulator is

$$C_{id} = \frac{R_n P^{-1}}{1 - R_n} = \frac{R_n(1 + sT_1)(1 + sT_2)}{A(1 - R_n)}; \quad T_1 > T_2 \tag{5.51}$$

Let the reference model R_n be of first order

$$R_n = \frac{1}{1 + sT_n} \tag{5.52}$$

which means that the first term of the regulator is an integrator

$$\frac{R_n}{1 - R_n} = \frac{\frac{1}{1+sT_n}}{1 - \frac{1}{1+sT_n}} = \frac{1}{1 + sT_n - 1} = \frac{1}{sT_n} \tag{5.53}$$

whose integrating time is equal to the time constant of the reference model. Thus, the resulting regulator corresponds to the design principle, i.e., it is an ideal *PID* regulator

$$C_{PID} = A_{PID}\frac{(1 + sT_I)(1 + sT_D)}{sT_I} = A_{PID}\frac{(1 + sT_1)(1 + sT_2)}{sT_1} \tag{5.54}$$

with

$$A_{PID} = \frac{T_1}{AT_n}; \quad T_I = T_1; \quad T_D = T_2 \tag{5.55}$$

The Youla parameter Q in the ideal regulator is

$$Q = R_n P^{-1} = \frac{1}{A}\frac{(1 + sT_1)(1 + sT_2)}{1 + sT_n} \tag{5.56}$$

It is not necessary, but desirable to ensure the realizability, i.e., it is reasonable to use

$$Q = R_n P^{-1} = \frac{1}{A} \frac{(1 + sT_1)(1 + sT_2)}{(1 + sT_n)(1 + sT)} \tag{5.57}$$

where T can be considered the time constant of the derivative action $(0.1T_D \leq T \leq 0.5T_D)$. The regulator $\widehat{C}'$ and the feedback term $\widehat{H}$ must be always realizable. In practice, the *PID* regulator and the Youla parameter are always model based, so

$$\widehat{C}_{PID}(\widehat{P}) = \widehat{A}_{PID} \frac{(1 + s\widehat{T}_1)(1 + s\widehat{T}_2)}{s\widehat{T}_1}; \quad \widehat{A}_{PID} = \frac{\widehat{T}_1}{\widehat{A}T_n} \tag{5.58}$$

$$\widehat{Q} = R_n \widehat{P}^{-1} = \frac{1}{\widehat{A}} \frac{(1 + s\widehat{T}_1)(1 + s\widehat{T}_2)}{1 + sT_n} \tag{5.59}$$

The scheme of the observer-based *PID* regulator is shown in Figure 5.6, where a simple *PI* regulator

$$\widehat{K}_l = A_l \frac{1 + sT_l}{sT_l} \tag{5.60}$$

is applied in the observer loop. Here T_l must be in the range of T, i.e., considerably smaller than T_1 and T_2.

Note that the frequency characteristic of $\widehat{H}$ cannot easily be designed to reach a proper error suppression. For example, it is almost impossible to design a good realizable high-cut filter in this architecture. The

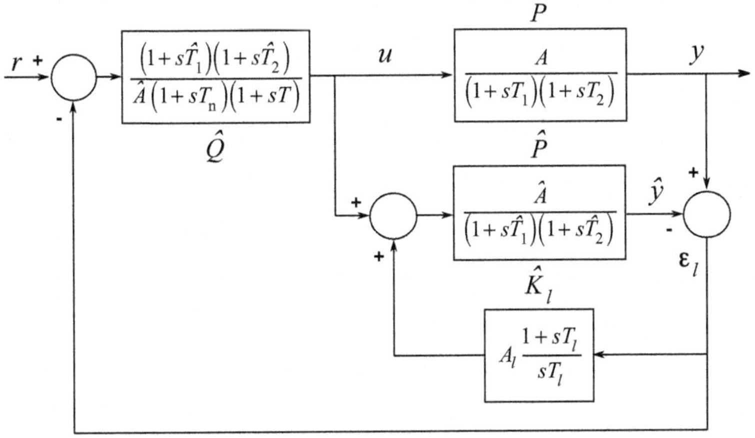

Figure 5.6 Observer-based *PID* regulator.

high-frequency domain is always more interesting to speed up a control loop, so the target of the future research is how to select $\widehat{K}_l$ for the desired shape of $\widehat{H}$.

Simulation Examples

The simulation experiments were performed using the observer-based *PID* scheme shown in Figure 5.6.

Example 5.1

The process parameters are $T_1 = 20$, $T_2 = 10$, and $A = 1$. The model parameters are $\widehat{T}_1 = 25$, $\widehat{T}_2 = 12$, and $\widehat{A} = 1.2$. The purpose of the regulation is to speed up the basic step response by 4, i.e., $T_n = 5$ is selected in the first order R_n. In the observer loop a simple proportional regulator $\widehat{K}_l = 0.01$ is applied. The ideal form of Q (5.56) was used. Figure 5.7 shows some step responses in the operation of the observer-based *PID* regulator.

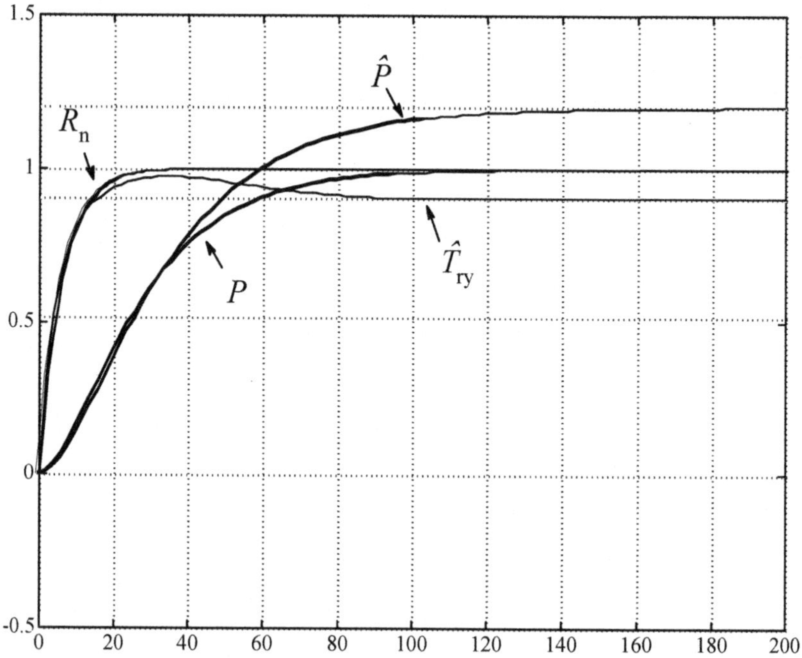

Figure 5.7 Step responses using the observer-based *PID* regulator.

It is clear that the $\widehat{T}'_{ry}$ very well approximates R_n in the high frequencies (for small time values) in spite of the very bad model $\widehat{P}$.

Example 5.2

The process parameters and the selected first order R_n are the same as in the previous example. The model parameters are $\widehat{T}_1 = 30$, $\widehat{T}_2 = 20$, and $\widehat{A} = 0.5$. In the observer loop, a *PI* regulator (5.60) is applied with $A_I = 0.001$ and $T_I = 2$. The ideal form of Q (5.56) was used. Figure 5.8 shows some step responses in the operation of the observer-based *PID* regulator.

It is clear that the $\widehat{T}'_{ry}$ well approximates R_n in the high frequencies (for small time values) in spite of the very bad model $\widehat{P}$.

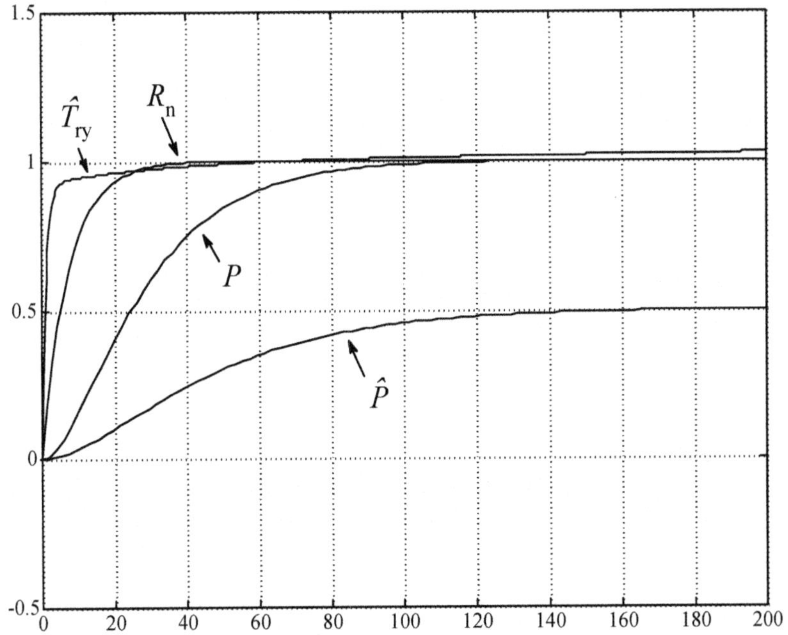

Figure 5.8 Step responses using the observer-based *PID* regulator.

CHAPTER 6

Control of Stochastic Processes

In the 1970s, computer-aided process control flourished thanks to the so-called *minimum variance* (*MV*) control devised by Åström and Wittenmark [11]. This discrete regulator was well suited to the discrete-time (DT) process identification techniques suggested earlier by Åström and Bohlin [6,7], and from the very beginning the regulator had a typical application scope in the paper industry. The method models the process together with the effecting stochastic noise, a very important approach both from perspective of identification and control.

MINIMUM VARIANCE (*MV*) REGULATOR

Let the DT system equation of the process be

$$y[k] = Gu[k] + H_n\varsigma[k] = \frac{\mathcal{G}z^{-d}}{\mathcal{F}}u[k] + \frac{C}{\mathcal{F}}\varsigma[k] = \frac{\mathcal{G}}{\mathcal{F}}u[k-d] + y_n[k] \tag{6.1}$$

where $u[k]$ and $y[k]$ are the sampled series of the process input and output signals, $y_n[k]$ is the sampled series of the additive output noise. $\varsigma[k]$ is the source noise from which $y_n[k]$ is derived. It is assumed that $\varsigma[k]$ is a white noise with zero mean value. Arbitrary colored noise can be derived from the white noise input by the model $H_n = \mathcal{C}/\mathcal{F}$.

The target of the *MV* regulator is to minimize the variance of the output. To this end, a method is followed which is similar to what was used to derive the predictive regulator in Section 2.4. Introduce the polynomial equation

$$\mathcal{C} = \mathcal{F}\mathcal{T} + \mathcal{P}z^{-d} \tag{6.2}$$

whose solution is unambiguous by seeking $(d-1)$ order and $(n-1)$ order polynomials $\mathcal{T} = 1 + t_1z^{-1} + \ldots + t_{d-1}z^{-(d-1)}$ and $\mathcal{P} = p_o + p_1z^{-1} + \ldots + p_{n-1}z^{-(n-1)}$, if $\mathcal{C}$ and $\mathcal{F}$ are of n order. Equation (6.2) is a special version of the Diophantine equation (*DE*) discussed in Chapter 4. The form (6.1) can be decomposed as

$$y[k] = \frac{\mathcal{G}\mathcal{T}}{\mathcal{C}}u[k-d] + \frac{\mathcal{P}}{\mathcal{C}}y[k-d] + \mathcal{T}\varsigma[k] = y[k|k-d] + \mathcal{T}\varsigma[k] \tag{6.3}$$

Here $y[k|k-d]$ is the predictor of the output, by means of which the $[k+d]$-th future value can be predicted in the $[k]$-th moment. Thus it is sensible to use the form

$$y[k+d] = y[k+d|k] + \mathcal{T}\varsigma[k+d] = y[k+d|k] + \varepsilon[k+d]$$
$$= \left\{\frac{\mathcal{GT}}{\mathcal{C}}u[k] + \frac{\mathcal{P}}{\mathcal{C}}y[k]\right\} + \varepsilon[k+d] \tag{6.4}$$

where the prediction error $\varepsilon[k+d] = \mathcal{T}\varsigma[k+d]$ is independent of $y[k+d|k]$. Do not forget that the order of $\mathcal{T}$ is $(d-1)$, and does not contain values or their functions which have appeared up to the time $[k]$. Since the elements of (6.4) are independent, the variance of the output is

$$V = \text{var}\{y[k+d]\} = E\{y^2[k+d]\} = E\{y^2[k+d|k]\} + E\{\varepsilon^2[k+d]\} \tag{6.5}$$

and it can be minimized by the trivial use of $y[k+d|k] = 0$. The MV is $E\{\varepsilon^2[k+d]\}$. The condition of the minimum is

$$y[k+d|k] = \frac{\mathcal{GT}}{\mathcal{C}}u[k] + \frac{\mathcal{P}}{\mathcal{C}}y[k] = 0 \tag{6.6}$$

and so the classical form of the MV regulator is

$$u[k] = -\frac{\mathcal{P}}{\mathcal{GT}}y[k] \tag{6.7}$$

If it is necessary to follow the output of the reference model R_r then from the condition $y[k+d|k] = R_r y_r[k]$ we obtain for the regulator

$$u[k] = \frac{R_r y_r[k] - \mathcal{P}y[k]}{\mathcal{GT}} \tag{6.8}$$

This formally corresponds to (2.4.7) but now $\mathcal{P}$ is computed from the DE of (6.2). The interpretation of Equation (6.8) can be seen in Figure 6.1(a). The equivalent block scheme of Figure 6.1(b) shows the conventional closed loop.

The MV regulator takes the form

$$C_{\text{MV}} = \frac{R_n}{1 - R_n z^{-d}}\frac{\mathcal{F}}{\mathcal{G}} = \frac{\mathcal{P}}{\mathcal{C} - \mathcal{P}z^{-d}}\frac{\mathcal{F}}{\mathcal{G}} = \frac{\mathcal{P}}{\mathcal{GT}} \quad \text{or} \quad C_{\text{MV}} = \frac{H_n - \mathcal{T}}{\mathcal{GT}} \tag{6.9}$$

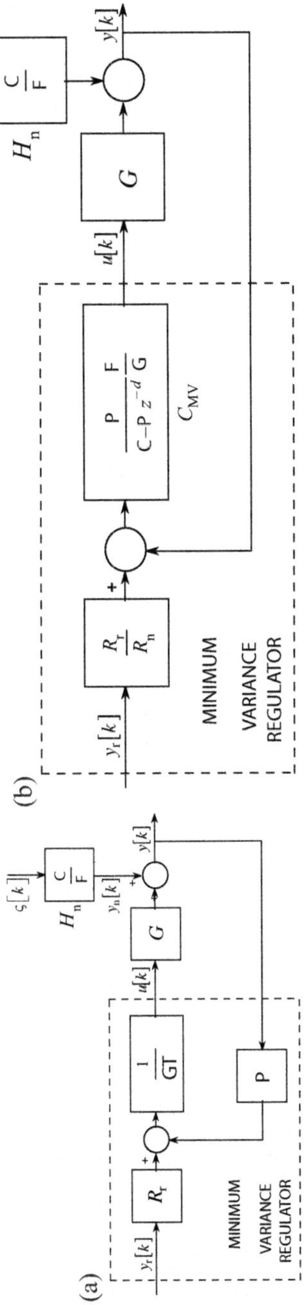

Figure 6.1 Forms of the *minimum variance* regulator.

i.e., the reference model referring to the noise rejection is now

$$R_n = \frac{\mathcal{P}}{\mathcal{C}} \tag{6.10}$$

and depends on the parameters of the noise model. It is obtained from (6.2). The result is quite obvious since the task is now to optimally compensate the effect of the output noise. Observe that the regulator C_{MV} is formally equal to the prediction regulator C_{pr} but $\mathcal{P}$ should be computed from (2.4.2) instead of (6.2). If different denominators are assumed in the pulse transfer functions (6.1) of the process and noise model

$$y[k] = Gu[k] + H_n\varsigma[k] = \frac{\mathcal{G}z^{-d}}{\mathcal{F}}u[k] + \frac{\mathcal{C}}{\mathcal{D}}\varsigma[k] = \frac{\mathcal{G}}{\mathcal{F}}u[k-d] + y_n[k] \tag{6.11}$$

then the DE in (6.2) changes to

$$\mathcal{C} = \mathcal{D}\mathcal{T} + \mathcal{P}z^{-d} \tag{6.12}$$

otherwise all other expressions (concerning the predictor, regulator, etc.) formally remain unchanged. Since (6.10) is the expression of the optimal predictor, the noise models $H_n = \mathcal{C}/\mathcal{F}$ and $H_n = \mathcal{C}/\mathcal{D}$ must be assumed to be inverse stable. (This follows from the realization theory of the stationary stochastic signals.)

It is easy to check that with the *MV* regulator the characteristic equation of the closed loop takes the form

$$\mathcal{D}\mathcal{T} + \mathcal{P}z^{-d} = \mathcal{C} = 0 \tag{6.13}$$

which also requires the stability of $\mathcal{C}$.

GENERALIZED MINIMUM VARIANCE REGULATOR

It can be seen from (6.9) that the *MV* regulator is actually a Youla regulator and corresponds to its ideal form. Obviously in this case the existence of the complete inverse of the process should be assumed. Since this condition is very rarely fulfilled, several recommendations have been made on how to modify the original *MV* regulator, i.e., how to *de-tune* it from its optimal form. The prominent representative of this school is the so-called *generalized minimum variance* (*GMV*) regulator of Clarke [34].

Clarke suggested [34,35] introducing an additional term, instead of the variance of the process output, which also penalizes the input variance. The most general form of the suggested criterion is

$$V = E\left\{ \left(W_y(z) y[k+d] \right)^2 + \left(W_u(z) u[k] \right)^2 \middle| k \right\} \tag{6.14}$$

whereby the stable filter $W_y(z)$ penalizes the corresponding frequency region of the output and the stable filter $W_u(z)$ does the same regarding the input. It is deduced in A.3 that the condition of the minimum is now

$$y^F[k+d|k] + \frac{w_{no}^u}{w_{no}^y g_o} W_u(z) u[k] = y^F[k+d|k] + W_u'(z) u[k] = 0 \tag{6.15}$$

Here the d-step ahead prediction of the filtered output y^F is

$$y^F[k+d|k] = \frac{\mathcal{G}'\mathcal{D}'\mathcal{T}'}{\mathcal{F}'\mathcal{C}'} u[k] + \frac{\mathcal{P}'}{\mathcal{C}'} y[k] = \frac{\mathcal{G}\mathcal{D}\mathcal{T}'}{\mathcal{F}\mathcal{C}} u[k] + \frac{\mathcal{P}'}{\mathcal{C}\mathcal{W}_d^y} y[k] \tag{6.16}$$

Based on these latter two equations, the equation of the *GMV* regulator takes the form

$$u[k] = -\frac{\mathcal{F}\mathcal{P}'}{\left(\mathcal{G}\mathcal{D}\mathcal{T}' + \mathcal{F}\mathcal{C}W_u'\right)\mathcal{W}_d^y} y[k] \quad \text{i.e.,}$$

$$C_{GMV} = \frac{\mathcal{F}\mathcal{P}'}{\left(\mathcal{G}\mathcal{D}\mathcal{T}' + \mathcal{F}\mathcal{C}W_u'\right)\mathcal{W}_d^y} \tag{6.17}$$

The polynomials $\mathcal{T}'$ and $\mathcal{P}'$ are derived from the special polynomial equation

$$\left(\mathcal{C}\mathcal{W}_n^y\right) = \left(\mathcal{D}\mathcal{W}_d^y\right)\mathcal{T}' + \mathcal{P}' z^{-d} \tag{6.18}$$

where it is assumed that the filter $W_y(z)$ is given by the following pulse transfer function

$$W_y(z) = \frac{\mathcal{W}_n^y}{\mathcal{W}_d^y} \tag{6.19}$$

Moreover, $W_u'(z)$ in (6.15) is a scaled form of $W_u(z)$, i.e.,

$$W_u'(z) = \frac{w_{no}^u}{w_{no}^y g_o} W_u(z) \tag{6.20}$$

Here g_o, w_{no}^y, and w_{no}^u are the zero-order terms in z^{-1} of the polynomials $\mathcal{G}$, $\mathcal{W}_n^y$, and $\mathcal{W}_n^u$, respectively.

With the *GMV* regulator, the characteristic equation of the closed system is

$$\left(\mathcal{W}_n^y \mathcal{G} + \mathcal{W}_d^y \mathcal{F} W_u'\right)\mathcal{C} = 0 \tag{6.21}$$

which is a significant change compared with (6.13). It is clear from the equation that the stability requirement for $\mathcal{C}$ still stands but not for the process. The consequence of the complex criterion can be also recognized. Namely, if $W'_{\rm u}$ is large compared with $W_{\rm y}$, then even in the case of unstable $\mathcal{G}$ the roots of the characteristic equation approach the roots of $\mathcal{F}$. Thus, if $\mathcal{F}$ is stable then the *GMV* regulator is able to control even inverse unstable processes, too.

Example 6.1

Choosing the following penalizing filters $W_{\rm y}(z) = 1$ and $W_{\rm u}(z) = \alpha(1 - z^{-1})$, we obtain the following loss function:

$$V = E\{y^2[k+d] + \alpha(u[k] - u[k-1])^2 | k\} \tag{6.22}$$

It is clear that this criterion penalizes the change of the actuating signal and so results in the control of a slow operation. If we choose an α of large enough value, the transfer function of the *GMV* regulator (6.17) will be

$$C_{\rm GMV} = \frac{g_{\rm o}\mathcal{P}'}{\alpha^2(1 - z^{-1})\mathcal{C}} \tag{6.23}$$

i.e., the loss function (6.22) makes it possible to approach the integrating effect. Here $w^{\rm u}_{\rm no} = \alpha$ is applied.

Example 6.2

Reshape the transfer function of the *GMV* regulator (6.17) to retain only polynomials in both its numerator and denominator, i.e.,

$$C_{\rm GMV} = \frac{\mathcal{F}\mathcal{P}'}{(\mathcal{G}\mathcal{D}\mathcal{T}' + \mathcal{F}\mathcal{C}W'_{\rm u})\mathcal{W}^{\rm y}_{\rm n}} = \frac{\mathcal{F}\mathcal{P}' w^{\rm u}_{\rm o}\mathcal{W}^{\rm u}_{\rm d}}{(w^{\rm u}_{\rm o}\mathcal{W}^{\rm u}_{\rm d}\mathcal{G}\mathcal{D}\mathcal{T}' + g_{\rm o}\mathcal{W}^{\rm u}_{\rm n}\mathcal{F}\mathcal{C})\mathcal{W}^{\rm y}_{\rm n}} \tag{6.24}$$

Note that the term $\mathcal{W}^{\rm u}_{\rm d}$ which appears in the numerator, i.e., the denominator of $W_{\rm u}(z)$, provides a good chance to include the differentiating effect in the regulator. For this purpose, the most reasonable and simplest choice is $\mathcal{W}^{\rm u}_{\rm d} = (1 - z^{-1})$.

Example 6.3

In many applications $W_{\rm y}(z) = 1$ and $W_{\rm u}(z) = \alpha$ are used and lead to the loss function

$$V = E\{y^2[k+d] + \alpha u^2[k] | k\} \tag{6.25}$$

The equation of the *GMV* regulator is now

$$C_{\mathrm{GMV}} = \frac{\mathcal{F}\mathcal{P}'}{\mathcal{G}\mathcal{D}\mathcal{T}' + \frac{\alpha^2}{g_o}\mathcal{F}\mathcal{C}} \overset{\mathcal{D}=\mathcal{F}}{\Rightarrow} \frac{\mathcal{P}'}{\mathcal{G}\mathcal{T}' + \frac{\alpha^2}{g_o}\mathcal{C}} \tag{6.26}$$

which is very simple and therefore preferred in industrial practice, too.

PREDICTION OF DETERMINISTIC SIGNALS

Certain disturbance signals affecting control systems are deterministic and have special features which correspond to the solution of a homogeneous difference equation. For these signals it holds that

$$\mathcal{K}_x(z) y_x[k] = 0; \quad k > l_x \tag{6.27}$$

where

$$\mathcal{K}_x(z) = 1 + k_1^x z^{-1} + k_2^x z^{-2} + \ldots + k_{l_x}^x z^{-l_x} \tag{6.28}$$

is the characteristic polynomial of the signal. The typical test signals used in deterministic design can be described by their characteristic polynomials: for the unit step signal $\mathcal{K}_x(z) = 1 + k_1^x z^{-1}$ and for the speed-like signal $\mathcal{K}_x(z) = 1 + k_1^x z^{-1} + k_2^x z^{-2}$, respectively. This determination, however, is valid only after a transient consisting of l_x steps, when all past values of $y_x[k]$ are available (6.27). Thus, the typical DT test signals have $(l_x + 1)$ nonregular points. Otherwise, the deterministic signals characterized by Equation (6.27) can be interpreted as a stationary component of the impulse function series of the transfer function $\mathcal{N}/\mathcal{K}_x(z)$.

The *d*-step prediction of signals with the characteristic polynomial $\mathcal{K}_x(z)$ can be simply computed on the basis of the following decomposition

$$1 = \mathcal{T}_x \mathcal{K}_x + \mathcal{P}_x z^{-d} \tag{6.29}$$

where

$$\mathcal{T}_x(z) = 1 + t_1^x z^{-1} + t_2^x z^{-2} + \ldots + t_{d-1}^x z^{-(d-1)} \tag{6.30}$$

and

$$\mathcal{P}_x(z) = p_o^x + p_1^x z^{-1} + p_2^x z^{-2} + \ldots + p_{l_x-1}^x z^{-(l_x-1)} \tag{6.31}$$

Given (6.26) and (6.28), we can write

$$y_x[k+d] = \mathcal{P}_x y_x[k] + \mathcal{T}_x \mathcal{K}_x y_x[k+d] = y_x[k+d|k] + \varepsilon_x[k+d|k] \tag{6.32}$$

It can be seen from the above expression that the prediction error $\varepsilon_x[k+d|k]$ is not zero in the (l_x+1) nonregular points of the signal or in the point $(d-1)$ corresponding to the memory of $\mathcal{T}_x(z)$. At any further points the prediction error is already zero.

Thus, given the above, it can be stated that with a reference signal of characteristic polynomial $\mathcal{K}_r(z)$, the *Youla parameterized* (*YP*) regulator computed by the optimal predictor (6.32) reacts with a dead-beat regulator (see Section 2.3) process containing (l_r+d) sampling points, and finally the tracking error regarding the reference signal is zero.

PREDICTION OF STOCHASTIC SIGNALS

Investigate the prediction possibility of the disturbance signal effect on the output of the DT process

$$y_n[k] = H_n\varsigma[k] = \frac{\mathcal{C}}{\mathcal{D}}\varsigma[k] \tag{6.33}$$

Following the derivation procedure of (6.4), we can write that

$$\begin{aligned} y_n[k+d] &= \frac{\mathcal{P}}{\mathcal{C}}y_n[k] + \mathcal{T}\varsigma[k+d] = R_n + \mathcal{T}\varsigma[k+d] \\ &= y_n[k+d|k] + \varepsilon_n[k+d|k] \end{aligned} \tag{6.34}$$

which is the d-step prediction of the output disturbance $y_n[k]$. Here the polynomial decomposition (6.12)

$$\mathcal{C} = \mathcal{D}\mathcal{T} + \mathcal{P}z^{-d} \tag{6.35}$$

is applied. The prediction error of the disturbance signal is

$$\varepsilon_n[k+d|k] = \mathcal{T}\varsigma[k+d] \tag{6.36}$$

which is uncorrelated with the d-step prediction $y_n[k+d|k]$, because the highest order of $\mathcal{T}$ is $(d-1)$. Therefore, the *MV* phrase is used not only for the regulator seen above but also for the predictor of the random signal. The block scheme of the d-step prediction of the output disturbance is shown in Figure 6.2.

The importance of the expression (6.34) is that the prediction of the stochastic output disturbance $y_n[k]$ can be determined by the signal itself. Namely, the *MV* predictor is

$$y_n[k+d|k] = \frac{\mathcal{P}}{\mathcal{C}}y_n[k] = R_n y_n[k] \tag{6.37}$$

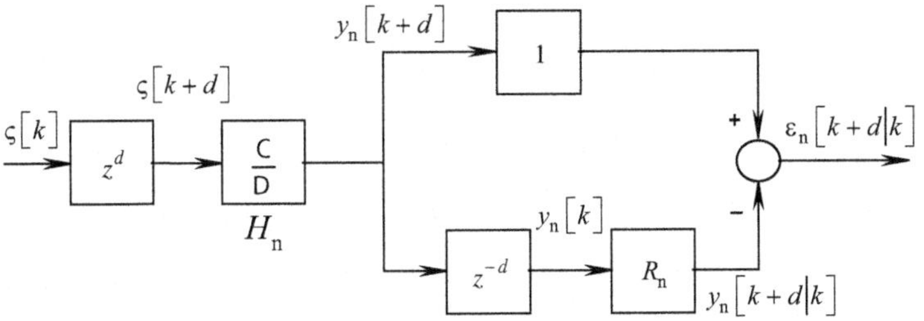

Figure 6.2 Block scheme of the *d*-step prediction of the output disturbance.

In the closed loop, however, $y_n[k]$ cannot be measured separately, and therefore the expression (6.36) has theoretical importance. For any stochastic signal, derived via (6.32), however, the *MV* *d*-step prediction can be simply computed from its measured values.

Given the scheme in Figure 6.2, the $\mathcal{H}_2$ norm of the prediction error $\varepsilon_n[k+d|k]$ can be simply computed by the following expression

$$J_n = \left\| \frac{\mathcal{C}}{\mathcal{D}} - \frac{\mathcal{C}}{\mathcal{D}} R_n z^{-d} \right\|_{\mathcal{H}_2} = \|\mathcal{T}\|_{\mathcal{H}_2} = 1 + \sum_{i=1}^{d-1} t_i^2 \tag{6.38}$$

Since in the closed loop $y_n[k]$ is an inner signal, the above prediction algorithm should be modified. The disturbance signal should be computed according to Figure 6.3, where $y_n[k] = -e[k] \quad \text{if} \quad r[k] \equiv 0$

If the regulator (2.1.25) is derived by *YP* design, then

$$C = \frac{R_n G_n G_+^{-1}}{1 - R_n G_n G_- z^{-d}} = \frac{R_n K_n}{1 - R_n K_n G} = \frac{\left(R_n G_n G_+^{-1}\right)}{1 - \left(R_n G_n G_+^{-1}\right) G} = \frac{X_n}{1 - X_n G} \tag{6.39}$$

and the prediction scheme shown in Figure 6.4 of the output disturbance can be simply obtained, since now

$$\varepsilon_n[k+d|k] = (1 - X_n G) y_n[k+d] \tag{6.40}$$

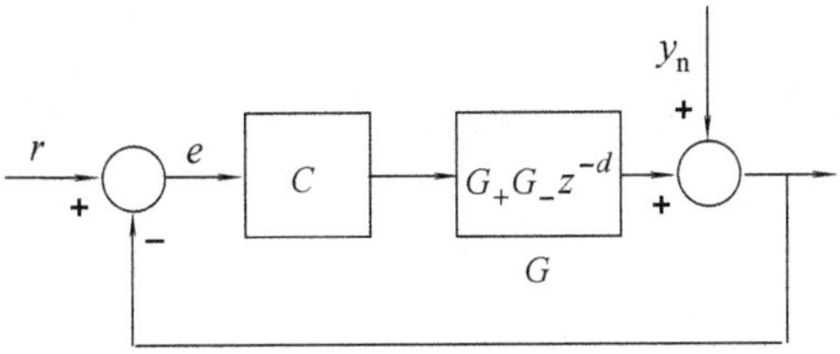

Figure 6.3 Closed-loop control for the derivation of $y_n[k]$

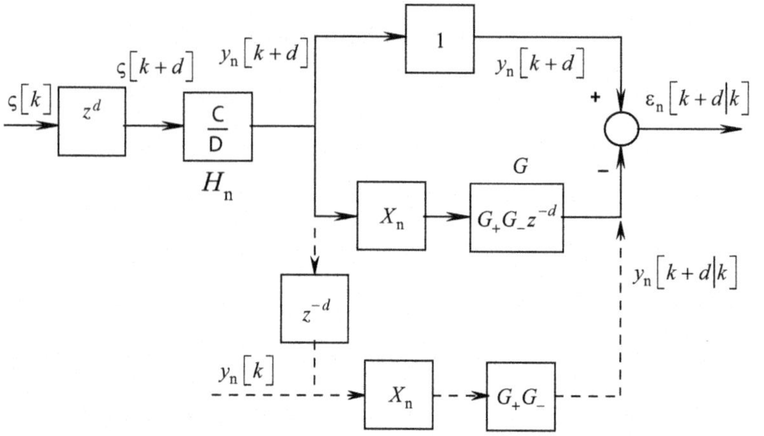

Figure 6.4 Prediction scheme of the output disturbance in *Youla parameterization* closed control loop.

Assume that the DT process is given by Equation (6.11) and determine the $\mathcal{H}_2$ norm for the error signal ε_n

$$
\begin{aligned}
J_n &= \left\| \frac{\mathcal{C}}{\mathcal{D}} \left(1 - X_n G_+ G_- z^{-d}\right) \right\|_{\mathcal{H}_2} \\
&= \left\| \frac{G_- z^{-d}}{G_-^-} \right\|_{\mathcal{H}_2} \left\| \frac{\mathcal{C}\mathcal{G}_-^-}{\mathcal{G}_- z^{-d} \mathcal{D}} - X_n \frac{\mathcal{C}\mathcal{G}_-^- \mathcal{G}_+}{\mathcal{D}} \right\|_{\mathcal{H}_2} \\
&= \left\| \frac{\mathcal{C}\mathcal{G}_-^-}{\mathcal{G}_- z^{-d} \mathcal{D}} - X_n \frac{\mathcal{C}\mathcal{G}_-^- \mathcal{G}_+}{\mathcal{D}} \right\|_{\mathcal{H}_2}
\end{aligned}
\tag{6.41}
$$

where $\mathcal{G}_-^-$, contrary to the CT systems, is now a polynomial derived by mirroring the zeros of $\mathcal{G}_-$ to outside the unit circle; thus the condition should be fulfilled for the new zeros z_i^*. In (6.41) $|z^{\pm jd\omega}| = 1$ and $\|\mathcal{G}_-/\mathcal{G}_-^-\|_{\mathcal{H}_2} = 1$ is taken into consideration, i.e., it is a norm-keeping factor which can be checked easily by simple examples. The minimization of (6.41) is solved by the Wiener method [158], i.e., the argument is decomposed to causal and noncausal parts

$$
\frac{\mathcal{C}\mathcal{G}_-^-}{\mathcal{D}\mathcal{G}_- z^{-d}} = \frac{\mathcal{T}}{\mathcal{G}_- z^{-d}} + \frac{\mathcal{P}}{\mathcal{D}} = \frac{\mathcal{D}\mathcal{T}}{\mathcal{D}\mathcal{G}_- z^{-d}} + \frac{\mathcal{P}\mathcal{G}_- z^{-d}}{\mathcal{D}\mathcal{G}_- z^{-d}} \tag{6.42}
$$

where $\mathcal{T}/\mathcal{G}_- z^{-d}$ is the noncausal term and $\mathcal{P}/\mathcal{D}$ the causal term, respectively. Then it can be written that

$$J_n = \left\| \frac{\mathcal{T}}{\mathcal{G}_- z^{-d}} \right\|_{\mathcal{H}_2} + \left\| \frac{\mathcal{P}}{\mathcal{D}} - X_n \frac{\mathcal{C}\mathcal{G}_-^- \mathcal{G}_+}{\mathcal{D}} \right\|_{\mathcal{H}_2} \geq \left\| \frac{\mathcal{T}}{\mathcal{G}_- z^{-d}} \right\|_{\mathcal{H}_2} \tag{6.43}$$

The loss function has its minimum when the second term is zero, and thus

$$\frac{\mathcal{P}}{\mathcal{D}} - X_n \frac{\mathcal{C}\mathcal{G}_-^- \mathcal{G}_+}{\mathcal{D}} = 0 \tag{6.44}$$

From this expression the optimality of the $\mathcal{H}_2$ norm can be ensured by the following serial compensator

$$X_n = \frac{\mathcal{P}}{\mathcal{C}\mathcal{G}_-^- \mathcal{G}_+} = \frac{\mathcal{P}}{\mathcal{C}} \frac{1}{\mathcal{G}_-^-} \frac{1}{\mathcal{G}_+} = R_n G_n G_+^{-1} = R_n K_n \tag{6.45}$$

Do not forget that now the serial compensator is directly computed, since the $\mathcal{H}_2$ optimal disturbance compensation did not give us the opportunity to design the predictor directly. After formal comparison with the parameters of the conventional *YP* regulator and the predictor of the *MV* regulator (6.10), the following parameters are obtained:

$$R_n = \frac{\mathcal{P}}{\mathcal{C}}; \quad K_n = G_n G_+^{-1} = \frac{1}{\mathcal{G}_-^-} \frac{1}{\mathcal{G}_+}; \quad G_n = \frac{1}{\mathcal{G}_-^-} \tag{6.46}$$

The value of $\mathcal{T}$ and $\mathcal{P}$ can be obtained by the solution of a special *DE*

$$\mathcal{C}\mathcal{G}_-^- = \mathcal{D}\mathcal{T} + \mathcal{P}\mathcal{G}_- z^{-d} \tag{6.47}$$

This expression can be derived from comparison of the two sides of (6.42). Using the expressions (2.1.25) and (6.43), we can obtain the $\mathcal{H}_2$ optimal *YP* regulator as

$$C_* = \frac{\mathcal{P}}{\mathcal{G}_+ \left(\mathcal{C}\mathcal{G}_-^- - \mathcal{P}\mathcal{G}_- z^{-d} \right)} = \frac{\mathcal{P}}{\mathcal{G}_+ \mathcal{D}\mathcal{T}} \quad \text{or} \quad C_* = \frac{H_n - \mathcal{T}'}{\mathcal{T}' G} \tag{6.48}$$

The minimum of J_n is

$$\min_{X_n} \{J_n\} = \left\| \frac{\mathcal{T}}{\mathcal{G}_- z^{-d}} \right\|_{\mathcal{H}_2} = \left\| \frac{\mathcal{G}_- z^{-d}}{\mathcal{G}_-^-} \right\|_{\mathcal{H}_2} \left\| \frac{\mathcal{T}}{\mathcal{G}_- z^{-d}} \right\|_{\mathcal{H}_2} = \left\| \frac{\mathcal{R}}{\mathcal{G}_-^-} \right\|_{\mathcal{H}_2}$$
$$= \left\| \frac{\mathcal{C}}{\mathcal{D}} (1 - X_n G) \right\|_{\mathcal{H}_2} \tag{6.49}$$

where using the *DE* (6.47), the form $\mathcal{T}' = \mathcal{T}/\mathcal{G}_-^-$ is reshaped as

$$\mathcal{T}' = \frac{\mathcal{T}}{\mathcal{G}_-^-} = \frac{\mathcal{C}\mathcal{G}_-^- - \mathcal{P}\mathcal{G}_- z^{-d}}{\mathcal{D}\mathcal{G}_-^-} = \frac{\mathcal{C}}{\mathcal{D}}\left(1 - \frac{\mathcal{P}}{\mathcal{C}\mathcal{G}_-^-\mathcal{G}_+}\mathcal{G}_+\mathcal{G}_- z^{-d}\right)$$
$$= \frac{\mathcal{C}}{\mathcal{D}}(1 - \mathrm{X_n G}) \tag{6.50}$$

This completely corresponds to the scheme shown in Figure 6.4. It is worth mentioning that the obtained regulator (6.48) is $\mathcal{H}_2$ optimal, but is not of *MV*, since the *DE* (6.47) does not provide independent prediction error. A similar algorithm was published by Peterka [137,138], who named this approach, unsettlingly, suboptimal solution.

The product $\mathcal{C}\mathcal{G}_-^-$ is on the left side of the *DE* (6.45). If the term $\mathcal{G}_-^-$, obtained by mirroring the zeros, is ordered according to the power of z^{-1} similarly to the other polynomials, then we obtain the form $\mathcal{G}_-^- = z^{m_-}\overline{\mathcal{G}}_-^-$, where m_- is the number of inverse unstable zeros. As far as the numerator of the noise model $\mathcal{C}/\mathcal{D}$ is extended by z^{-m_-}, then the absolute value of the noise model spectrum does not change, because $|z^{-m_-}\mathcal{C}| = |\mathcal{C}|$. Therefore, the same procedure can be followed as that recommended at the end of Section 4.2. It is therefore reasonable to assume the term $\mathcal{C} = \overline{\mathcal{C}}z^{-m_-}$ in the numerator of the noise model, in order to ensure the solvability of the *DE*.

CHAPTER 7

Control of Multivariable Processes

The state equation *multiple input multiple output* (*MIMO*), i.e., a multivariable linear dynamic system is formally the same as the single variable system given by (1.1.10). The main difference is that now both the input $\boldsymbol{u}$ and the output $\boldsymbol{y}$ are p-dimensional vectors. For simplicity, let the number of the input and output variables be the same and denoted by p (quadratic systems). Thus the state equation of the process takes the form

$$\frac{\mathrm{d}\boldsymbol{x}}{\mathrm{d}t} = \dot{\boldsymbol{x}} = \mathbf{A}\boldsymbol{x} + \mathbf{B}\boldsymbol{u} \qquad \boldsymbol{y} = \mathbf{C}\boldsymbol{x} + \mathbf{D}\boldsymbol{u} \tag{7.1}$$

where $\mathbf{A}$, $\mathbf{B}$, $\mathbf{C}$, and $\mathbf{D}$ are an $(n\times n)$, an $(n\times p)$, a $(p\times n)$, and a $(p\times p)$ matrix, respectively. The $(n\times n)$-dimensional *transfer function matrix* (*TFM*) of the *MIMO* processes is

$$\boldsymbol{P}(s) = \mathbf{C}(s\mathbf{I}-\mathbf{A})^{-1}\mathbf{B} + \mathbf{D} = \mathbf{C}\boldsymbol{\Phi}(s)\mathbf{B} + \mathbf{D} = \frac{1}{\mathcal{A}(s)}\boldsymbol{\mathcal{B}}(s) \tag{7.2}$$

where

$$\boldsymbol{\Phi}(s) = (s\mathbf{I}-\mathbf{A})^{-1} = \frac{1}{\mathcal{A}(s)}\boldsymbol{\Psi}(s) \quad \text{and} \quad \boldsymbol{\Psi}(s) = \mathbf{adj}(s\mathbf{I}-\mathbf{A}) \tag{7.3}$$

The scalar denominator

$$\mathcal{A}(s) = \det(s\mathbf{I}-\mathbf{A}) \tag{7.4}$$

is the n-th-degree characteristic polynomial of the process. $\boldsymbol{\Phi}$ and $\boldsymbol{\Psi}$ are also $(n\times n)$-dimensional. The form (7.2) means the simplest *MIMO* process model, though $\boldsymbol{P}(s)$ is not necessarily minimal and may be reduced. The right side of (7.2) is usually also called the "naïve" model of the *MIMO* process. (In this chapter the parameter matrices of the state equation are denoted by bold upright fonts and cursive bold fonts denote the *TFM*.)

The *TFM* $\boldsymbol{P}(s)$ of the *MIMO* process can always be transformed into the Smith form [64]

$$\boldsymbol{P}(s) = \boldsymbol{\mathcal{U}}(s)\boldsymbol{D}(s)\boldsymbol{\mathcal{V}}(s) \tag{7.5}$$

where $\mathcal{U}(s)$ and $\mathcal{V}(s)$ are unimodular polynomial matrices. (A matrix is called unimodular, if its determinant is scalar constant.) The inverse of the unimodular polynomial matrices is also a polynomial matrix. The $\mathbf{D}(s)$ in (7.5) is a special *TFM*

$$\begin{aligned}\mathbf{D}(s) &= \mathbf{D}_{\mathrm{D}}(s)\mathcal{D}_{\mathrm{N}}(s) = \mathcal{D}_{\mathrm{N}}(s)\mathbf{D}_{\mathrm{D}}(s) = \mathcal{D}_{\mathrm{D}}^{-1}(s)\mathcal{D}_{\mathrm{N}}(s) = \mathcal{D}_{\mathrm{N}}(s)\mathcal{D}_{\mathrm{D}}^{-1}(s)\\ &= \mathbf{diag}\left\langle \frac{\mathcal{E}_1}{\psi_1}, \frac{\mathcal{E}_2}{\psi_2}, \ldots, \frac{\mathcal{E}_{\mathrm{r}}}{\psi_{\mathrm{r}}}, 0, \ldots, 0 \right\rangle = \mathbf{diag}\langle \mathcal{L}_1, \mathcal{L}_2, \ldots, \mathcal{L}_{\mathrm{r}}, 0, \ldots, 0\rangle\end{aligned} \tag{7.6}$$

where the multiplication of the diagonal matrices is commutative. In (7.6)

$$\mathcal{D}_{\mathrm{N}}(s) = \mathbf{diag}\langle \mathcal{E}_1(s), \mathcal{E}_2(s), \ldots, \mathcal{E}_{\mathrm{r}}(s), 0, \ldots, 0\rangle; \quad \mathcal{L}_i(s) = \frac{\mathcal{E}_i(s)}{\psi_i(s)} \tag{7.7}$$

and

$$\mathbf{D}_{\mathrm{D}}(s) = \mathcal{D}_{\mathrm{D}}^{-1}(s); \quad \mathcal{D}_{\mathrm{D}}(s) = \mathbf{diag}\langle \psi_1(s), \psi_2(s), \ldots, \psi_{\mathrm{r}}(s), 0, \ldots, 0\rangle \tag{7.8}$$

The polynomials $\mathcal{E}_i(s)$ and $\psi_i(s)$ are relative primes. The expression obtained by the rearrangement of (7.6)

$$\mathbf{D}(s) = \mathcal{U}^{-1}(s)\mathbf{P}(s)\mathcal{V}^{-1}(s) \tag{7.9}$$

is called the Smith–McMillan form of the *MIMO* process [30,31]. In the diagonal there are r nonzero elements, where $r \leq n$ is the so-called McMillan degree. (Here the *bold* fonts note the polynomial matrices.) Note that, in general, the multiplication of the polynomial matrices is not commutative.

Based on the Smith form (7.5), the *MIMO* process can be rearranged to two equivalent, so-called *matrix fractional descriptions* (*MFDs*)

$$\mathbf{P}(s) = \mathcal{N}_{\mathrm{R}}(s)\mathcal{D}_{\mathrm{R}}^{-1}(s) = \mathcal{D}_{\mathrm{L}}^{-1}(s)\mathcal{N}_{\mathrm{L}}(s) \tag{7.10}$$

where $\mathcal{N}_{\mathrm{R}}(s)$, $\mathcal{D}_{\mathrm{R}}(s)$ and $\mathcal{N}_{\mathrm{L}}(s)$, $\mathcal{D}_{\mathrm{L}}(s)$ are the polynomial matrices of the right- and the left-*MFD*, respectively. These polynomial matrices, at least formally, can be simply computed

$$\mathcal{D}_{\mathrm{L}}(s) = \mathcal{D}_{\mathrm{D}}(s)\mathcal{U}^{-1}(s); \quad \mathcal{N}_{\mathrm{L}}(s) = \mathcal{D}_{\mathrm{N}}(s)\mathcal{V}(s) \tag{7.11}$$

and

$$\mathcal{D}_{R}(s) = \mathcal{V}^{-1}(s)\mathcal{D}_{\mathrm{D}}(s); \quad \mathcal{N}_{\mathrm{R}}(s) = \mathcal{U}(s)\mathcal{D}_{\mathrm{N}}(s) \tag{7.12}$$

The roots of the scalar polynomials are $\mathcal{E}_i(s)$ in $\mathcal{D}_{\mathrm{N}}(s)$, and thus the roots of $\prod_{i=1}^{r} \mathcal{E}_i(s)$ are the zeros of the *TFM* $\mathbf{P}(s)$.

In the control design of the *single input single output* (*SISO*) processes, the decomposition of the process into *inverse stable* (*IS*) and *inverse unstable* (*IU*) factors is usually a requirement. The *TFM* $\boldsymbol{P}$ of a *MIMO* process can be decomposed in a similar way

$$\boldsymbol{P} = \boldsymbol{P}_{-}\boldsymbol{P}_{+} \neq \boldsymbol{P}_{+}\boldsymbol{P}_{-} \tag{7.13}$$

where $\boldsymbol{P}_{+}$ and $\boldsymbol{P}_{-}$ are the *IS* and *IU* matrix operators (*TFM*), respectively. Obviously $\boldsymbol{P}$ can always be written in another equivalent form

$$\boldsymbol{P} = \bar{\boldsymbol{P}}_{+}\bar{\boldsymbol{P}}_{-} \neq \bar{\boldsymbol{P}}_{-}\bar{\boldsymbol{P}}_{+} \tag{7.14}$$

With the left- and right-*MFD* representations of (7.13) and (7.14), the following equivalent form is obtained

$$\boldsymbol{P}(s) = \boldsymbol{\mathcal{N}}_{\mathrm{R}}^{-}(s)\boldsymbol{\mathcal{N}}_{\mathrm{R}}^{+}(s)\boldsymbol{\mathcal{D}}_{\mathrm{R}}^{-1}(s) = \boldsymbol{\mathcal{D}}_{\mathrm{L}}^{-1}(s)\boldsymbol{\mathcal{N}}_{\mathrm{L}}^{+}(s)\boldsymbol{\mathcal{N}}_{\mathrm{L}}^{-}(s) \tag{7.15}$$

where the factorized factors are

$$\begin{aligned} &\boldsymbol{P}_{-}(s) = \boldsymbol{\mathcal{N}}_{\mathrm{R}}^{-}(s); \quad \boldsymbol{P}_{+}(s) = \boldsymbol{\mathcal{N}}_{\mathrm{R}}^{+}(s)\boldsymbol{\mathcal{D}}_{\mathrm{R}}^{-1}(s); \\ &\bar{\boldsymbol{P}}_{-}(s) = \boldsymbol{\mathcal{N}}_{\mathrm{L}}^{-}(s); \quad \bar{\boldsymbol{P}}_{+}(s) = \boldsymbol{\mathcal{D}}_{\mathrm{L}}^{-1}(s)\boldsymbol{\mathcal{N}}_{\mathrm{L}}^{+}(s) \end{aligned} \tag{7.16}$$

It is assumed that the factorization can always be done in the numerator matrix polynomials, and thus

$$\mathcal{N}_{\mathrm{R}}(s) = \boldsymbol{\mathcal{N}}_{\mathrm{R}}^{-}(s)\boldsymbol{\mathcal{N}}_{\mathrm{R}}^{+}(s) \quad \text{and} \quad \mathcal{N}_{\mathrm{L}}(s) = \boldsymbol{\mathcal{N}}_{\mathrm{L}}^{+}(s)\boldsymbol{\mathcal{N}}_{\mathrm{L}}^{-}(s) \tag{7.17}$$

YOULA-PARAMETERIZED *MIMO* CLOSED-LOOP CONTROL

Formally it is very easy to extend the *Youla parameterization* (*YP*) to *MIMO* processes by introducing the *TFM*

$$\mathbf{Q} = \mathbf{C}(\mathbf{I} - \mathbf{PC})^{-1} = (\mathbf{I} - \mathbf{CP})^{-1}\mathbf{C} \tag{7.18}$$

which results in the following *YP MIMO* regulator

$$\mathbf{C} = (\mathbf{I} - \mathbf{QP})^{-1}\mathbf{Q} = \mathbf{Q}(\mathbf{I} - \mathbf{PQ})^{-1} \tag{7.19}$$

(see Figure 7.1). Here $\boldsymbol{P}$ is assumed to be stable. It can be easily verified that the two sides of (7.18) and (7.19) are the same.

The following identities have important roles in the investigation of the *TFM* of the *MIMO* closed-loop control

$$\begin{aligned} (\mathbf{I} + \mathbf{PC})^{-1} &= \left[\mathbf{I} + \mathbf{PQ}(\mathbf{I} - \mathbf{PQ})^{-1}\right]^{-1} = \mathbf{I} - \mathbf{PQ} \\ (\mathbf{I} + \mathbf{CP})^{-1} &= \left[\mathbf{I} + (\mathbf{I} - \mathbf{QP})^{-1}\mathbf{QP}\right]^{-1} = \mathbf{I} - \mathbf{QP} \end{aligned} \tag{7.20}$$

and can be proved by using the expressions (see (A.1.17))

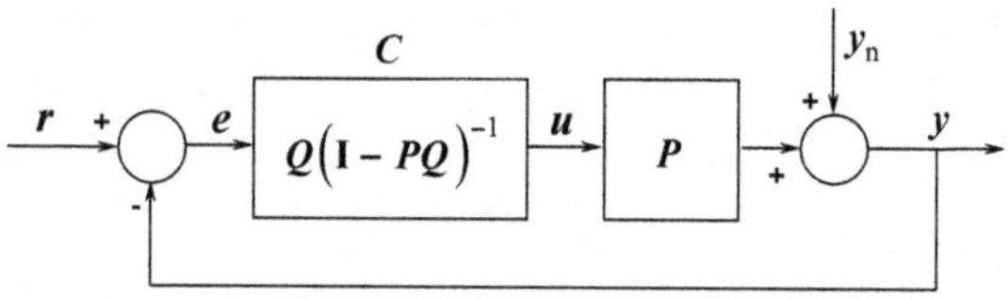

Figure 7.1 Generalization of *Youla parameterization* for *multiple input multiple output* processes.

$$\left[\mathbf{I}+(\mathbf{I}-\mathbf{A})^{-1}\mathbf{A}\right]^{-1}=\mathbf{I}-\mathbf{A} \quad \text{and} \quad \left[\mathbf{I}+\mathbf{B}(\mathbf{I}-\mathbf{B})^{-1}\right]^{-1}=\mathbf{I}-\mathbf{B} \tag{7.21}$$

The overall transfer characteristics of the *YP* closed system shown in Figure 7.1 can be obtained by simple calculations

$$y=\boldsymbol{PQ}r+(\mathbf{I}-\boldsymbol{PQ})y_{\mathrm{n}} \tag{7.22}$$

but it should be taken into account that the multiplication of the *TFM* is not commutative. Here the *Keviczky–Bányász* (*KB*) parameterization introduced in *SISO* processes can also be applied, so multiplication by the prefilter $\mathbf{Q}^{-1}$ results in the *two-degree-of-freedom* (*TDOF*) *MIMO* closed system of Figure 7.2, where the overall transfer characteristic is

$$y=\boldsymbol{P}r+(\mathbf{I}-\boldsymbol{PQ})y_{\mathrm{n}} \tag{7.23}$$

and virtually opens the closed loop. Note that the *KB* parameterization can be applied for all closed-loop controls, not only for the *YP* loops and it always virtually opens the loop, thus ensuring $\boldsymbol{Pr}$ tracking properties. The noise rejection property ($\mathbf{I}-\boldsymbol{PQ}$), however, appears only in the case of *YP*.

Generalizing the *generic TDOF* closed system of (2.1.3) and Figure 2.1.7 of the *SISO* processes to *MIMO* processes, we obtain the closed loop shown in Figure 7.3.

The overall characteristics of the generic closed system are

$$y=\boldsymbol{PQ}_{\mathrm{r}}y_{\mathrm{r}}+(\boldsymbol{I}-\boldsymbol{PQ}_{\mathrm{n}})y_{\mathrm{n}}=\boldsymbol{PK}_{\mathrm{r}}\boldsymbol{R}_{\mathrm{r}}y_{\mathrm{r}}+(\boldsymbol{I}-\boldsymbol{PK}_{\mathrm{n}}\boldsymbol{R}_{\mathrm{n}})y_{\mathrm{n}}=y_{\mathrm{t}}+y_{\mathrm{d}} \tag{7.24}$$

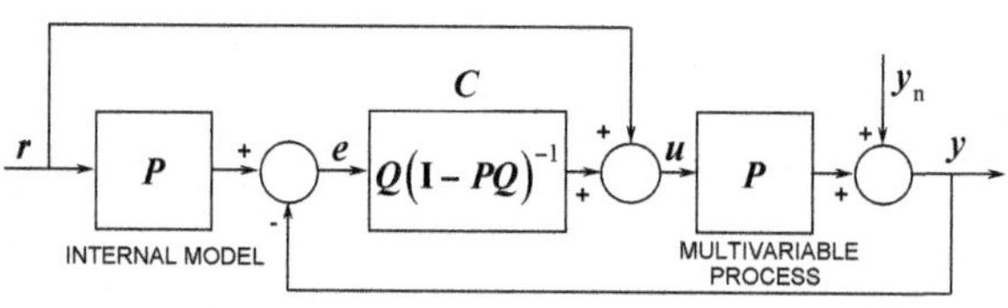

Figure 7.2 *KB*-parameterized *multiple input multiple output* closed-loop control.

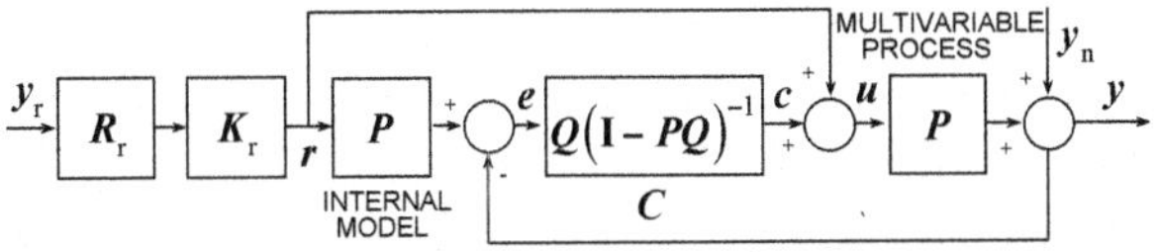

Figure 7.3 *Generic two-degree-of-freedom* closed loop of *MIMO* processes.

Assume that the stable *MIMO* process $\boldsymbol{P}$ can be decomposed according to (7.13). Then the *MIMO* Youla parameters are

$$\mathbf{Q} = \mathbf{Q}_n = \boldsymbol{K}_n\boldsymbol{R}_n = \boldsymbol{P}_+^{-1}\mathbf{G}_n\boldsymbol{R}_n \tag{7.25}$$

and

$$\mathbf{Q}_r = \boldsymbol{K}_r\boldsymbol{R}_r = \boldsymbol{P}_+^{-1}\mathbf{G}_r\boldsymbol{R}_r; \quad \boldsymbol{K}_n = \boldsymbol{P}_+^{-1}\mathbf{G}_n; \quad \boldsymbol{K}_r = \boldsymbol{P}_+^{-1}\mathbf{G}_r \tag{7.26}$$

The Youla-parametcrized *MIMO* regulator is

$$\begin{aligned} \boldsymbol{C} &= \mathbf{Q}(\mathbf{I}-\boldsymbol{P}\mathbf{Q})^{-1} = \boldsymbol{K}_n\boldsymbol{R}_n(\mathbf{I}-\boldsymbol{P}\boldsymbol{K}_n\boldsymbol{R}_n)^{-1} \\ &= \boldsymbol{P}_+^{-1}\mathbf{G}_n\boldsymbol{R}_n(\mathbf{I}-\boldsymbol{P}_-\mathbf{G}_n\boldsymbol{R}_n)^{-1} \end{aligned} \tag{7.27}$$

Using the expressions (7.25)–(7.27), the obtained closed system takes the form

$$\boldsymbol{y} = \boldsymbol{P}_-\mathbf{G}_r\boldsymbol{R}_r\boldsymbol{y}_r + (\mathbf{I}-\boldsymbol{P}_-\mathbf{G}_n\boldsymbol{R}_n)\boldsymbol{y}_n = \boldsymbol{y}_t + \boldsymbol{y}_d \tag{7.28}$$

where, similar to the *SISO* case, $\boldsymbol{y}_t$ and $\boldsymbol{y}_d$ mean the tracking and noise rejection properties, respectively. Here $\boldsymbol{K}_r$ and $\boldsymbol{K}_n$ contain the inverse $\boldsymbol{P}_+^{-1}$ of the invertible part $\boldsymbol{P}_+$ of $\boldsymbol{P}$, furthermore $\mathbf{G}_r$ and $\mathbf{G}_n$ attenuate the effect of the invariant factor $\boldsymbol{P}_-$.

If the process $\boldsymbol{P}$ is decomposed according to (7.14), then we obtain the Youla-parameterized *MIMO* regulator as

$$\bar{\boldsymbol{C}} = \left(\mathbf{I}-\bar{\mathbf{Q}}\bar{\boldsymbol{P}}\right)^{-1}\bar{\mathbf{Q}} = \left(\mathbf{I}-\boldsymbol{R}_n\bar{\boldsymbol{K}}_n\bar{\boldsymbol{P}}\right)^{-1}\boldsymbol{R}_n\bar{\boldsymbol{K}}_n = \left(\mathbf{I}-\boldsymbol{R}_n\bar{\mathbf{G}}_n\bar{\boldsymbol{P}}_-\right)^{-1}\boldsymbol{R}_n\bar{\mathbf{G}}_n\bar{\boldsymbol{P}}_+^{-1} \tag{7.29}$$

where

$$\bar{\boldsymbol{K}}_n = \bar{\mathbf{G}}_n\bar{\boldsymbol{P}}_+^{-1} \tag{7.30}$$

Now the equation of the closed system becomes

$$\boldsymbol{y} = \boldsymbol{P}_-\mathbf{G}_r\boldsymbol{R}_r\boldsymbol{y}_r + \left(\mathbf{I}-\boldsymbol{R}_n\bar{\mathbf{G}}_n\bar{\boldsymbol{P}}_-\right)\boldsymbol{y}_n = \boldsymbol{y}_t + \bar{\boldsymbol{y}}_d \tag{7.31}$$

It is clear that the tracking property $\boldsymbol{y}_t$ is the same for the two types of the decomposition but the noise rejection properties $\boldsymbol{y}_d$ and $\bar{\boldsymbol{y}}_d$, however, may be different.

Note that whereas for the *SISO* case the realizability of the Youla-parameterized regulator can be simply ensured by the reasonable choice of the pole access of the reference models R_r and R_n, the same cannot be said for the *MIMO* case. It is true that in many cases, raising the pole access of the elements in the main diagonal of $\boldsymbol{R}_r$ and $\boldsymbol{R}_n$ helps the realizability, if they are given in *TFM* form. The general condition, however, always needs further, thorough, investigation.

YOULA-PARAMETERIZED *MIMO* REGULATOR FOR THE "NAÏVE" PROCESS MODEL

The derivation of the regulators (7.27) and (7.29) requires complex operations between the *TFM*. This computation demand can be slightly decreased by using the "naïve" model shown in (7.2). In this case, the decomposition (7.13) takes the form

$$\boldsymbol{P}(s) = \boldsymbol{P}_{-}\boldsymbol{P}_{+} = \frac{1}{\mathcal{A}(s)}\boldsymbol{\mathcal{B}}(s) = \frac{1}{\mathcal{A}(s)}\boldsymbol{\mathcal{B}}_{-}(s)\boldsymbol{\mathcal{B}}_{+}(s) \tag{7.32}$$

where the designated operation with the polynomial $\mathcal{A}(s)$ in the denominator can be replaced by matrix polynomials.

For the model (7.31), we get

$$\boldsymbol{P} = \frac{1}{\mathcal{A}(s)}\boldsymbol{\mathcal{B}}_{+}(s); \quad \boldsymbol{\mathcal{B}}_{+} = \boldsymbol{\mathcal{B}}; \quad \boldsymbol{\mathcal{B}}_{-} = \mathbf{I} \tag{7.33}$$

Let the reference models be given in the "naïve" form, i.e.,

$$\boldsymbol{R}_r = \frac{1}{\mathcal{A}_r(s)}\boldsymbol{\mathcal{B}}_r(s) \quad \text{and} \quad \boldsymbol{R}_n = \frac{1}{\mathcal{A}_n(s)}\boldsymbol{\mathcal{B}}_n(s) \tag{7.34}$$

If $\boldsymbol{\mathcal{B}}_{-} = \mathbf{I}$ and $\boldsymbol{\mathcal{B}}_{+} = \boldsymbol{\mathcal{B}}$, then further optimization is impossible, so it is reasonable to choose $\mathbf{G}_r = \mathbf{G}_n = \mathbf{I}$. In this case, the *MIMO* Youla parameter is

$$\mathbf{Q} = \mathbf{Q}_n = \mathcal{A}(s)\boldsymbol{\mathcal{B}}^{-1}(s)\boldsymbol{R}_n = \frac{\mathcal{A}(s)}{\mathcal{A}_n(s)}\boldsymbol{\mathcal{B}}^{-1}(s)\boldsymbol{\mathcal{B}}_n(s) \tag{7.35}$$

and the Youla-parameterized *MIMO* regulator becomes

$$\begin{aligned}\boldsymbol{C}(s) &= \mathcal{A}(s)\boldsymbol{\mathcal{B}}^{-1}(s)\boldsymbol{R}_n(s)[\mathbf{I} - \boldsymbol{R}_n(s)]^{-1} \\ &= \mathcal{A}(s)\boldsymbol{\mathcal{B}}_{+}^{-1}(s)\boldsymbol{\mathcal{B}}_n(s)[\mathcal{A}_n(s)\mathbf{I} - \boldsymbol{\mathcal{B}}_n(s)]^{-1}\end{aligned} \tag{7.36}$$

CONTROL OF INVERSE STABLE *MIMO* PROCESS MODELS

In many practical cases, the *MIMO* process model is given in a special *IS* form. This is especially valid for sampled processes

$$\mathbf{G} = \mathbf{G}_{-}\mathbf{G}_{+} = z^{-d}\bar{\mathbf{G}}_{+} = \bar{\mathbf{G}}_{+}\bar{\mathbf{G}}_{-} = \bar{\mathbf{G}}_{+}z^{-d}; \quad \mathbf{G}_{+} = \bar{\mathbf{G}}_{+} \tag{7.37}$$

For all inputs in the main diagonal, the time delay is z^{-d}. All other variants can be taken into account in $\mathbf{G}_{+}$. In this case, the Youla parameter is

$$\mathbf{Q} = \mathbf{G}_{+}^{-1}\mathbf{R}_{\mathrm{n}}; \quad \bar{\mathbf{Q}} = \mathbf{R}_{\mathrm{n}}\bar{\mathbf{G}}_{+}^{-1} \tag{7.38}$$

With these parameters, the regulator (7.27) and (7.29) becomes

$$\begin{aligned} \mathbf{C} &= \mathbf{Q}(\mathbf{I} - \mathbf{P}\mathbf{Q})^{-1} = \mathbf{G}_{+}^{-1}\mathbf{R}_{\mathrm{n}}\left(\mathbf{I} - \mathbf{R}_{\mathrm{n}}z^{-d}\right)^{-1} = \mathbf{G}_{+}^{-1}\left(\mathbf{I} - \mathbf{R}_{\mathrm{n}}z^{-d}\right)^{-1}\mathbf{R}_{\mathrm{n}} \\ \bar{\mathbf{C}} &= \left(\mathbf{I} - \bar{\mathbf{Q}}\bar{\mathbf{P}}\right)^{-1}\bar{\mathbf{Q}} = \left(\mathbf{I} - \mathbf{R}_{\mathrm{n}}z^{-d}\right)^{-1}\mathbf{R}_{\mathrm{n}}\mathbf{G}_{+}^{-1} = \mathbf{R}_{\mathrm{n}}\left(\mathbf{I} - \mathbf{R}_{\mathrm{n}}z^{-d}\right)^{-1}\mathbf{G}_{+}^{-1} \end{aligned} \tag{7.39}$$

Here $\mathbf{G}_{+} = \bar{\mathbf{G}}_{+}$ is considered and the identity

$$\mathbf{R}_{\mathrm{n}}\left(\mathbf{I} - \mathbf{R}_{\mathrm{n}}z^{-d}\right)^{-1} = \left(\mathbf{I} - \mathbf{R}_{\mathrm{n}}z^{-d}\right)^{-1}\mathbf{R}_{\mathrm{n}} \tag{7.40}$$

can be easily checked. The closed systems for the two types of regulators are exactly the same

$$\begin{aligned} \mathbf{y} &= \mathbf{R}_{\mathrm{r}}z^{-d}\mathbf{y}_{\mathrm{r}} + \left(\mathbf{I} - \mathbf{R}_{\mathrm{n}}z^{-d}\right)\mathbf{y}_{\mathrm{n}} = \mathbf{y}_{\mathrm{t}} + \mathbf{y}_{\mathrm{d}} \\ \mathbf{y} &= \mathbf{R}_{\mathrm{r}}z^{-d}\mathbf{y}_{\mathrm{r}} + \left(\mathbf{I} - \mathbf{R}_{n}z^{-d}\right)\mathbf{y}_{\mathrm{n}} = \mathbf{y}_{\mathrm{t}} + \bar{\mathbf{y}}_{\mathrm{d}} \end{aligned} \tag{7.41}$$

and thus $\mathbf{y}_{\mathrm{d}} = \bar{\mathbf{y}}_{\mathrm{d}}$. Note that in this case, $\mathbf{G}_{\mathrm{r}} = \mathbf{G}_{\mathrm{n}} = \mathbf{I}$ has been chosen since the effect of the invariant factor $\mathbf{G}_{-} = z^{-d}\mathbf{I}$ cannot be attenuated.

DECOUPLING CONTROL OF *MIMO* PROCESS MODELS

Generally, in *MIMO* process models each input signal has an effect on each output signal. The same is valid for all the elements of the output disturbance. It is an important practical task to construct a control system where each reference signal has effect only on the corresponding output signal. Similarly, it is a favorable case when a certain output disturbance has an effect on a given output signal and has no effect at all on the other outputs. The joint solution of the above tasks is known as decoupling or decoupling control. The practical solutions available in the literature usually apply two approaches [131,139,144].

The first approach applies state feedback whereby the decoupling vector can be chosen by algebraic methods for partial or complete decoupling. These methods are very complicated, do not illustrate well how the decoupling operates, and therefore are not widely used in engineering practice [139].

The other approach introduces process model structures (P and V structures), which handle the feed-forward and feedback elements of the *TFM* separately. The analysis of these elements makes the design of the necessary control easier though it does not provide a systematic solution and does not give the theoretical limits of the decoupling [144].

Let us investigate the decoupling in sampled systems where the *TFM* of the process is assumed to be

$$\mathbf{G} = \mathbf{G}_{\mathrm{D}} + \mathbf{G}_{\mathrm{A}} = \mathbf{G}_{\mathrm{D}}\left(\boldsymbol{I} + \mathbf{G}_{\mathrm{D}}^{-1}\mathbf{G}_{\mathrm{A}}\right) = \left(\boldsymbol{I} + \mathbf{G}_{\mathrm{A}}\mathbf{G}_{\mathrm{D}}^{-1}\right)\mathbf{G}_{\mathrm{D}} \tag{7.42}$$

Here $\mathbf{G}_{\mathrm{D}}$ contains the diagonal elements, i.e., it is a diagonal matrix, and $\mathbf{G}_{\mathrm{A}}$ contains the elements outside the diagonal (antidiagonal elements) in the original structure. The block scheme of the *MIMO* processes is usually feed-forward as shown in Figure 7.4 for the two-variable case. The operation of the decoupling regulators is usually demonstrated on two-input/two-output simple *MIMO* systems where the essence of the method can be understood in the simplest way. In industrial practice the input and output variables are usually considered in pairs if the technology allows. These kinds of schema are used below to illustrate the methods.

The decoupler, serially connected with the *MIMO* process and providing the decoupling effect, is denoted by $\boldsymbol{D}$. One of the most natural decouplings can be provided by the compensator $\boldsymbol{D} = \boldsymbol{D}_{\mathrm{o}} = \mathbf{G}^{-1}$, i.e., by

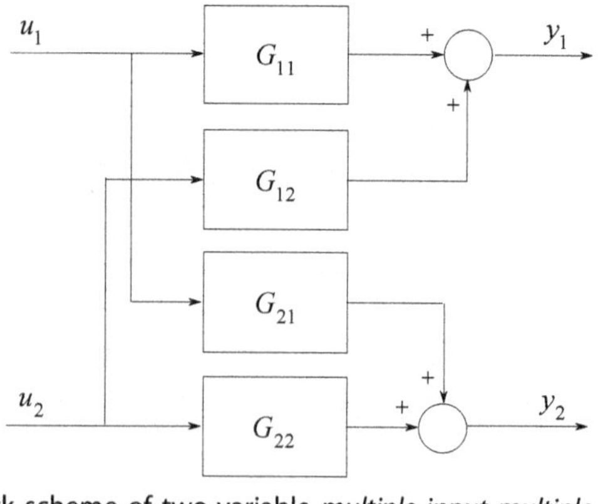

Figure 7.4 Block scheme of two-variable *multiple input multiple output* process.

the inverse of the process, which would mean complete decoupling $\boldsymbol{D}_o\boldsymbol{G} = \boldsymbol{I}$. But the inverse is usually not realizable and there is almost never any need to eliminate the complete dynamics of the process. In general, the structure of the decoupler $\boldsymbol{D}$ corresponds to the process model shown in Figure 7.4 if the elements G_{ij} are simply substituted by D_{ij}.

Considering the engineering aspects, the ideal decoupling would be when the main diagonal contains the process dynamics $\mathbf{G}_D$ provided by the following compensator

$$\boldsymbol{D} = \boldsymbol{D}_i = \mathbf{G}^{-1}\mathbf{G}_D \tag{7.43}$$

Observe that this case also requires the inverse of the process, though in certain cases there are more opportunities to realize the elements of $\mathbf{G}^{-1}\mathbf{G}_D$ than those of $\mathbf{G}^{-1}$.

There are models for decoupling wherein the feed-forward and feed-back effects appear to be mixed. Such topology is shown in Figure 7.5. This structure is called V-topology or inverse (inverted) structure [57,144].

The relationships of the resulting model can be written as

$$\begin{bmatrix} u_1 \\ u_2 \end{bmatrix} = \begin{bmatrix} V_{11} & 0 \\ 0 & V_{22} \end{bmatrix} \begin{bmatrix} e_1 \\ e_2 \end{bmatrix} + \begin{bmatrix} V_{11} & 0 \\ 0 & V_{22} \end{bmatrix} \begin{bmatrix} 0 & V_{12} \\ V_{21} & 0 \end{bmatrix} \begin{bmatrix} u_1 \\ u_2 \end{bmatrix} \tag{7.44}$$

Analogously with the notations introduced in (7.42), we can now write that

$$\boldsymbol{u} = \boldsymbol{V}_D\boldsymbol{e} + \boldsymbol{V}_D\boldsymbol{V}_A\boldsymbol{u} \tag{7.45}$$

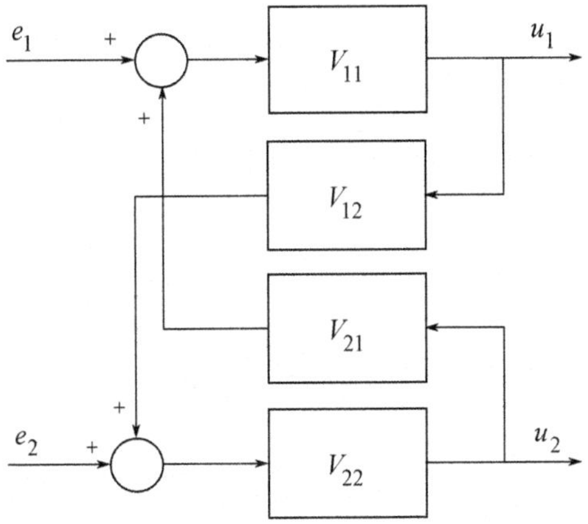

Figure 7.5 Block scheme of the decoupler of V-topology.

where the decomposition

$$\boldsymbol{V} = \boldsymbol{V}_{\mathrm{D}} + \boldsymbol{V}_{\mathrm{A}} \tag{7.46}$$

for diagonal and antidiagonal components is similar to what was used for the process model. Given (7.45) we can write that

$$\boldsymbol{u} = (\boldsymbol{I} - \boldsymbol{V}_{\mathrm{D}}\boldsymbol{V}_{\mathrm{A}})^{-1}\boldsymbol{V}_{\mathrm{D}}\boldsymbol{e} = \boldsymbol{D}_{\mathrm{V}}\boldsymbol{e} \tag{7.47}$$

The V-topology can be simply used for the design of a decoupling compensator. Let us use the following equation for the design of the decoupling

$$\boldsymbol{G}\boldsymbol{D}_{\mathrm{V}} = \boldsymbol{G}_{\mathrm{D}}\left(\boldsymbol{I} + \boldsymbol{G}_{\mathrm{D}}^{-1}\boldsymbol{G}_{\mathrm{A}}\right)(\boldsymbol{I} - \boldsymbol{V}_{\mathrm{D}}\boldsymbol{V}_{\mathrm{A}})^{-1}\boldsymbol{V}_{\mathrm{D}} \tag{7.48}$$

Observe that if in the decoupler $\boldsymbol{V}_{\mathrm{D}}\boldsymbol{V}_{\mathrm{A}} = -\boldsymbol{G}_{\mathrm{D}}^{-1}\boldsymbol{G}_{\mathrm{A}}$ is chosen then the ideal decoupling $\boldsymbol{G}\boldsymbol{D}_{\mathrm{V}} = \boldsymbol{G}_{\mathrm{D}}\boldsymbol{V}_{\mathrm{D}}$ is ensured. Note that the elements of $\boldsymbol{V}_{\mathrm{D}}$ can provide the decoupled single variable regulator in the control loop. The realization of the above compensation is ensured by the following choices

$$\boldsymbol{V}_{\mathrm{A}} = -\boldsymbol{G}_{\mathrm{A}}; \quad \boldsymbol{V}_{\mathrm{D}} = \boldsymbol{G}_{\mathrm{D}}^{-1} \tag{7.49}$$

These relationships explain the introduction of the V-topology since the prescribed operations are so simple that they can be performed manually.

From the design relationships (7.49) it can be seen that the V-topology shown here corresponds to the following decoupling compensator

$$\boldsymbol{D}_{\mathrm{V}} = \boldsymbol{G}^{-1}\boldsymbol{G}_{\mathrm{D}}\boldsymbol{V}_{\mathrm{D}} \tag{7.50}$$

The decoupling tasks inspired the introduction of a very useful structure which can be seen on the right side of Figure 7.6 between the variables $\boldsymbol{c}$ and $\boldsymbol{u}$. Let us call this U-topology, where U (unity) refers to the channels with unity transfer. It is clear from the comparison with the V-topology

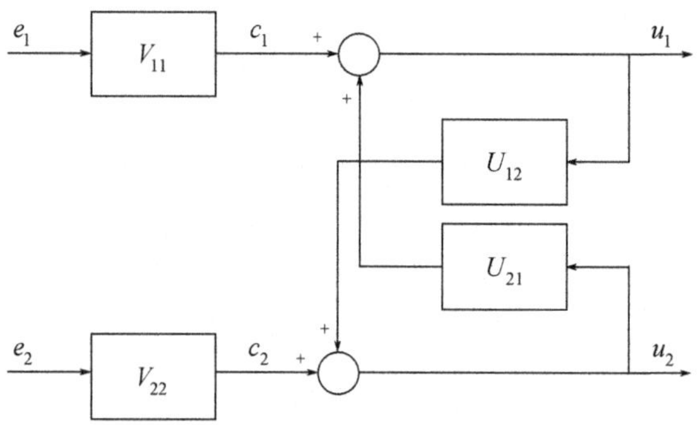

Figure 7.6 Block scheme of the decoupler of the U-topology.

that the U-topology can be obtained by choosing $U_{11} = 1$ and $U_{22} = 1$, and thus it corresponds to $\boldsymbol{V}_{\mathrm{D}} = \boldsymbol{I}$. Let us write $\boldsymbol{D}_{\mathrm{U}}$ for this case. Substituting $\boldsymbol{V}_{\mathrm{D}} = \boldsymbol{I}$ into $\boldsymbol{D}_{\mathrm{V}}$, we get

$$\boldsymbol{D}_{\mathrm{U}} = \boldsymbol{D}_{\mathrm{V}}\big|_{\boldsymbol{V}_{\mathrm{D}}=\boldsymbol{I}} = (\boldsymbol{I} - \boldsymbol{U}_{\mathrm{A}})^{-1} \tag{7.51}$$

Here, it is assumed that $\boldsymbol{U} = \boldsymbol{I} + \boldsymbol{U}_{\mathrm{A}}$ complies with the notations in (7.42) and (7.46). After identical rearrangements, we get

$$\boldsymbol{D}_{\mathrm{U}} = (\boldsymbol{I} - \boldsymbol{U}_{\mathrm{A}})^{-1} = \left[\boldsymbol{G}_{\mathrm{D}}(\boldsymbol{I} - \boldsymbol{U}_{\mathrm{A}})\right]^{-1}\boldsymbol{G}_{\mathrm{D}} = (\boldsymbol{G}_{\mathrm{D}} - \boldsymbol{G}_{\mathrm{D}}\boldsymbol{U}_{\mathrm{A}})^{-1}\boldsymbol{G}_{\mathrm{D}} \tag{7.52}$$

It is clear that with $\boldsymbol{U}_{\mathrm{A}} = -\boldsymbol{G}_{\mathrm{D}}^{-1}\boldsymbol{G}_{\mathrm{A}}$ the final form of the decoupler becomes

$$\boldsymbol{D}_{\mathrm{U}} = (\boldsymbol{G}_{\mathrm{D}} + \boldsymbol{G}_{\mathrm{A}})^{-1}\boldsymbol{G}_{\mathrm{D}} = \boldsymbol{G}^{-1}\boldsymbol{G}_{\mathrm{D}} \tag{7.53}$$

Using the compensator, we obtain decoupling as follows:

$$\boldsymbol{G}\boldsymbol{D}_{\mathrm{U}} = (\boldsymbol{G}_{\mathrm{D}} + \boldsymbol{G}_{\mathrm{A}})(\boldsymbol{G}_{\mathrm{D}} + \boldsymbol{G}_{\mathrm{A}})^{-1}\boldsymbol{G}_{\mathrm{D}} = \boldsymbol{G}_{\mathrm{D}} \tag{7.54}$$

Compared with the V-topology, the effect of the main diagonal elements is still missing. It can be easily substituted if a diagonal element $\boldsymbol{V}_{\mathrm{D}}$ is serially connected to the compensator $\boldsymbol{D}_{\mathrm{U}}$. This effect is illustrated on the left side of Figure 7.6 between the variables $\boldsymbol{e}$ and $\boldsymbol{c}$. This means at the same time that the relation between the two compensators can be simply written as

$$\boldsymbol{G}_{\mathrm{V}} = \boldsymbol{G}_{\mathrm{U}}\boldsymbol{V}_{\mathrm{D}} = \boldsymbol{G}^{-1}\boldsymbol{G}_{\mathrm{D}}\boldsymbol{V}_{\mathrm{D}} \tag{7.55}$$

The following simple relationship exists between the V- and U-topology

$$\boldsymbol{V} = \boldsymbol{V}_{\mathrm{D}} + \boldsymbol{V}_{\mathrm{A}} = \boldsymbol{V}_{\mathrm{D}}\left(\boldsymbol{I} + \boldsymbol{V}_{\mathrm{D}}^{-1}\boldsymbol{V}_{\mathrm{A}}\right) = \boldsymbol{V}_{\mathrm{D}}(\boldsymbol{I} + \boldsymbol{U}_{\mathrm{A}}) = \boldsymbol{V}_{\mathrm{D}}\boldsymbol{U} \tag{7.56}$$

which explains all the above results.

Figure 7.7 summarizes the joint block scheme of the process and the decoupler of the U-topology. The cross effects can be eliminated by the equations $G_{12} = -U_{21}G_{11}$ and $G_{21} = -U_{12}G_{22}$, whence the equations $U_{12} = -G_{21}/G_{22}$ and $U_{21} = -G_{12}/G_{11}$ are obtained for the decoupler. Thanks to its simplicity, this method is widely used in the industrial practice of the decoupling by pairs.

This structure is beloved in practical applications because the two inputs (V_{11} and U_{21} or V_{22} and U_{12}) of the summing elements allow us to use standard *programmable logic controller (PLC)* elements where the regulator (now V_{11} and V_{22}) appears together with feed-forward elements (now U_{21} and U_{12}),

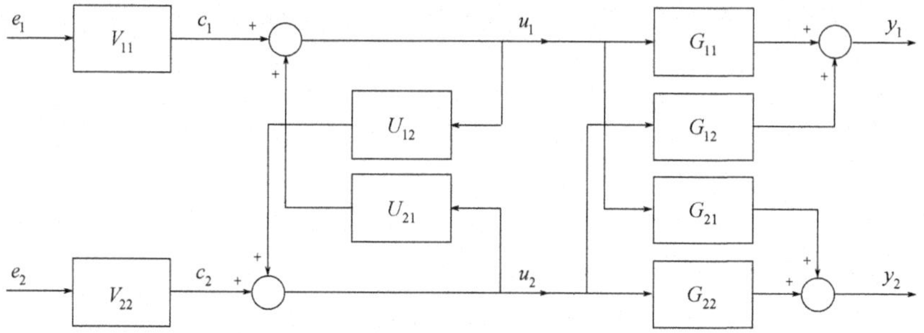

Figure 7.7 Joint block scheme of the decoupler and the process.

which is usually the conventional tool of the classical solution of noise compensation.

Also, decoupling can be performed by further simple topologies. The unity values of the diagonal elements can be used also for feed-forward structures. This method is used to be called the *simple* or *simplified* decoupling method [131]. The corresponding S-topology of the decoupling block scheme is shown in Figure 7.8.

The basic relationships of the model can be written as

$$\boldsymbol{u} = \begin{bmatrix} u_1 \\ u_2 \end{bmatrix} = \begin{bmatrix} 1 & S_{12} \\ S_{21} & 1 \end{bmatrix} \begin{bmatrix} c_1 \\ c_2 \end{bmatrix} = (\boldsymbol{I} + \boldsymbol{S}_A)\boldsymbol{c} \tag{7.57}$$

Introduce the following notation for the inverse of the process

$$\mathbf{G}^{-1} = (\mathbf{G}_\mathrm{D} + \mathbf{G}_\mathrm{A})^{-1} = \bar{\mathbf{G}}_\mathrm{D} + \bar{\mathbf{G}}_\mathrm{A} \tag{7.58}$$

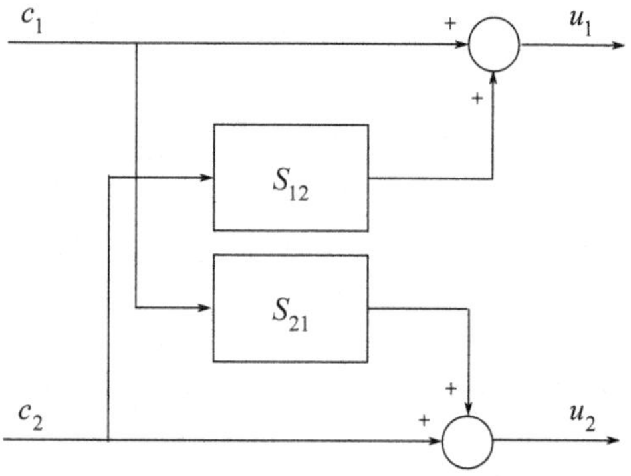

Figure 7.8 Block scheme of the decoupler of the S-topology.

Let $S_A = \bar{G}_A \bar{G}_D^{-1}$ be, then

$$(I + S_A) = I + \bar{G}_A \bar{G}_D^{-1} = \left(I + \bar{G}_A \bar{G}_D^{-1}\right) \bar{G}_D \bar{G}_D^{-1} \\ = \left(\bar{G}_D + \bar{G}_A\right) \bar{G}_D^{-1} = G^{-1} \bar{G}_D^{-1} \tag{7.59}$$

With the S-compensator the decoupling becomes

$$GG^{-1} \bar{G}_D^{-1} = \bar{G}_D^{-1} \tag{7.60}$$

Thus, the decoupling is fulfilled but there are very complicated transfer functions in the diagonals, namely the reciprocals of the main diagonal of G^{-1}.

It is worth noting that the decouplers of feedback topology are welcome in sampled time applications because the actuator signal can be easily computed and programmed using (7.47) from the following expression:

$$u = V_D e - V_D V_A u = V_D (e - V_A u) = G_D^{-1} (e + G_A u) \tag{7.61}$$

DECOUPLING CONTROL USING YOULA-PARAMETERIZED *MIMO* REGULATORS

The Youla-parameterized *MIMO* regulator introduced above also makes it possible to solve the decoupling problem. The real advantage of this approach is that it is clearly observable whether the decoupling is possible or not.

According to (7.28) and with $G_r = I$ and $G_n = I$, the overall transfer characteristics of the closed system obtained by *YP* for *MIMO* systems take the form

$$y = P_- R_r y_r + (I - P_- R_n) y_n = y_t + y_d \tag{7.62}$$

It is clear that if the invariant *MIMO* process factor P_- is nondiagonal, then it is impossible to apply a decoupling regulator. If P_- is diagonal or $P_- = I$, then tracking and noise rejection decoupling can be performed with the diagonal R_r or R_n. If P_- is diagonal, then the diagonal inner matrix filters G_r or G_n can also be applied for the optimal compensation of the invariant factors. In the case of diagonal reference models providing decoupling, the design of the main diagonal elements of the inner filters is completely the same as in the optimization methods shown for scalar (*SISO*) systems.

Sampled Data Systems

The discrete-time (DT) "naïve" model of the *MIMO* process is

$$\mathbf{G}(z) = \frac{1}{\mathcal{A}(z)}\boldsymbol{\mathcal{B}}(z); \quad \boldsymbol{\mathcal{B}}_{+} = \boldsymbol{\mathcal{B}}; \quad \boldsymbol{\mathcal{B}}_{-} = z^{-d}\mathbf{I} \tag{7.63}$$

and the sampled Youla-parameterized *MIMO* regulator is obtained by performing the analogous computations providing (7.36)

$$\begin{aligned}\mathbf{C}(z) &= \mathcal{A}(z)\boldsymbol{\mathcal{B}}^{-1}(z)\boldsymbol{\mathcal{B}}_{\mathrm{n}}(z)\left[\mathcal{A}_{\mathrm{n}}(z)\mathbf{I} - z^{-d}\boldsymbol{\mathcal{B}}_{\mathrm{n}}(z)\right]^{-1} \\ &= \mathcal{A}(z)\boldsymbol{\mathcal{B}}_{+}^{-1}(z)\boldsymbol{\mathcal{B}}_{\mathrm{n}}(z)\left[\mathcal{A}_{\mathrm{n}}(z)\mathbf{I} - z^{-d}\boldsymbol{\mathcal{B}}_{\mathrm{n}}(z)\right]^{-1}\end{aligned} \tag{7.64}$$

It is worth checking that the similarly computed sampled Youla-parameterized *SISO* regulator takes the form

$$\begin{aligned}C &= \mathcal{A}(z)\mathcal{B}^{-1}(z)\mathcal{B}_{\mathrm{n}}(z)\left[\mathcal{A}_{\mathrm{n}}(z) - z^{-d}\mathcal{B}_{\mathrm{n}}(z)\right]^{-1} \\ &= \mathcal{A}(z)\mathcal{B}_{+}^{-1}(z)\mathcal{B}_{\mathrm{n}}(z)\left[\mathcal{A}_{\mathrm{n}}(z) - z^{-d}\mathcal{B}_{\mathrm{n}}(z)\right]^{-1}\end{aligned} \tag{7.65}$$

In these expressions the reference model also has the "naïve" form. Let us assume now that $\boldsymbol{R}_{\mathrm{n}}$ is given in left-*MFD* form, i.e.,

$$\boldsymbol{R}_{\mathrm{n}} = \boldsymbol{\mathcal{A}}_{\mathrm{n}}^{-1}(z)\boldsymbol{\mathcal{B}}_{\mathrm{n}}(z) = \left[\mathbf{I}+\tilde{\boldsymbol{\mathcal{A}}}_{\mathrm{n}}(z^{-1})\right]^{-1}\boldsymbol{\mathcal{B}}_{\mathrm{n}}(z^{-1}) \tag{7.66}$$

In this case, the output of the regulator can be computed by a two-step algorithm. Let us denote the output by the vector $\boldsymbol{c}[k]$. It is reasonable to use (7.29) according to which the necessary computation takes the form

$$\boldsymbol{c} = \bar{\mathbf{C}}\boldsymbol{e} = \left(\boldsymbol{I} - \mathbf{R}_{\mathrm{n}}\bar{\mathbf{G}}_{-}\right)^{-1}\mathbf{R}_{\mathrm{n}}\bar{\mathbf{G}}_{+}^{-1}\boldsymbol{x}; \quad \boldsymbol{x} = \bar{\mathbf{G}}_{+}^{-1}\boldsymbol{e} \tag{7.67}$$

Here the auxiliary variable $x[k]$ is introduced. Using these equations the regulator can be written in the form of vector difference equation form linear in parameters

$$\boldsymbol{c} = \left(\boldsymbol{\mathcal{B}}_{\mathrm{n}}\boldsymbol{\mathcal{N}}_{\mathrm{L}}^{-} - \tilde{\boldsymbol{\mathcal{A}}}_{\mathrm{n}}\right)\boldsymbol{c} + \boldsymbol{\mathcal{B}}_{\mathrm{n}}\boldsymbol{x} \tag{7.68}$$

where $\boldsymbol{x}[k]$ can also be given in similar form

$$\boldsymbol{x} = \boldsymbol{\mathcal{D}}_{\mathrm{L}}\boldsymbol{e} - \tilde{\boldsymbol{\mathcal{N}}}_{\mathrm{L}}^{+}\boldsymbol{x} \tag{7.69}$$

In the equations of the regulator the following simple notations are used

$$\bar{\mathbf{G}}_{+}^{-1} = \left[\boldsymbol{\mathcal{N}}_{\mathrm{L}}^{+}(z)\right]^{-1}\boldsymbol{\mathcal{D}}_{\mathrm{L}}(z); \quad \bar{\mathbf{G}}_{-} = \boldsymbol{\mathcal{N}}_{\mathrm{L}}^{-}; \quad \boldsymbol{\mathcal{N}}_{\mathrm{L}}^{+} = \mathbf{I} + \tilde{\boldsymbol{\mathcal{N}}}_{\mathrm{L}}^{+} \tag{7.70}$$

DECOUPLING EXAMPLES

Example 7.1

Consider a very simple *MIMO* process, whose *TFM* is

$$\boldsymbol{P}(s) = \frac{1}{\mathcal{A}(s)}\boldsymbol{\mathcal{B}}(s) = \begin{bmatrix} \dfrac{1}{1+s} & \dfrac{1}{1+2s} \\ 0 & \dfrac{1}{1+4s} \end{bmatrix}$$
$$= \frac{1}{(1+s)(1+2s)(1+4s)}\begin{bmatrix} (1+2s)(1+4s) & (1+s)(1+4s) \\ 0 & (1+s)(1+2s) \end{bmatrix} \tag{7.71}$$

Choose reference models which can meet both the speeding-up and decoupling design goals

$$\boldsymbol{R}_{\mathrm{n}}(s) = \frac{1}{\mathcal{A}_{\mathrm{n}}(s)}\boldsymbol{\mathcal{B}}_{\mathrm{n}}(s) = \frac{1}{(1+0.5s)}\begin{bmatrix} 1 & 0 \\ 0 & 1 \end{bmatrix} = \frac{1}{(1+0.5s)}\mathbf{I} \tag{7.72}$$

After the calculations of (7.36), the following regulator is obtained

$$\mathbf{C}(s) = \begin{bmatrix} \dfrac{1+s}{0.5s} & -\dfrac{(1+s)(1+4s)}{0.5s(1+2s)} \\ 0 & \dfrac{1+4s}{0.5s} \end{bmatrix} \tag{7.73}$$

whose elements contain signal forming of *proportional-integral (PI)* and *proportional-integral-derivative* (*PID*) character.

Example 7.2

Investigate a DT process where the impulse *TFM* of the process is

$$\mathbf{G}(z) = \begin{bmatrix} \dfrac{0.5z^{-1}}{1-0.5z^{-1}} & \dfrac{0.2z^{-1}}{1-0.8z^{-1}} \\ 0 & \dfrac{z^{-1}-0.5z^{-2}}{1-1.7z^{-1}+0.2z^{-1}} \end{bmatrix} \tag{7.74}$$

Reapply the speeding-up and decoupling design goals using the following reference model:

$$\boldsymbol{R}_{\mathrm{n}}(z) = \begin{bmatrix} \dfrac{0.8z^{-1}}{1-0.2z^{-1}} & 0 \\ 0 & \left(\dfrac{0.9z^{-1}}{1-0.1z^{-1}}\right)^2 \end{bmatrix} = \frac{\begin{bmatrix} 0.8z^{-1}\left(1-0.1z^{-1}\right)^2 & 0 \\ 0 & \left(0.9z^{-1}\right)^2\left(1-0.2z^{-1}\right) \end{bmatrix}}{(1-0.2z^{-1})(1-0.1z^{-1})^2} \tag{7.75}$$

After the calculations given by (7.36), the impulse *TFM* of the obtained matrix regulator is

$$\mathbf{C}(z) = \begin{bmatrix} C_{11}(z) & C_{12}(z) \\ C_{21}(z) & C_{22}(z) \end{bmatrix} \tag{7.76}$$

where

$$C_{11}(z) = \frac{1.6(1-0.5z^{-1})}{1-z^{-1}}; \quad C_{12}(z) = \frac{-0.32(1-1.7z^{-1}+0.2z^{-2})}{(1-z^{-1})(1-0.8z^{-1})} \tag{7.77}$$

$$C_{21}(z) = 0; \quad C_{22}(z) = \frac{0.81z^{-1}(1-1.7z^{-1}+0.2z^{-2})}{(1-z^{-1})(1+0.8z^{-1})(1-0.5z^{-1})} \tag{7.78}$$

All elements of the regulator can be realized, which is the consequence of the specially chosen reference model *TFM*. Since all nontrivial elements of $\boldsymbol{R}_{\mathrm{n}}$ have unity gain, the scalar regulators have an integrating character (i.e., all elements have the pole $z = 1$).

Example 7.3

An aircraft obviously has a very complex dynamics [3,128,133], which can be described by many state, input, and/or output variables. Experts state that the vital lateral dynamics, however, can be described by relatively simple models with four state variables and two major input signals. The input variables are the aileron δ_a and the rudder δ_r. For the small changes

$\Delta\delta_a$ and $\Delta\delta_r$ in the vicinity of a working point, we can introduce the following input vector

$$u = \begin{bmatrix} \Delta\delta_a & \Delta\delta_r \end{bmatrix}^T \tag{7.79}$$

so the next state equation describes the dynamics [43,133]

$$\dot{x} = \begin{bmatrix} Y_\beta & (\approx 0) & (\approx -1) & \frac{g}{V} \\ L_\beta & L_p & L_r & 0 \\ N_\beta & N_p & N_r & 0 \\ 0 & 1 & 0 & 0 \end{bmatrix} x + \begin{bmatrix} 0 & Y_{\delta_r} \\ L_{\delta_a} & L_{\delta_r} \\ N_{\delta_a} & N_{\delta_r} \\ 0 & 0 \end{bmatrix} u = \mathbf{A}x + \mathbf{B}u$$

$$y = \mathbf{C}x \tag{7.80}$$

Here **A** and **B** contain the so-called dimensional derivatives typical of a given aircraft. The subscripts δ_a and δ_r refer to the aileron and rudder input, respectively. Introduce the following variables: the sideslip angle β, the roll rate p, the yaw rate r, and the roll angle ϕ. The small changes in the above variables produce the elements of the state vector, i.e.,

$$x = \begin{bmatrix} \Delta\beta & \Delta p & \Delta r & \Delta\phi \end{bmatrix}^T \tag{7.81}$$

The output variables depend on the selection of the structure for matrix **C**. The following, for example,

$$\mathbf{C} = \begin{bmatrix} 0 & 1 & 0 & 0 \\ 0 & 0 & 0 & 1 \end{bmatrix} \tag{7.82}$$

means that the output variables are the roll rate p and the roll angle ϕ, i.e., for their small changes

$$y = \begin{bmatrix} \Delta p & \Delta\phi \end{bmatrix}^T \tag{7.83}$$

It is an interesting task to design a simple decoupling regulator encompassing independent roll rate and roll angle or other selected output variables. The parameter matrices of the above state equation are available for different types of aircraft in the literature. First, choose an aircraft where this model is stable. A possible model according to [131] is

$$\dot{x} = \begin{bmatrix} -0.099593 & 0 & -1 & 0.1056796 \\ -1.700982 & -1.184647 & 0.223908 & 0 \\ 0.407420 & -0.056276 & -0.188010 & 0 \\ 0 & 1 & 0 & 0 \end{bmatrix} x + \begin{bmatrix} 0 & 0.740361 \\ 0.531304 & 0.049766 \\ 0.005685 & -0.106592 \\ 0 & 0 \end{bmatrix} u = \mathbf{A}x + \mathbf{B}u \tag{7.84}$$

Of the eigenvalues $\{-0.0603 + 0.7555i; -0.0603 - 0.7555i; -1.3198; -0.0319\}$ of the matrix $\mathbf{A}$, two are complex conjugates and one is very slow. Let the output variables be the sideslip angle β and the yaw rate r, i.e.,

$$y = [\Delta\beta \quad \Delta r]^{\mathrm{T}} \tag{7.85}$$

This task can be solved by choosing

$$C = \begin{bmatrix} 1 & 0 & 0 & 0 \\ 0 & 0 & 1 & 0 \end{bmatrix} \tag{7.86}$$

For the decoupling, choose the diagonal reference models

$$\mathbf{R}_{\mathrm{n}}(s) = \frac{1}{\mathcal{A}_{\mathrm{n}}(s)}\mathcal{B}_{\mathrm{n}}(s) = \frac{1}{(1+0.5s)}\begin{bmatrix} 1 & 0 \\ 0 & 1 \end{bmatrix} = \frac{2}{s+2}\mathbf{I} = \mathbf{R}_{\mathrm{r}} \tag{7.87}$$

Using (7.36), we can compute the decoupling regulator as

$$\mathcal{C}(s) = \begin{bmatrix} \mathcal{C}_{11}(s) & \mathcal{C}_{12}(s) \\ \mathcal{C}_{21}(s) & \mathcal{C}_{22}(s) \end{bmatrix} \tag{7.88}$$

where

$$\mathcal{C}_{11}(s) = \frac{50.6501(s-2.862)(s+1.383)(s-0.04035)}{s(s+3.687)(s+3.687)} \tag{7.89}$$

$$C_{12}(s) = \frac{351.803(s+1.154)(s+0.3676)(s-0.004883)}{s(s+3.687)(s+3.687)} \tag{7.90}$$

$$C_{21}(s) = \frac{2.7014(s-3.79)(s-1.15)(s+0.9645)}{s(s+3.687)(s+3.687)} \tag{7.91}$$

$$C_{22}(s) = \frac{2.7014(s-14.08)(s+0.1335)}{s(s+3.687)(s+3.687)} \tag{7.92}$$

It can be checked by simple calculations that the overall characteristics of the closed system are

$$y = \begin{bmatrix} \Delta\beta \\ \Delta r \end{bmatrix} = \frac{2}{s+2}\mathbf{I}\begin{bmatrix} \Delta\delta_a \\ \Delta\delta_r \end{bmatrix} + \frac{2}{s+2}\mathbf{I}y_n \tag{7.93}$$

i.e., the decoupling is realized both for tracking and for noise rejection. Each element of the *MIMO* regulator is a realizable integrating regulator with third-order transfer functions. Of course, depending on the nature of the task, different reference models can be chosen for $\boldsymbol{R}_r$ and $\boldsymbol{R}_n$.

On the basis of [133], the state equation of an unstable aircraft can be obtained by linearization around the working point

$$\dot{\boldsymbol{x}} = \begin{bmatrix} -0.05 & -0.003 & -0.98 & 0.2 \\ -1.0 & -0.75 & 1.0 & 0 \\ 0.3 & -0.3 & -0.15 & 0 \\ 0 & 1 & 0 & 0 \end{bmatrix}\boldsymbol{x} + \begin{bmatrix} 0 & 0 \\ 1.7 & -0.2 \\ 0.3 & -0.6 \\ 0 & 0 \end{bmatrix}\boldsymbol{u} = \mathbf{A}\boldsymbol{x} + \mathbf{B}\boldsymbol{u} \tag{7.94}$$

where the relative gains for the aileron and rudder are $g_1 = 1.0$ and $g_2 = \delta_r/\delta_a$, respectively.

The dynamic model of most of the aircraft with the above state variables, however, is unstable. *YP*-based regulators can be applied only for stable processes. The solution may be the two-step method mentioned in Section 2.1, where first an inner control loop is applied to stabilize the system.

The eigenvalues of the matrix $\mathbf{A}$ are $\{-0.0035 \pm 0.8834i;\ -0.9821;\ -0.0391\}$. The two complex conjugate poles and one of the real poles are stable; the other pole is unstable. This latter one corresponds to the instability of the so-called spiral dynamics. Different types of stabilizing regulators can be applied. The simplest case is when the stabilization is solved by state feedback. Choose the following design poles: $\{-0.0035 \pm 0.8834i;\ -0.9821;\ -0.0391\}$, i.e., mirror the unstable pole on the complex axis. This pole-assignment task can be solved by the following state feedback matrix

$$\boldsymbol{K} = \begin{bmatrix} -0.5606 & -0.3848 & 0.5529 & 0.5071 \\ -0.7622 & 0.2099 & -1.0143 & 0.6824 \end{bmatrix} \tag{7.95}$$

Let the output variables be the sideslip β and roll angle ϕ, i.e.,

$$y = [\Delta\beta \quad \Delta\phi]^{\mathrm{T}} \tag{7.96}$$

and the corresponding control matrix is

$$\boldsymbol{C} = \begin{bmatrix} 1 & 0 & 0 & 0 \\ 0 & 0 & 0 & 1 \end{bmatrix} \tag{7.97}$$

Similarly to the previous case, the elements of the *MIMO* regulator are

$$\mathcal{C}_{11}(s) = \frac{0.42517(s^2 + 0.5416s + 0.6217)}{s} \tag{7.98}$$

$$\mathcal{C}_{12}(s) = \frac{1.2513(s^2 - 0.0332s + 0.7768)}{s} \tag{7.99}$$

$$\mathcal{C}_{21}(s) = \frac{3.6139(s + 0.8423)(s + 0.107)}{s} \tag{7.100}$$

$$\mathcal{C}_{22}(s) = \frac{0.63584(s^2 - 1.395s + 1.124)}{s} \tag{7.101}$$

Here, we obtain *PID* regulators (ideal but not realizable) in each element of the matrix regulator. The overall characteristics of the closed system are

$$y = \begin{bmatrix} \Delta\beta \\ \Delta\phi \end{bmatrix} = \frac{2}{s+2}\mathbf{I}\begin{bmatrix} \Delta\delta_{\mathrm{a}} \\ \Delta\delta_{\mathrm{r}} \end{bmatrix} + \frac{2}{s+2}\mathbf{I}\,y_{\mathrm{n}} \tag{7.102}$$

MIMO PROCESS MODELS LINEAR IN PARAMETER MATRICES

The "naïve" models of the DT *MIMO* process and the Youla-parameterized regulator have been shown above. It is important to note that if, on the basis of (7.10), we construct a left-*MFD* representation of a DT process

$$\mathbf{G}(z) = \mathcal{D}_{\mathrm{L}}^{-1}(z)\mathcal{N}_{\mathrm{L}}(z) \tag{7.103}$$

then it is easy to formulate a differential vector equation

$$\begin{aligned}\mathcal{D}_{\mathrm{L}}\left(z^{-1}\right) &= \mathbf{I} + \mathbf{D}_{\mathrm{L}}^{1} z^{-1} + \mathbf{D}_{\mathrm{L}}^{2} z^{-2} + \ldots + \mathbf{D}_{\mathrm{L}}^{n_{\mathrm{d}}} z^{-n_{\mathrm{d}}} = \mathbf{I} + \tilde{\mathcal{D}}_{\mathrm{L}}\left(z^{-1}\right) \\ \mathcal{N}_{\mathrm{L}}\left(z^{-1}\right) &= \mathbf{N}_{\mathrm{L}}^{1} z^{-1} + \mathbf{N}_{\mathrm{L}}^{2} z^{-2} + \ldots + \mathbf{N}_{\mathrm{L}}^{n_{\mathrm{n}}} z^{-n_{\mathrm{n}}}\end{aligned} \tag{7.104}$$

where the expression is normalized by the coefficient $\mathbf{D}_{\mathrm{L}}^{\mathrm{o}}$. Given the introduced matrix polynomials we can write

$$\begin{aligned}\boldsymbol{y}[k] &= \mathcal{N}_{\mathrm{L}}\left(z^{-1}\right)\boldsymbol{u}[k] - \tilde{\mathcal{D}}_{\mathrm{L}}\left(z^{-1}\right)\boldsymbol{y}[k] \\ &= \mathbf{N}_{\mathrm{L}}^{1}\boldsymbol{u}[k-1] + \mathbf{N}_{\mathrm{L}}^{2}\boldsymbol{u}[k-2] + \ldots + \mathbf{N}_{\mathrm{L}}^{n_{\mathrm{n}}}\boldsymbol{u}[k-n_{n}] \\ &\quad - \mathbf{D}_{\mathrm{L}}^{1}\boldsymbol{y}[k-1] + \mathbf{D}_{\mathrm{L}}^{2}\boldsymbol{y}[k-2] + \ldots + \mathbf{D}_{\mathrm{L}}^{n_{\mathrm{d}}}\boldsymbol{y}[k-n_{\mathrm{d}}]\end{aligned} \tag{7.105}$$

This equation is linear in the matrix parameters $\mathbf{N}_{\mathrm{L}}^{i}$ and $\mathbf{D}_{\mathrm{L}}^{i}$, and offers a very good opportunity to apply a simple *MIMO* process identification method.

The time-delay version of the process (7.46), which corresponds to (7.37), is

$$\mathbf{G}(z) = \mathcal{D}_{\mathrm{L}}^{-1}(z)\mathcal{N}_{\mathrm{L}}(z)z^{-d}; \quad \mathbf{G}_{+} = \mathcal{D}_{\mathrm{L}}^{-1}\mathcal{N}_{\mathrm{L}} \tag{7.106}$$

Based on (7.27), (7.29), and (7.39), the regulators are

$$\begin{aligned}\mathbf{C} &= \mathcal{N}_{\mathrm{L}}^{-1}\mathcal{D}_{\mathrm{L}}\mathbf{R}_{\mathrm{n}}\left(\mathbf{I} - \mathbf{R}_{\mathrm{n}} z^{-d}\right)^{-1} = \mathcal{N}_{\mathrm{L}}^{-1}\mathcal{D}_{\mathrm{L}}\left(\mathbf{I} - \mathbf{R}_{\mathrm{n}} z^{-d}\right)^{-1}\mathbf{R}_{\mathrm{n}} \\ \bar{\mathbf{C}} &= \left(\mathbf{I} - \mathbf{R}_{\mathrm{n}} z^{-d}\right)^{-1}\mathbf{R}_{\mathrm{n}}\mathcal{N}_{\mathrm{L}}^{-1}\mathcal{D}_{\mathrm{L}} = \mathbf{R}_{\mathrm{n}}\left(\mathbf{I} - \mathbf{R}_{\mathrm{n}} z^{-d}\right)^{-1}\mathcal{N}_{\mathrm{L}}^{-1}\mathcal{D}_{\mathrm{L}}\end{aligned} \tag{7.107}$$

MIMO PREDICTIVE REGULATORS

In the sampled process with time-delay (7.86), the time-delay d is supposed to be the same for each input. Any other variant can be represented by $\mathcal{N}_{\mathrm{L}}$. Following the procedure presented in Section 2.4, let us search again for an expression by means of which the output value of the process in the $[k+d]$-th sampling time can be estimated on the basis of the information available until the $[k]$-th sampling time. Introduce the following special polynomial equation

$$\mathbf{I} = \mathcal{T}_{\mathrm{L}}\mathcal{D}_{\mathrm{L}} + \mathcal{P}_{\mathrm{L}} z^{-d} \tag{7.108}$$

whose solution is unambiguously seeking $\mathcal{T}_\mathrm{L} = \mathbf{I} + \mathbf{T}_1 z^{-1} + \ldots + \mathbf{T}_{d-1} z^{-(d-1)}$ as $(d-1)$-order polynomial and $\mathcal{P}_\mathrm{L} = \mathbf{P}_\mathrm{o} + \mathbf{P}_1 z^{-1} + \ldots + \mathbf{P}_{n_\mathrm{d}-1} z^{-(n_\mathrm{d}-1)}$ as $(n_\mathrm{d}-1)$-order polynomial, if the matrix polynomial is n_d-order. Equation (7.108) is a special matrix polynomial version of the Diophantine equation (*DE*) discussed in Chapter 4, and can be considered as the *MIMO* generalization of (2.4.2). With equivalent manipulations, the process $\mathbf{G}(z)$ can be decomposed in the form

$$\begin{aligned}\mathbf{G} &= \mathcal{D}_\mathrm{L}^{-1}\mathcal{N}_\mathrm{L} z^{-d} = (\mathbf{I})\mathcal{D}_\mathrm{L}^{-1}\mathcal{N}_\mathrm{L} z^{-d} = \left(\mathcal{T}_\mathrm{L}\mathcal{D}_\mathrm{L} + \mathcal{P}_\mathrm{L} z^{-d}\right)\mathcal{D}_\mathrm{L}^{-1}\mathcal{N}_\mathrm{L} z^{-d} \\ &= \mathcal{T}_\mathrm{L}\mathcal{N}_\mathrm{L} z^{-d} + \mathcal{P}_\mathrm{L} z^{-d}\left(\mathcal{D}_\mathrm{L}^{-1}\mathcal{N}_\mathrm{L} z^{-d}\right)\end{aligned} \tag{7.109}$$

Apply both sides for the input series $\boldsymbol{u}[k]$

$$\begin{aligned}\boldsymbol{y}[k] &= \mathcal{T}_\mathrm{L}\mathcal{N}_\mathrm{L} z^{-d}\boldsymbol{u}[k] + \mathcal{P}_\mathrm{L} z^{-d}\left(\mathcal{D}_\mathrm{L}^{-1}\mathcal{N}_\mathrm{L} z^{-d}\boldsymbol{u}[k]\right) = \mathcal{P}_\mathrm{L} z^{-d}\boldsymbol{y}[k] \\ &= \mathcal{T}_\mathrm{L}\mathcal{N}_\mathrm{L}\boldsymbol{u}[k-d] + \mathcal{P}_\mathrm{L}\boldsymbol{y}[k-d]\end{aligned} \tag{7.110}$$

The vector differential equation can be rewritten for the $[k+d]$-th sampling time as

$$\boldsymbol{y}[k+d] = \mathcal{T}_\mathrm{L}\mathcal{N}_\mathrm{L}\boldsymbol{u}[k] + \mathcal{P}_\mathrm{L}\boldsymbol{y}[k] = \boldsymbol{y}[k+d|k] \tag{7.111}$$

where the vector $\boldsymbol{y}[k+d|k]$ is the estimated/predicted value of the series $\boldsymbol{y}[k]$ in the $[k+d]$-*th* sampling time. Note that the prediction is error-free and uses only the values of both the input and output signals available until the $[k]$-th sampling time. Both $\mathcal{T}_\mathrm{L}\mathcal{N}_\mathrm{L}$ and $\mathcal{P}_\mathrm{L}$ are matrix polynomials of z^{-1}, and they only weight the values available in the $[k]$-th and the former sampling times in (7.111). Based on the d-step predictor, a special so-called predictive regulator can be constructed. If the process is required to follow the output of the reference model $\boldsymbol{R}_\mathrm{r}$, then the equation of the regulator becomes

$$\boldsymbol{R}_\mathrm{r}\boldsymbol{y}_\mathrm{r}[k] = \boldsymbol{y}[k+d|k] = \mathcal{T}_\mathrm{L}\mathcal{N}_\mathrm{L}\boldsymbol{u}[k] + \mathcal{P}_\mathrm{L}\boldsymbol{y}[k] \tag{7.112}$$

and the fundamental equation of the *MIMO* predictive regulator can be formulated for the input of the process as

$$\boldsymbol{u}[k] = \mathcal{N}_\mathrm{L}^{-1}\mathcal{T}_L^{-1}\left\{\boldsymbol{R}_\mathrm{r}\boldsymbol{y}_\mathrm{r}[k] - \mathcal{P}_\mathrm{L}\boldsymbol{y}[k]\right\} \tag{7.113}$$

It can easily be seen that the regulator corresponding to Equation (7.92) is

$$\begin{aligned} \boldsymbol{C} &= \mathcal{N}_{\mathrm{L}}^{-1}\mathcal{T}_{\mathrm{L}}^{-1}\mathcal{P}_{\mathrm{L}} = \mathcal{N}_{\mathrm{L}}^{-1}\mathcal{D}_{\mathrm{L}}\mathcal{P}_{\mathrm{L}}\left(\mathbf{I} - \mathcal{P}_{\mathrm{L}}z^{-d}\right)^{-1} \\ &= \mathcal{N}_{\mathrm{L}}^{-1}\mathcal{D}_{\mathrm{L}}\left(\mathbf{I} - \mathcal{P}_{\mathrm{L}}z^{-d}\right)^{-1}\mathcal{P}_{\mathrm{L}} \end{aligned} \tag{7.114}$$

Thus, the *MIMO* predictive regulator is basically a Youla-parameterized regulator, where $\boldsymbol{R}_{\mathrm{n}} = \mathcal{P}_{\mathrm{L}}$ is the reference model referring to the noise rejection, which does not depend on us but comes from the *DE* of (7.108). The equation of the overall closed system is

$$\boldsymbol{y} = \boldsymbol{R}_{\mathrm{r}}z^{-d}\boldsymbol{y}_{\mathrm{r}} + \left(\mathbf{I} - \boldsymbol{R}_{\mathrm{n}}z^{-d}\right)\boldsymbol{y}_{\mathrm{n}} = \boldsymbol{R}_{\mathrm{r}}z^{-d}\boldsymbol{y}_{\mathrm{r}} + \left(\mathbf{I} - \mathcal{P}_{\mathrm{L}}z^{-d}\right)\boldsymbol{y}_{\mathrm{n}} \tag{7.115}$$

MIMO MINIMUM VARIANCE (MV) REGULATOR

For simplicity and on the basis of (7.106), let the DT system equation describing the *MIMO* process be the same as the generalization of (6.1)

$$\begin{aligned} \boldsymbol{y}[k] &= \mathcal{D}_{\mathrm{L}}^{-1}(z)\mathcal{N}_{\mathrm{L}}(z)z^{-d}\boldsymbol{u}[k] + \mathcal{D}_{\mathrm{L}}^{-1}(z)\mathcal{C}_{\mathrm{L}}(z)\boldsymbol{\varsigma}[k] \\ &= \mathcal{D}_{\mathrm{L}}^{-1}(z)\mathcal{N}_{\mathrm{L}}(z)z^{-d}\boldsymbol{u}[k] + \boldsymbol{y}_{\mathrm{n}}[k] \end{aligned} \tag{7.116}$$

where $\boldsymbol{u}[k]$ and $\boldsymbol{y}[k]$ are the vectors of the process input and output signals, and $\boldsymbol{y}_{\mathrm{n}}[k]$ is the sampled series of the additive output noise vector. $\boldsymbol{\varsigma}[k]$ is the so-called source noise vector from which $\boldsymbol{y}_{\mathrm{n}}[k]$ is derived. $\boldsymbol{\varsigma}[k]$ is also assumed to be a white noise vector with zero mean. The noise model $\boldsymbol{H}_{\mathrm{n}} = \mathcal{D}_{\mathrm{L}}^{-1}\mathcal{C}_{\mathrm{L}}$ given by left-*MFD* can produce a so-called "color" noise with an arbitrary spectrum from the white noise input. Note that here the assumption of the common matrix polynomial $\mathcal{D}_{\mathrm{L}}(z)$ in both the process and noise model is much stronger restriction than in the denominator of Equation (6.1) in the *SISO* case. The process model is a left-*MFD*.

The goal of the *MIMO MV* regulator is to minimize the variance of the output. To this end, a similar method can be followed to that in Section 2.4 for the *SISO* case and in Chapter 7 for *MIMO* predictive regulators. Introduce the polynomial equation

$$\mathcal{C}_{\mathrm{L}} = \mathcal{D}_{\mathrm{L}}\mathcal{T} + \mathcal{P}z^{-d} \tag{7.117}$$

whose solution is unambiguously seeking $\mathcal{T} = \mathbf{I} + \mathbf{T}_1 z^{-1} + \ldots + \mathbf{T}_{d-1}z^{-(d-1)}$ as $(d-1)$-order polynomial, and $\mathcal{P} = \mathbf{P}_{\mathrm{o}} + \mathbf{P}_1 z^{-1} + \ldots + \mathbf{P}_{n_{\mathrm{d}}-1}z^{-(n_{\mathrm{d}}-1)}$ $(n_{\mathrm{d}}-1)$-order polynomial, if the matrix polynomials

$\mathcal{C}_L$ and $\mathcal{D}_L$ are of n_d-order (this can always be done for $\mathcal{C}_L$ by introducing additional zero matrices). Equation (7.117) is a special matrix polynomial version of the *DE* discussed in Chapter 4.

In the derivation, the reverse order product of $\mathcal{T}$ and $\mathcal{P}$ obtained from (7.117) is required, so introduce the matrix polynomials $\bar{\mathcal{T}}$ and $\bar{\mathcal{P}}$

$$\mathcal{P}\mathcal{T}^{-1} = \bar{\mathcal{T}}^{-1}\bar{\mathcal{P}} \tag{7.118}$$

Equation (7.118) always has a solution, if it is assumed that $\det\left[\bar{\mathcal{T}}(z)\right] = \det[\mathcal{T}(z)]$ and $\bar{\mathcal{T}}(0) = \mathbf{T}_o = \mathbf{I}$. The identical values of the determinants can always be reached by normalization since these *MFDs* are unimodular matrices, too. With the help of the obtained matrix polynomials $\bar{\mathcal{T}}$ and $\bar{\mathcal{P}}$, let us define a new matrix polynomial

$$\bar{\mathcal{C}}_L = \bar{\mathcal{T}}\mathcal{D}_L + \bar{\mathcal{P}}z^{-d} \tag{7.119}$$

Substituting the expression (7.118) into (7.117) and (7.119), and comparing the two equations we can check that the identity

$$\mathcal{T}\mathcal{C}_L^{-1} = \bar{\mathcal{C}}_L^{-1}\bar{\mathcal{T}} \tag{7.120}$$

exists.

The form (7.116) of the process for the $[k+d]$-th sampling time can be written by using the decomposition (7.117) as

$$\boldsymbol{y}[k+d] = \mathcal{D}_L^{-1}\mathcal{N}_L\boldsymbol{u}[k] + \mathcal{D}_L^{-1}\mathcal{P}\boldsymbol{\varsigma}[k] + \mathcal{T}\boldsymbol{\varsigma}[k] \tag{7.121}$$

Let

$$\boldsymbol{\varepsilon}[k+d|k] = \mathcal{T}\boldsymbol{\varsigma}[k] \tag{7.122}$$

and express $\boldsymbol{\varsigma}[k]$ from (7.116)

$$\boldsymbol{\varsigma}[k] = \mathcal{C}_L^{-1}\left\{\mathcal{D}_L\boldsymbol{y}[k] - \mathcal{N}_L z^{-d}\boldsymbol{u}[k]\right\} \tag{7.123}$$

then substitute it into (7.121)

$$\begin{aligned}\boldsymbol{y}[k+d] &= \mathcal{D}_L^{-1}\mathcal{N}_L\boldsymbol{u}[k] + \mathcal{D}_L^{-1}\mathcal{P}\mathcal{C}_L^{-1}\left\{\mathcal{D}_L\boldsymbol{y}[k] - \mathcal{N}_L z^{-d}\boldsymbol{u}[k]\right\} + \mathcal{T}\boldsymbol{\varsigma}[k]\\ &= \left\{\mathcal{D}_L^{-1}\mathcal{N}_L - \mathcal{D}_L^{-1}\mathcal{P}\mathcal{C}_L^{-1}\mathcal{N}_L z^{-d}\right\}\boldsymbol{u}[k] + \mathcal{D}_L^{-1}\mathcal{P}\mathcal{C}_L^{-1}\mathcal{D}_L\boldsymbol{y}[k]\\ &\quad + \boldsymbol{\varepsilon}[k+d|k]\\ &= \boldsymbol{y}[k+d|k] + \boldsymbol{\varepsilon}[k+d|k]\end{aligned} \tag{7.124}$$

Here $\boldsymbol{y}[k+d|k]$ is the predictor of the process output in the $[k]$-th sampling time, by means of which the future value in the $[k+d]$-th sampling time can be predicted. The prediction error $\boldsymbol{\varepsilon}[k+d|k] = \mathcal{T}\varsigma[k+d]$ is independent of $\boldsymbol{y}[k+d|k]$. Do not forget that $\mathcal{T}$ is of $d-1$ order, so it does not contain values or their functions which occurred until the $[k]$-th sampling time. The first part of the following identity is obtained from (7.117) by simple rearrangement

$$\mathcal{D}_{\mathrm{L}}^{-1}\mathcal{P}\mathcal{C}_{\mathrm{L}}^{-1}\mathcal{D}_{\mathrm{L}} = z^{d}\left(\mathbf{I} - \mathcal{T}\mathcal{C}_{\mathrm{L}}^{-1}\mathcal{D}_{\mathrm{L}}\right) = z^{d}\left(\mathbf{I} - \bar{\mathcal{C}}_{\mathrm{L}}^{-1}\bar{\mathcal{T}}\mathcal{D}_{\mathrm{L}}\right) = \bar{\mathcal{C}}_{\mathrm{L}}^{-1}\bar{\mathcal{P}} \tag{7.125}$$

and the second part comes from (7.120). Using the above expressions, the predictor $\boldsymbol{y}[k+d|k]$ can be further simplified

$$\boldsymbol{y}[k+d|k] = \mathcal{T}\mathcal{C}_{\mathrm{L}}^{-1}\mathcal{N}_{\mathrm{L}}\boldsymbol{u}[k] + \mathcal{D}_{\mathrm{L}}^{-1}\mathcal{P}\mathcal{C}_{\mathrm{L}}^{-1}\mathcal{D}_{\mathrm{L}}\boldsymbol{y}[k] \tag{7.126}$$

Taking (7.120) and (7.125) into account, (7.126) can be further modified as

$$\boldsymbol{y}[k+d|k] = \bar{\mathcal{C}}_{\mathrm{L}}^{-1}\bar{\mathcal{T}}\mathcal{N}_{\mathrm{L}}\boldsymbol{u}[k] + \bar{\mathcal{C}}_{\mathrm{L}}^{-1}\bar{\mathcal{P}}\boldsymbol{y}[k] = \bar{\mathcal{C}}_{\mathrm{L}}^{-1}\left\{\bar{\mathcal{T}}\mathcal{N}_{\mathrm{L}}\boldsymbol{u}[k] + \bar{\mathcal{P}}\boldsymbol{y}[k]\right\} \tag{7.127}$$

Since the elements of (7.124) are independent, so the variance of the input signal is

$$\begin{aligned} V = \operatorname{var}\{\boldsymbol{y}[k+d]\} &= E\left\{\boldsymbol{y}^{\mathrm{T}}[k+d]\boldsymbol{y}[k+d]\right\} \\ &= E\left\{\boldsymbol{y}^{\mathrm{T}}[k+d|k]\boldsymbol{y}[k+d|k]\right\} + E\left\{\boldsymbol{\varepsilon}^{\mathrm{T}}[k+d|k]\boldsymbol{\varepsilon}[k+d|k]\right\} \end{aligned} \tag{7.128}$$

which can be minimized by the trivial choice $\boldsymbol{y}[k+d|k] = \mathbf{0}$. The minimum variance is $E\left\{\boldsymbol{\varepsilon}^{\mathrm{T}}[k+d|k]\boldsymbol{\varepsilon}[k+d|k]\right\}$. Thus, the condition of the minimum is to ensure the equality

$$\boldsymbol{y}[k+d|k] = \bar{\mathcal{C}}_{L}^{-1}\left\{\bar{\mathcal{T}}\mathcal{N}_{\mathrm{L}}\boldsymbol{u}[k] + \bar{\mathcal{P}}\boldsymbol{y}[k]\right\} = \mathbf{0} \tag{7.129}$$

from which the classical form of the *MIMO MV* regulator is obtained as

$$\boldsymbol{u}[k] = -\mathcal{N}_{L}^{-1}\bar{\mathcal{T}}^{-1}\bar{\mathcal{P}}\boldsymbol{y}[k] \tag{7.130}$$

If the process is required to follow the output of the reference model $\boldsymbol{R}_{\mathrm{r}}$, then the control equation is obtained from the condition $\boldsymbol{y}[k+d|k] = \boldsymbol{R}_{\mathrm{r}}\boldsymbol{y}_{\mathrm{r}}[k]$ as

$$\boldsymbol{u}[k] = \mathcal{N}_{\mathrm{L}}^{-1}\bar{\mathcal{T}}^{-1}\left\{\boldsymbol{R}_{\mathrm{r}}\boldsymbol{y}_{\mathrm{r}}[k] - \bar{\mathcal{P}}\boldsymbol{y}[k]\right\} \tag{7.131}$$

This expression formally corresponds to (7.113) with the difference that now $\bar{\mathcal{P}}$ is computed from (7.119).

After long but not very complicated calculation, it can be seen that the regulator corresponding to (7.130) is

$$\begin{aligned} C &= \mathcal{N}_{\mathrm{L}}^{-1}\bar{\mathcal{T}}^{-1}\bar{\mathcal{P}} = \mathcal{N}_{\mathrm{L}}^{-1}\mathcal{D}_{\mathrm{L}}\bar{\mathcal{C}}_{\mathrm{L}}^{-1}\bar{\mathcal{P}}\left(\mathbf{I} - \bar{\mathcal{C}}_{\mathrm{L}}^{-1}\bar{\mathcal{P}}z^{-d}\right)^{-1} \\ &= \mathcal{N}_{\mathrm{L}}^{-1}\mathcal{D}_{\mathrm{L}}\boldsymbol{R}_{\mathrm{n}}\left(\mathbf{I} - \boldsymbol{R}_{\mathrm{n}}z^{-d}\right)^{-1} \end{aligned} \tag{7.132}$$

Thus, the *MIMO* predictive regulator is practically a Youla-parameterized regulator whereby the reference model $\boldsymbol{R}_{\mathrm{n}} = \bar{\mathcal{C}}_{\mathrm{L}}^{-1}\bar{\mathcal{P}}$ refers to the noise rejection and does not depend on us but comes from Equation (7.119).

CHAPTER 8

Control of Nonlinear Cascade Processes

Nonlinear dynamic systems can have infinite versions though some well-defined classifications are found [56]. The purpose of this chapter is not to solve the control problems of these systems in general. Our goal here is very simple: to show some models where the Youla parameterization-based control design methods, discussed in detail earlier for linear systems, can be applied.

Generic two-degrees-of-freedom control systems, obtained by *KB* parameterization for linear processes, can be simply extended formally to nonlinear processes, as can be seen in Figure 8.1.

The equation of the closed control loop is

$$y = \mathsf{NQ_r}y_r + (1 - \mathsf{NQ_n})y_n = \mathsf{NK_r}R_r y_r + (1 - \mathsf{NK_n R_n})y_n = y_t + y_d \tag{8.1}$$

where N is the nonlinear operator of the process, $\mathsf{Q_n}$ and $\mathsf{K_r}$ are the nonlinear extension of the Youla parameter Q_n and the serial compensator K_r, respectively, introduced in Section 2.1 for linear processes. The computation of the inverse shown in the figure, if it exists at all, can be performed according to the conventional definition of the inverse functions.

Let the nonlinear process be factorable (decomposable for products)

$$\mathsf{N} = \mathsf{N_-N_+} \neq \mathsf{N_+N_-} \tag{8.2}$$

where it is assumed that $\mathsf{N_+}$ is inverse stable and $\mathsf{N_-}$ is inverse unstable. Given (8.2), it should be noted that, except for very special cases, the multiplication of nonlinear operators is not commutative. Based on the

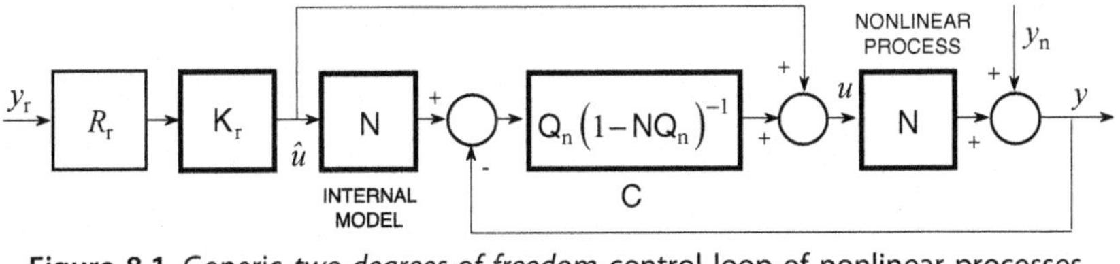

Figure 8.1 Generic *two-degrees-of-freedom* control loop of nonlinear processes.

decomposition (8.2), the nonlinear extension of the relevant linear Youla parameters can be simply written as

$$\begin{aligned} \mathsf{Q} &= \mathsf{Q}_n = \mathsf{K}_n R_n = \mathsf{N}_+^{-1}\mathsf{G}_n R_n \\ \mathsf{Q}_r &= \mathsf{K}_r R_r = \mathsf{N}_+^{-1}\mathsf{G}_r R_r \end{aligned} \tag{8.3}$$

where

$$\mathsf{K}_n = \mathsf{N}_+^{-1}\mathsf{G}_n \quad \text{and} \quad \mathsf{K}_r = \mathsf{N}_+^{-1}\mathsf{G}_r \tag{8.4}$$

Finally, the Youla-parameterized regulator belonging to the factorization (8.2) takes the form

$$\begin{aligned} \mathsf{C} &= \mathsf{Q}_n(1-\mathsf{N}\mathsf{Q}_n)^{-1} = \mathsf{K}_n R_n(1-\mathsf{N}\mathsf{K}_n R_n)^{-1} \\ &= \mathsf{N}_+^{-1}\mathsf{G}_n R_n(1-\mathsf{N}_-\mathsf{G}_n R_n)^{-1} \end{aligned} \tag{8.5}$$

The equation describing the closed system is

$$y = \mathsf{N}_-\mathsf{G}_r R_r y_r + (1-\mathsf{N}_-\mathsf{G}_n R_n)y_n = y_t + y_d \tag{8.6}$$

In the above equations, K_r and K_n contain the functional inverse N_+^{-1} of N_+; the operators G_r and G_n, however, after suitable optimization, can attenuate the effect of the invariant factor. Linear reference models R_r and R_n are chosen, because they can be defined more easily than nonlinear versions. It is important to note that if the nonlinear process can be factorized in the form $\mathsf{N} = \mathsf{N}_+\mathsf{N}_-$, then the above general scheme cannot be derived yet. In the literature, both forms of the factorization (8.2) are known as cascade models.

Based on the Figures 2.1.8 shown for linear processes, the block scheme in Figure 8.2 is introduced whose importance is that the regulator is

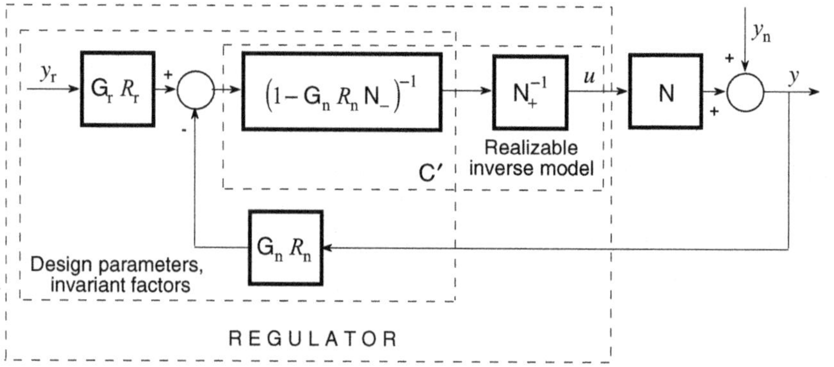

Figure 8.2 Simplified form of the nonlinear generic *two-degrees-of-freedom* regulator.

decomposed into two important parts. The first part contains the design parameters and the invariant factors, and the other part contains the realizable inverse model.

SIMPLE NONLINEAR CASCADE MODELS

The two best-known cascade models are the Wiener and Hammerstein models [56], shown in Figure 8.3. In the simple Wiener model N_W (Figure 8.3(a)), a static nonlinear subsystem $\mathsf{N}^{W,stat}$ is serially connected with a linear dynamic subsystem $Y^{W,dyn}$

$$\mathsf{N}_W = \mathsf{N}^{W,stat} Y^{W,dyn} \quad \text{or simply} \quad \mathsf{N}_W = \mathsf{N}^W Y^W \tag{8.7}$$

In the simple Hammerstein model N_H (Figure 8.3(b)), a linear dynamic subsystem $Y^{H,dyn}$ is serially connected with a nonlinear static subsystem $\mathsf{N}^{H,stat}$

$$\mathsf{N}_H = Y^{H,dyn} \mathsf{N}^{H,stat} \quad \text{or simply} \quad \mathsf{N}_H = Y^H \mathsf{N}^H \tag{8.8}$$

Observe that in the equations the order of the nonlinear operators is reverse compared with what is shown in the block scheme. (This used to be a very common mistake in the computations of nonlinear systems.)

Wiener Model

Assume that the transfer characteristics of the process is represented by the following equations:

$$\begin{aligned} \mathsf{N}_W &= \mathsf{N}^W Y_-^W Y_+^W = \mathsf{N}_+^W Y_-^W = \mathsf{N}^W Y_+^W \overline{Y}_-^W z^{-d} = \mathsf{N}_+ \mathsf{N}_- \\ \mathsf{N}_+ &= \mathsf{N}_+^W = \mathsf{N}^W Y_+^W \quad ; \quad \mathsf{N}_- = Y_-^W = \overline{Y}_-^W z^{-d} \end{aligned} \tag{8.9}$$

This structure apparently does not suit the decomposition form (8.2), except if $\overline{Y}_-^W = 1$ or $Y_-^W = z^{-d}$, whose factors are commutative in the order of operators. Hence, $Y_+^W = Y^W$. Here Y_+^W and Y_-^W mean the inverse stable and the inverse unstable factors, respectively. Furthermore, it is assumed that

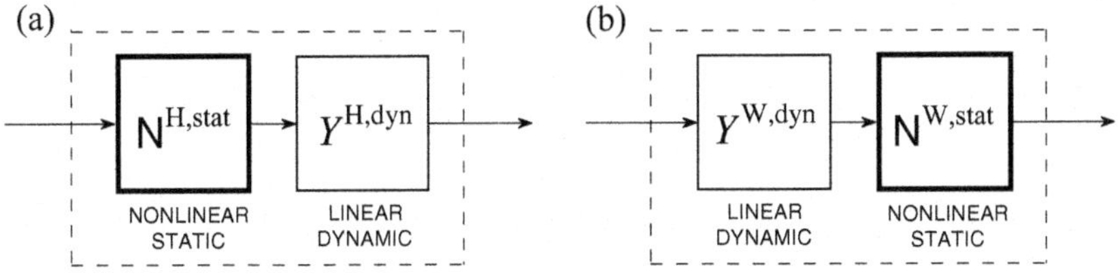

Figure 8.3 Topology of the simple Wiener and Hammerstein models.

the inverse static characteristic $(\mathsf{N}^{\mathrm{W}})^{-1}$ is realizable in the operating region, i.e., $(\mathsf{N}^{\mathrm{W}})^{-1} \neq \infty$. For this special Wiener model, it can be written that

$$\begin{aligned}\mathsf{N}_{\mathrm{W}} &= (\mathsf{N}^{\mathrm{W}} Y^{\mathrm{W}}) z^{-d} = \mathsf{N}_{+}^{\mathrm{W}} \mathsf{N}_{-} = z^{-d} (\mathsf{N}^{\mathrm{W}} Y^{\mathrm{W}}) = \mathsf{N}_{-}\mathsf{N}_{+} \\ \mathsf{N}_{+} &= \mathsf{N}^{\mathrm{W}} Y^{\mathrm{W}} \quad ; \quad \mathsf{N}_{-} = z^{-d} = N_{-}\end{aligned} \tag{8.10}$$

Since $\overline{Y}_{-}^{\mathrm{W}} = 1$, the following choice can be made: $\mathsf{G}_{\mathrm{n}}^{\mathrm{W}} = G_{\mathrm{n}}^{\mathrm{W}} = 1$ and $\mathsf{G}_{\mathrm{r}}^{\mathrm{W}} = G_{\mathrm{r}}^{\mathrm{W}} = 1$. The nonlinear Youla-parameterized regulator is now

$$\mathsf{C}_{\mathrm{W}} = (\mathsf{N}^{\mathrm{W}} Y^{\mathrm{W}})^{-1} \mathsf{G}_{\mathrm{n}}^{\mathrm{W}} R_{\mathrm{n}} (1 - \mathsf{N}_{-} \mathsf{G}_{\mathrm{n}}^{\mathrm{W}} R_{\mathrm{n}})^{-1} = \frac{1}{Y^{\mathrm{W}}} (\mathsf{N}^{\mathrm{W}})^{-1} \frac{R_{\mathrm{n}}}{1 - R_{\mathrm{n}} z^{-d}} \tag{8.11}$$

and the necessary nonlinear compensators are

$$\begin{aligned}\mathsf{K}_{\mathrm{r}}^{\mathrm{W}} &= (\mathsf{N}_{+}^{\mathrm{W}})^{-1} \mathsf{G}_{\mathrm{r}}^{\mathrm{W}} = \frac{1}{Y^{\mathrm{W}}} (\mathsf{N}^{\mathrm{W}})^{-1} \\ \mathsf{K}_{\mathrm{n}}^{\mathrm{W}} &= (\mathsf{N}_{+}^{\mathrm{W}})^{-1} \mathsf{G}_{\mathrm{n}}^{\mathrm{W}} = \frac{1}{Y^{\mathrm{W}}} (\mathsf{N}^{\mathrm{W}})^{-1}\end{aligned} \tag{8.12}$$

The simplified regulator form of Figure 8.2 becomes

$$\mathsf{C}'_{\mathrm{W}} = (\mathsf{N}^{\mathrm{W}} Y^{\mathrm{W}})^{-1} (1 - \mathsf{G}_{\mathrm{n}}^{\mathrm{W}} R_{\mathrm{n}} \mathsf{N}_{-})^{-1} = \frac{1}{Y^{\mathrm{W}}} (\mathsf{N}^{\mathrm{W}})^{-1} \frac{1}{1 - R_{\mathrm{n}} z^{-d}} \tag{8.13}$$

The nonlinear regulator, using the inverse $(\mathsf{N}^{\mathrm{W}})^{-1}$, apparently compensates completely the static nonlinearity N_{W} and the equation of overall closed system corresponds to the operation of a linear system

$$\begin{aligned}y &= z^{-d} \overline{Y}_{-}^{\mathrm{W}} G_{\mathrm{r}}^{\mathrm{W}} R_{\mathrm{r}} y_{\mathrm{r}} - \left(1 - z^{-d} \overline{Y}_{-}^{\mathrm{W}} G_{\mathrm{n}}^{\mathrm{W}} R_{\mathrm{n}}\right) y_{\mathrm{n}} \\ &= R_{\mathrm{r}} z^{-d} y_{\mathrm{r}} - (1 - R_{\mathrm{n}} z^{-d}) y_{\mathrm{n}} = y_{\mathrm{t}} + y_{\mathrm{d}}\end{aligned} \tag{8.14}$$

Example 8.1

Let the Wiener model of a nonlinear dynamic process be

$$\mathsf{N}_{\mathrm{W}} = \mathsf{N}^{\mathrm{W}} Y_{+}^{\mathrm{W}} Y_{-}^{\mathrm{W}} = (-0.125u^2 + u) \frac{0.2z^{-1}}{1 - 0.8z^{-1}} z^{-5} \tag{8.15}$$

where

$$\begin{aligned}\mathsf{N}^{\mathrm{W}} &= -0.125u^2 + u \quad ; \quad Y_{+}^{\mathrm{W}} = \frac{0.2z^{-1}}{1 - 0.8z^{-1}} \\ Y_{-}^{\mathrm{W}} &= \overline{Y}_{-}^{\mathrm{W}} z^{-5} = z^{-5} \quad ; \quad \overline{Y}_{-}^{\mathrm{W}} = 1\end{aligned} \tag{8.16}$$

Applying the above design method for the nonlinear regulator, we obtain

$$\mathsf{C}_{\mathrm{W}} = \left(5 - 4z^{-1}\right)\left(\mathsf{N}^{\mathrm{W}}\right)^{-1}\frac{0.8}{1 - 0.2z^{-1} - 0.8z^{-6}}$$
$$\left(\mathsf{N}^{\mathrm{W}}\right)^{-1} = 4\left(1 - \sqrt{1 - 0.5u}\right) \tag{8.17}$$

and the necessary nonlinear compensators are obtained as

$$\mathsf{K}_{\mathrm{r}}^{\mathrm{W}} R_{\mathrm{r}} = \left(5 - 4z^{-1}\right)\left(\mathsf{N}^{\mathrm{W}}\right)^{-1}\frac{0.7}{1 - 0.3z^{-1}} \tag{8.18}$$

In this task, the following reference models are applied as design goals

$$R_{\mathrm{r}} = \frac{0.7z^{-1}}{1 - 0.3z^{-1}} \quad \text{and} \quad R_{\mathrm{n}} = \frac{0.8z^{-1}}{1 - 0.2z^{-1}} \tag{8.19}$$

Figure 8.4 shows the operation of the closed system, if a square reference signal is used as excitation.

The transient processes show clearly the linear property expected after the complete compensation of the nonlinearity.

Hammerstein Model

Assume that the transfer characteristic of the process is represented by the equations

$$\mathsf{N}_{\mathrm{H}} = \mathrm{Y}_{-}^{\mathrm{H}}\mathrm{Y}_{+}^{\mathrm{H}}\mathsf{N}^{\mathrm{H}} = \mathrm{Y}_{-}^{\mathrm{H}}\mathsf{N}_{+}^{\mathrm{H}} = z^{-d}\overline{\mathrm{Y}}_{-}^{\mathrm{H}}\mathsf{N}_{+}^{\mathrm{H}} = \mathsf{N}_{-}\mathsf{N}_{+}$$
$$\mathsf{N}_{+} = \mathsf{N}_{+}^{\mathrm{H}} = \mathrm{Y}_{+}^{\mathrm{H}}\mathsf{N}^{\mathrm{H}} \quad ; \quad \mathsf{N}_{-} = \mathrm{Y}_{-}^{\mathrm{H}} = z^{-d}\overline{\mathrm{Y}}_{-}^{\mathrm{H}} \tag{8.20}$$

This structure apparently corresponds to the decomposition in (8.2). Here $\mathrm{Y}_{+}^{\mathrm{H}}$ and $\mathrm{Y}_{-}^{\mathrm{H}}$ denote the inverse stable and inverse unstable factors, respectively. Furthermore, it is assumed that the inverse static

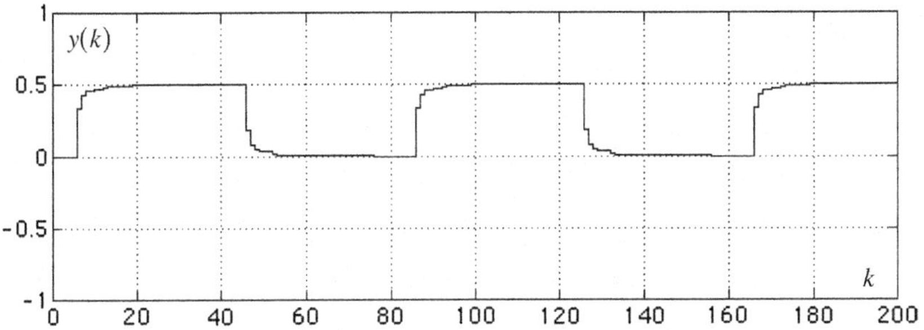

Figure 8.4 Operation of the nonlinear Youla-parameterized regulator for a simple Wiener model.

characteristic $(\mathsf{N}^{\mathrm{H}})^{-1}$ can be realized in the operating region, i.e., $(\mathsf{N}^{\mathrm{H}})^{-1} \neq \infty$. For the special Hammerstein model, it can be written that

$$\mathsf{C}'_{\mathrm{H}} = \left(Y_{+}^{\mathrm{H}}\mathsf{N}^{\mathrm{H}}\right)^{-1}\left(1 - \mathsf{G}_{\mathrm{n}}^{\mathrm{H}} R_{\mathrm{n}} \mathsf{N}_{-}\right)^{-1} = \left(\mathsf{N}^{\mathrm{H}}\right)^{-1} \frac{1}{\left(1 - G_{\mathrm{n}}^{\mathrm{H}} R_{\mathrm{n}} \overline{Y}_{-}^{\mathrm{H}} z^{-d}\right) Y_{+}^{\mathrm{H}}} \tag{8.21}$$

and the necessary nonlinear compensators become

$$\begin{aligned} \mathsf{K}_{\mathrm{r}}^{\mathrm{H}} &= \left(\mathsf{N}_{+}^{\mathrm{H}}\right)^{-1} \mathsf{G}_{\mathrm{r}}^{\mathrm{H}} = \left(\mathsf{N}^{\mathrm{H}}\right)^{-1} \frac{G_{\mathrm{r}}^{\mathrm{H}}}{Y_{+}^{\mathrm{H}}} \\ \mathsf{K}_{\mathrm{n}}^{\mathrm{H}} &= \left(\mathsf{N}_{+}^{\mathrm{H}}\right)^{-1} \mathsf{G}_{\mathrm{n}}^{\mathrm{H}} = \left(\mathsf{N}^{\mathrm{H}}\right)^{-1} \frac{G_{\mathrm{n}}^{\mathrm{H}}}{Y_{+}^{\mathrm{H}}} \end{aligned} \tag{8.22}$$

Since the invariant factor Y_{-}^{H} is linear, the following simplification can be made: $\mathsf{G}_{\mathrm{n}}^{\mathrm{H}} = G_{\mathrm{n}}^{\mathrm{H}}$ and $\mathsf{G}_{\mathrm{r}}^{\mathrm{H}} = G_{\mathrm{r}}^{\mathrm{H}}$. The simplified regulator form C' of Figure 8.2 is

$$C'_{\mathrm{H}} = \left(Y_{+}^{\mathrm{H}} N^{\mathrm{H}}\right)^{-1}\left(1 - G_{\mathrm{n}}^{\mathrm{H}} R_{\mathrm{n}} N_{-}\right)^{-1} = \left(N^{\mathrm{H}}\right)^{-1} \frac{1}{\left(1 - G_{\mathrm{n}}^{\mathrm{H}} R_{\mathrm{n}} \overline{Y}_{-}^{\mathrm{H}} z^{-d}\right) Y_{+}^{\mathrm{H}}} \tag{8.23}$$

Using the inverse $(\mathsf{N}^{\mathrm{H}})^{-1}$, the nonlinear regulator completely compensates the static nonlinearity N_{H} and the equation of the overall closed system corresponds to the operation of a linear system.

$$\begin{aligned} y &= z^{-d}\overline{Y}_{-}^{\mathrm{H}} G_{\mathrm{r}}^{\mathrm{H}} R_{\mathrm{r}} y_{\mathrm{r}} - \left(1 - z^{-d}\overline{Y}_{-}^{\mathrm{H}} G_{\mathrm{n}}^{\mathrm{H}} R_{\mathrm{n}}\right) y_{\mathrm{n}} \\ &= R_{\mathrm{r}} G_{\mathrm{r}}^{\mathrm{H}} \overline{Y}_{-}^{\mathrm{H}} z^{-d} y_{\mathrm{r}} - \left(1 - R_{\mathrm{n}} G_{\mathrm{n}}^{\mathrm{H}} \overline{Y}_{-}^{\mathrm{H}} z^{-d}\right) y_{\mathrm{n}} = y_{\mathrm{t}} + y_{\mathrm{d}} \end{aligned} \tag{8.24}$$

Example 8.2

Let the Hammerstein model of the nonlinear dynamic process be

$$\mathsf{N}_{\mathrm{H}} = Y_{-}^{\mathrm{H}} Y_{+}^{\mathrm{H}} \mathsf{N}^{\mathrm{H}} = z^{-5} \frac{0.2z^{-1}}{1 - 0.8z^{-1}} [\ln(u+1)] \tag{8.25}$$

where

$$\begin{aligned} &\mathsf{N}^{\mathrm{H}} = \ln(u+1) \quad ; \quad Y_{+}^{\mathrm{H}} = \frac{0.2z^{-1}}{1 - 0.8z^{-1}} \\ &Y_{-}^{\mathrm{H}} = \overline{Y}_{-}^{\mathrm{H}} z^{-5} = z^{-5} \quad ; \quad \overline{Y}_{-}^{\mathrm{H}} = 1 \end{aligned} \tag{8.26}$$

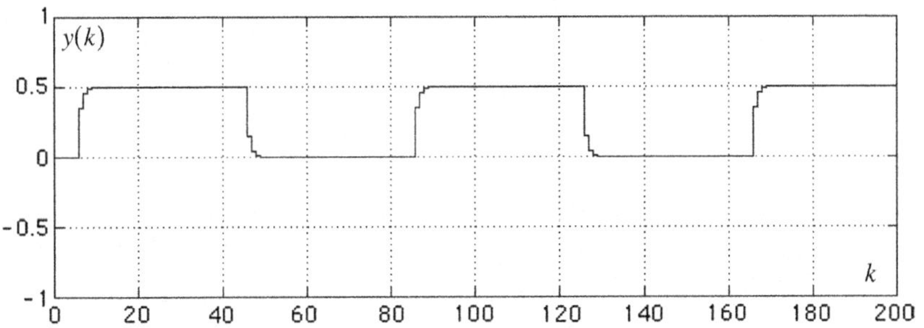

Figure 8.5 Operation of the nonlinear Youla-parameterized regulator for a simple Hammerstein model.

Applying the above design method for the nonlinear regulator, we obtain

$$\mathsf{C}_{\mathrm{H}} = \left(\mathsf{N}^{\mathrm{H}}\right)^{-1} \frac{4 - 3.2z^{-1}}{1 - 0.2z^{-1} - 0.8z^{-5}} \qquad (8.27)$$

$$\left(\mathsf{N}^{\mathrm{H}}\right)^{-1} = \exp(u) - 1$$

and the nonlinear compensator is obtained as

$$\mathsf{K}_{\mathrm{r}}^{\mathrm{H}} R_{\mathrm{r}} = \left(\mathsf{N}^{\mathrm{H}}\right)^{-1} \frac{3.5 - 2.8z^{-1}}{1 - 0.3z^{-1}} \qquad (8.28)$$

In this task the reference models of (8.19) are used.

Figure 8.5 shows the operation of the closed system, where a square signal is used as excitation.

The transients clearly show the linear properties as expected after the complete compensation of the nonlinearity.

NONLINEAR *PROPORTIONAL-INTEGRAL-DERIVATIVE* (*PID*) REGULATOR FOR NONLINEAR CASCADE MODELS

Consider the closed control system seen in Figure 8.6 where the process can be represented by a nonlinear cascade model. Investigate now whether this process can be controlled by a properly constructed, possibly a cascade,

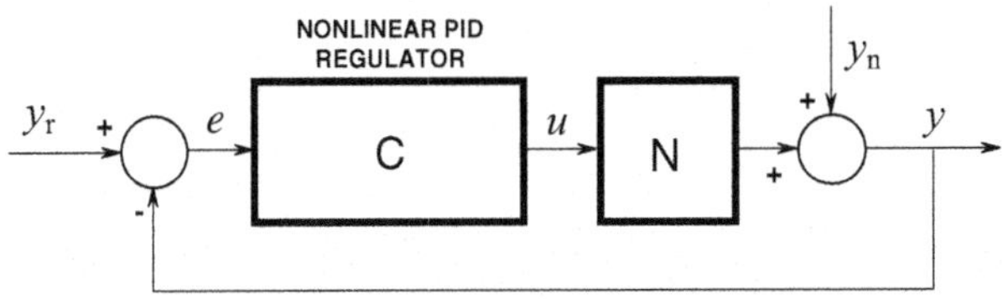

Figure 8.6 Nonlinear *proportional-integral-derivative* regulator and the process.

conventional *proportional-integral-derivative* (*PID*) regulator. To compute the sensitivity or complementary sensitivity of nonlinear operators of the closed system, the nonlinear operator $(1 + \mathsf{NC})^{-1}$ should be determined, which is a complex task, as seen at the beginning of this section. There is only one exception, that is, if the overall closed loop can be linearized. This is possible by means of the cascade models shown above, if decomposition (8.2) exists for the nonlinear process.

In our investigations, assume that the linear factor in the cascade models of the nonlinear process N corresponds to a second-order time-delay structure including the pulse transfer functions of (5.46) and (5.47)

$$G(z) = \frac{\mathcal{G}(z^{-1})}{\mathcal{F}(z^{-1})} = \frac{g_o(1+\gamma z^{-1})}{1+f_1 z^{-1}+f_2 z^{-2}} z^{-d} \tag{8.29}$$

for which the transfer function of a practical, modified *PID* regulator was given in (5.39) as

$$C(z) = C_{PID} = \frac{q_o + q_1 z^{-1} + q_2 z^{-2}}{1 - z^{-1}} G_F(z) = \frac{q_o \mathcal{F}(z)}{1 - z^{-1}} G_F(z) = \frac{K_I \mathcal{F}(z)}{g_o(1 - z^{-1})} G_F(z), \tag{8.30}$$

Hammerstein Model

Thus, given (8.20), the expressions for the Hammerstein model are now

$$\begin{aligned} \mathsf{N}_H &= Y_-^H Y_+^H \mathsf{N}^H = Y_-^H \mathsf{N}_+^H = z^{-d}\overline{Y}_-^H \mathsf{N}_+^H = \mathsf{N}_- \mathsf{N}_+ \\ \mathsf{N}_+ &= \mathsf{N}_+^H = Y_+^H \mathsf{N}^H \quad ; \quad \mathsf{N}_- = Y_-^H = z^{-d}\overline{Y}_-^H \end{aligned} \tag{8.31}$$

where

$$\begin{aligned} Y_-^H &= \mathcal{G}(z^{-1}) z^{-d} = g_o(1+\gamma z^{-1}) z^{-d} \quad ; \quad |\gamma| < 1 \\ Y_-^H &= z^{-d} \quad ; \quad |\gamma| \geq 1 \end{aligned} \tag{8.32}$$

and

$$\begin{aligned} Y_+^H &= \frac{\mathcal{G}(z^{-1})}{\mathcal{F}(z^{-1})} \quad ; \quad |\gamma| < 1 \\ Y_+^H &= \frac{1}{\mathcal{F}(z^{-1})} \quad ; \quad |\gamma| \geq 1 \end{aligned} \tag{8.33}$$

It is clear that the Hammerstein model fulfills the factorization condition (8.2). Assume that the inverse static characteristic $(\mathsf{N}^H)^{-1}$ is realizable and exists in the operation region, i.e., $(\mathsf{N}^{\mathrm{H}})^{-1} \neq \infty$. Following the derivation of (8.23) and using (8.30), we can cite the transfer operator of the "nonlinear" *PID* regulator as

$$\mathsf{C}_{\mathrm{H}} = \mathsf{C}_{\mathrm{N}}^{\mathrm{H}} C_{PID}^{\mathrm{H}} = (\mathsf{N}^{\mathrm{H}})^{-1} \frac{K_{\mathrm{I}} \mathcal{F}(z^{-1})}{g_{\mathrm{o}}(1 - z^{-1})} G_F(z^{-1})$$

$$C_{PID}^{\mathrm{H}}(z) = \frac{q_{\mathrm{o}} + q_1 z^{-1} + q_2 z^{-2}}{1 - z^{-1}} G_{\mathrm{F}}(z) = \frac{q_{\mathrm{o}} \mathcal{F}(z)}{1 - z^{-1}} G_{\mathrm{F}}(z) \qquad (8.34)$$

$$= \frac{K_{\mathrm{I}} \mathcal{F}(z)}{g_{\mathrm{o}}(1 - z^{-1})} G_{\mathrm{F}}(z)$$

Here, $C_{\mathrm{PID}}^{\mathrm{H}} = C_{\mathrm{PID}}$ and it is exactly the same as the regulator (8.30) used in the linear case, and furthermore

$$\mathsf{C}_{\mathrm{N}}^{\mathrm{H}} = (\mathsf{N}^{\mathrm{H}})^{-1} \qquad (8.35)$$

is the inverse of the static process characteristic.

Wiener Model

The Wiener model can be written as

$$\begin{aligned} \mathsf{N}_{\mathrm{W}} &= \mathsf{N}^{\mathrm{W}} Y_{-}^{\mathrm{W}} Y_{+}^{\mathrm{W}} = \mathsf{N}_{+}^{\mathrm{W}} Y_{-}^{\mathrm{W}} = \mathsf{N}^{\mathrm{W}} Y_{+}^{\mathrm{W}} \overline{Y}_{-}^{\mathrm{W}} z^{-d} = \mathsf{N}_{+} \mathsf{N}_{-} \\ \mathsf{N}_{+} &= \mathsf{N}_{+}^{\mathrm{W}} = \mathsf{N}^{\mathrm{W}} Y_{+}^{\mathrm{W}} \quad ; \quad \mathsf{N}_{-} = Y_{-}^{\mathrm{W}} = \overline{Y}_{-}^{\mathrm{W}} z^{-d} \end{aligned} \qquad (8.36)$$

which clearly does not meet the factorization condition (8.2), except when $\overline{Y}_{-}^{\mathrm{W}} = 1$, i.e., the factor $Y_{+}^{\mathrm{W}} = Y^{\mathrm{W}}$ is *inverse stable* (*IS*). Here again Y_{+}^{W} and Y_{-}^{W} denote the inverse stable and inverse unstable factors, respectively. In the case of this special Wiener model

$$\begin{aligned} \mathsf{N}_{\mathrm{W}} &= (\mathsf{N}^{\mathrm{W}} Y^{\mathrm{W}}) z^{-d} = \mathsf{N}_{+}^{\mathrm{W}} \mathsf{N}_{-} = z^{-d} (\mathsf{N}^{\mathrm{W}} Y^{\mathrm{W}}) = \mathsf{N}_{-} \mathsf{N}_{+} \\ \mathsf{N}_{+} &= \mathsf{N}^{\mathrm{W}} Y^{\mathrm{W}} \quad ; \quad \mathsf{N}_{-} = z^{-d} = N_{-} \end{aligned} \qquad (8.37)$$

where the unrealizable invariant factor $Y_{-}^{\mathrm{W}} = z^{-d}$ can be arbitrarily interchanged in the operator order.

Assume that the inverse static characteristic $(\mathsf{N}^{\mathrm{W}})^{-1}$ is realizable and exists in the required operating region, i.e., $(\mathsf{N}^{\mathrm{W}})^{-1} \neq \infty$. Following the derivation of (8.23) and using (8.30), we can cite the transfer operator of the "nonlinear" *PID* regulator as

$$\mathsf{C}_{\mathrm{W}} = \mathsf{C}_{\mathrm{N}}^{\mathrm{W}} C_{\mathrm{PID}}^{\mathrm{W}} = \frac{\mathcal{F}(z^{-1})}{g_{\mathrm{o}}(1 + \gamma z^{-1})} (\mathsf{N}^{\mathrm{W}})^{-1} \frac{K_{\mathrm{I}}}{g_{\mathrm{o}}(1 - z^{-1})} \qquad (8.38)$$

Now, contrary to (8.34),

$$C_{PID}^{W}(z) = \frac{K_I}{g_o(1 - z^{-1})} \tag{8.39}$$

is linear, but has no *PID* property and

$$C_N^W = \left[N^W \frac{g_o(1 + \gamma z^{-1})}{\mathcal{F}(z^{-1})} \right]^{-1} = (G_D)^{-1} (N^W)^{-1} \tag{8.40}$$

is not the inverse of the static process model but the inverse of the overall model. This requires the inverse stability of the process to be assumed.

Example 8.3

Let the nonlinear dynamic process be of Wiener type, i.e.,

$$N_W = N^W Y_+^W Y_-^W \tag{8.41}$$

where

$$N^W = -0.125u^2 + u \quad ; \quad Y_+^W = \frac{0.2z^{-1}}{1 - 1.5z^{-1} + 0.7z^{-2}}$$
$$Y_-^W = \overline{Y}_-^W z^{-5} = z^{-5} \quad ; \quad \overline{Y}_-^W = 1 \tag{8.42}$$

This nonlinear process corresponds to a situation when the output of the linear time-delay process $Y = Y_-^W Y_+^W = Y_+^W z^{-5}$ is measured via a sensor N^W without nonlinear dynamics. The nonlinear characteristic N^W of the sensor and its inverse are presented in Figure 8.7.

The reference signal $y_r = 1[k-1]$ and the outer noise $y_n = 1[k-100]$ take effect on the closed control loop. (Here $1[k]$ is the classical unity jump

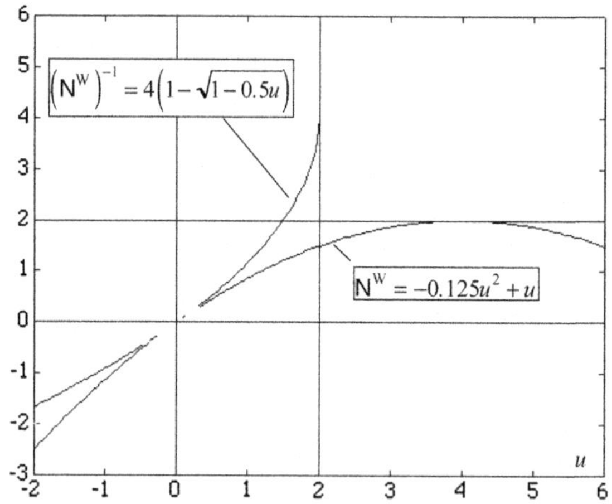

Figure 8.7 Nonlinear static characteristics of the process and the regulator.

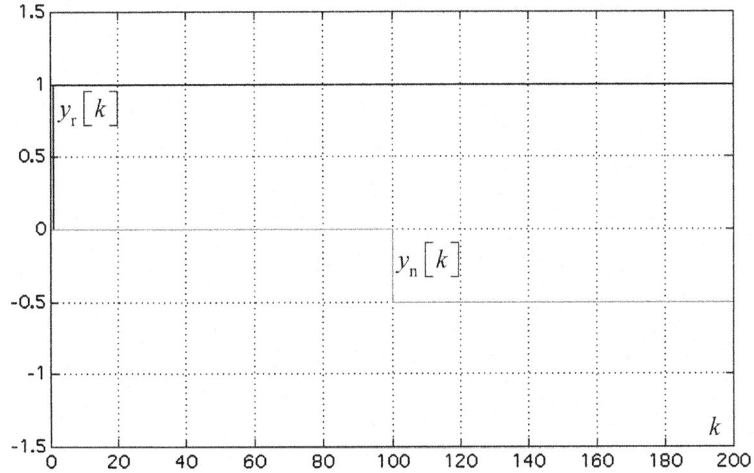

Figure 8.8 Time functions of the applied outer disturbing signals $y_r[k]$ and $y_n[k]$.

signal, i.e., $1[k] = 1$ if $k \geq 0$ and $1[k] = 0$, if $k < 0$.) The applied signals $y_r[k]$ and $y_n[k]$ can be seen in Figure 8.8. Figure 8.9 shows the case when the process is linear ($\mathbf{N}^W \equiv 1$) and the original regulator C_{PID} of (8.39) is applied, where, using the expression (5.44), the optimal integral gain is

$$K_I^o = \frac{1}{2d - 1} = 0.09 \tag{8.43}$$

Figure 8.10 shows the quality loss if the linear regulator regulates the nonlinear process. It is clear from Figure 8.11 that the nonlinear regulator of (8.38) completely reinstated the transient processes of the linear operation.

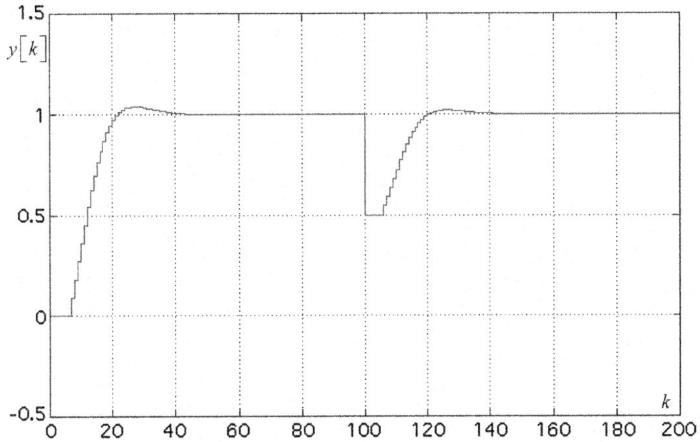

Figure 8.9 Time function of the output $y[k]$ if both the process and the regulator are linear.

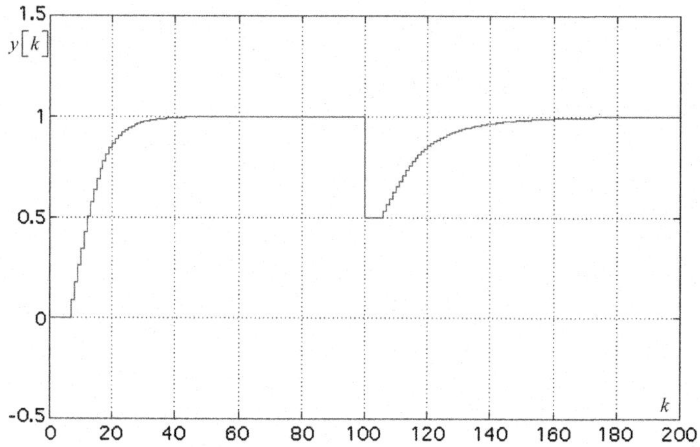

Figure 8.10 Time function of the output $y[k]$ if the regulator is linear and the process is nonlinear.

Example 8.4

Let the nonlinear dynamic process be of Hammerstein type, i.e.,

$$\mathsf{N}_{\mathrm{H}} = \mathcal{Y}_{-}^{\mathrm{H}} \mathcal{Y}_{+}^{\mathrm{H}} \mathsf{N}^{\mathrm{H}} \tag{8.44}$$

where

$$\begin{aligned} \mathsf{N}^{\mathrm{H}} &= -0.125u^2 + u \quad ; \quad \mathcal{Y}_{+}^{\mathrm{H}} = \frac{0.2z^{-1}}{1 - 1.5z^{-1} + 0.7z^{-2}} \\ \mathcal{Y}_{-}^{\mathrm{H}} &= \overline{\mathcal{Y}}_{-}^{\mathrm{H}} z^{-5} = z^{-5} \quad ; \quad \overline{\mathcal{Y}}_{-}^{\mathrm{H}} = 1 \end{aligned} \tag{8.45}$$

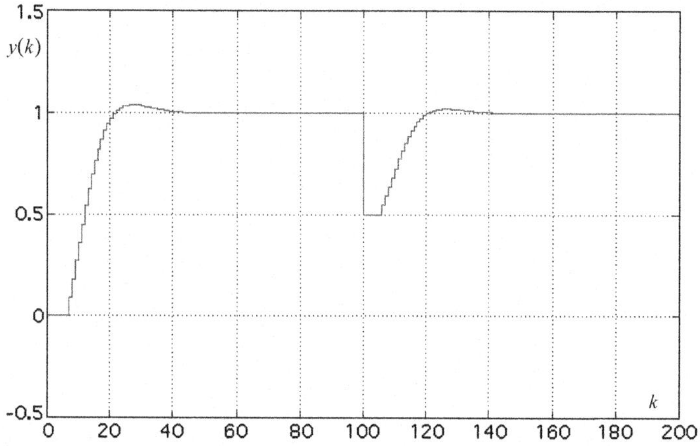

Figure 8.11 Time function of the output $y[k]$ if the regulator of (8.38) is nonlinear and the process is nonlinear.

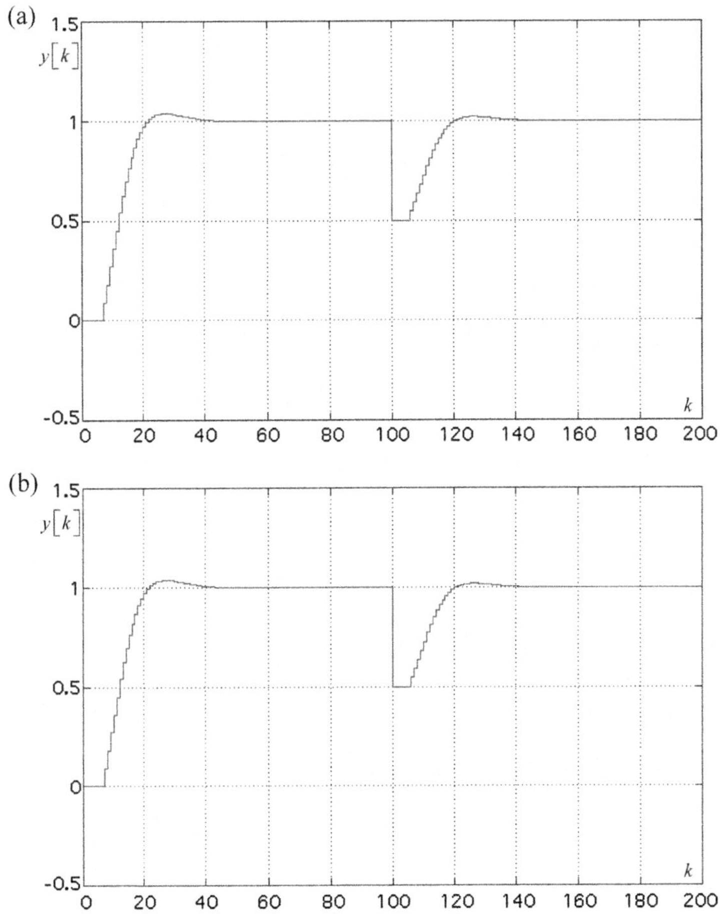

Figure 8.12 Time function of the output *y*[*k*] in the linear closed system (a) and in the case of the nonlinear process compensated by a nonlinear regulator (b).

This nonlinear process corresponds to a situation where a nonlinear actuator N^{H}, serially connected to the linear time-delay process $Y = Y_{+}^{\mathrm{H}}Y_{-}^{\mathrm{H}} = Y_{+}^{\mathrm{H}}z^{-5}$, is applied to the output of the regulator. The nonlinear characteristic N^{H} of the actuator and its inverse are the same as in Figure 8.7. It is clear from Figure 8.12 that the nonlinear *PID* regulator $\mathsf{C}_{\mathrm{H}} = \mathsf{C}_{\mathrm{N}}^{\mathrm{H}}C_{\mathrm{PID}}^{\mathrm{H}}$ of (8.34) completely reinstated the transient processes of the linear operation.

CHAPTER 9

Robust Control

In control engineering other concepts apart from optimality have recently appeared, such as the design methodology and robustness (insensitiveness) of regulators, closed-loop controls and stability conditions, etc.

In the literature, nature of robustness is not always clearly distinguished. It can refer to stability, or system properties (e.g., performance) or regulators, that is, to their insensitivity to parameter changes. Making a distinction is not easy.

The process parameters are never known precisely and the process is subject to change. The environment can change, which can in turn change the parameters of the process in a given region. Negative feedback reduces the sensitivity of the system to parameter changes. Therefore regulator design needs to take possible parameter changes into account. The required behavior of the control loop must be fulfilled not only for the nominal parameters but also for the possible parameter changes.

Let us now investigate the behavior of the control system if the transfer function of the process changes from the (real) value $P(s)$ to the nominal (model) value $\hat{P}(s)$. The overall transfer function of the open loop (Figure 1.2.1) is $L = CP$. For small changes in the process

$$\Delta L = \frac{\partial L}{\partial P}\Delta P = C\Delta P \tag{9.1}$$

Applying the relative changes we obtain

$$\frac{\Delta L}{L} = \ell_{\mathrm{L}} = \frac{C\Delta P}{CP} = \frac{\Delta P}{P} = \ell \quad \text{where} \quad \Delta P(s) = P(s) - \hat{P}(s) = \Delta \tag{9.2}$$

The overall transfer function of the negative feedback closed loop (Figure 1.2.3) is

$$T = \frac{CP}{1 + CP} \tag{9.3}$$

For small changes

$$\Delta T = \frac{\partial T}{\partial P}\Delta P = \frac{C}{(1 + CP)^2}\Delta P \tag{9.4}$$

For relative changes:

$$\frac{\Delta T}{T} = \ell_{\mathrm{T}} = \frac{1}{1+CP}\frac{\Delta P}{P} = S\frac{\Delta P}{P} = S\ell \tag{9.5}$$

where S is the sensitivity function of the closed loop

$$S = \frac{\Delta T/T}{\Delta P/P} = \frac{1}{1+CP} \tag{9.6}$$

The sensitivity function shows how the relative change of a term influences the relative change of the overall transfer function (the name derives from this property). In the frequency region where $|L(j\omega)| \to \infty$, the sensitivity function has small values, and big parameter changes in the terms have less effect on the overall transfer function and on the output of the closed system.

For infinitesimally small changes ($\Delta P \to 0$):

$$\frac{\partial T}{T} = S\frac{\partial P}{P} \tag{9.7}$$

whence

$$S = \frac{\partial T/T}{\partial P/P} = \frac{\partial \ln T}{\partial \ln P} \tag{9.8}$$

The overall transfer function T of closed systems is also called the complementary sensitivity function since the following expression holds

$$S + T = 1 \tag{9.9}$$

The typical frequency functions of the loop transfer function $|L(j\omega)|$ of the open loop, the sensitivity function $|S(j\omega)|$, and the complementary sensitivity function $|T(j\omega)|$ of the closed system are presented in Figure 9.1.

Let us consider the sensitivity of the control loop to parameter changes in the term in the feedback loop (Figure 9.2). In this case the sensitivity function can be defined as

$$S_{\mathrm{H}} = \frac{\Delta T/T}{\Delta H/H} \tag{9.10}$$

Now

$$T = \frac{CP}{1+CPH} \quad \text{so} \quad \Delta T = \frac{\partial T}{\partial H}\Delta H = -\frac{(CP)^2}{(1+CPH)^2}\Delta H, \tag{9.11}$$

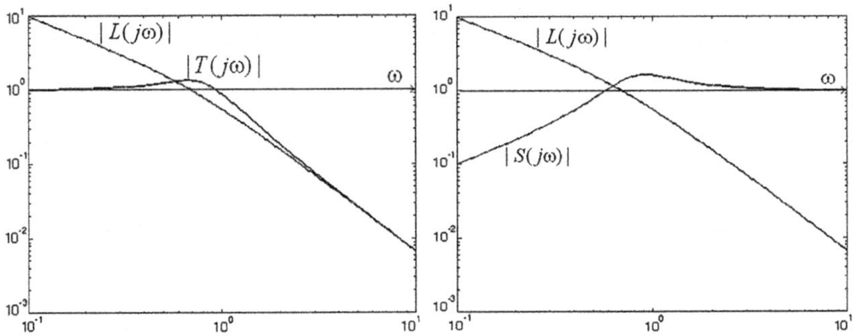

Figure 9.1 The loop transfer function, the sensitivity function, and the complementary sensitivity function.

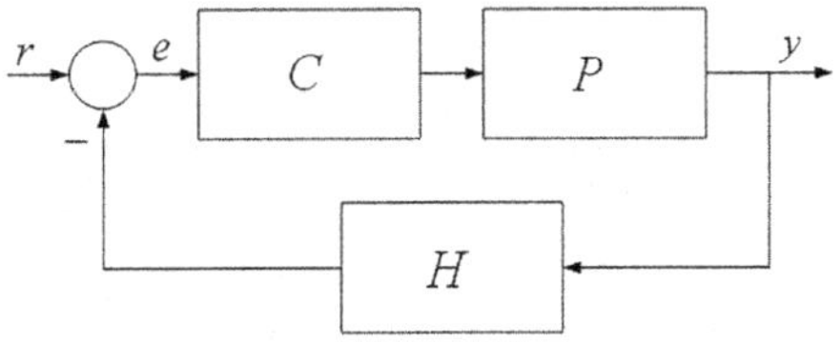

Figure 9.2 Feedback control loop showing the sensor has some dynamics.

and the relative error of T is

$$\frac{\Delta T}{T} = \ell_{\text{T}} = S_{\text{H}} \frac{\Delta H}{H} = -\frac{CPH}{1 + CPH} \frac{\Delta H}{H} = -\frac{L}{1 + L} \frac{\Delta H}{H} = -\frac{L}{1 + L} \ell_{\text{H}} \tag{9.12}$$

Since good reference signal tracking requires a value of $S_{\text{H}} = -L/(1 + L) = -T$ close to unity in a very wide frequency region, the parameter changes of the feedback term can significantly influence the output. Therefore we aim to make very precise measurements, or apply unity (rigid) feedback. Hereafter the simplified notation $\Delta P(s) = \Delta$ will be used.

Section 1.3 discussed the stability of the closed loop and the different properties of stability. The necessary and sufficient conditions of the so-called robust stability were given by (1.3.18), that is,

$$|\ell(j\omega)| < \left| \frac{1 + \hat{L}(j\omega)}{\hat{L}(j\omega)} \right| = \frac{1}{\left| \hat{T}(j\omega) \right|} \quad \forall \omega \tag{9.13}$$

where ℓ is the relative error of the process model and $\hat{T} = \hat{L}/(1 + \hat{L})$ is the nominal complementary sensitivity function. The expression (9.13) is very specific because it does not depend directly on the real closed system, but only on the model and $\hat{T}$, computed by the regulator designed on the basis of the model [32]. Equation (9.13) can be written in the form of inequality as

$$\left|\hat{T}(j\omega)\right| \left|\ell(j\omega)\right| < 1 \quad \forall\omega \tag{9.14}$$

which can also be interpreted as follows

$$\left|\begin{matrix}\text{design}\\ \text{factor}\end{matrix}\right| \times \left|\begin{matrix}\text{modeling}\\ \text{factor}\end{matrix}\right| < 1 \tag{9.15}$$

that is, the design and modeling factors can be improved at the expense of each other. This dialectic condition has great significance. In practice the inequality (9.14) does not mean very much. In low frequencies it is generally true that $\left|\hat{T}\right| \cong 1$, so the condition $|\ell|<1$ is sufficient for robust stability. In high frequencies, where $\left|\hat{T}\right| \ll 1$ in general, relatively bad models can be accepted for which $|\ell|>1$. In control systems the medium frequency region has significant importance when $\left|\hat{T}\right|$ is usually maximal, so $|\ell|$ must be minimal here, that is, the most accurate model is required.

In terms of robust stability the distance $\rho = |1 + L(j\omega)|$ has significant importance, as it gives the distance of a point $L(j\omega)$ of the Nyquist diagram in all frequencies from the point $-1 + j0$. Its lowest value is known as the modulus margin (1.3.13), that is,

$$\rho_{\mathrm{m}} = \frac{1}{\max\limits_{\omega} |S(j\omega)|} = \min_{\omega}\left|S^{-1}(j\omega)\right| = \min_{\omega}|1 + L(j\omega)| \tag{9.16}$$

ρ_{m} also used to be called the Nyquist stability measure (see Figure 9.3).

Thus the maximum of the absolute value of the sensitivity function appears on the frequency ω_{E} where the closest point of the Nyquist diagram is to $-1 + j0$. The maximum value is the reciprocal of this distance.

It is not immaterial how the curve $E(\omega) = |S(j\omega)|$ behaves in the vicinity of the maximum ω_{E}. Let us therefore investigate the second derivate in the vicinity of the maximum

$$\frac{\partial^2|S(j\omega)|}{\partial\omega^2} = 2|S(j\omega)|^3\left[\frac{\partial|S(j\omega)|}{\partial\omega}\right]^2 - S^2(j\omega)\frac{\partial^2|1 + L(j\omega)|}{\partial\omega^2}\Bigg|_{\omega=\omega_S}$$
$$= -S^2(j\omega)\frac{\partial^2|1 + L(j\omega)|}{\partial\omega^2} = -S^2(j\omega)\frac{\partial^2\rho}{\partial\omega^2} \tag{9.17}$$

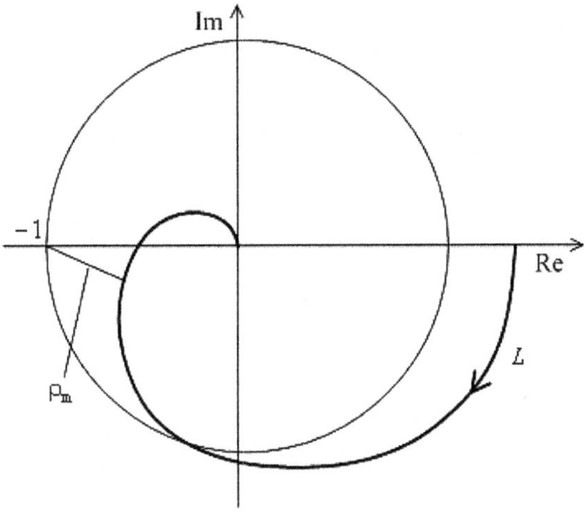

Figure 9.3 Nyquist stability measure.

since the first term is zero. The second derivate is the reciprocal of the radius of the circle fitted in the place of the extremum. Let r_S and r_ρ be the radii of the circles fitted to the curves $|S(j\omega)|$ and ρ, respectively, so

$$r_S = -\left[S^2(j\omega)\frac{\partial^2|1+L(j\omega)|}{\partial\omega^2}\right]^{-1}\Bigg|_{\omega=\omega_S} = -\rho_m^2\left[\frac{\partial^2\rho}{\partial\omega^2}\right]^{-1}_{\omega=\omega_S} = -\rho_m^2 r_\rho \tag{9.18}$$

The reciprocal of the fitting radius is also called the radius of curvature. The negative sign means the minimum and the positive the maximum, and thus $r_S = \rho_m^2 r_\rho$.

The regulator design procedure maximizing ρ_m, increases, at the same time, the radius of the fitting circle in the maximum of the curve $|S(j\omega)|$, but more precise investigations exploit the maximization possibilities of the radius r_ρ. Geometrically the largest radius is obtained when the two neighboring corner points of the Bode diagram around the extremum are as far away as possible.

Consider the following three simple closed control loops which can be used in model-based regulator design. The first closed loop can be seen in Figure 9.4. Here it is assumed that the regulator C is computed from the theoretical real process P and is placed together with the real process in the closed loop. Obviously this closed loop is not realistic and represents an ideal case.

The next version can be seen in Figure 9.5, and is usually applied in design tasks namely, when the regulator $\hat{C}$ is determined on the basis of the

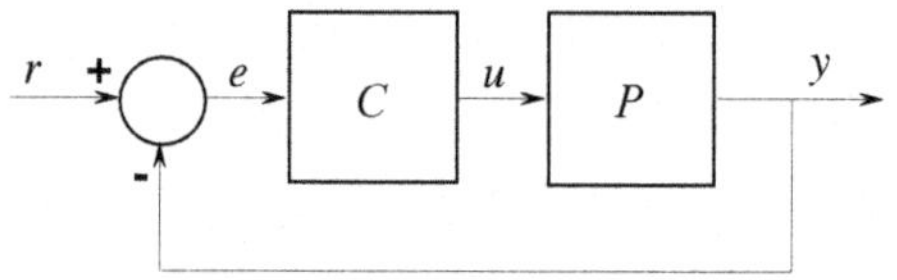

Figure 9.4 The theoretical closed system.

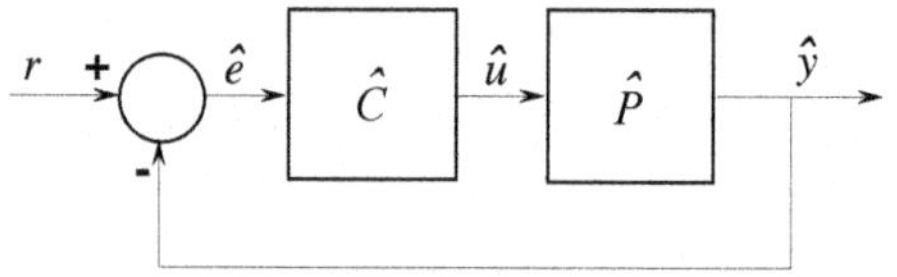

Figure 9.5 The nominal closed system.

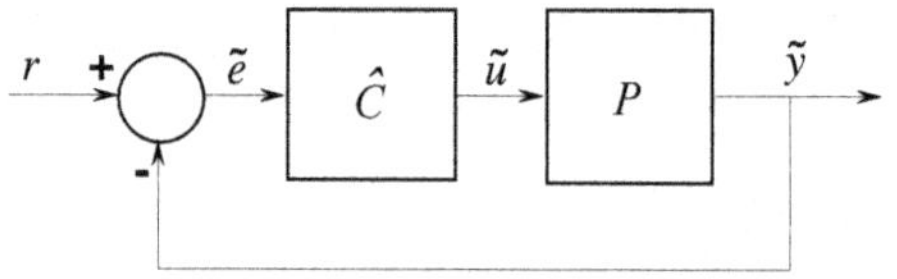

Figure 9.6 The real closed system appearing in practice.

process model $\hat{P}$ the whole closed loop is model based. This case is usually called the nominal system. This closed loop depends only on the designer, knowledge of the process, and the suggested regulator. The scheme can be used in simulation, optimization, and design tasks.

The third version of the closed system is what operates in reality. A model-based regulator is used together with the real process in the closed loop as in Figure 9.6. Usually measurements, verifications, and application of identification methods take place in this kind of closed loops.

The sensitivity and complementary sensitivity functions for the above three closed systems are summarized in Table 9.1. The computation of each element is very different and they must not be mixed. Obviously, in the ideal case when $\hat{P} = P$ the elements in the same rows are equal.

Table 9.1 The sensitivity and complementary sensitivity functions of the three systems

System function	Ideal	Nominal	Real
T	$T = \dfrac{CP}{1+CP}$	$\hat{T} = \dfrac{\hat{C}\hat{P}}{1+\hat{C}\hat{P}}$	$\tilde{T} = \dfrac{\hat{C}P}{1+\hat{C}P}$
S	$S = \dfrac{1}{1+CP}$	$\hat{S} = \dfrac{1}{1+\hat{C}\hat{P}}$	$\tilde{S} = \dfrac{1}{1+\hat{C}P}$

9.1 ROBUSTNESS OF YOULA-PARAMETERIZED REGULATOR

The functions of the regulator, of the Youla parameter, and of the loop transfer function are summarized in Table 9.1.1 for the three closed systems discussed in the previous section, and the three most important signals, that is, y, u, e can be found in Table 9.1.2.

In the last column $\tilde{C}$, $\tilde{Q}$, and $\tilde{L}$ have theoretical importance because, similarly to the ideal case, they can never be computed precisely. In practice the model-based regulator $\hat{C}$ is used. Obviously the following expression should always be taken into account for the three basic cases

$$\tilde{C} = \frac{\hat{Q}}{1 - \hat{Q}P} \neq C = \frac{Q}{1 - QP} \neq \hat{C} = \frac{\hat{Q}}{1 - \hat{Q}\hat{P}} \tag{9.1.1}$$

It is interesting to observe that the following identity holds

$$\frac{1}{\hat{Q}} - \frac{1}{\tilde{Q}} = \frac{\tilde{Q} - \hat{Q}}{\tilde{Q}\hat{Q}} = \Delta \tag{9.1.2}$$

Table 9.1.1 The regulator, the Youla parameter, and the loop transfer function of the three systems in the *YP* control loop

System function	Theoretical	Nominal	Real
C	$C = \frac{Q}{1 - QP} = C_o$	$\hat{C} = \frac{\hat{Q}}{1 - \hat{Q}\hat{P}}$	$\tilde{C} = \frac{\hat{Q}}{1 - \hat{Q}P}$
Q	$Q = \frac{C}{1 + CP} = Q_o$	$\hat{Q} = \frac{\hat{C}}{1 + \hat{C}\hat{P}}$	$\tilde{Q} = \frac{\hat{C}}{1 + \hat{C}P}$
L	$L = CP$	$\hat{L} = \hat{C}\hat{P}$	$\tilde{L} = \hat{C}P$

Table 9.1.2 The most important signals of the three systems

System signal	Theoretical	Nominal	Real
y	$y = CPe = QPr$	$\hat{y} = \hat{C}\hat{P}e = \hat{Q}\hat{P}r$	$\tilde{y} = \hat{C}P\tilde{e} = \frac{\hat{Q}\hat{P}(\ell + 1)}{1 + \hat{Q}\hat{P}\ell} r$
u	$u = Ce = Qr$	$\hat{u} = \hat{C}e = \hat{Q}r$	$\tilde{u} = \hat{C}\tilde{e} = \frac{\hat{Q}}{1 + \hat{Q}\hat{P}\ell} r$
e	$e = (1 - QP)r$	$e = (1 - \hat{Q}\hat{P})r$	$\tilde{e} = \frac{1}{1 + \hat{C}P} r = \frac{1 - \hat{Q}\hat{P}}{1 + \hat{Q}\hat{P}\ell} r$

All the elements of the process factorization (2.1.9) can have uncertainty. These can be denoted by

$$\Delta = P - \hat{P}; \quad \Delta_+ = P_+ - \hat{P}_+; \quad \Delta_- = P_- - \hat{P}_-; \quad \Delta_d = T_d - \hat{T}_d \tag{9.1.3}$$

for the absolute errors and by

$$\ell = \frac{\Delta}{\hat{P}} = \frac{P - \hat{P}}{\hat{P}}; \quad \ell_+ = \frac{\Delta_+}{\hat{P}_+}; \quad \ell_- = \frac{\Delta_-}{\hat{P}_-}; \quad \ell_d = \frac{\Delta_d}{\hat{T}_d} \tag{9.1.4}$$

for the relative errors. It is clearly seen that in the case of the ideal *Youla parameterized* (*YP*) regulator of (2.1.5), the factor $\hat{P}_-$ is included in the characteristic equation $\hat{P}_+\hat{P}_- = 0$; thus the invariant (unstable) zeros cannot be eliminated by the regulator. In the case of the realizable regulator (2.1.10) and using the above uncertainty notations, we obtain the characteristic equation

$$\left(\hat{P}_+ + R_n\Delta_+ + R_n\Delta_-\right)\Big|_{(\Delta_+,\Delta_-)\to 0} = \hat{P}_+ = 0 \tag{9.1.5}$$

for the simplified case $\Delta_d = 0$; thus it does not depend on invariant (unstable) factors.

The condition of robust stability (1.3.17) for the *YP* control loops can be further simplified so the expression (9.14) becomes

$$\begin{aligned}\left|\hat{Q}\hat{P}\ell\right| &= \left|R_n G_n \hat{P}_+^{-1}\hat{P}\ell\right| = \left|R_n G_n \hat{P}_- e^{-s\hat{T}_d}\ell\right| = \left|R_n G_n \hat{P}_- \ell\right| \\ &= \left|R_n G_n \hat{P}_-\right|\left|\ell\right| < 1 \quad \forall\omega\end{aligned} \tag{9.1.6}$$

where $\hat{T}_d$ is the dead-time of the model and $\left|e^{-s\hat{T}_d}\right| = 1$. The inequality (9.13), limiting the relative error, is now

$$|\ell(j\omega)| < \frac{1}{|R_n|\left|G_n\hat{P}_-\right|} \quad \forall\omega \tag{9.1.7}$$

If the process is *inverse stable* (*IS*), that is, $\hat{P}_- = 1$, then $G_n = 1$ can be chosen and the condition of robust stability can be further simplified as

$$|\ell(j\omega)| < \frac{1}{|R_n|} \quad \forall\omega \tag{9.1.8}$$

that is, it does not depend on the model $\hat{P}$ but only on the reference model or the design goal.

The reference model is an important parameter of the general Youla design, by means of which the condition of robust stability (9.1.8) can be guaranteed.

Let

$$R_n = \frac{1}{1 + sT_n} \tag{9.1.9}$$

then the constraining condition of the right side of (9.1.8) can be seen in Figure 9.1.1. Given the latter condition and choosing first-order reference model R_n, we see that robust stability can be ensured even in the case of 100% relative model error.

If the process is *inverse unstable (IU)*, even the factor $\left|G_n \hat{P}_-\right|$ appears in (9.1.6), and can significantly modify (9.1.7). The worst case is when this factor has a large value in the region of the cutoff frequency.

Investigate the sensitivity of the ideal *YP* regulator

$$\ell_C^{id} = \frac{\Delta C_{id}}{C_{id}} = \frac{C_{id} - \hat{C}}{C_{id}} = \frac{P^{-1} - \hat{P}^{-1}}{P^{-1}} = -\frac{P - \hat{P}}{\hat{P}} = -\ell \tag{9.1.10}$$

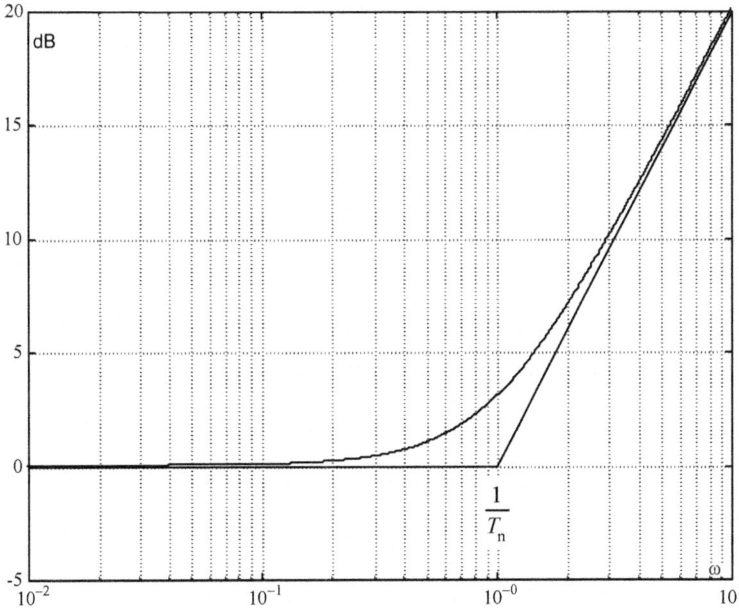

Figure 9.1.1 Condition constraining the relative model error in the case of the first-order reference model.

This is a very important relationship, which means that any identification method what minimizes, for example, the norm $\|\ell\|_\infty$ at the same time ensures the least sensitive regulator for the parameter changes.

On the basis of Table 9.1.1, further interesting statements can be made regarding the robust stability of the real *YP* closed-loop control. Using the following expressions

$$\tilde{C} = \frac{\hat{Q}}{1-\hat{Q}P} = \frac{\hat{C}}{1-\hat{C}\Delta} \Rightarrow \tilde{C} - \tilde{C}\hat{C}\Delta = \hat{C} \Rightarrow \tilde{C} - \hat{C} = \tilde{C}\hat{C}\Delta \tag{9.1.11}$$

finally we obtain

$$\Delta = \frac{\tilde{C} - \hat{C}}{\tilde{C}\hat{C}} \tag{9.1.12}$$

for the absolute error, and the relative error becomes

$$\ell = \frac{\Delta}{\hat{P}} = \frac{\tilde{C} - \hat{C}}{\tilde{C}} \frac{1}{\hat{C}\hat{P}} = \frac{\tilde{C} - \hat{C}}{\tilde{C}} \frac{1}{\hat{L}}; \quad \hat{L} = \hat{C}\hat{P} \tag{9.1.13}$$

Then the robust stability condition (9.1.8) takes the form

$$\left| \frac{\tilde{C} - \hat{C}}{\tilde{C}} \frac{1}{\hat{L}} R_{\mathrm{n}} \right| = \left| \ell_{\mathrm{C}} \frac{1}{\hat{L}} R_{\mathrm{n}} \right| < 1 \quad \text{or} \quad \ell_{\mathrm{C}} = \left| \frac{\tilde{C} - \hat{C}}{\tilde{C}} \right| < \left| \frac{\hat{L}}{R_{\mathrm{n}}} \right| = \frac{|\hat{L}|}{|R_{\mathrm{n}}|} \quad \forall\omega \tag{9.1.14}$$

The result is very interesting with regard to the constraining right-side condition, as shown in Figure 9.1.2. In the usual case (when $|\hat{L}(j\omega)|$

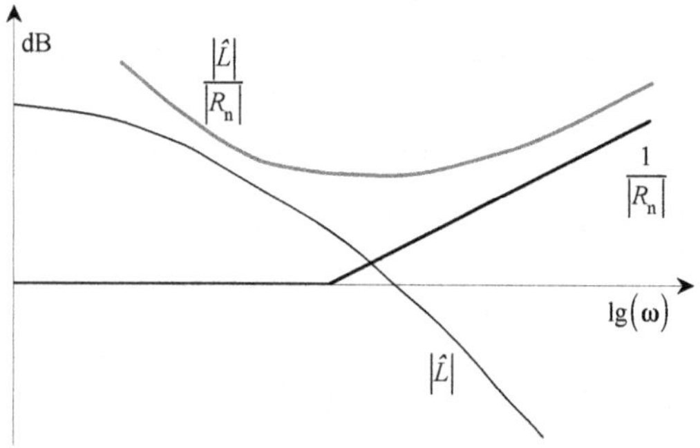

Figure 9.1.2 Constraint of the robust stability of the *Youla parameterized* control loop.

and $|R_n(j\omega)|$ is monotonously decreasing) it is clear that the regulator must be most accurate in the medium frequency region, where $|\hat{L}|/|R_n|$ is minimal.

Given (9.1.13) the robust stability condition (1.3.17) becomes

$$\left|\hat{T}\ell\right| = \left|\frac{\hat{C}\hat{P}}{1+\hat{C}\hat{P}}\frac{\tilde{C}-\hat{C}}{\tilde{C}}\frac{1}{\hat{C}\hat{P}}\right| = \left|\hat{S}\,\ell_C\right| < 1 \quad \forall\omega \tag{9.1.15}$$

Here it is assumed that the dead-time T_d is known

$$\hat{S} = \frac{1}{1+\hat{C}\hat{P}} = \frac{1}{1+\frac{R_n\hat{G}_n\hat{P}_+^{-1}}{1-R_n\hat{G}_n\hat{P}_-z^{-d}}\hat{P}_+\hat{P}_-e^{-sT_d}} = 1 - R_n\hat{G}_n\hat{P}_-e^{-sT_d} \tag{9.1.16}$$

Thus

$$\ell_C = \left|\frac{\tilde{C}-\hat{C}}{\tilde{C}}\right| < \left|\frac{1}{1-R_n\hat{G}_n\hat{P}_-e^{-sT_d}}\right| = \left|1+\hat{C}\hat{P}\right| = \hat{\rho} \quad \forall\omega \tag{9.1.17}$$

This means that in the nominal closed system the maximization of $\hat{\rho}$ maximizes the stability region. The worst-case optimization paradigm $\max\left\{\min\left|\hat{S}^{-1}\right|\right\} = \max\{\min|\hat{\rho}|\}$, which maximizes the Nyquist stability measure $\hat{\rho}_m$, results in the most robust and most precise regulator (i.e., when the difference $\tilde{C}-\hat{C}$ is the lowest). (Note that, in general, $\left|\hat{S}(j\omega)\right|$ has its maximum in the medium frequency region.)

It is easy to calculate the stability measure $\hat{\rho} = \left|1+\hat{C}\hat{P}\right|$ or its maximum $\hat{\rho}_m$ in the nominal system, since all factors are known. This is not the case for the measure $\rho = \left|1+\hat{C}P\right|$ in the real system. It is possible, however, to give lower and upper limits if the supremum of the relative error is known. Let

$$\rho_m(\hat{C}) = \min_{\omega}\left|\rho(\omega,\hat{C})\right| = \min_{\omega}\left|1+\hat{C}P\right| \tag{9.1.18}$$

which is the distance between the point $-1+0j$ and the closest point of the real loop transfer function $\tilde{L} = \hat{C}P$. (Note that in the most common case, when $\left|\tilde{L}(\omega=\infty)\right| = 0$, the measure ρ_m is in the region

$0 \le \rho_m \le 1$ for stable closed loops.) Let us denote the largest value of ρ_m regarding $\hat{C}$ by

$$\widehat{\rho}_m = \max_{\hat{C}}\left\{\min_{\omega}\left|\rho(\omega, \hat{C})\right|\right\} = \max_{\hat{C}}\left\{\min_{\omega}(\rho)\right\} \tag{9.1.19}$$

With simple computations the inner function becomes

$$\rho_m(\hat{C}) = \min_{\omega}\left|1 + \hat{C}P\right| = \left\|\frac{1 - R_n\hat{G}_n\hat{P}_-}{1 + R_n\hat{G}_n\hat{P}_-\ell}\right\|_\infty^{-1} \tag{9.1.20}$$

whence the following upper limit can be derived

$$\frac{\rho_m(\hat{C})}{\widehat{\rho}_m^{\,o}} \le \left\|1 + R_n\hat{G}_n\hat{P}_-\ell\right\|_\infty \le 1 + \|R_n\|_\infty\|\ell\|_\infty \tag{9.1.21}$$

Here $\widehat{\rho}_m^{\,o}$ is theoretically the best (maximum) robust measure in the case of $\|\ell\|_\infty \to 0$; furthermore it is also taken into account that the optimization of $\left|e^{-sT_d}\right| = 1$ and $\hat{G}_n$ can provide the all-pass condition $\left|\hat{G}_n\hat{P}_-\right| = 1$. It can be checked, obviously, that

$$\widehat{\rho}_m^{\,o} = \widehat{\rho}_m(\ell = 0) = \frac{1}{\left\|1 - R_n\hat{G}_n\hat{P}_-\right\|_\infty} = \min_{\omega}\left|\frac{1}{1 - R_n\hat{G}_n\hat{P}_-}\right| \tag{9.1.22}$$

In the special inverse stable case

$$\widehat{\rho}_{m,IS}^{\,o} = \widehat{\rho}_{m,IS}(\ell = 0) = \frac{1}{\|1 - R_n\|_\infty} = \min_{\omega}\left|\frac{1}{1 - R_n}\right| \tag{9.1.23}$$

which depends only on the reference model. In the continuous-time (CT) case the first-order reference model (9.1.9) provides the trivial value $\widehat{\rho}_{m,IS}^{\,o} = 1$. Observe that this is not so the discrete-time (DT) case, when the first-order reference model takes the form

$$R_n = \frac{b_{n1}z^{-1}}{1 + a_{n1}z^{-1}} \tag{9.1.24}$$

and

$$\widehat{\rho}_{m,IS}^{\,o} = \frac{|a_{n1} - 1|}{2} \tag{9.1.25}$$

This gives a robust measure of $\widehat{\rho}^{o}_{m,IS} = 0.9$ under the parameters $b_{n1} = 0.2$ and $a_{n1} = -0.8$ chosen for reference models with unity gain.

As with the upper limit (9.1.21) it is easy to determine a lower limit also, that is,

$$1 - \|R_n\|_\infty \|\ell\|_\infty \leq \min_{\omega}\left(1 + R_n \hat{G}_n \hat{P}_{-}\ell\right) \leq \frac{\rho_m(\hat{C})}{\widehat{\rho}^{o}_m} \tag{9.1.26}$$

Combining the two conditions we obtain

$$1 - \|R_n\|_\infty \|\ell\|_\infty \leq \min_{\omega}(1 + R_n \hat{G}_n \hat{P}_{-}\ell) \leq \frac{\rho_m(\hat{C})}{\widehat{\rho}^{o}_m} \leq \left\|1 + R_n \hat{G}_n \hat{P}_{-}\ell\right\|_\infty \leq 1 + \|R_n\|_\infty \|\ell\|_\infty \tag{9.1.27}$$

Comparison of the lower and upper limits clearly shows that those identification methods, where the condition $\|\ell\|_\infty \to 0$ is fulfilled, ensure the following convergence for the *YP* regulator

$$\rho_m(\hat{C}) \to \widehat{\rho}^{o}_m = \max_{\hat{C}}\left\{\min_{\omega}\left|\rho(\omega, \hat{C})\right|\right\}, \quad \text{if } \|\ell\|_\infty \to 0 \tag{9.1.28}$$

that is, they provide the most robust regulator in the real closed loop. Next, C_o and Q_o denote the regulator obtained in the theoretical case $\|\ell\|_\infty \to 0$ and the Youla parameter, respectively. (See Table 9.1.1.)

Minimum Sensitivity of the *YP* Regulator

The sensitivity of the regulator regarding the uncertainties of the process and noise models can be optimized. According to Gevers [48,49], importantly, in sampled systems a regulator can be given for the measuring case of (6.1) whose variance is minimum, if N samples are used for the model identification. In this case the variance $E\left\{\left|\Delta \hat{C}_N\right|^2\right\}$ is minimum, that is, the optimal regulator has the least possible sensitivity if the regulator takes the form

$$C^{\min}_{\text{opt}} = -\frac{F_1}{F_1 G - F_2 H_n} \tag{9.1.29}$$

Here G and H_n are the process and noise models, respectively; furthermore F_1 and F_2 are the following sensitivity functions

$$F_1 = \frac{\partial C}{\partial G} \quad \text{and} \quad F_2 = \frac{\partial C}{\partial H_n} \tag{9.1.30}$$

According to (6.46) and in the case of the assumed stochastic noise, the $\mathcal{H}_2$ optimal *YP* regulator is

$$C_* = \frac{H_n - \mathcal{T}'}{\mathcal{T}' G} \tag{9.1.31}$$

This form also includes the minimum variance regulator of (6.9), since in this case $\mathcal{G}_-^- = 1$, that is, $\mathcal{T}' = \mathcal{T}$. The derivates of the sensitivity can be obtained by simple calculations such as

$$F_1 = \frac{\partial C}{\partial G} = \frac{1}{G_+} \frac{\partial C}{\partial G_-} = -C\frac{1}{G} \tag{9.1.32}$$

$$F_2 = \frac{\partial C}{\partial H_n} = \frac{\partial}{\partial H_n} \frac{H_n - \mathcal{T}'}{\mathcal{T}' G} = \frac{1}{G\mathcal{T}'} \tag{9.1.33}$$

Using the results in (9.1.29), we obtain the minimum sensitivity optimal regulator

$$C_{\text{opt}}^{\text{min}} = -\frac{C_*\frac{1}{G}}{C_*\frac{1}{G}G - \frac{1}{G\mathcal{T}'}H_n} = \frac{C_*}{\frac{H_n}{\mathcal{T}'} - C_* G} = C_* \tag{9.1.34}$$

that is, it is equal to the $\mathcal{H}_2$ optimal *YP* regulator.

9.2 LIMITS OF REGULATOR ROBUSTNESS

It has already been seen that during the realization of the regulator several limiting circumstances can appear which significantly influence achievement of the best reachable control. The most important of these is the amplitude constraint of the actuator. In this section we will investigate how this constraint influences the robustness of the control loop. Since this constraint is basically a feature of the regulator it is discussed in this section. In the case of step response reference signal or output noise the following coefficient

$$u_t = \frac{u(0)}{u_\infty} \tag{9.2.1}$$

is called dynamic overexcitation in the step response function of the regulator (see, e.g., Section 4.1).

Here the first-order reference model (9.1.9) is applied and the process is assumed to be a first-order lag with dead-time (which is quite a good approximation used in several industrial processes), that is,

$$P = \frac{e^{-sT_d}}{1 + sT} \tag{9.2.2}$$

In this case the optimal *YP* regulator is (see Section 2.1)

$$C_o = \frac{R_n P_+^{-1}}{1 - P_n e^{-sT_d}} = \frac{1}{1 - e^{-sT_d}/(1 + sT_n)} \frac{1 + sT}{1 + T_n} \tag{9.2.3}$$

For this regulator the value of the overexcitation is

$$u_t = \frac{u(0)}{u_\infty} = C_o(\omega = \infty) = T/T_n \tag{9.2.4}$$

which also allows us to increase the speed from the original bandwidth $\omega_{b,o} = 1/T$ of the open loop to the bandwidth $\omega_{b,c} = 1/T_n$ of the closed loop. Thus we can write

$$u_t = \omega_{b,c}/\omega_{b,o} = T/T_n \tag{9.2.5}$$

The sensitivity function of the closed loop is

$$S = 1 - R_n e^{-sT_d} = 1 - \frac{1}{1 + sT_n} e^{-sT_d} = \frac{1 + sT_n - e^{-sT_d}}{1 + sT_n} \tag{9.2.6}$$

from which the Nyquist stability measure can be computed numerically.

$$\rho_m = \rho_{min}(C) = \min_{\omega} |\rho(\omega, C)| = \min_{\omega} |1 + CP| = \frac{1}{\|S\|_\infty} \tag{9.2.7}$$

Introducing $x = T_n/T_d$ and the process parameters

$$\nu = T/T_d = T/T_n \;\; T_n/T_d = u_t x \tag{9.2.8}$$

we can easily draw the dependence of $\rho_m = \rho_{min}$ on the variables x and ν (see Figure 9.2.1).

The interpretation of the curve $\rho_m(x)$ is very important because it gives the theoretically reachable best robustness measure for an arbitrary *IS* process. The measure ρ_m very noticeably tends to the limit $\rho_m(0) \rightarrow 0.5$ if the reference model R_n requires a very fast transient from the process with dead-time. If the dead-time T_d is negligibly small compared with the time

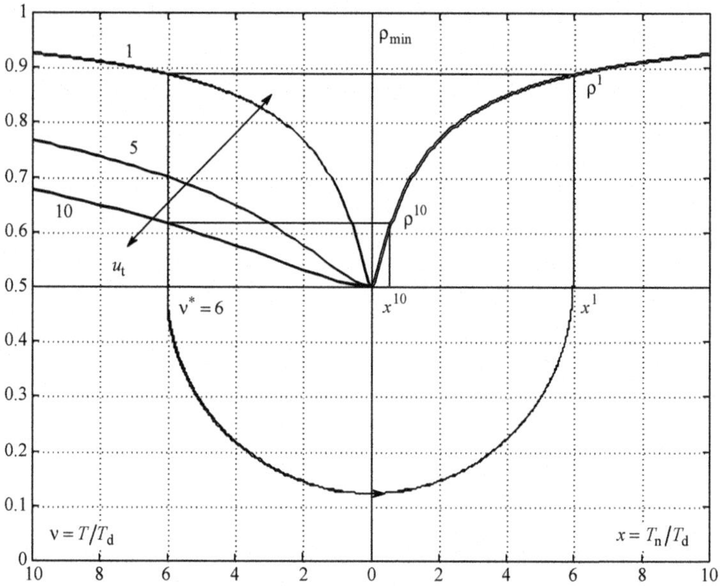

Figure 9.2.1 Dependence of ρ_m on the variables x and $u_t = \nu = \omega_{b,c}/\omega_{b,o} = T/T_n$.

constant T in R_n, then the curve ρ_m tends to the limit $\rho_m(\infty) \to 1$. The curves $\rho_m(x)$ are given for the most important overexcitation values ($u_t = 1$, 5, 10 in the figure).

Assume that $\nu^* = T/T_d = 6$ is given for the process. Then, for all those open loops where the condition

$$x < x^1 = \nu^* = T/T_d = 6 \tag{9.2.9}$$

holds, the response of the closed loop can be accelerated. It is very rare for the condition $u_t > 10$ to be fulfilled, and therefore the limit parameter of the process is x^{10}, which corresponds to $u_t = 10$. The reachable qualitative parameter (acceleration) can lie only in the region $x^{10} < x = T_n/T_d < x^1$. On the basis of the figure, for the process $\nu^* = T/T_d = 6$ we obtain the interval $\rho^{10} < \rho_m < \rho^1$. Note that, in a given case, the values x^1, x^{10}, ρ^1, and ρ^{10} depend on the magnitude of ν^*. Similarly the expressions must be recalculated for another u_t.

It may happen that the dominant dynamics is a second-order, oscillating lag corresponding to a complex conjugate pole pair instead of a first-order lag. In this case

$$P = \frac{e^{-sT_d}}{1 + 2\xi Ts + T^2 s^2} = P_+ e^{-sT_d}; \quad 0 < \xi < 1 \tag{9.2.10}$$

To make the *YP* regulator realizable, it seems to be reasonable to use a second-order reference model

$$R_n = \frac{1}{(1+sT_n)^2} \tag{9.2.11}$$

In this case the optimal *YP* regulator is (see Section 2.1)

$$C_o = \frac{R_n P_+^{-1}}{1 - R_n e^{-sT_d}} = \frac{1}{1 - e^{-sT_d}/(1+sT_n)^2} \frac{1 + 2\xi Ts + T^2 s^2}{(1+sT_n)^2} \tag{9.2.12}$$

Hence the initial step of the step response function is

$$u_t = \frac{u(0)}{u_\infty} = C_o(\omega = \infty) = T^2/T_n^2 \neq \omega_{b,c}/\omega_{b,o} \tag{9.2.13}$$

Note that u_t does not depend on ξ but, at the same time, the bandwidth improvement $\omega_{b,c}/\omega_{b,o}$ is not equal to u_t since it depends on ξ.

Observe that the cutoff frequency ω_c determined by R_n changed to ω_c' because of the time delay e^{-sT_d}, which is shown in Figure 9.2.2. The measure of $\|S\|_\infty$ is also interpreted graphically in the figure.

The sensitivity function of the closed loop is

$$S = 1 - R_n e^{-sT_d} = 1 - \frac{1}{(1+sT_n)^2} e^{-sT_d} = \frac{(1+sT_n)^2 - e^{-sT_d}}{(1+sT_n)^2} \tag{9.2.14}$$

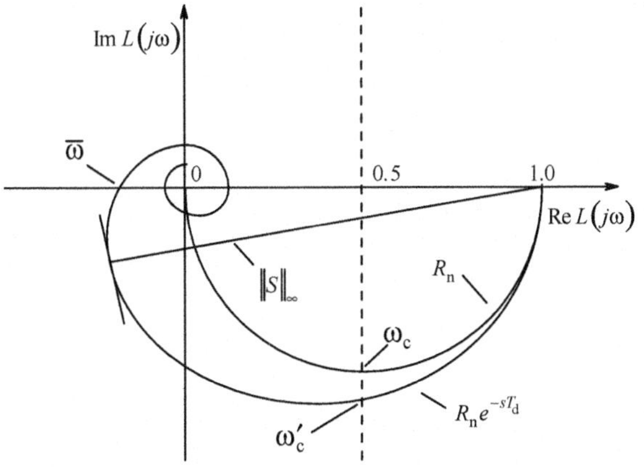

Figure 9.2.2 Change in the cutoff frequency from ω_c to ω_c'.

As seen in Figure 9.2.2, $\|S\|_{\infty}$ is the distance between the point $1 + 0j$ and the farthest point of $R_n(j\omega)e^{-j\omega T_d}$, from which ρ_m can be calculated by (9.2.7). The process parameter (9.2.8) is now

$$\nu = \frac{T}{T_d} = \frac{T}{T_n}\frac{T_n}{T_d} = \sqrt{u_t}\,\frac{T_n}{T_d} = \sqrt{u_t}x \qquad (9.2.15)$$

The dependence of $\rho_m = \rho_{min}$ on the variables ν and x, parameterized in u_t, is shown in Figure 9.2.3.

Note that the limit value x^{10} for the same process $\nu^* = T/T_d = 6$ is larger than in the previous example. The reason for this is that both the lower and the upper parts of the curve series are squeezed. Consequently the permitted region $x^{10} < x < x^1$ is smaller and, as a consequence, the robustness factor $\rho^{10} < \rho_m < \rho^1$ is smaller in the regulator design. The upper part of the figure is also shown separately in Figure 9.2.4.

Here an upper limit $\rho^* = 0.866$ can be observed for ρ_m if $T_d \to 0$. This limit can be simply calculated from (9.2.14), since

$$|S(j\omega)|_{T_d=0} = \sqrt{\left(\frac{T_n\omega}{1+T_n^2\omega^2}\right)^2 (4 + T_n^2\omega^2)} \qquad (9.2.16)$$

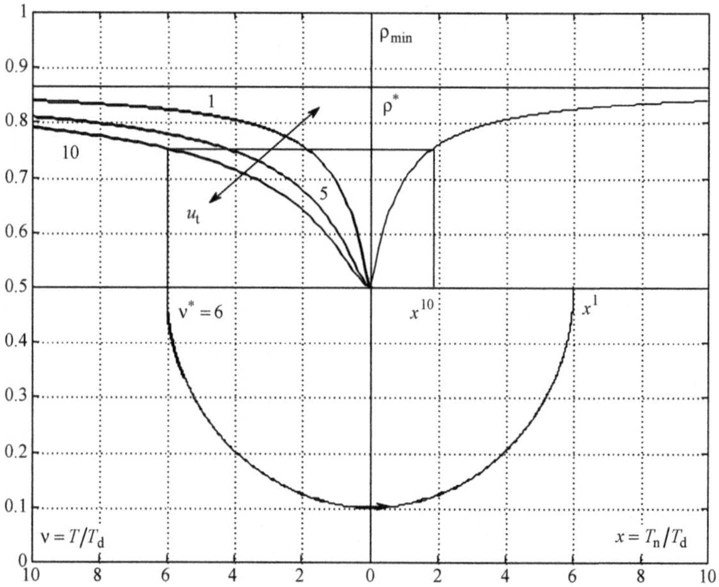

Figure 9.2.3 Dependence of ρ_m on the variables ν and x, parameterized in u_t.

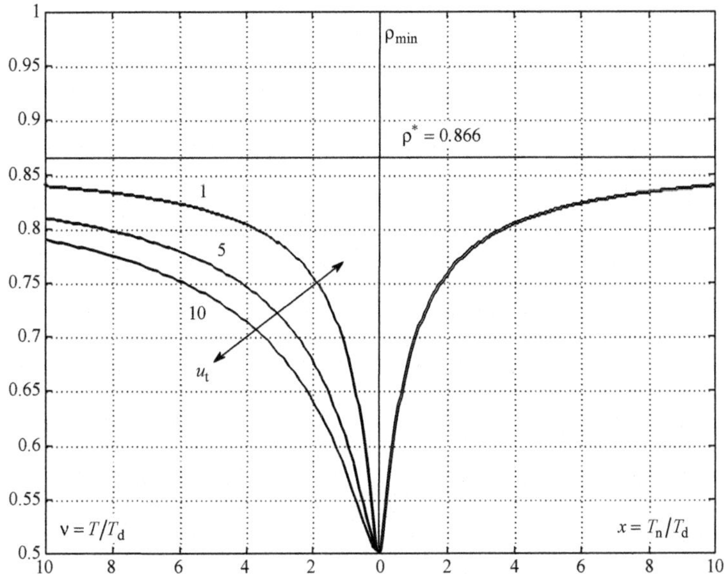

Figure 9.2.4 Relationship between ρ_m, ν, and x parameterized in u_t.

whose maximum is in the point

$$\omega^* = \frac{\sqrt{2}}{T_n} \tag{9.2.17}$$

where

$$\|S\|_\infty = |S(j\omega^*)|_{T_d=0} = \frac{2}{\sqrt{3}} = 1.155 \tag{9.2.18}$$

As a consequence

$$\rho^* = \frac{1}{\|S\|_\infty} = \frac{\sqrt{3}}{2} = 0.866 \tag{9.2.19}$$

Investigate the robustness measure $\rho^*(n)$ as a function of n as shown in Figure 9.2.5 if the process is of nth order with dead-time and the reference model is of nth order

$$R_n = \frac{1}{(1+sT_n)^n}; \quad P = \frac{1}{(1+sT)^n} e^{-sT_d} \tag{9.2.20}$$

The figure shows that $\rho^*(n)$ drastically decreases for higher-order processes, and since the compressions of the curve series have been

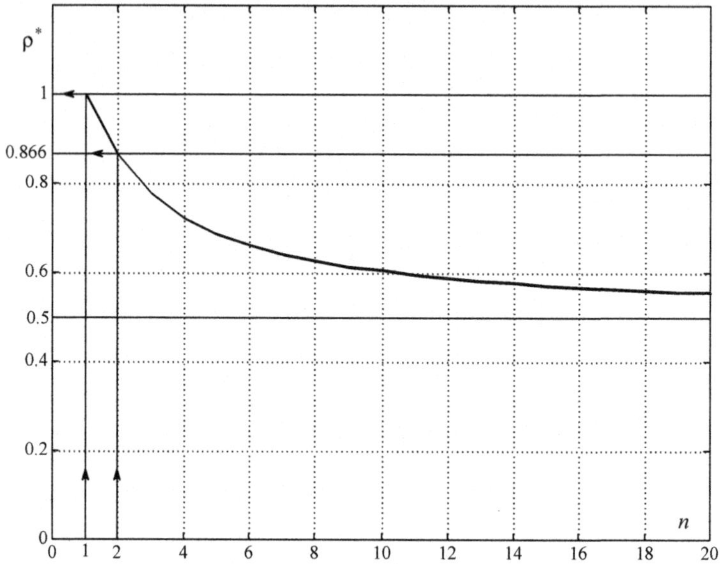

Figure 9.2.5 Change in the upper limit of the robustness measure $\rho^*(n)$ as a function of n.

recognized even in the second-order case, their joint consequence is that the acceptable design goal region $x^{10} < x < x^1$ decreases more and more. If the numerator of the process has a stable zero, this can improve the situation.

At the intersection frequency $\overline{\omega}$ of the characteristic $R_n(j\omega)e^{-j\omega T_d}$ and the real axis, a lower limit can be obtained for $\|S\|_\infty$ with the real part of the curve. Hence the upper limit $\rho_m = \rho_{min}$ becomes

$$\rho_m \leq \frac{1}{1 - \mathrm{Re}\left\{R_n(j\overline{\omega})e^{-j\overline{\omega}T_d}\right\}} = \overline{\rho}_m \tag{9.2.21}$$

Dependence of the real ρ_m and upper limit $\overline{\rho}_m$ on $x = T_n/T_d$ is shown in Figures 9.2.6 and 9.2.7 for $n = 1$ and $n = 2$, respectively.

Clearly, in the second-order case the upper limit provides a much worse approximation than in the first-order case.

If the numerator of the process has an unstable zero, it is an *IU* case and the robustness deteriorates substantially. To investigate this situation further, consider the following process

$$P = \frac{1 - sT_1}{1 + sT} e^{-sT_d} \tag{9.2.22}$$

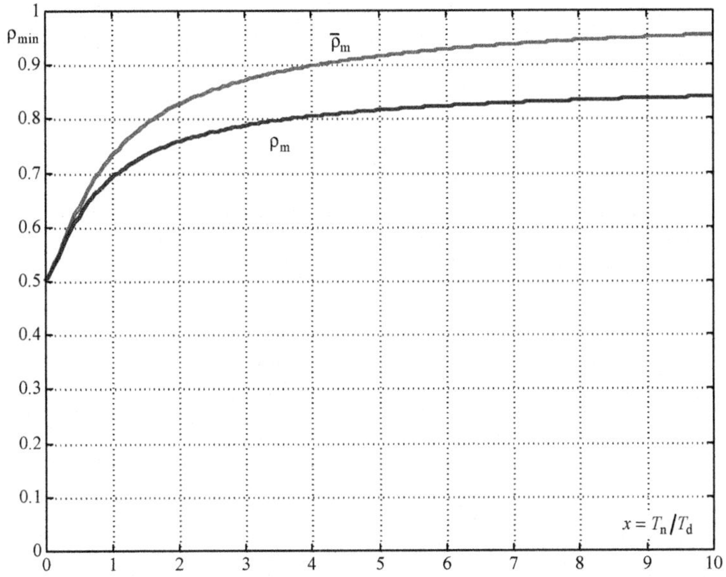

Figure 9.2.6 Dependence of the real ρ_m and upper limit $\bar{\rho}_m$ on $x = T_n/T_d$ $(n = 1)$.

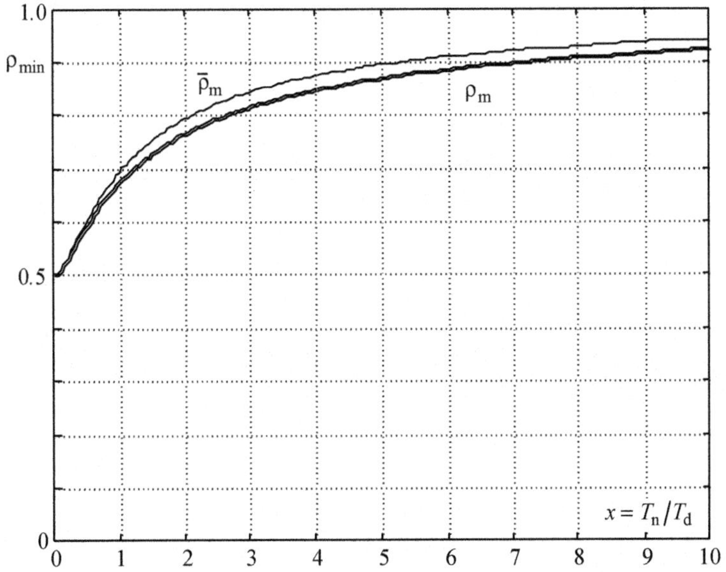

Figure 9.2.7 Dependence of the real ρ_m and upper limit $\bar{\rho}_m$ on $x = T_n/T_d$ $(n = 2)$.

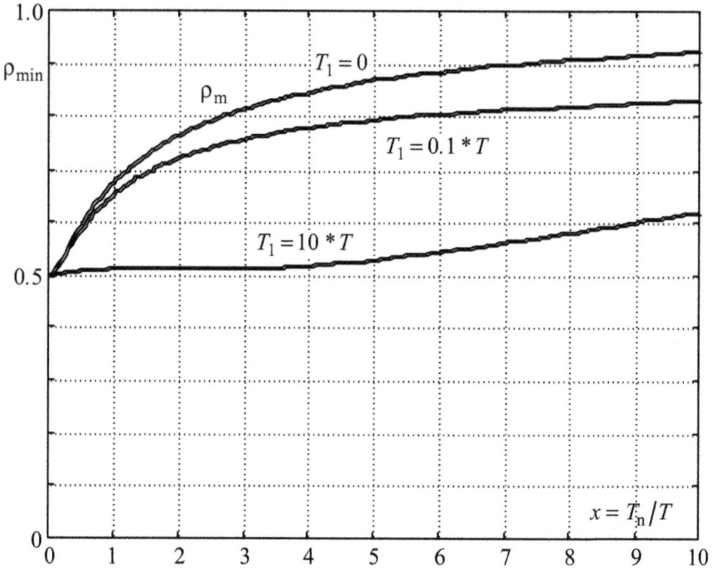

Figure 9.2.8 Values of $\rho_m(x)$ parameterized in T_1/T.

and let the reference model be first order according to (9.1.9). The function $\rho_m(x)$ parameterized in T_1/T is shown in Figure 9.2.8. Note that a large unstable zero has a much worse effect on the reachable robustness than the higher order of the process mentioned earlier.

9.3 GAP METRICS

ρ_m is an excellent measure for the characterization of stability and also gives a very good interpretation of the optimization of the $\mathcal{H}_\infty$ norm, since this requires the maximization of ρ_m according to (9.16): $\|S(j\omega)\|_\infty = 1/\rho_m$. Thus the maximization of the $\mathcal{H}_\infty$ norm maximizes the distance between the point $-1 + j0$ and the nearest point of the Nyquist diagram to the stability limit.

ρ_m does not, however, perform quite so well if the goal is to compare the stability robustness of two frequency functions, that is, to measure a very small distance (gap). Of course, the differences between the loop transfer functions L_1 and L_2 can be formulated as

$$\Delta\rho_m(L_1, L_2) = \rho_m(L_1) - \rho_m(L_2) \tag{9.3.1}$$

This difference is not normalized for the region [0,1], but the condition of the robust stability can be formulated as following (1.3.17)

$$\Delta\rho_{\mathrm{m}}(L_1, L_2) < \hat{\rho}_{\mathrm{m}} \tag{9.3.2}$$

The metric $\hat{\rho}_{\mathrm{m}}$ should be determined for $\hat{L}(j\omega)$ calculated on the basis of the model. Thus, because of (9.3.2), the $\Delta\rho_{\mathrm{m}}$ resulting from the parameter changes must be always less than $\hat{\rho}_{\mathrm{m}}$ calculated from the model to the robust stability.

It is possible to define a normalized metric, that is, when it is always in the region [0,1]. For this purpose, Vinnicombe recommended a new metric [155–157]. Its original name was the "ν-gap metric." Vinnicombe applied the classical Riemann spherical method, which maps the whole complex plane to a sphere. If the sphere chosen is of unity diameter then its equator corresponds to a circle of unity radius; thus this metric is most sensitive around the cutoff frequency ω_c for all frequency characteristics.

Denote a point of the frequency function $L(j\omega)$ of the open loop by

$$\xi = \mathrm{Re}\, L(j\omega); \quad \eta = \mathrm{Im}\, L(j\omega); \quad \zeta = 0; \quad |L|^2 = \xi^2 + \eta^2 \tag{9.3.3}$$

and map this point to the Riemann (R) sphere by a line, that is, map the original $L(j\omega)$ to a function $L'(j\omega)$

$$\xi' = \frac{\xi}{1+\xi^2+\eta^2} = \frac{\xi}{1+|L|^2}; \quad \eta' = \frac{\eta}{1+\xi^2+\eta^2} = \frac{\eta}{1+|L|^2};$$
$$\zeta' = \frac{|L|^2}{1+|L|^2} \tag{9.3.4}$$

From the perpendicular projection of $L'(j\omega)$ we obtain points where

$$|L'|^2 = (\xi')^2 + (\eta')^2 = \frac{|L|}{1+|L|^2} \tag{9.3.5}$$

The distance between two transfer functions L_1 and L_2 on the plane is

$$|\Delta_L| = |L_1 - L_2| \tag{9.3.6}$$

The "chord" distance between two transfer functions L_1 and L_2 on the $R-$ sphere is

$$\kappa_V(L_1, L_2) = \frac{|L_1 - L_2|}{\sqrt{1+\xi_1^2+\eta_1^2}\sqrt{1+\xi_2^2+\eta_2^2}} = \frac{|\Delta_L|}{\sqrt{1+|L_1|^2}\sqrt{1+|L_2|^2}} = \frac{|\Delta_L|}{\sqrt{\left(1+|L_1|^2\right)\left(1+|L_2|^2\right)}} \tag{9.3.7}$$

The definition of the ν-distance suggested by Vinnicombe is

$$\delta_V = \delta_V(L_1, L_2) = \begin{cases} \max \kappa_V(L_1, L_2); & \delta_V(L_1, L_2) \le 1 \\ 1 \end{cases} \tag{9.3.8}$$

The value of δ_V is between zero and one, and the mapping is obviously most sensitive on the equator of the sphere concerning the distinction of two points on the plane, that is, in the vicinity of the frequency ω_c.

During the evolution of control engineering the place of the relevant frequency regions has changed. First law frequency region was considered important as it enabled the static properties of the control to be determined. The investigation of this region led to the introduction of the type 0, 1, and 2 regulators, which resulted in zero steady-state error in the case of type 0, 1, and 2 (i.e., Dirac delta, step response, and impulse response) external disturbances. Later, the so-called medium frequency region, especially the vicinity of the cutoff frequency ω_c, became more important. The design methods of a given phase access or the most important transient behaviors are connected to this region. Robustness is related to a different frequency region, that is, to point ω_E, where $E(\omega) = |S(j\omega)|$ of (1.2.17) has its maximum (if the extremum exists at all).

Now shift the $R-$ sphere left to have its center in the stability point $-1 + j0$. Let us denote the resulting sphere by $K-$ sphere. It is clear that, similarly to $R-$ sphere, the whole complex plane can be mapped also to the $K-$ sphere. The expressions of the transformation $L'_K(j\omega)$ are also very simple and can be calculated from the coordinates of $L'(j\omega)$ as

$$\xi_K = \xi + 1; \quad \eta'_K = \eta'; \quad \zeta'_K = \zeta' \tag{9.3.9}$$

and using (9.3.4) we obtain

$$\xi'_K = \frac{\xi}{1+\xi_K^2+\eta^2} - 1; \quad \eta'_K = \frac{\eta}{1+\xi_K^2+\eta^2}; \quad \zeta'_K = \frac{\xi_K^2+\eta^2}{1+\xi_K^2+\eta^2} \tag{9.3.10}$$

The "chord" distance between two transfer functions L_1 and L_2 in the $K-$ sphere is

$$\begin{aligned}\kappa_K(L_{K,1}, L_{K,2}) = \kappa_K(L_1, L_2) &= \frac{|(1+L_1)-(1+L_2)|}{\sqrt{1+|1+L_1|^2}\sqrt{1+|1+L_2|^2}} \\ &= \frac{|L_1 - L_2|}{\sqrt{1+|1+L_1|^2}\sqrt{1+|1+L_2|^2}}\end{aligned} \tag{9.3.11}$$

The new gap metric can be defined by the chord distance measured in the $K-$ sphere

$$\delta_K = \delta_K(L_1, L_2) = \begin{cases} \max \kappa_K(L_1, L_2); & \delta_K(L_1, L_2) \le 1 \\ 1 \end{cases} \tag{9.3.12}$$

The most important expressions are presented in Figure 9.3.1. The upper part of the figure shows the front pictures of the $R-$ and $K-$ spheres with projected frequency functions $L'(j\omega)$ and $L'_K(j\omega)$. The front borderline of the two cones corresponds to the geometrical places $|L(j\omega)| = 1$ and $E(\omega) = 1$ (i.e., to circles). The lower part of the figure shows the conventional Nyquist curves L_1 and L_2, and the lower half ($\mathrm{Im} \le 0$) of the two above-mentioned circles. The equator of the $R-$ sphere corresponds to the cutoff frequency ω_c, the equator of the $K-$ sphere, however, to the frequency deriving from the condition $E(\omega) = |E(\omega)| = |S(j\omega)| = 1$.

Similarly, the frequency ω_M can also be very important, as it is where $M(\omega) = |T(j\omega)|$ has its maximum (if the extreme exists). Unfortunately the determination of this maximum is more complicated than that of ω_E, and significantly less method can be connected to it. The values of ω_E and ω_M are usually greater than ω_c, but this is not always true and it is difficult to formulate system-independent statements.

Shift the $R-$ sphere left to have its center in the point $-0.5 + j0$. Let us denote the resulting sphere by $KM-$ sphere. It is clear that, similarly to

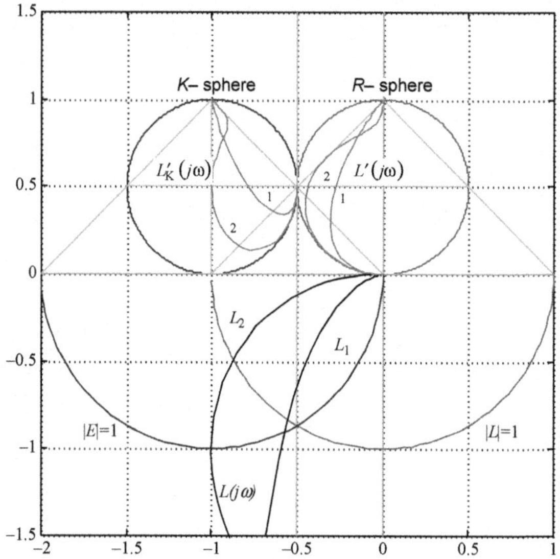

Figure 9.3.1 Joint figure showing the $R-$ and $K-$ spheres.

$R-$ sphere, the whole complex plane can be mapped also to the $KM-$ sphere. The expressions of the transformation $L'_K(j\omega)$ are also very simple now and can be calculated from the coordinates of $L'(j\omega)$ as

$$\xi_{KM} = \xi + 0.5; \quad \eta'_{KM} = \eta'; \quad \zeta'_{KM} = \zeta' \tag{9.3.13}$$

and using (9.3.4) we obtain

$$\begin{aligned} \xi'_{KM} &= \frac{\xi}{1+\xi_{KM}^2+\eta^2} - 0.5; \quad \eta'_K = \frac{\eta}{1+\xi_{KM}^2+\eta^2}; \\ \zeta'_{KM} &= \frac{\xi_{KM}^2+\eta^2}{1+\xi_{KM}^2+\eta^2} \end{aligned} \tag{9.3.14}$$

The "chord" distance between two transfer functions L_1 and L_2 on the $KM-$ sphere is

$$\begin{aligned} \kappa_{KM}(L_{KM,1}, L_{KM,2}) &= \frac{|(0.5+L_1)-(0.5+L_2)|}{\sqrt{1+|0.5+L_1|^2}\sqrt{1+|0.5+L_2|^2}} \\ &= \frac{|L_1-L_2|}{\sqrt{1+|0.5+L_1|^2}\sqrt{1+|0.5+L_2|^2}} \end{aligned} \tag{9.3.15}$$

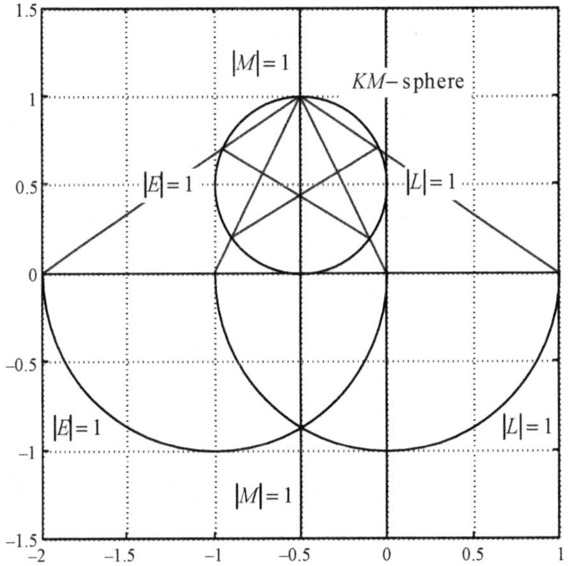

Figure 9.3.2 Figure showing the *KM*− sphere.

The new gap metric can be defined by the chord distance measured in the *KM*− sphere

$$\delta_{\mathrm{KM}} = \delta_{\mathrm{KM}}(L_1, L_2) = \begin{cases} \max\ \kappa_{\mathrm{KM}}(L_1, L_2); & \delta_{\mathrm{KM}}(L_1, L_2) \leq 1 \\ 1 \end{cases} \tag{9.3.16}$$

The upper part of Figure 9.3.2 shows the front picture of the *KM*− spheres. The front borderline of the cones correspond to the geometrical places $|L(j\omega)| = 1$ and $|E(j\omega)| = 1$ (i.e., to circles). A vertical axis on the *KM*− sphere corresponds to the condition $|M(j\omega)| = 1$.

Example 9.3.1

Consider a simple type 1 closed-loop control, in which

$$L_1(j\omega) = \frac{K_{\mathrm{I}}}{s(1+sT)} = \frac{b_o}{a_2 s^2 + a_1 s + a_o} = \frac{b_o}{a_2 s^2 + a_1 s} \tag{9.3.17}$$

$$b_o = 0.8; \quad a_2 = 0.8; \quad a_1 = 1; \quad a_o = 0$$

Suppose that another loop transfer function $L_2(j\omega)$ is determined by the conditions

$$L_2(j\omega):\ \ 0.1 \leq b_o \leq 1.0 \tag{9.3.18}$$

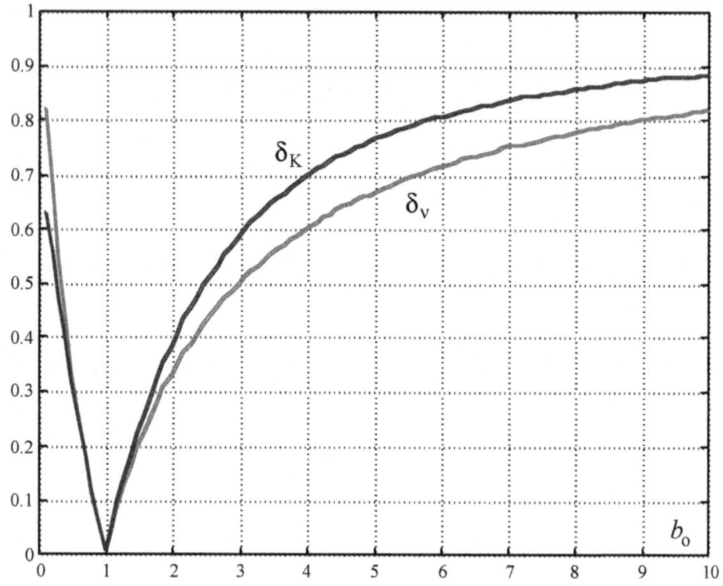

Figure 9.3.3 Sensitivity of δ_K and δ_V to change in the coefficient b_o.

The sensitivity of δ_K and δ_V to the change of the coefficient b_o is shown in Figure 9.3.3.

Assume that now the coefficient a_2 is changing in $L_2(j\omega)$

$$L_2(j\omega): \quad 0.1 \leq a_2 \leq 1.0 \tag{9.3.19}$$

The sensitivity of δ_K and δ_V to the change in the coefficient a_2 is shown in Figure 9.3.4. The above two figures clearly show that the gap metric δ_K suggested by us is much more sensitive to parameter changes than δ_V.

This greater sensitivity is not necessarily true in all cases, however. It is fair to say that each distance maps are different for different regions. A summary of this region mapping is shown in Figure 9.3.5. It is well known that in the design of control loops regions 1, 2, and 3 are the most important. Compared with them the importance of regions 4, 5, and 6 is almost negligible. (Perhaps the only exception is the loop transfer function in the vicinity of the zero frequency, which determines the static behavior of the closed loop.) Note that region 1 is much the largest, and region 2 is larger in the $K-$ sphere, whereas region 3 is smaller than in the $R-$ sphere. That is why it cannot be stated that δ_K is always more sensitive than δ_V, but it is true in the majority of the design tasks, and can be proved by several numerical examples.

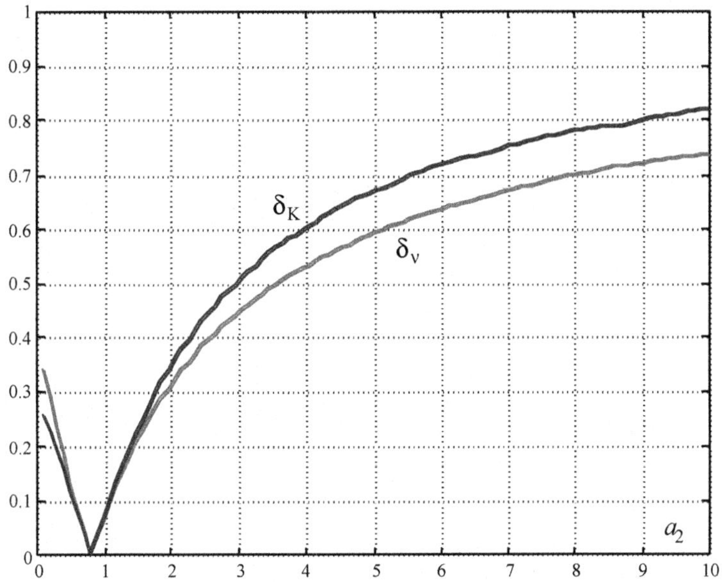

Figure 9.3.4 Sensitivity of δ_K and δ_V to change in the coefficient a_2.

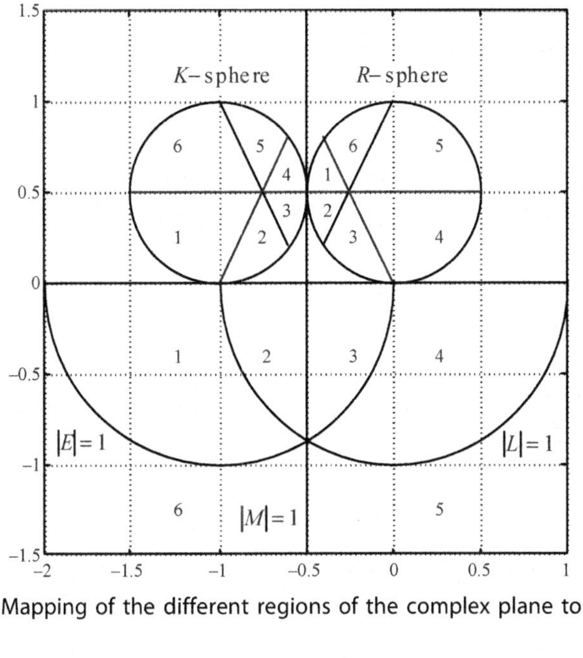

Figure 9.3.5 Mapping of the different regions of the complex plane to the $R-$ and $K-$ spheres.

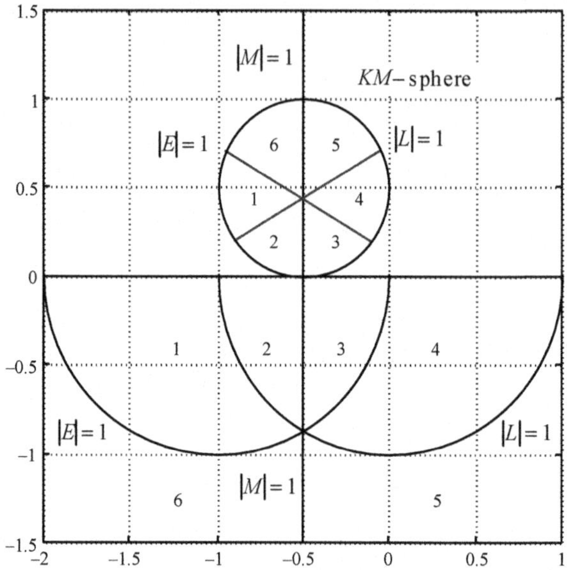

Figure 9.3.6 Mapping of the different regions of the complex plane to the *KM*− sphere.

The mapping of the different regions of the complex plane to the *KM*− sphere is shown in Figure 9.3.6. Note that both regions 1 and 2 are larger, but region 3 is approximately the same as in the *R*− sphere. (Further advantages of the introduced new metrics are discussed at the end of Section 10.3.)

9.4 DIALECTIC BETWEEN PERFORMANCE AND ROBUSTNESS

The dialectic of the quality and robustness has already been seen in the condition of robust stability (9.15). Further important observations can be made via specific tasks. First investigate the robustness of the *YP* control loop, if the model error derives from inaccurate knowledge of the dead-time. The exact knowledge of the dead-time is an important condition of the application of these regulators. The investigations are first completed for CT systems. Assume that the delay-free part of the process is exactly known to be $\hat{P}_+ = P_+$, $\hat{\overline{P}}_- = 1$ and the relative model error is

$$\ell = \ell_{T_d} = \frac{\Delta}{\hat{P}} = \frac{P - \hat{P}}{\hat{P}} = \frac{P_+ e^{-sT_d} - P_+ e^{-s\hat{T}_d}}{P_+ e^{-s\hat{T}_d}} = e^{-\Delta T_d s} - 1 \tag{9.4.1}$$

where $\varDelta T_{\mathrm{d}} = T_{\mathrm{d}} - \hat{T}_{\mathrm{d}}$. Assuming the most rigorous case in the inequality expression (9.1.8) of the robust stability we obtain

$$\sup_{\omega} \left|\ell_{T_{\mathrm{d}}}\right| = \sup_{\omega} \left|e^{-j\varDelta T_{\mathrm{d}}\omega} - 1\right| \leq \frac{1}{|R_{\mathrm{n}}(\omega)|} \tag{9.4.2}$$

Applying the first-order reference model (9.1.9), we can solve Equation (9.4.2) to obtain the following inequality

$$\ell_{T_{\mathrm{d}}} = \left|\frac{\varDelta T_{\mathrm{d}}}{T_{\mathrm{d}}}\right| = \left|1 - \frac{\hat{T}_{\mathrm{d}}}{T_{\mathrm{d}}}\right| < \frac{\pi}{\sqrt{3}} \frac{T_{\mathrm{n}}}{T_{\mathrm{d}}} = 1.82 \frac{T_{\mathrm{n}}}{T_{\mathrm{d}}} \tag{9.4.3}$$

The derivation of the inequality can be simply followed via the geometry shown in Figure 9.4.1.

If the exponential term in (9.4.2) is expanded into Taylor series then we get a sufficient condition simpler than (9.4.3)

$$\ell_{T_{\mathrm{d}}} = \left|\frac{\varDelta T_{\mathrm{d}}}{T_{\mathrm{d}}}\right| = \left|1 - \frac{\hat{T}_{\mathrm{d}}}{T_{\mathrm{d}}}\right| < \frac{T_{\mathrm{n}}}{T_{\mathrm{d}}} \tag{9.4.4}$$

If the condition of robust stability is chosen a higher-order (sharper cutting) transfer function as

$$R_{\mathrm{n}} = \frac{1}{(1 + sT_{\mathrm{n}})^n} \Rightarrow R_{\mathrm{n}}(j\omega) = \frac{1}{(1 + j\omega T_{\mathrm{n}})^n} \tag{9.4.5}$$

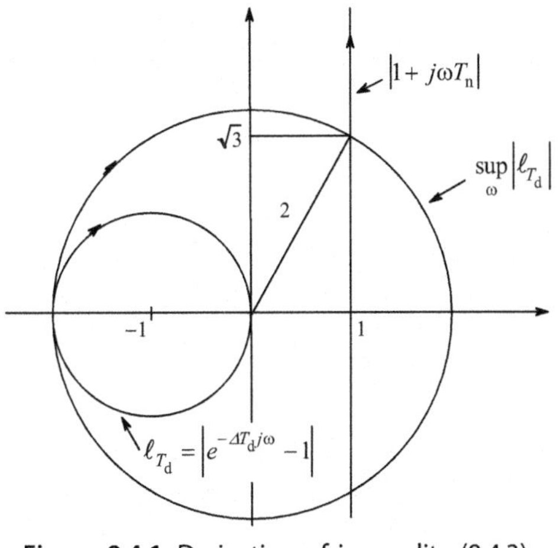

Figure 9.4.1 Derivation of inequality (9.4.3).

then we get

$$\ell_{T_d} = \left|\frac{\Delta T_d}{T_d}\right| = \left|1 - \frac{\hat{T}_d}{T_d}\right| < a(n)\frac{T_n}{T_d} \tag{9.4.6}$$

The coefficient $a(n)$ is shown in Figure 9.4.2 as a function of n.

Investigate now how the stability measure $\rho_m = \rho_{min}$ is shaping. The sensitivity function for the above example can be calculated as

$$S = \frac{1 - R_n e^{-s\hat{T}_d}}{1 + \ell R_n e^{-s\hat{T}_d}} = \frac{1 - R_n e^{-s\hat{T}_d}}{1 + \ell_{T_d} R_n e^{-s\hat{T}_d}} \tag{9.4.7}$$

and applying the first-order reference model (9.1.9) we obtain

$$S = \frac{1 + sT_n - e^{-s\hat{T}_d}}{1 + sT_n + e^{-sT_d} - e^{-s\hat{T}_d}} \tag{9.4.8}$$

There is no analytical solution, but the complex function $\rho_m = \rho_{min}(\hat{T}_d/T_d, T_n/T_d)$ can be calculated numerically. The function is presented in Figure 9.4.3 for the different values $\hat{T}_d = 0.5T_d, T_d, 2T_d$. Here the quotient $\hat{T}_d/T_d$ is used instead of the model error $\Delta T_d = T_d - \hat{T}_d$.

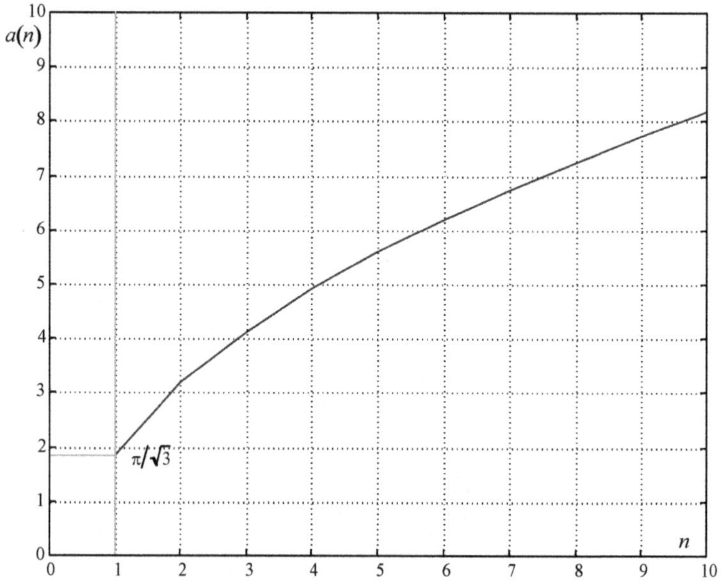

Figure 9.4.2 Plot of the function $a(n)$.

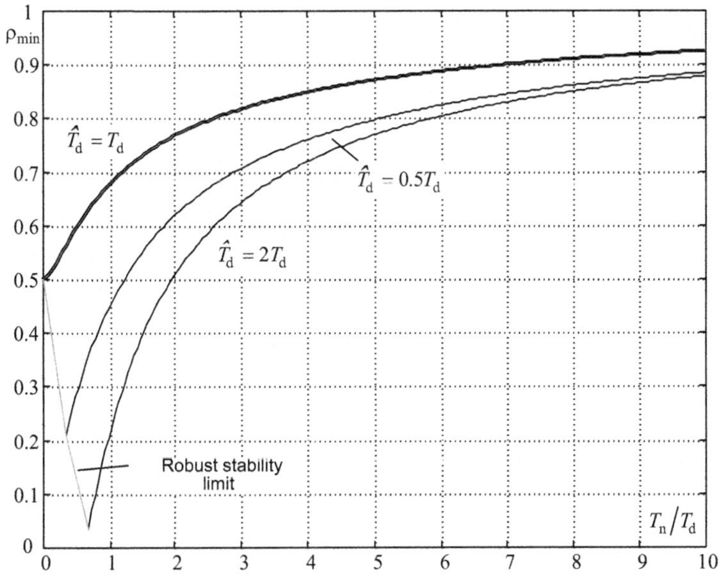

Figure 9.4.3 Plot of the curve $\rho_{min}(T_n/T_d)$ for the values $\hat{T}_d = 0.5T_d, T_d, 2T_d$.

Implicitly the quotient $\hat{T}_d/T_d = 1$ corresponds to the value $\Delta T_d = 0$ and T_n/T_d is the process parameter. In the ideal case $\hat{T}_d = T_d$, ρ_m depends only on the design goal T_n and the dead-time of the process T_d or, more precisely, only on the quotient T_n/T_d. The best robustness measure is $\rho_{min}(0) = 0.5$ if R_n requires a fast transient from the dead-time process and $\rho_{min}(\infty) \approx 1$ if T_d can be neglected compared with the time constant T_n of the reference model. (The left side of the curves corresponds to the robust stability limit.) Here the zero model error, independently of T_n/T_d, always ensures stability. Note that the under- or overestimation of the dead-time significantly decreases the robustness measure. Perhaps ρ_{min} is more sensitive to the overestimation of the dead-time. In the case of fast design, the violation of the robust stability limit should almost always be taken into account.

These observations are better illustrated in Figure 9.4.4, where the curve $\rho_{min}(T_n/T_d)$ is parameterized in T_n/T_d.

It is clear from the figure how the robustness measure is sensitive if the value T_n/T_d is small, and how this sensitivity decreases for the higher parameter T_n/T_d values. It is worth noting that in the case of smaller uncertainty regarding knowledge of the dead-time the overestimation of the dead-time gives higher ρ_{min} values, and furthermore, ρ_m is more

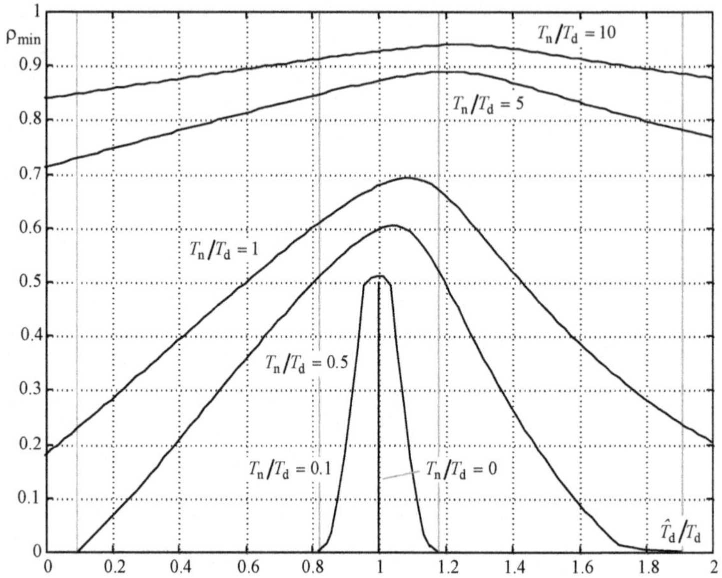

Figure 9.4.4 Plot of the curve $\rho_{\min}(T_n/T_d)$ parameterized in T_n/T_d.

sensitive to higher model errors. In a very wide region of the ρ_m, the overestimation of the time delay $\hat{T}_d/T_d = T_d^*/T_d$ (here T_d^* is higher than T_d by definition) changes $\rho_{\min}$ to $\rho_{\min}^*$ owing to the maxima of the curves in Figure 9.4.4. This improvement is illustrated in Figure 9.4.5. If the overestimation is less than 25%, then this improvement is smaller than 5% or negligible. If the dead-time error is less than 20% (which is the case in many practical tasks), then the decrease of the robustness in the region $T_n/T_d \geq 0.5$ is always less than 10%. If the task is to speed up the process by a time constant significantly smaller than the dead-time, then the dead-time must be known very precisely. Vice versa, if the dead-time error is high, then only modest improvement (speeding up) can be expected from the robust stable regulator.

The complex relationships between quality, robustness, and dead-time uncertainty are summarized in Figure 9.4.6, where the region acceptable for the design is designated.

Investigate now an example where the dead-time is known exactly ($\hat{T}_d = T_d$) and the model error comes from the time-constant uncertainty of a first-order lag process. Let

$$P_+ = \frac{1}{1+sT}; \quad \hat{P}_+ = \frac{1}{1+s\hat{T}} \tag{9.4.9}$$

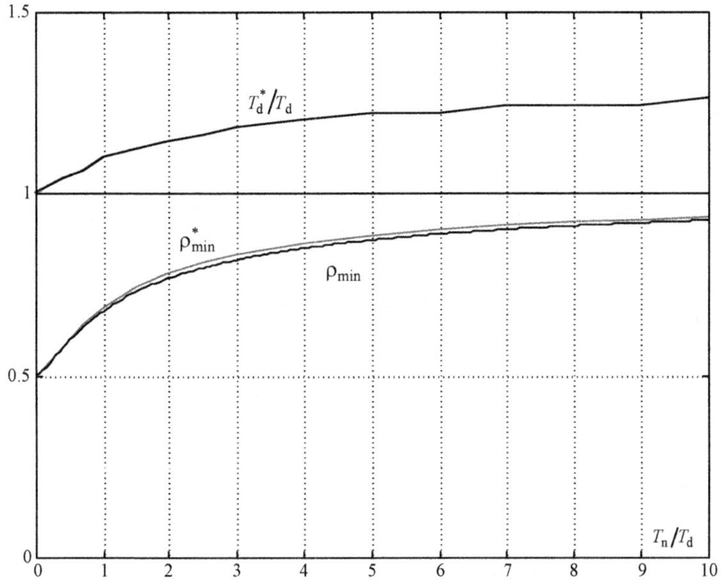

Figure 9.4.5 Effect of the overestimation of dead-time.

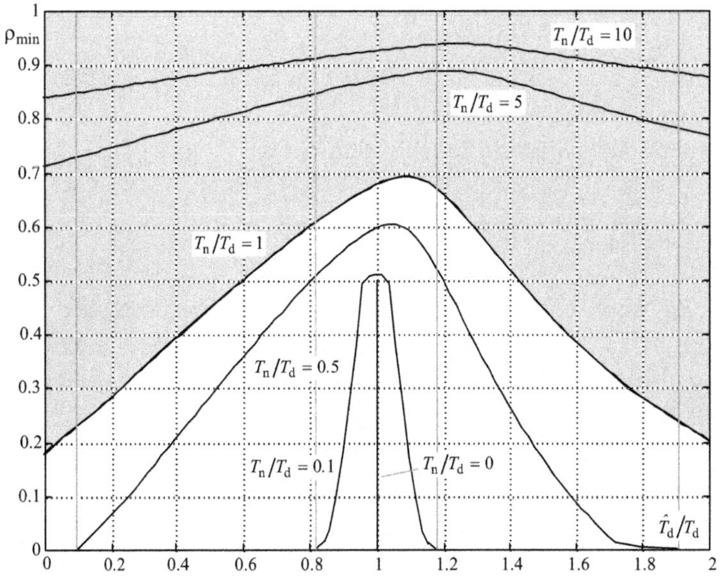

Figure 9.4.6 Plot of the curve $\rho_{min}(T_n/T_d)$ parameterized in T_n/T_d.

so the relative model error becomes

$$\ell = \ell_T = \frac{\Delta}{\hat{P}} = \frac{P - \hat{P}}{\hat{P}} = \frac{1 + s\hat{T}}{1 + sT} - 1 = \frac{s(\hat{T} - T)}{1 + sT} = \frac{s\Delta T}{1 + sT} \tag{9.4.10}$$

With $\Delta T = \hat{T} - T$ the robust stability condition (9.13) can be formulated as

$$|\ell P_n| = \left|\frac{s\Delta\hat{T}}{1 + sT}\frac{1}{1 + sT_n}\right| = \left|\frac{\Delta\hat{T}}{T}\right|\left|\frac{sT}{(1 + sT)(1 + sT_n)}\right| < 1 \tag{9.4.11}$$

The second term has its maximum

$$\left|\frac{T}{T + T_n}\right| = \frac{1}{|1 + T_n/T|} \tag{9.4.12}$$

at the frequency

$$\omega^* = \frac{1}{\sqrt{TT_n}} \tag{9.4.13}$$

The robust stability condition analogous with (9.4.2) is obtained in the inequality

$$\ell_T = \left|\frac{\Delta\hat{T}}{T}\right| \leq \left|1 + \frac{T_n}{T}\right| \tag{9.4.14}$$

which is very similar to (9.4.4).

The sensitivity function for this example is

$$S = \frac{1 - R_n}{1 + \ell R_n} = \frac{sT_n(1 + sT)}{sT_n(1 + sT) + (1 + s\hat{T})} \tag{9.4.15}$$

where R_n now corresponds to (9.1.9). Here the complex function $\rho_m = \rho_{min}(\hat{T}/T, T_n/T_d)$ can be calculated by numerical methods. The resulting function $\rho_{min}(\hat{T}/T)$ parameterized in T_n/T_d is illustrated in Figure 9.4.7 for the values $T_d/T = 0.1$ and $T_d/T = 1$.

It is clear from the figure that the effect of the dead-time uncertainty significantly depends on the process parameter T_d/T, but the maximum of ρ_m depends on the quality property T_n/T_d (speeding up). Thus it is not possible to find such unambiguous and simple expressions as those found in the investigation of dead-time uncertainty. Perhaps the only similar

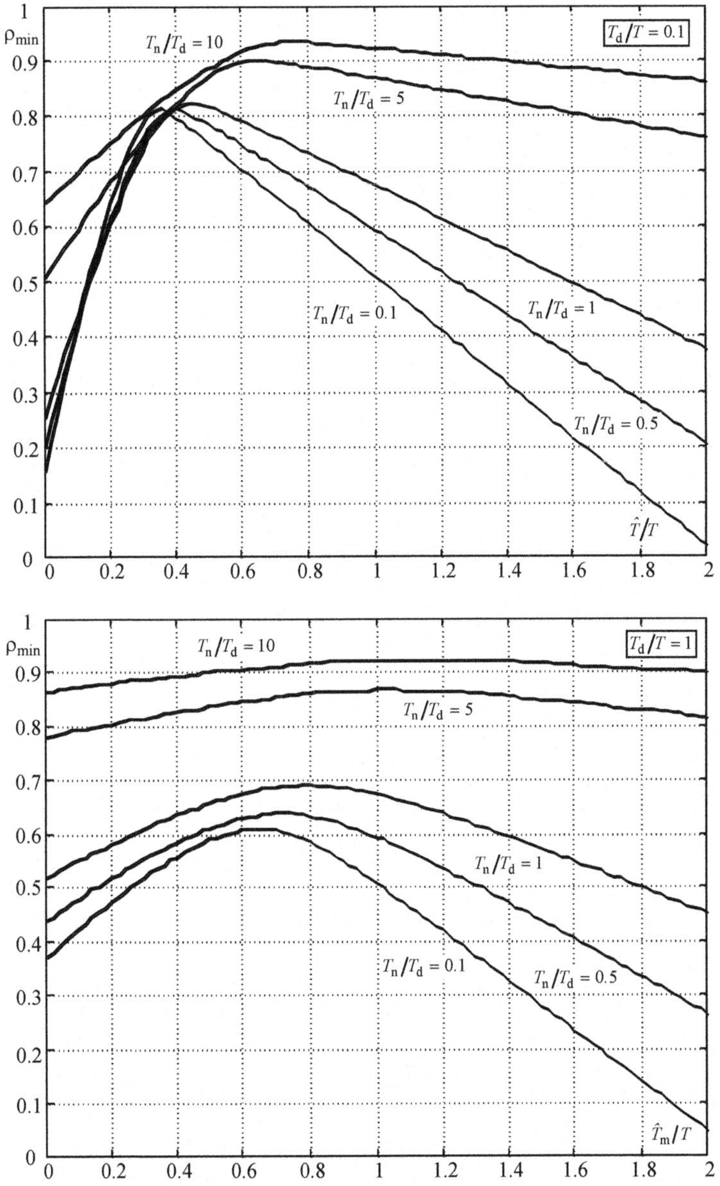

Figure 9.4.7 Plot of the array of curves $\rho_{min}(\hat{T}/T)$ for the parameter values $\hat{T}_d/T = 0.1$ and $\hat{T}/T = 1$.

conclusion can be drawn is that if the quality requirement is not so serious, that is, T_n/T_d is large, then the sensitivity of the robustness is less.

That is why, in many cases, a general design method is applied for the maximization of the robustness measure. This means the maximization of $\|S\|_\infty$, that is, the methodology of the $\mathcal{H}_\infty$ design. (This subject was discussed in detail in Section 4.2.)

Sampled Data Systems

The above results are also valid for DT systems in many respects. In sampled-data systems the time delay of the process $G(z)$ and the model $\hat{G}(z)$ is denoted by z^{-d} and $z^{-\hat{d}}$, respectively. The first-order reference model equivalent to (9.1.9) is now

$$R_n = \frac{(1+a_1)z^{-1}}{1+a_1 z^{-1}} \tag{9.4.16}$$

where $a_1 = -e^{-T_s/T}$ and T_s is the sampling time. The relative model error is

$$\ell = \ell_d \stackrel{\Delta}{=} \frac{\tilde{G}}{\hat{G}} = \frac{G - \hat{G}}{\hat{G}} = \frac{G_+ z^{-d} - G_+ z^{-d_m}}{G_+ z^{-d_m}} = z^{-(d-d_m)} - 1 \tag{9.4.17}$$

The following inequality is obtained for the robust stability condition

$$\sup_\omega |\ell_d| = \sup_\omega \left| z^{-(d-d_m)} - 1 \right| \le \frac{1}{|R_n|} = \left| \frac{1+a_1 z^{-1}}{(1+a_1)z^{-1}} \right| \tag{9.4.18}$$

which, in a stable case, takes the following form in the frequency region

$$\begin{aligned} \sup_\omega |\ell_d| = \sup_\omega \left| e^{-j\omega(d-d_m)T_s} - 1 \right| &\le \frac{\left|1 - e^{-T_s/T} e^{-j\omega T_s}\right|}{\left(1 - e^{-T_s/T}\right)\left|e^{-j\omega T_s}\right|} \\ &= \frac{\left|1 - e^{-T_s/T} e^{-j\omega T_s}\right|}{\left(1 - e^{-T_s/T}\right)} \end{aligned} \tag{9.4.19}$$

9.5 PRODUCT INEQUALITIES

Section 9.4 discussed the dialectics of the quality and robustness for some special cases, especially for dead-time systems. It is possible to derive more general relationships than can be given in the form of the so-called product inequalities.

The conditions of robust stability (1.3.20), (9.14), (9.15) already contain a product inequality. Here $\left|\hat{T}(j\omega)\right|$ (although it is usually called a design factor) can be considered as the quality factor of the control. The other factor, however, can be considered as the relative correctness of the applied model. In the light of practical experience control, engineers favor applying a mostly heuristic expression

(quality of the control) × (robustness of the control) ≤ limit

This product inequality can be simply demonstrated by the integral criteria of classical control engineering. Let I_2 be a square integral criterion (*integral square of error, ISE*) whose optimum is I_2^* when the regulator is properly set, and the Nyquist stability limit (i.e., robustness measure) is ρ_m. The well-known empirical, heuristics formula is

$$\frac{I_2^*}{I_2}\rho_m \leq \text{limit} \tag{9.5.1}$$

The inequality is illustrated in Figure 9.5.1.

The fact that the quality of the identification (which is the inverse of the model correctness) can have a certain relationship with the robustness of the control is not very trivial. This can be observed only in a special case, namely in the identification technique based on *Keviczky–Bányász* (*KB*) parameterization, as described in Section 10.3, when $\varepsilon_{ID} = -\tilde{e}$.

Introduce a new relationship for the characterization of the quality of the control

$$\delta = \delta(\omega, \hat{C}) = \frac{|-\tilde{e}(j\omega)|}{|y_n(j\omega)|} = \left|\frac{1}{1+\hat{C}P}\right| = \frac{1}{|1+\tilde{L}|} \tag{9.5.2}$$

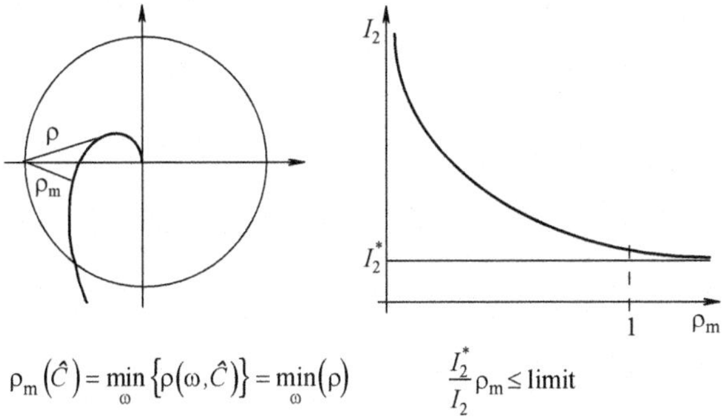

Figure 9.5.1 Illustration of the inequality of (9.5.1).

Notice that δ is the absolute value of the sensitivity function. Obviously, $\delta\rho = 1$ for all frequencies (here $\rho = |1 + \tilde{L}|$). Of course, the same equalities are valid for the minimum and maximum values, i.e.,

$$\begin{aligned} \rho_m(\hat{C}) &= \min_{\omega}\left[\rho(\omega, \hat{C})\right] = \min_{\omega}(\rho) \\ & \qquad\qquad ;\ \text{i.e., their product is unity: } \delta_M\rho_m = 1 \\ \delta_M(\hat{C}) &= \max_{\omega}\left[\delta(\omega, \hat{C})\right] = \max_{\omega}(\delta) \end{aligned} \tag{9.5.3}$$

Denote the worst value of these measures by

$$\begin{aligned} \frown{\rho}_m &= \max_{\hat{C}}\left\{\min_{\omega}\left[\rho(\omega, \hat{C})\right]\right\} = \max_{\hat{C}}\left[\min_{\omega}(\rho)\right] \\ & \qquad\qquad ;\quad \smile{\delta}_M\frown{\rho}_m = 1 \\ \smile{\delta}_M &= \min_{\hat{C}}\left\{\max_{\omega}\left[\delta(\omega, \hat{C})\right]\right\} = \min_{\hat{C}}\left[\max_{\omega}(\delta)\right] \end{aligned} \tag{9.5.4}$$

The above three basic relationships can be summarized in the inequalities below

$$\delta\rho = 1; \quad \delta_M\rho_m = 1; \quad \smile{\delta}_M\frown{\rho}_m = 1 \tag{9.5.5}$$

where the following simple calculations prove the existence of (9.5.3) and (9.5.4)

$$\rho_m(\hat{C}) = \min_{\omega}\left|1 + \hat{C}P\right| = \frac{1}{\max_{\omega}\left|\frac{1}{1+\hat{C}P}\right|} = \frac{1}{\left\|\frac{1}{1+\hat{C}P}\right\|_{\infty}} = \frac{1}{\delta_M(\hat{C})} \tag{9.5.6}$$

$$\begin{aligned} \smile{\delta}_M &= \min_{\hat{C}}\left\{\max_{\omega}\left|\frac{1}{1+\hat{C}P}\right|\right\} = \frac{1}{\max_{\hat{C}}\left\{\max_{\omega}\left|\frac{1}{1+\hat{C}P}\right|\right\}} \\ &= \frac{1}{\max_{\hat{C}}\left\{\min_{\omega}\left|1 + \hat{C}P\right|\right\}} = \frac{1}{\frown{\rho}_m} \end{aligned} \tag{9.5.7}$$

Given (9.5.3), (9.5.4), and (9.5.5) further basic, almost trivial, inequalities can also be simply formulated

$$\breve{\delta}_{\mathrm{M}} \le \delta_{\mathrm{M}}(\hat{C}); \quad \rho_{\mathrm{m}}(\hat{C}) \le \overset{\frown}{\rho}_{\mathrm{m}}$$

$$\frac{1}{\overset{\frown}{\rho}_{\mathrm{m}}} = \breve{\delta}_{\mathrm{M}} \le \delta_{\mathrm{M}}(\hat{C}) = \frac{1}{\rho_{\mathrm{m}}(\hat{C})}; \qquad \frac{1}{\delta_{\mathrm{M}}(\hat{C})} = \rho_{\mathrm{m}}(\hat{C}) \le \overset{\frown}{\rho}_{\mathrm{m}} = \frac{1}{\breve{\delta}_{\mathrm{M}}} \tag{9.5.8}$$

The above results are not surprising. The fact that they are valid even for the modeling error in the case of *KB*-parameterized identification methods makes them special. So it can be clearly seen that when the modeling error decreases, the robustness of the control increases. Namely, if the minimum of the modeling error $\breve{\delta}_{\mathrm{M}}$ is decreased, then the maximum of the minimum robustness measure $\overset{\frown}{\rho}_{\mathrm{m}}$ is increased, since $\breve{\delta}_{\mathrm{M}} \overset{\frown}{\rho}_{\mathrm{m}} = 1$.

Similar relationships can be obtained if the $\mathcal{H}_2$ norm of the "joint" modeling and control error is used instead of the absolute values. Introduce the following relative fidelity measure

$$\sigma = \frac{\|\varepsilon_{\mathrm{ID}}\|_2}{\|y_{\mathrm{n}}\|_2} = \frac{\|\tilde{e}\|_2}{\|y_{\mathrm{n}}\|_2}; \quad \|y_{\mathrm{n}}\|_2 \neq 0 \tag{9.5.9}$$

The upper limit for this measure can be formulated as

$$\sigma = \frac{\|\varepsilon_{\mathrm{ID}}\|_2}{\|y_{\mathrm{n}}\|_2} = \frac{\|\tilde{e}\|_2}{\|y_{\mathrm{n}}\|_2} \le \left\| \frac{1}{1 + \hat{C}P} \right\| = \delta_{\mathrm{M}}(\hat{C}) \tag{9.5.10}$$

so it is very easy to find similar equations for σ. Let $\sigma_{\mathrm{M}}(\hat{C}) = \max_{\ell}[\sigma(\ell, \hat{C})]$ and $\breve{\sigma}_{\mathrm{M}} = \min_{\hat{C}}\left\{ \max_{\ell} \left[\sigma(\ell, \hat{C})\right]\right\}$. Using these definitions and the former equations we obtain the following interesting relationship

$$\breve{\sigma}_{\mathrm{M}} \le \breve{\sigma}_{\mathrm{M}}(\hat{C}) \le \delta_{\mathrm{M}}(\hat{C}) = \frac{1}{\rho_{\mathrm{m}}(\hat{C})} \tag{9.5.11}$$

for the relative quadratic identification error.

Example 9.5.1

Use again the first-order reference model (9.1.23) for the design of the noise rejection in the *IS* process. Here the maximum of the robustness measure is $\overset{\frown}{\rho}{}^{\mathrm{o}}_{\mathrm{m}} = \overset{\frown}{\rho}{}^{\mathrm{o}}_{\mathrm{m,IS}} = 0.9$ according to (9.1.25). The values of the typical variables (see above) are

$$\breve{\delta}{}^{\mathrm{o}}_{\mathrm{M}} = \frac{1}{\overset{\frown}{\rho}{}^{\mathrm{o}}_{\mathrm{m}}} = \frac{2}{|a_{\mathrm{n}1} - 1|} = 1.1111 \Rightarrow \breve{\delta}{}^{\mathrm{o}}_{\mathrm{M}} \overset{\frown}{\rho}{}^{\mathrm{o}}_{\mathrm{m}} = 1 \tag{9.5.12}$$

$$\sigma_{\mathrm{M}}^{\mathrm{o}}=\frac{1}{\sqrt{\widehat{\rho}_{\mathrm{m}}^{\mathrm{o}}}}=\sqrt{\frac{2}{|a_{\mathrm{n}1}-1|}}=1.054\Rightarrow\sigma_{\mathrm{M}}^{\mathrm{o}}\widehat{\rho}_{\mathrm{m}}^{\mathrm{o}}=0.9486\leq\breve{\delta}_{\mathrm{M}}^{\mathrm{o}}\widehat{\rho}_{\mathrm{m}}^{\mathrm{o}}=1 \tag{9.5.13}$$

Considering the data of (9.5.1) and applying again the relative sampling time $x=T_{\mathrm{s}}/T_{\mathrm{n}}$, the different measures in (9.5.11) are illustrated in Figure 9.5.2. Here T_{n} is the time constant of the CT first-order reference model.

Introduce the following coefficient for the excitation caused by the reference signal

$$\xi=\frac{\|y_{\mathrm{r}}\|_2}{\|y_{\mathrm{n}}\|_2};\quad \|y_{\mathrm{n}}\|_2\neq 0 \tag{9.5.14}$$

which represents a signal/noise ratio. Investigate the product $\sigma\rho$ (which is called the uncertainty product) in an iterative procedure where the relative error ℓ of the model is improved gradually. For simplicity, let us assume an *IS* process. It can be simply derived that

$$\sigma\rho_{\mathrm{m}}\leq\sigma\widehat{\rho}_{\mathrm{m}}^{\mathrm{o}}\leq\frac{1+\xi\|R_{\mathrm{n}}\ell\|_\infty}{\min|1+R_{\mathrm{n}}\ell|}\leq\left.\frac{1+\xi\|R_{\mathrm{n}}\|_\infty\|\ell\|_\infty}{1-\xi\|R_{\mathrm{n}}\|_\infty\|\ell\|_\infty}\right|_{\ell\to 0}=\sigma_{\mathrm{o}}\widehat{\rho}_{\mathrm{m}}^{\mathrm{o}}=1 \tag{9.5.15}$$

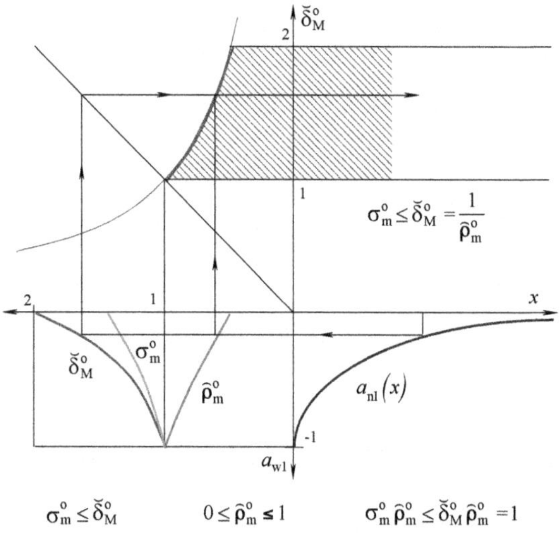

Figure 9.5.2 Illustration of uncertainty relationships (9.5.11).

i.e.,

$$\sigma\rho_m \le \sigma\overset{\frown}{\rho}{}^o_m \underset{\ell\to 0}{\le} \sigma_o\overset{\frown}{\rho}{}^o_m = 1 \quad \text{or} \quad \sigma \le \sigma_o = \frac{1}{\overset{\frown}{\rho}{}^o_m} = \breve{\delta}{}^o_M \le \delta_M = \frac{1}{\rho} \tag{9.5.16}$$

where $\sigma_o = \sigma(\ell = 0)$. Similarly to the notations $\sigma_M(\hat{C})$ and $\breve{\sigma}_M$ applied above, the notations $\sigma_m(\ell) = \min_{\ell}[\sigma(\ell, \hat{C})]$ and $\sigma^o_m = \sigma_m(\ell = 0)$ can also be introduced. It is not an easy task, however, to derive the relationship between σ^o_m and σ_o or $\breve{\sigma}_M$ and $\sigma_M(\hat{C})$. The simplest case to investigate (9.5.15) is when $\ell = 0$, since then

$$\sigma^o_m \le \sigma^o_M(\hat{C}) \le \delta^o_M(\hat{C}) = \frac{1}{\rho^o_m(\hat{C})} \tag{9.5.17}$$

This equation gives a new uncertainty relationship, according to which

$$\frac{\|\tilde{e}\|_2}{\|y_n\|_2} \min_{\hat{C}} \left|1 + \hat{C}P\right| \underset{\ell\to 0}{\le} 1 \tag{9.5.18}$$

The product of the modeling accuracy and the robustness measure of the control must not be greater than one, when the optimality condition $\ell = 0$ is reached. The obtained uncertainty relation can be written in another form, since

$$\sup\left\{\frac{\|\tilde{e}\|_2}{\|y_n\|_2}\right\} \min_{\hat{C}} \left|1 + \hat{C}P\right|_{\ell=0} = 1 \tag{9.5.19}$$

The earlier results of control engineering referred only for the statement that the quality of the control cannot be improved, only at the expense of the robustness, so this result, which connects the quality of the identification and the robustness of the control, can be considered, by all mean, novel.

This phenomenon can arguably be considered as the Heisenberg uncertainty relation of control engineering, according to which

$$\frac{1}{\Delta z}\frac{1}{\Delta p} \le 1 \tag{9.5.20}$$

Here Δz and Δp are the alterations of the canonical coordinate and the impulse variables, respectively, and thus their inverse corresponds to the generalized accuracy and "rigidity" which are known as performance and robustness in control engineering.

The consequence of the new uncertainty relation is very simple: *KB*-parameterized identification is the only method where the improvement of the modeling error also increases the robustness of the control. With other methods, and other identification topology, modeling and control errors are interrelated in a very complex way, and in many cases this relation cannot be given in an explicit form. This is the main reason why it is difficult to elaborate a method which guarantees, or at least forces, similar behavior by the two errors, though some results can be found in the literature [4,50].

There is a myth in the literature concerning the antagonistic conflict between control and identification. A "good" regulator minimizes the internal signal changes in the closed loop and therefore most of the identification methods, which use these inner signals provide worse modeling error, if the regulator is better. The exciting signal of *KB*-parameterized identification is an outer signal and therefore the phenomenon does not exist. The relevant feature of this relationship is shown in Figures 9.5.3 and 9.5.4 for a general identification method and a *KB*-parameterized technique.

In Figure 9.5.3, there is no clear relation between δ_{ID} and δ, or σ_{ID} and σ, and therefore there is no guarantee that minimizing δ_M increases ρ_m.

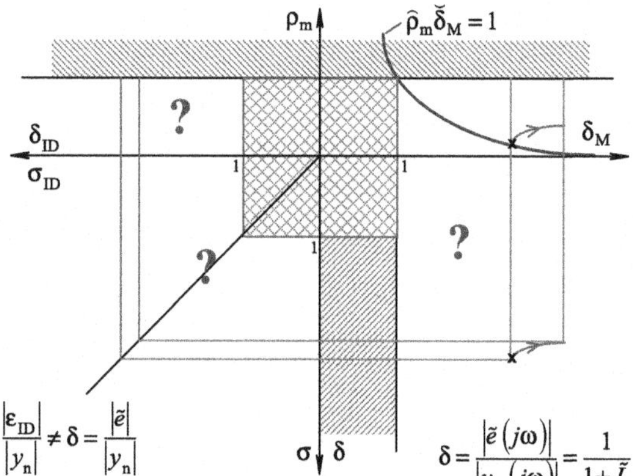

$$\delta_{ID} = \frac{|\varepsilon_{ID}|}{|y_n|} \neq \delta = \frac{|\tilde{e}|}{|y_n|} \qquad \delta = \frac{|\tilde{e}(j\omega)|}{|y_n(j\omega)|} = \frac{1}{1+\tilde{L}}$$

$$\sigma_{ID} = \sqrt{\frac{\operatorname{var}\{\varepsilon_{ID}\}}{\operatorname{var}\{y_n\}}} = \frac{|\varepsilon_{ID}|_2}{|y_n|_2} \neq \sigma = \frac{|\tilde{e}|_2}{|y_n|_2} \qquad \tilde{e} \neq \varepsilon_{ID} \qquad \rho_m = \min_{\omega} |1+\tilde{L}|$$

Figure 9.5.3 Relationship between the control and identification error in the general case.

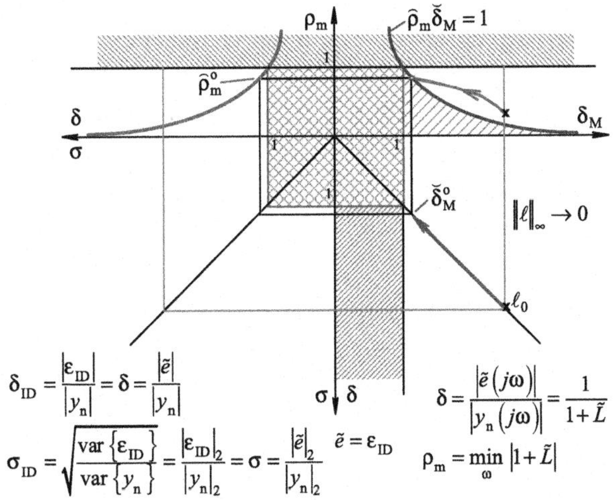

Figure 9.5.4 Relationship between the control and identification error in the case of the *Keviczky–Bányász*-parameterized identification method.

In Figure 9.5.4 $\delta_{ID} = \delta$ and $\sigma_{ID} = \sigma$, and thus the minimization of δ_M directly maximizes ρ_m. Thus if during the iterative identification the condition $\|\ell_k\|_\infty \underset{k\to\infty}{=} 0$ is guaranteed then, at the same time, the convergences $\breve{\delta}_M^k \underset{k\to\infty}{=} \breve{\delta}_M^o$ and $\widehat{\rho}_m^k \underset{k\to\infty}{=} \widehat{\rho}_m^o$ are ensured.

CHAPTER 10

Process Identification

Model creation is an important element of the investigation of control systems. The model describes the signal transfer features of the system in mathematical form. The static and dynamic behavior of the system can be analyzed with the help of the model without performing real tests on the system. Based on the system model, computations can be performed and the behavior of the system can be numerically simulated. The system model is used to design the regulator. To do this, the models of the elements of the control loop have to be determined. The signal transfer behavior P of each element should be determined from the physical relationships describing the given element or from its measured input and output signals. This latter procedure is called process identification (Figure 10.1).

The signal transfer behavior of an element can be given by mathematical relationships describing the physical operation. In order to provide this mathematical description it is necessary to understand the physical operation. The parameter values in these equations can be determined by computations or measurements, so these values might be inaccurate, but the extent of the uncertainty or the domain of the parameters can usually be given.

The static and dynamic behavior of the system can be determined by investigating the input and output signals. To gain relevant information for this investigation the input signals must be chosen properly. This procedure assumes a system model, and the parameters of this model are those which best fit the output signals of the system and the model for a given excitation. This procedure is called process identification.

In many cases the physical modeling and identification are used jointly to determine the system model (Figure 10.2).

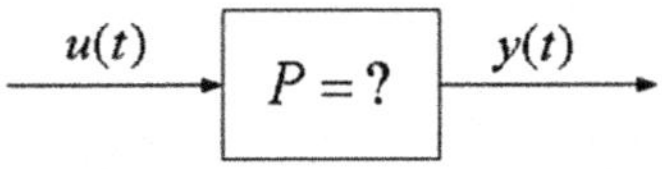

Figure 10.1 Process identification.

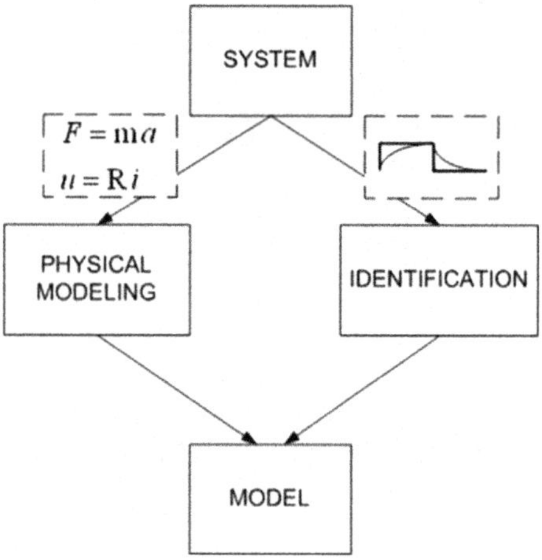

Figure 10.2 Determination of the system model.

TYPES OF MODELS

During the solution of a control problem, one of the basic tasks is the *process identification* (ID) when the model $\hat{P}$ of the controlled process P should be determined from the measured input and output data. The process identification, initiated by simple grapho-analytical methods, has become a substantive discipline, and its results and methods are discussed in numerous books. In modern computer libraries there are almost standard instruments to solve the most important tasks. Here only some of the most important methods are used to illustrate the applied algorithms and techniques.

The general nonlinear state equation of a process can have different forms, so the classification, which could help the overview, is not an easy task. The possible classification of the process models has been discussed in detail in Section 1.1. Further details will be given during the discussion and classification of the process identification tasks. For example, the process identification methods substantially depend on what kind of task should be solved: the determination of the static characteristic or the dynamic model of the process.

Since the process models are mainly determined by ID methods from experimental measurements, the classification of the models follows the applicability of the identification methods widely used in practice. This approach is followed below.

MODEL VALIDATION

A model of acceptable accuracy can usually be determined by an iterative procedure whose main steps are illustrated in Figure 10.3.

Identification is started from initial/prior/preliminary information. First, the so-called *design of experiments* is performed, that is, in the case of static modeling the optimal allocation of the measurement points is determined; in the dynamic modeling case, optimal input signal is generated which assumes the vital frequency region. This is called *input design*. The accuracy of the final model strongly depends on this step, and therefore numerous theories deal with the optimal execution of this task (see, e.g., Section 4.3).

The effect of the optimal measurement points and the input signal is realized by active experiments and the measured data collected.

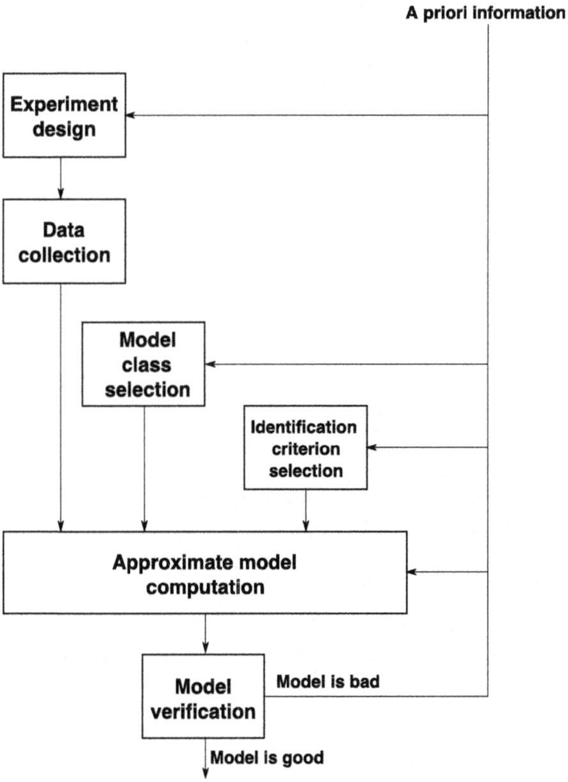

Figure 10.3 Block scheme of the complete process identification.

Preliminary information facilitates the choice of model class and identification criterion, and the approximate model is determined (by parameter estimation).

Then the output of the model and the measured output of the process are compared for the same input signals. Qualitative measures are constructed for *model validation* from the deviations.

If the accuracy of the model is not satisfactory, then the iteration is continued by a new set of experiments. If the accuracy is satisfactory, then the procedure is stopped.

PARAMETER ESTIMATION

The goal of the process identification is always to approximate best the process P by its model $\hat{P}$. The process is never known completely; only a close model class P^* can be applied what is known. If the expected value of the deviations is to be minimized, then the criterion of the approximation should be changed as follows [123]

$$V\left(P,\hat{P}\right)=E\left\{\left(P-\hat{P}\right)^2\right\}=\underbrace{E\{(P-P^*)^2\}}_{\text{bias}}+\underbrace{E\left\{\left(P^*-\hat{P}\right)^2\right\}}_{\text{fitting}} \quad (10.1)$$

Here the first term means the average square error between the real process and the chosen model class, which is called implicit bias. The second term is a square fitting error, and can be minimized during the parameter estimation. In practice, it is almost always assumed that $P\approx P^*=g(u,\boldsymbol{p})$, and thus the process belongs to the applied model class. The further technique of parameter estimation depends only on the structure of $g(u, \boldsymbol{p})$ and the assumption made regarding the noise arising from the measurements.

The following system equation is generally used for the parametric identification of *single input single output* (*SISO*) models

$$y=g(u,\boldsymbol{p})+\varsigma \quad (10.2)$$

where u and y are the input and output signals of the process, respectively, and the parameter vector is denoted by $\boldsymbol{p}$. It is assumed that the structure is estimated precisely, and the expected value of ς is zero, so only accidental (i.e., not systematic) error components appear in the deviation $y-g(u, \boldsymbol{p})$. The classical parameter estimation problem is to determine an estimation $\hat{\boldsymbol{p}}$ of the parameter vector $\boldsymbol{p}$, which ensures the best fitting of $g(u,\hat{\boldsymbol{p}})$ to the observed y in the context of a certain optimality criterion. The different statistical estimation methods are usually classified on the basis of this quality criterion and preliminary knowledge concerning the residual (measurement noise) [44].

Given the system Equation (10.2), the errors (residuals) $\hat{\varsigma}[1], \ldots, \hat{\varsigma}[N]$ can be computed from the N samples of the input and output signals for any parameter vector $\hat{\boldsymbol{p}}$, where $\hat{\varsigma} = y - g(u, \hat{\boldsymbol{p}})$. (As it will be seen at the ID of the discrete-time (DT) process models, the arguments of all variables and the errors $\hat{\varsigma}[k]$ will be DT.) Arrange these values into a vector $\hat{\varsigma} = [\hat{\varsigma}[1], \ldots, \hat{\varsigma}[N]]^{\mathrm{T}}$. By means of the *maximum likelihood* (*ML*) method, the estimate $\hat{\boldsymbol{p}}$ is determined by maximizing the joint probability density of the N elements of the error vector $\hat{\varsigma}$, that is, maximizing the probability that $\hat{\varsigma}$, computed from the value pairs $u[k]$, $y[k]$, and $\hat{\boldsymbol{p}}$, is equal to the vector $\boldsymbol{e}$ of the real accidental error. Thus in the *ML* method the conditional probability density function $d_N(\hat{\varsigma}|\hat{\boldsymbol{p}})$ should be maximized. In many cases, however, instead of $d_N(\hat{\varsigma}|\hat{\boldsymbol{p}})$ its natural base logarithm, that is, the so-called likelihood function $L(\hat{\boldsymbol{p}}) = \ln[d_N(\hat{\varsigma}|\hat{\boldsymbol{p}})]$, is maximized. Thus, with this method the likelihood function can be constructed only with the a priori knowledge (or assumption) of the density function. In most cases, $\varsigma[k]$ is assumed to be of normal distribution (this is the reason why the ln function can be used), which is valid for the majority of practical cases. (Otherwise, the normality investigation of the residuals (i.e., the values $\hat{\varsigma}[k]$ computed after the parameter estimation), and thus the checking of our assumption, means a routine statistical task [44,123,147].)

If ς is assumed to be a vector variable of normal distribution, its probability density function can be given as [123]

$$d_N(\varsigma|\boldsymbol{p}) = (2\pi)^{-N/2}|\boldsymbol{Z}|^{-1/2}\exp\left\{-\frac{1}{2}\varsigma^{\mathrm{T}}\boldsymbol{Z}^{-1}\varsigma\right\} \tag{10.3}$$

where

$$\boldsymbol{Z} = \boldsymbol{Z}_e = E\{\varsigma\varsigma^{\mathrm{T}}\} \tag{10.4}$$

that is, it is the covariance matrix of the noise. Here $|\ldots|$ notes the determinant. Thus, in the case of normal distribution the likelihood function is $L(\boldsymbol{p}) = \ln[d_N(\varsigma \boldsymbol{p})]$ and it can be similarly computed for $\hat{\varsigma}$, too.

$$L(\hat{\boldsymbol{p}}) = \ln[d_N(\hat{\varsigma}|\hat{\boldsymbol{p}})] = -\frac{N}{2}\ln 2\pi - \frac{1}{2}\ln|\boldsymbol{Z}| - \frac{1}{2}\hat{\varsigma}^{\mathrm{T}}\boldsymbol{Z}^{-1}\hat{\varsigma} \tag{10.5}$$

For a given $\boldsymbol{Z}$, the maximum of $L(\hat{\boldsymbol{p}})$ can also be ensured by the following equivalent assumption

$$\min_{\hat{p}}\{V(\hat{\boldsymbol{p}})\} = \min_{\hat{p}}\left\{\frac{1}{2}\hat{\varsigma}^{\mathrm{T}}\boldsymbol{Z}^{-1}\hat{\varsigma}\right\} \tag{10.6}$$

Of course, in this case the covariance matrix $\boldsymbol{Z}$ of the noise should be known. If, however, the elements of $\boldsymbol{Z}$ are assumed to be unknown, the minimization should be performed by these variables.

If the noise $\varsigma[k]$ is uncorrelated, its variance $E\{\varsigma^2\} = \sigma_\varsigma^2$ is constant independently of the time (this condition means stationary source noise in the case of DT processes), that is, the measurements have identical accuracies, and

$$\boldsymbol{Z} = \lambda^2 \boldsymbol{I} \tag{10.7}$$

In this case the *ML* estimation can be obtained by solving the minimization task

$$\min_{\hat{p}}\{V(\hat{\boldsymbol{p}})\} = \min_{\hat{p}}\left\{\frac{1}{2}\hat{\varsigma}^{\mathrm{T}}\hat{\varsigma}\right\} \tag{10.8}$$

Note that this criterion means the minimization of the following loss function by $\hat{\boldsymbol{p}}$

$$V(\hat{\boldsymbol{p}}, N) = \frac{1}{2}\sum_{j=1}^{N}\left[y_j - g(u, \hat{\boldsymbol{p}})\right]^2 \tag{10.9}$$

which corresponds to the classical so-called *least squares* (*LS*) method. Note that the *LS* algorithm can always be applied but it can provide *ML* estimation only if the measurement error has normal distribution [31]. The method can be applied even in the case when λ is a priori not known.

The tasks (10.5) or (10.7) can be generally considered a nonlinear extremum-seeking problem. In the next sections, this task is further simplified for the special classes of the models which are linear in parameters.

If the system has multiple-input and multiple-output, then the *multiple input multiple output* (*MIMO*) process model takes the form

$$\boldsymbol{y} = \boldsymbol{g}(\boldsymbol{u}, \boldsymbol{P}) + \boldsymbol{e} \tag{10.10}$$

where it is assumed that $\varsigma[k]$ has zero mean value and its covariance matrix is $\boldsymbol{\Lambda}$. (Note that $\boldsymbol{\Lambda}$ corresponds to the generalization of σ_ς^2.) If N

measurement value pairs of $\boldsymbol{u}[k]$ and $\boldsymbol{y}[k]$ are available, then the joint density function of $\boldsymbol{\varsigma}[k]$ can be given as

$$d_N(\boldsymbol{\varsigma}_\mathrm{M}|\boldsymbol{P}) = \prod_{j=1}^{N} d(\boldsymbol{\varsigma}[j]|\boldsymbol{P}) = (2\pi)^{-pN/2}|\boldsymbol{\Lambda}|^{-N/2}\exp\left\{-\frac{1}{2}\sum_{j=1}^{N}\boldsymbol{\varsigma}^\mathrm{T}[j]\boldsymbol{\Lambda}\boldsymbol{\varsigma}[j]\right\} \tag{10.11}$$

where

$$\boldsymbol{\varsigma}_\mathrm{M} = \left[\boldsymbol{\varsigma}^\mathrm{T}[1], \ldots, \boldsymbol{\varsigma}^\mathrm{T}[N]\right] \tag{10.12}$$

and

$$d(\boldsymbol{\varsigma}[j]|\boldsymbol{P}) = (2\pi)^{-p/2}|\boldsymbol{\Lambda}|^{-1/2}\exp\left\{-\frac{1}{2}\boldsymbol{\varsigma}^\mathrm{T}[j]\boldsymbol{\Lambda}\boldsymbol{\varsigma}[j]\right\} \tag{10.13}$$

is the conditional probability function of $\boldsymbol{\varsigma}[j]$, $\boldsymbol{P}$ is the parameter matrix of the *MIMO* system, and p is the dimension of $\boldsymbol{\varsigma}$, that is, the number of the outputs. On the basis of (10.11), the likelihood function of the error of the parameter estimation for *MIMO* systems can be given as

$$L(\hat{\boldsymbol{P}}) = \ln\left[d_N\left(\boldsymbol{\varsigma}_\mathrm{M}|\hat{\boldsymbol{P}}\right)\right] = -\frac{pN}{2}\ln 2\pi - \frac{N}{2}\ln|\boldsymbol{\Lambda}| - \frac{1}{2}\sum_{j=1}^{N}\boldsymbol{\varsigma}^\mathrm{T}[j]\boldsymbol{\Lambda}^{-1}\boldsymbol{\varsigma}[j] \tag{10.14}$$

According to the derivation shown in A.5.3 of Appendix A.5, the maximization of the likelihood function (10.14) is an equivalent task to the minimization of the loss function

$$\min_{\hat{\boldsymbol{P}}}\left\{V\left(\hat{\boldsymbol{P}}\right)\right\} = \min_{\hat{\boldsymbol{P}}}\left|\frac{1}{N}\sum_{j=1}^{N}\hat{\boldsymbol{\varsigma}}^\mathrm{T}[j]\hat{\boldsymbol{\varsigma}}[j]\right| \tag{10.15}$$

(i.e., minimization of the determinant of a matrix).

10.1 OFF-LINE PROCESS IDENTIFICATION METHODS

Methods of process identification strongly depend on the kind of model selected: one with static characteristics or a dynamic model.

Static Process Identification

Assume that the static characteristic of a process is a straight line $p_\mathrm{o} + p_1 u$, which is measured by an additive measurement error ς

$$y = p_\mathrm{o} + p_1 u + \varsigma \tag{10.1.1}$$

It is assumed that the input signal u is measured noise-free (or set by us; see active experiment). The input and output signals are measured in N points, and the measured samples are approximated by a linear model

$$\hat{y} = \hat{p}_o + \hat{p}_1 u = \boldsymbol{f}^T(u)\hat{\boldsymbol{p}}; \quad \boldsymbol{f}^T(u) = [1 \quad u]; \quad \hat{\boldsymbol{p}} = [\hat{p}_o \quad \hat{p}_1]^T \qquad (10.1.2)$$

as shown in Figure 10.1.1. Here $\boldsymbol{f}(u)$ is the vector of the approximating function elements.

If the additive measurement error is a noise of zero mean value then the so-called *LS* method provides the unbiased parameter estimation of the static process. The *LS* method summarizes the squares of the differences between the measured value and the output in each sample point and minimizes the criterion

$$V(\hat{\boldsymbol{p}}, N) = \frac{1}{2}\sum_{j=1}^{N}\left[y_j - \boldsymbol{f}^T(u_j)\hat{\boldsymbol{p}}\right]^2 = \frac{1}{2}[\boldsymbol{y} - \boldsymbol{F}_u\hat{\boldsymbol{p}}]^T[\boldsymbol{y} - \boldsymbol{F}_u\hat{\boldsymbol{p}}] \qquad (10.1.3)$$

where

$$\boldsymbol{F}_u = \begin{bmatrix} 1 & u_1 \\ 1 & u_2 \\ \vdots & \vdots \\ 1 & u_N \end{bmatrix} = \begin{bmatrix} \boldsymbol{f}^T(u_1) \\ \boldsymbol{f}^T(u_2) \\ \vdots \\ \boldsymbol{f}^T(u_N) \end{bmatrix} \quad \text{and} \quad \boldsymbol{y} = \begin{bmatrix} y_1 \\ y_2 \\ \vdots \\ y_N \end{bmatrix} \qquad (10.1.4)$$

The vector system equation for N samples is

$$\boldsymbol{y} = \boldsymbol{F}_u\boldsymbol{p} + \varsigma \qquad (10.1.5)$$

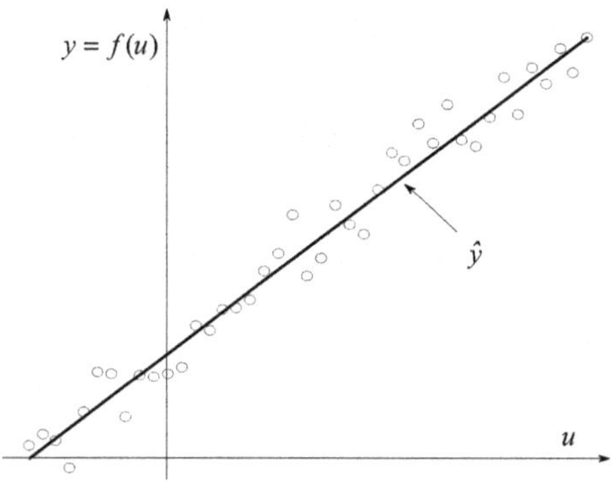

Figure 10.1.1 Identification of a linear static model.

where $\boldsymbol{\varsigma} = \begin{bmatrix} \varsigma_1 & \varsigma_1 & \dots & \varsigma_N \end{bmatrix}^{\mathrm{T}}$ is the vector of the measurement error. The parameter estimation minimizing the sum of squares (10.1.3) is obtained according to Section A.6.1 of Appendix A.6 as

$$\hat{\boldsymbol{p}} = \left[\boldsymbol{F}_{\mathrm{u}}^{\mathrm{T}}\boldsymbol{F}_{\mathrm{u}}\right]^{-1}\boldsymbol{F}_{\mathrm{u}}^{\mathrm{T}}\boldsymbol{y} \tag{10.1.6}$$

Based on the classical regression computation, this relationship is the solution of the Gaussian normal equation system. This estimation is unbiased, that is, $E\{\hat{\boldsymbol{p}}\} = \boldsymbol{p}$, so the expected value of $\hat{\boldsymbol{p}}$ gives the unknown original parameter vector $\boldsymbol{p}$. If ς has normal distribution, then $\hat{\boldsymbol{p}}$, which results from the *LS* estimation, is the *minimum variance* (*MV*) (best) estimation of $\boldsymbol{p}$.

The accuracy of the estimated parameter vector $\hat{\boldsymbol{p}}$ can be characterized by the following covariance matrix

$$\boldsymbol{Z}_{\hat{p}} = E\left\{(\hat{\boldsymbol{p}} - \boldsymbol{p})(\hat{\boldsymbol{p}} - \boldsymbol{p})^{\mathrm{T}}\right\} = \left[\boldsymbol{F}_{\mathrm{u}}^{\mathrm{T}}\boldsymbol{F}_{\mathrm{u}}\right]^{-1}\sigma_{\varsigma}^2 \tag{10.1.7}$$

where σ_{ς}^2 is the variance of the measurement error. (The inverse of the covariance matrix is called the information matrix.)

The input signal of the process, however, is always set but simply observed, that is, measured (passive experiment). If u is a random variable then the independence of ς and u needs to be assumed to obtain an unbiased estimation by the *LS* method.

The solution (10.1.6) can be simplified by taking the following relationships into account

$$\boldsymbol{F}_{\mathrm{u}}^{\mathrm{T}}\boldsymbol{F}_{\mathrm{u}} = \sum_{j=1}^{N}\boldsymbol{f}(u_j)\boldsymbol{f}^{\mathrm{T}}(u_j) \quad \text{and} \quad \boldsymbol{F}_{\mathrm{u}}^{\mathrm{T}}\boldsymbol{y} = \sum_{j=1}^{N}\boldsymbol{f}(u_j)y_j \tag{10.1.8}$$

Assume that the static characteristic of the process is a parabola $p_{\mathrm{o}} + p_1 u + p_2 u^2$, measured with additive measurement error ς (see Figure 10.1.2)

$$y = p_{\mathrm{o}} + p_1 u + p_2 u^2 + \varsigma \tag{10.1.9}$$

The input u is also assumed to be measured noise-free (or is set, see active experiment). The input and output are measured in N points and the measurements are approximated by nonlinear (quadratic) model as shown in Figure 10.1.2.

$$\begin{aligned} &\hat{y} = \hat{p}_{\mathrm{o}} + \hat{p}_1 u + \hat{p}_2 u^2 = \boldsymbol{f}^{\mathrm{T}}(u)\hat{\boldsymbol{p}}; \quad \boldsymbol{f}^{\mathrm{T}}(u) = \begin{bmatrix} 1 & u & u^2 \end{bmatrix}; \\ &\hat{\boldsymbol{p}} = \begin{bmatrix} \hat{p}_{\mathrm{o}} & \hat{p}_1 & \hat{p}_2 \end{bmatrix}^{\mathrm{T}} \end{aligned} \tag{10.1.10}$$

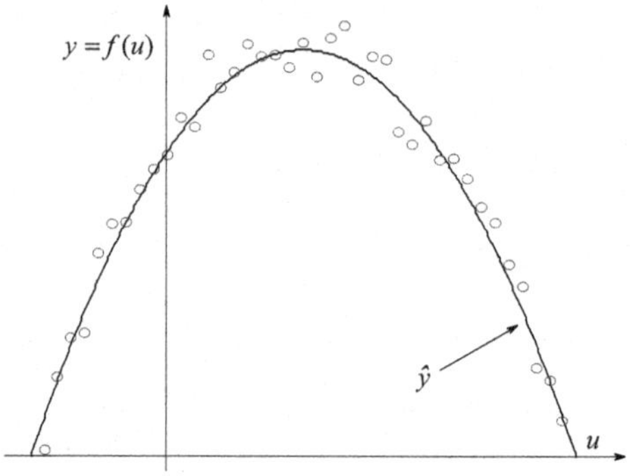

Figure 10.1.2 Identification of a nonlinear (quadratic) static model.

Observe that the model $\hat{y} = \boldsymbol{f}^{\mathrm{T}}(u)\hat{\boldsymbol{p}}$ is still linear in terms of parameters (shortly: linear in parameters). Thus the *LS* method can be applied in unchanged form, if the matrix $\boldsymbol{F}_{\mathrm{u}}$ is derived from the function component vector $\boldsymbol{f}(u)$ of the quadratic model (10.1.9)

$$\boldsymbol{F}_{\mathrm{u}} = \begin{bmatrix} 1 & u_1 & u_1^2 \\ 1 & u_2 & u_2^2 \\ \vdots & \vdots & \vdots \\ 1 & u_N & u_N^2 \end{bmatrix} = \begin{bmatrix} \boldsymbol{f}^{\mathrm{T}}(u_1) \\ \boldsymbol{f}^{\mathrm{T}}(u_2) \\ \vdots \\ \boldsymbol{f}^{\mathrm{T}}(u_N) \end{bmatrix} \tag{10.1.11}$$

To obtain $\hat{\boldsymbol{p}}$, Equation (10.1.6) needs to be used again.

Note that quite a wide range of the function classes can be written as linear in parameters.

If the static characteristic is nonlinear then an extreme seeking method needs to be applied for the minimization of $V(\hat{\boldsymbol{p}}, N)$.

Identification of Linear Dynamic Processes

The identification of a linear dynamic process almost always means the determination of the DT model. It is shown in Appendix A.3 that a linear DT system given in the so-called filter form

$$y[k] = G(z^{-1})u[k] = \frac{\mathcal{B}(z^{-1})z^{-d}}{\mathcal{A}(z^{-1})}u[k] = \frac{\mathcal{B}(z^{-1})z^{-d}}{1 + \tilde{\mathcal{A}}(z^{-1})}u[k] \tag{10.1.12}$$

can always be written in the form of a difference equation that is linear in parameters [7,44]

$$\begin{aligned} y[k] &= \mathcal{B}(z^{-1})z^{-d}u[k] - \tilde{\mathcal{A}}(z^{-1})y[k] = \boldsymbol{f}^{\mathrm{T}}[u, y, k]\boldsymbol{p}_{\mathrm{ba}} \\ &= b_1 u[k-d-1] + b_2 u[k-d-2] + \ldots + b_n u[k-d-n] \\ &\quad - a_1 y[k-1] - \ldots - a_n y[k-n] \end{aligned} \tag{10.1.13}$$

where

$$f^{\mathrm{T}}[u,y,k]=[u[k-d-1]\; u[k-d-2]\; \ldots \; u[k-d-n]\; -y[k-1]\; \ldots \; -y[k-n]]$$

$$p_{\mathrm{ba}}=\begin{bmatrix} b_1 & b_2 & \ldots & b_n & a_1 & \ldots & a_n \end{bmatrix}^{\mathrm{T}} \tag{10.1.14}$$

This technique, making the difference equation of the dynamic DT system "quasi-linear," paves the way for process identification methods similar to the *LS* method. The measurement error problem of the DT systems, however, needs to be discussed rather differently from how it is in the case of static characteristics. The measurement situation of the identification is assumed to be as shown in Figure 10.1.3.

Here $u[k]$ is assumed to be without noise, but the noise-free output $v[k]$ of the process is measured with the additive error $y_{\mathrm{n}}[k]$. This output noise $y_{\mathrm{n}}[k]$ can be derived from an independent, zero mean value, white, so-called source noise $\varsigma[k]$ by the noise model $H_{\mathrm{n}}(z^{-1})=\mathcal{C}(z^{-1})/\mathcal{D}(z^{-1})$. In this case, process identification means the estimation of two models: the process and the noise model. The task can be substantially simplified if certain assumptions are made for the noise model. If the noise model has the form $H_{\mathrm{n}}(z^{-1})=1/\mathcal{A}(z^{-1})$, then

$$\begin{aligned} y[k] &= \frac{\mathcal{B}(z^{-1})z^{-d}}{\mathcal{A}(z^{-1})}u[k]+y_{\mathrm{n}}[k]=\frac{\mathcal{B}(z^{-1})z^{-d}}{\mathcal{A}(z^{-1})}u[k]+\frac{1}{\mathcal{A}(z^{-1})}\varsigma[k] \\ &= G(z^{-1})u[k]+H_{\mathrm{n}}(z^{-1})\varsigma[k] \end{aligned} \tag{10.1.15}$$

and, with the above technique, the original process can be rearranged as

$$y[k]=f^{\mathrm{T}}[u,y,k]p_{\mathrm{ba}}+\varsigma[k] \tag{10.1.16}$$

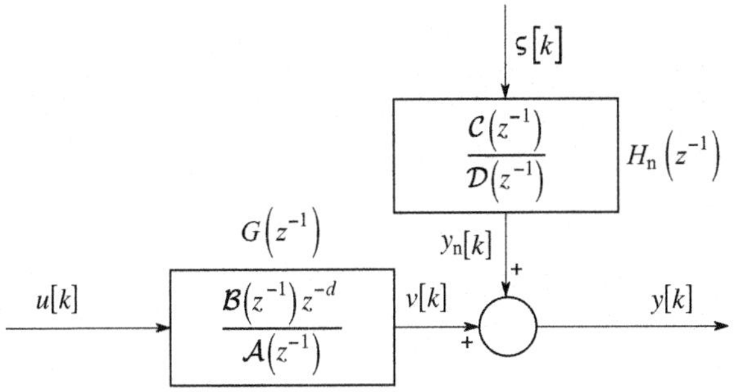

Figure 10.1.3 Measuring arrangement of a linear dynamic DT system.

Note that this form essentially corresponds to Equations (10.1.2) or (10.1.10) as shown in the identification of static characteristics, so the *LS* method can be directly applied if the matrix $\boldsymbol{F}(u)$ is constructed from $\boldsymbol{f}^{\mathrm{T}}[u, y, k]$ instead of $\boldsymbol{f}(u)$, and the parameter vector is $\hat{\boldsymbol{p}}_{\mathrm{ba}}$. Construct first the vectors

$$\begin{aligned} \boldsymbol{y} &= \begin{bmatrix} y[1] & y[2] & \dots & y[N] \end{bmatrix}^{\mathrm{T}} \\ \boldsymbol{u} &= \begin{bmatrix} u[1] & u[2] & \dots & u[N] \end{bmatrix}^{\mathrm{T}} \end{aligned} \tag{10.1.17}$$

then the matrix

$$\boldsymbol{F}_{\mathrm{uy}} = \begin{bmatrix} \boldsymbol{f}^{\mathrm{T}}(u, y, 1) \\ \boldsymbol{f}^{\mathrm{T}}(u, y, 2) \\ \vdots \\ \boldsymbol{f}^{\mathrm{T}}(u, y, N) \end{bmatrix} = \left[\boldsymbol{S}^{d+1}\boldsymbol{u}, \dots, \boldsymbol{S}^{d+n}\boldsymbol{u}, -\boldsymbol{S}^{1}\boldsymbol{y}, \dots, -\boldsymbol{S}^{n}\boldsymbol{y}\right] = \boldsymbol{F}(\boldsymbol{u}, \boldsymbol{y}) \tag{10.1.18}$$

Here $\boldsymbol{S}$ means the shift matrix introduced in A.1.4 of Appendix A.1. The equation of the process can be written by vectors containing N measurements as

$$\begin{aligned} \boldsymbol{y} &= \boldsymbol{S}^{d}\boldsymbol{A}^{-1}\boldsymbol{B}\boldsymbol{u} + \boldsymbol{A}^{-1}\boldsymbol{\varsigma} + \boldsymbol{A}^{-1}\boldsymbol{E}\boldsymbol{z}_{\mathrm{o}} = \boldsymbol{S}^{d}\boldsymbol{B}\boldsymbol{u} - \tilde{\boldsymbol{A}}\boldsymbol{y} + \boldsymbol{\varsigma} + \boldsymbol{E}\boldsymbol{z}_{\mathrm{o}}\big|_{\boldsymbol{z}_{\mathrm{o}}=0} \\ &= \boldsymbol{S}^{d}\boldsymbol{B}\boldsymbol{u} - \tilde{\boldsymbol{A}}\boldsymbol{y} + \boldsymbol{\varsigma} \end{aligned} \tag{10.1.19}$$

Here $\boldsymbol{\varsigma}$, similarly to $\boldsymbol{u}$ and $\boldsymbol{y}$, is a vector of N elements, which contains the values of the source noise. $\boldsymbol{E}$ is an $N \times n$ matrix, whose upper part is an $n \times n$ unity matrix; all its other elements are zero

$$\boldsymbol{E} = \begin{bmatrix} \boldsymbol{I} \\ \boldsymbol{0} \end{bmatrix} \tag{10.1.20}$$

$\boldsymbol{z}_{\mathrm{o}}$ is an $n \times 1$ vector containing the initial conditions (before the observation). Since the parameter estimations are asymptotically independent of the initial conditions [123], it can be assumed that $\boldsymbol{z}_{\mathrm{o}} = 0$ without losing the generality. The dimensions of the matrices in (10.1.17) are $N \times N$ and have the form

$$\begin{aligned} \boldsymbol{A} &= \boldsymbol{I} + a_1\boldsymbol{S} + \dots + a_n\boldsymbol{S}^{n} = \boldsymbol{I} + \tilde{\boldsymbol{A}} \\ \boldsymbol{B} &= b_1\boldsymbol{S} + \dots + b_n\boldsymbol{S}^{n} \end{aligned} \tag{10.1.21}$$

where $\boldsymbol{S}^g$ is an $N \times N$ Toeplitz matrix, in this case a shift matrix (see A.1.4. of Appendix A.1), whose elements in the g-th row under the main diagonal have the value one, and all other elements are zero. The last form of Equation (10.1.19) shows the quasi-linearization since the vector system equation has the form

$$\boldsymbol{y} = \boldsymbol{S}^d \boldsymbol{B} \boldsymbol{u} - \tilde{\boldsymbol{A}} \boldsymbol{y} + \boldsymbol{\varsigma} = \boldsymbol{F}(\boldsymbol{u}, \boldsymbol{y}) \boldsymbol{p}_{\mathrm{ba}} + \boldsymbol{\varsigma} \tag{10.1.22}$$

The *LS* parameter estimation is obtained again according to (10.1.6), that is,

$$\hat{\boldsymbol{p}}_{\mathrm{ba}} = \left[\boldsymbol{F}_{\mathrm{uy}}^{\mathrm{T}} \boldsymbol{F}_{\mathrm{uy}} \right]^{-1} \boldsymbol{F}_{\mathrm{uy}}^{\mathrm{T}} \boldsymbol{y} = \left[\boldsymbol{F}^{\mathrm{T}}(\boldsymbol{u}, \boldsymbol{y}) \boldsymbol{F}(\boldsymbol{u}, \boldsymbol{y}) \right]^{-1} \boldsymbol{F}^{\mathrm{T}}(\boldsymbol{u}, \boldsymbol{y}) \boldsymbol{y} \tag{10.1.23}$$

This estimation is asymptotically unbiased, that is, $\operatorname*{plim}_{N \to \infty} \{\hat{\boldsymbol{p}}\} = \boldsymbol{p}$, so the probability limit of $\hat{\boldsymbol{p}}$ gives the unknown original parameter vector $\boldsymbol{p}$. The independence and white noise character of $\varsigma[k]$ need to be assumed because $\boldsymbol{f}^{\mathrm{T}}[u, y, k]$ has measured components depending on $\varsigma[k]$. If $\varsigma[k]$ has normal distribution, then $\hat{\boldsymbol{p}}$, obtained by the *LS* estimation, is the *MV* (best) estimation of the parameter vector $\boldsymbol{p}$. The accuracy of the estimated parameter vector $\hat{\boldsymbol{p}}_{\mathrm{ba}}$ can be described by a covariance matrix

$$\boldsymbol{K} = E\left\{ \left(\hat{\boldsymbol{p}}_{\mathrm{ba}} - \boldsymbol{p} \right) \left(\hat{\boldsymbol{p}}_{\mathrm{ba}} - \boldsymbol{p} \right)^{\mathrm{T}} \right\} = \left[\boldsymbol{F}_{\mathrm{uy}}^{\mathrm{T}} \boldsymbol{F}_{\mathrm{uy}} \right]^{-1} \sigma_{\varsigma}^2 \tag{10.1.24}$$

where σ_{ς}^2 is the variance of the source noise. (Assuming the stationary character of the source noise series $\varsigma[k]$, the probability limit is equal to the expected value, and therefore the expected value is used to compute $\boldsymbol{K}$.)

Unfortunately, the applied noise model $1/\mathcal{A}(z^{-1})$ is very specific and cannot be used for the general case. A more general form is $H_{\mathrm{n}}(z^{-1}) = \mathcal{C}(z^{-1})/\mathcal{A}(z^{-1})$ [4], when

$$\begin{aligned} y[k] &= \frac{\mathcal{B}(z^{-1}) z^{-d}}{\mathcal{A}(z^{-1})} u[k] + \frac{\mathcal{C}(z^{-1})}{\mathcal{A}(z^{-1})} \varsigma[k] = \frac{\mathcal{B}(z^{-1}) z^{-d}}{1 + \tilde{\mathcal{A}}(z^{-1})} u[k] + \frac{1 + \tilde{\mathcal{C}}(z^{-1})}{\mathcal{A}(z^{-1})} \varsigma[k] \\ &= G(z^{-1}) u[k] + H_{\mathrm{n}}(z^{-1}) \varsigma[k] \end{aligned} \tag{10.1.25}$$

This form does not lose the generality of the noise model $H_{\mathrm{n}}(z^{-1}) = \mathcal{C}(z^{-1})/\mathcal{D}(z^{-1})$, but because of the common denominators, it probably contains a large number of redundant parameters. The quasi-linearization by (10.1.13) can easily be done formally as

$$\begin{aligned} y[k] &= \mathcal{B}(z^{-1})z^{-d}u[k] - \tilde{\mathcal{A}}(z^{-1})y[k] + \tilde{\mathcal{C}}(z^{-1})\varsigma[k] + \varsigma[k] \\ &= b_1 u[k-d-1] + b_2 u[k-d-2] + \ldots + b_n u[k-d-n] \\ &\quad - a_1 y[k-1] - \ldots - a_n y[k-n] + c_1\varsigma[k-1] \\ &\quad + \ldots + c_n\varsigma[k-n] + \varsigma[k] \\ &= \boldsymbol{f}^{\mathrm{T}}[u, y, \varsigma, k]\boldsymbol{p}_{\mathrm{bac}} + \varsigma[k] \end{aligned} \tag{10.1.26}$$

Similarly to (10.1.22), the vector equation of the process for N measurements can be given as

$$\boldsymbol{y} = \boldsymbol{S}^d\boldsymbol{A}^{-1}\boldsymbol{B}\boldsymbol{u} + \boldsymbol{A}^{-1}\boldsymbol{C}\boldsymbol{\varsigma} = \boldsymbol{S}^d\boldsymbol{B}\boldsymbol{u} - \tilde{\boldsymbol{A}}\boldsymbol{y} + \tilde{\boldsymbol{C}}\boldsymbol{\varsigma} + \boldsymbol{\varsigma} = \boldsymbol{F}(\boldsymbol{u}, \boldsymbol{y}, \boldsymbol{\varsigma})\boldsymbol{p}_{\mathrm{bac}} + \boldsymbol{\varsigma} \tag{10.1.27}$$

where vector $\boldsymbol{\varsigma}$, similarly to $\boldsymbol{u}$ and $\boldsymbol{y}$, has N elements and contains the values of the source noise and

$$\boldsymbol{C} = \boldsymbol{I} + c_1\boldsymbol{S} + \ldots + c_n\boldsymbol{S}^n = \boldsymbol{I} + \tilde{\boldsymbol{C}} \tag{10.1.28}$$

and

$$\begin{aligned} \boldsymbol{F}_{\mathrm{uye}} &= \begin{bmatrix} \boldsymbol{f}^{\mathrm{T}}(u, y, \varsigma, 1) \\ \boldsymbol{f}^{\mathrm{T}}(u, y, \varsigma, 2) \\ \vdots \\ \boldsymbol{f}^{\mathrm{T}}(u, y, \varsigma, N) \end{bmatrix} \\ &= \left[\boldsymbol{S}^{d+1}\boldsymbol{u}, \ldots, \boldsymbol{S}^{d+n}\boldsymbol{u}, -\boldsymbol{S}^1\boldsymbol{y}, \ldots, -\boldsymbol{S}^n\boldsymbol{y}, -\boldsymbol{S}^1\boldsymbol{\varsigma}, \ldots, -\boldsymbol{S}^n\boldsymbol{\varsigma}\right] = \boldsymbol{F}(\boldsymbol{u}, \boldsymbol{y}, \boldsymbol{\varsigma}) \end{aligned} \tag{10.1.29}$$

The main disadvantage of this model form is that the past values $\varsigma[k]$ in the vector $\boldsymbol{f}[u, y, \varsigma, k]$ are not known. If an estimation $\hat{\boldsymbol{p}}_{\mathrm{bac}}$ of $\boldsymbol{p}_{\mathrm{bac}}$ is known then the estimation $\hat{\varsigma}[k]$ of the source noise can always be computed as

$$\hat{\varsigma}[k] = y[k] - \boldsymbol{f}^{\mathrm{T}}[u, y, \hat{\varsigma}, k]\hat{\boldsymbol{p}}_{\mathrm{bac}} \tag{10.1.30}$$

where now the computed (estimated) values $\hat{\varsigma}[k]$ are in $\boldsymbol{f}[u, y, \hat{\varsigma}, k]$ or in the vector form

$$\hat{\boldsymbol{\varsigma}} = \hat{\boldsymbol{A}}\boldsymbol{y} - \boldsymbol{S}^d\hat{\boldsymbol{B}}\boldsymbol{u} - \hat{\tilde{\boldsymbol{C}}}\hat{\boldsymbol{\varsigma}} = \boldsymbol{y} - \boldsymbol{F}(\boldsymbol{u}, \boldsymbol{y}, \hat{\boldsymbol{\varsigma}})\hat{\boldsymbol{p}}_{\mathrm{bac}} = \begin{bmatrix} \hat{\varsigma}[1] & \hat{\varsigma}[2] & \ldots & \hat{\varsigma}[N] \end{bmatrix}^{\mathrm{T}} \tag{10.1.31}$$

Here the vector $\hat{\varsigma}$ has N elements and contains the estimated values of the source noise. Construct the matrix of (10.1.29) but with the estimated source noise as

$$\boldsymbol{F}_{\mathrm{uy}\hat{\varsigma}} = \begin{bmatrix} \boldsymbol{f}^{\mathrm{T}}(u, y, \hat{\varsigma}, 1) \\ \boldsymbol{f}^{\mathrm{T}}(u, y, \hat{\varsigma}, 2) \\ \vdots \\ \boldsymbol{f}^{\mathrm{T}}(u, y, \hat{\varsigma}, N) \end{bmatrix}$$

$$= \left[\boldsymbol{S}^{d+1}\boldsymbol{u}, \ldots, \boldsymbol{S}^{d+n}\boldsymbol{u}, -\boldsymbol{S}^{1}\boldsymbol{y}, \ldots, -\boldsymbol{S}^{n}\boldsymbol{y}, -\boldsymbol{S}^{1}\hat{\varsigma}, \ldots, -\boldsymbol{S}^{n}\hat{\varsigma}\right] = \boldsymbol{F}(\boldsymbol{u}, \boldsymbol{y}, \hat{\varsigma}) \tag{10.1.32}$$

An LS estimation can now be obtained by applying (10.1.6) and (10.1.23) in the form

$$\hat{\boldsymbol{p}}_{\mathrm{bac}} = \left[\boldsymbol{F}_{\mathrm{uy}\hat{\varsigma}}^{\mathrm{T}} \boldsymbol{F}_{\mathrm{uy}\hat{\varsigma}}\right]^{-1} \boldsymbol{F}_{\mathrm{uy}\hat{\varsigma}}^{\mathrm{T}} \boldsymbol{y} = \left[\boldsymbol{F}^{\mathrm{T}}(\boldsymbol{u}, \boldsymbol{y}, \hat{\varsigma})\boldsymbol{F}(\boldsymbol{u}, \boldsymbol{y}, \hat{\varsigma})\right]^{-1} \boldsymbol{F}^{\mathrm{T}}(\boldsymbol{u}, \boldsymbol{y}, \hat{\varsigma})\boldsymbol{y} \tag{10.1.33}$$

Since the series $\hat{\varsigma}[k]$ $(k = 1,\ldots,N)$ can be computed only for a given $\hat{\boldsymbol{p}}_{\mathrm{bac}}$, so only iterative method can be realized, where the series $\hat{\varsigma}[k]$, that is, the vector $\hat{\varsigma}$ is computed again during each estimation step according to (10.1.31). The iteration is continued until the difference between the consecutive estimations of the parameter vectors becomes less than a given error. (This is known as a relaxation-type iteration.) The solution (10.1.24), which belongs to Equation (10.1.21), is called the *extended least squares* (*ELS*) method. Several versions of this method are known, which, together with the different noise models, account for the large number of methods described in the literature [33,147].

It is important to note that only a so-called quasi-linearity is ensured by (10.1.26) since $\hat{\varsigma}[k]$ needs to be computed by an a priori known parameter estimation. This complex relationship is already nonlinear, so only an iterative procedure can be applied.

It is not a definite requirement that the model equation is linear in parameters, though its advantage can be clearly seen, but then different versions of the *LS* and *ELS* methods have to be applied. In this case the estimation $\hat{\varsigma}[k]$ of the source noise in (10.1.25) is computed by a model nonlinear in parameters

$$\hat{\varsigma}[k] = \frac{\hat{\mathcal{A}}(z^{-1})y[k] - \hat{\mathcal{B}}(z^{-1})z^{-d}u[k]}{\hat{\mathcal{C}}(z^{-1})} = \frac{1}{\hat{H}_{\mathrm{n}}(z^{-1}, \hat{\boldsymbol{p}})}\left[y[k] - \hat{G}(z^{-1}, \hat{\boldsymbol{p}})u[k]\right]$$

$$\hat{\boldsymbol{p}} = \hat{\boldsymbol{p}}_{\mathrm{bac}} = \begin{bmatrix} \hat{b}_1 & \hat{b}_2 & \ldots & \hat{b}_n & \hat{a}_1 & \hat{a}_2 & \ldots & \hat{a}_n & \hat{c}_1 & \hat{c}_2 & \ldots & \hat{c}_n \end{bmatrix}^{\mathrm{T}} \tag{10.1.34}$$

which constitutes to the theoretically most accurate method, if the minimization of the loss function

$$V(\hat{\boldsymbol{p}}_{\mathrm{bac}}, N) = \frac{1}{2}\sum_{j=1}^{N}\hat{\varsigma}^2[j] = \frac{1}{2}\sum_{j=1}^{N}\left\{\frac{1}{\hat{H}_{\mathrm{n}}(z^{-1},\hat{\boldsymbol{p}})}\left[y[j] - \hat{G}(z^{-1},\hat{\boldsymbol{p}})u[j]\right]\right\}^2$$
$$= \frac{1}{2}\sum_{j=1}^{N}\left\{\frac{\hat{\mathcal{A}}(z^{-1})y[j] - \hat{\mathcal{B}}(z^{-1})z^{-d}u[j]}{\hat{\mathcal{C}}(z^{-1})}\right\}^2 \tag{10.1.35}$$

is performed directly in the parameter vector space of $\hat{\boldsymbol{p}}_{\mathrm{bac}}$, which can be considered as a general minimum-seeking problem. Even in the case of different noise models $\hat{H}_{\mathrm{n}}$ it is relatively simple to compute the first- and second-order derivates of $V(\hat{\boldsymbol{p}}_{\mathrm{bac}}, N)$ by the parameters, and effective extremum-seeking algorithms can be constructed. The method, directly minimizing (10.1.35), is called the *ML* method. All methods assume the zero mean value, whiteness, and normal distribution of $e[k]$ and provide asymptotically unbiased *MV* estimation. This method, developed by Åström and Bohlin [7], can be considered the first basic technique of identification methods for DT processes.

The best approximating order and structure of the DT process is basically determined by investigating the highest order term of the parameter vector $\hat{\boldsymbol{p}}_{\mathrm{bac}}$. A more effective method is suggested by Bokor and Keviczky [29], which estimates the residua of the process model. The simplest version of the method assumes first-order real and second-order complex conjugate poles, and thus

$$\hat{G}(z^{-1},\hat{\boldsymbol{p}}) = \left(\sum_{i=1}^{n_{\mathrm{v}}}\frac{\hat{r}_i z^{-1}}{1+\hat{q}_i z^{-1}} + \sum_{i=1}^{n_{\mathrm{c}}}\frac{\hat{s}_{1i}z^{-1} + \hat{s}_{2i}z^{-2}}{1+\hat{t}_{1i}z^{-1} + \hat{t}_{2i}z^{-1}}\right)z^{-d}$$
$$\hat{\boldsymbol{p}} = \hat{\boldsymbol{p}}_{\mathrm{rqst}} = \left[\hat{r}_1 \;\ldots\; \hat{r}_{n_{\mathrm{v}}} \;\; \hat{q}_1 \;\ldots\; \hat{q}_{n_{\mathrm{v}}} \;\; \hat{s}_{11} \;\ldots\; \hat{s}_{1n_{\mathrm{c}}} \;\; \hat{s}_{21} \;\ldots\; \hat{s}_{2n_{\mathrm{c}}} \ldots\right.$$
$$\left.\ldots\hat{t}_{11} \;\ldots\; \hat{t}_{1n_{\mathrm{c}}} \;\; \hat{t}_{21} \;\ldots\; \hat{t}_{2n_{\mathrm{c}}}\right]^{\mathrm{T}}; \quad n = n_{\mathrm{v}} + 2n_{\mathrm{c}} \tag{10.1.36}$$

where n_{v} and n_{c} mean the number of the real poles and complex conjugate pole pairs, respectively. The same structure needs to be used for the noise model. If the absolute value of any estimation $\hat{r}_i$ can be statistically neglected then that pole is not included in the model. Similarly, if the absolute value of both estimations $\hat{s}_{1i}$ and $\hat{s}_{2i}$ can be statistically neglected then that complex conjugate pole pair is not included in the model. The model (10.1.36) is nonlinear in parameters so the same statements are valid as in (10.1.35). This model is called the *elementary subsystem* method [29].

In many cases there is no need to identify the noise model and, moreover, this case can be considered general. For this case the *instrumental variables (IVs)* method provides a good solution [147,163], according to which the estimation

$$\hat{\boldsymbol{p}}_{\mathrm{ba}} = \left[\mathbf{G}_{\mathrm{uv}}^{\mathrm{T}} \boldsymbol{F}_{\mathrm{uy}}\right]^{-1} \mathbf{G}_{\mathrm{uv}}^{\mathrm{T}} \boldsymbol{y} = \left[\mathbf{G}^{\mathrm{T}}(\boldsymbol{u}, \boldsymbol{v}) \boldsymbol{F}(\boldsymbol{u}, \boldsymbol{y})\right]^{-1} \mathbf{G}^{\mathrm{T}}(\boldsymbol{u}, \boldsymbol{v}) \boldsymbol{y} \qquad (10.1.37)$$

gives an asymptotically unbiased solution, if the conditions

$$\operatorname*{plim}_{N\to\infty}\left\{\frac{1}{N}\mathbf{G}_N^{\mathrm{T}}\boldsymbol{F}_N\right\} = \boldsymbol{Z}_{\mathrm{IV}} < \infty \quad \text{and} \quad \operatorname*{plim}_{N\to\infty}\left\{\frac{1}{N}\mathbf{G}_N^{\mathrm{T}}\boldsymbol{r}_N\right\} = \mathbf{0} \qquad (10.1.38)$$

are fulfilled. For simplicity the index N refers to the number of samples. Here

$$\boldsymbol{r}_N = \boldsymbol{y}_N - \boldsymbol{F}_N \boldsymbol{p} = \boldsymbol{y} - \boldsymbol{F}(\boldsymbol{u}, \boldsymbol{y}) \boldsymbol{p}_{\mathrm{ba}} \qquad (10.1.39)$$

is the so-called equation error. The second condition in (10.1.38) clearly shows that the asymptotical unbiasedness depends on whether or not the elements of $\mathbf{G}$ are correlated with the equation error. This is the key element of the method: the auxiliary variable $\boldsymbol{v}$ needs to be found which corresponds to the condition.

Equation (10.1.16) of the process can also be written in the following form which is linear in parameters

$$y[k] = \boldsymbol{f}^{\mathrm{T}}(u, y, k) \boldsymbol{p}_{\mathrm{ba}} + \mathcal{A}\left(z^{-1}\right) y_{\mathrm{n}}[k] \qquad (10.1.40)$$

where the equation error now is $\mathcal{A}(z^{-1}) y_{\mathrm{n}}[k]$, or for N samples

$$\boldsymbol{y} = \boldsymbol{S}^d \boldsymbol{B} \boldsymbol{u} - \tilde{\boldsymbol{A}} \boldsymbol{y} + \boldsymbol{A}^{-1} \boldsymbol{y}_{\mathrm{n}} = \boldsymbol{F}(\boldsymbol{u}, \boldsymbol{y}) \boldsymbol{p}_{\mathrm{ba}} + \boldsymbol{A}^{-1} \boldsymbol{y}_{\mathrm{n}} \qquad (10.1.41)$$

With this method, originally suggested in [124], the noise-free output of the process is chosen as auxiliary variable, so in Equation (10.1.37) of the parameter estimation

$$\mathbf{G}_{\mathrm{uv}} = \begin{bmatrix} \boldsymbol{g}^{\mathrm{T}}(u, v, 1) \\ \boldsymbol{g}^{\mathrm{T}}(u, v, 2) \\ \vdots \\ \boldsymbol{g}^{\mathrm{T}}(u, v, N) \end{bmatrix} = \left[\boldsymbol{S}^{d+1} \boldsymbol{u}, \ldots, \boldsymbol{S}^{d+n} \boldsymbol{u}, -\boldsymbol{S}^1 \boldsymbol{v}, \ldots, -\boldsymbol{S}^n \boldsymbol{v}\right] = \mathbf{G}(\boldsymbol{u}, \boldsymbol{v}) \qquad (10.1.42)$$

where

$$v[k] = \frac{\mathcal{B}(z^{-1})z^{-d}}{\mathcal{A}(z^{-1})} u[k] = \boldsymbol{f}^{\mathrm{T}}[u, v, k]\boldsymbol{p}_{\mathrm{ba}} \tag{10.1.43}$$

or for N samples

$$\boldsymbol{v} = \boldsymbol{A}^{-1}\boldsymbol{B}\boldsymbol{u} = \boldsymbol{B}\boldsymbol{u} - \tilde{\boldsymbol{A}}\boldsymbol{v} = \mathbf{G}(\boldsymbol{u}, \boldsymbol{v})\boldsymbol{p}_{\mathrm{ba}} \tag{10.1.44}$$

The noise-free output cannot be measured and can only be estimated

$$\hat{v}[k] = \frac{\mathcal{B}(z^{-1})z^{-d}}{\mathcal{A}(z^{-1})} u[k] = \boldsymbol{f}^{\mathrm{T}}[u, \hat{v}, k]\hat{\boldsymbol{p}}_{\mathrm{ba}} \tag{10.1.45}$$

or in vector form

$$\hat{\boldsymbol{v}} = \boldsymbol{A}^{-1}\boldsymbol{B}\boldsymbol{u} = \boldsymbol{B}\boldsymbol{u} - \tilde{\boldsymbol{A}}\hat{\boldsymbol{v}} = \mathbf{G}(\boldsymbol{u}, \hat{\boldsymbol{v}})\hat{\boldsymbol{p}}_{\mathrm{ba}} \tag{10.1.46}$$

Since the equation error $\mathcal{A}(z^{-1})y_{\mathrm{n}}[k]$ depends only on the uncorrelated source noise $\varsigma[k]$, the elements of $\mathbf{G}(\boldsymbol{u}, \hat{\boldsymbol{v}}) = \mathbf{G}(\boldsymbol{u}, \hat{\boldsymbol{A}}^{-1}\hat{\boldsymbol{B}}\boldsymbol{u}) = \mathbf{G}(\boldsymbol{u}, \hat{\boldsymbol{p}}_{\mathrm{ba}})$ are uncorrelated with the equation error, as those are functions of only $u[k]$ uncorrelated with $e[k]$. As a consequence of Equations (10.1.45) and (10.1.46) here only the iterative method can be applied.

Identification of Multivariable Linear Dynamic Processes

In Section 7, it was shown for the left-*MFD* (*matrix fractional description*) representation (7.83) that it can be rewritten in the form (7.85), which is linear in parameter matrices. This makes it possible to extend the identification methods discussed above to *MIMO* processes. The reasonable extension of (10.1.25) is

$$\boldsymbol{y}[k] = z^{-d}\boldsymbol{\mathcal{A}}^{-1}(z^{-1})\boldsymbol{\mathcal{B}}(z^{-1})\boldsymbol{u}[k] + \boldsymbol{\mathcal{A}}^{-1}(z^{-1})\boldsymbol{\mathcal{C}}(z^{-1})\boldsymbol{\varsigma}[k] \tag{10.1.47}$$

where the process parameters can be given by the matrix polynomials

$$\begin{aligned}
\boldsymbol{\mathcal{A}}(z^{-1}) &= \mathbf{I} + \mathbf{A}_1 z^{-1} + \mathbf{A}_2 z^{-2} + \ldots + \mathbf{A}_n z^{-n} = \mathbf{I} + \tilde{\boldsymbol{\mathcal{A}}}(z^{-1}) \\
\boldsymbol{\mathcal{B}}(z^{-1}) &= \mathbf{B}_1 z^{-1} + \mathbf{B}_2 z^{-2} + \ldots + \mathbf{B}_m z^{-m} \\
\boldsymbol{\mathcal{C}}(z^{-1}) &= \mathbf{I} + \mathbf{C}_1 z^{-1} + \mathbf{C}_2 z^{-2} + \ldots + \mathbf{C}_n z^{-n}
\end{aligned} \tag{10.1.48}$$

Here the input $\boldsymbol{u}[k]$ is assumed to be noise-free, but the output is measured with an additive measurement error. The output noise is the vector $\boldsymbol{\varsigma}[k]$ which derives from an independent, zero mean value, white source noise given by the noise model $\mathcal{A}^{-1}(z^{-1})\mathcal{C}(z^{-1})$. In this case, process identification requires the estimation of two models: the process and the noise models. Observe that here all channels have a common time-delay z^{-d}. Since this is an unreal assumption, certain elements of matrix $\mathbf{B}_i$ become zero for a small value of i. It needs to be underlined that the "order" of these models, that is, the values $n = n_{\mathrm{d}}$ and $m = n_{\mathrm{n}}$, has no connection with the classical orders but derive from the left-*MFD* representations (see (7.85)).

It is clear that the *MIMO* system (10.1.47) is linear in parameter matrices, since

$$
\begin{aligned}
\boldsymbol{y}[k] &= \sum_{i=1}^{m} \mathbf{B}_i \boldsymbol{u}[k-d-i] - \sum_{i=1}^{n} \mathbf{A}_i \boldsymbol{y}[k-i] + \sum_{i=1}^{n} \mathbf{C}_i \boldsymbol{\varsigma}[k-i] + \boldsymbol{\varsigma}[k] \\
&= \boldsymbol{P}_{\mathrm{BA}} \boldsymbol{f}[\boldsymbol{u}, \boldsymbol{y}, k] + \mathcal{C}\left(z^{-1}\right)\boldsymbol{\varsigma}[k] = \boldsymbol{P}_{\mathrm{BAC}} \boldsymbol{f}[\boldsymbol{u}, \boldsymbol{y}, \boldsymbol{\varsigma}, k] + \boldsymbol{\varsigma}[k]
\end{aligned}
\tag{10.1.49}
$$

Here

$$
\begin{aligned}
\boldsymbol{P}_{\mathrm{BA}} &= [\mathbf{B}_1, \mathbf{B}_2, \ldots, \mathbf{B}_m, \mathbf{A}_1, \ldots, \mathbf{A}_n] \\
\boldsymbol{P}_{\mathrm{BAC}} &= [\mathbf{B}_1, \mathbf{B}_2, \ldots, \mathbf{B}_m, \mathbf{A}_1, \ldots, \mathbf{A}_n, \mathbf{C}_1, \ldots, \mathbf{C}_n]
\end{aligned}
\tag{10.1.50}
$$

or

$$
\begin{aligned}
\boldsymbol{f}^{\mathrm{T}}[\boldsymbol{u}, \boldsymbol{y}, k] &= \left[\boldsymbol{u}^{\mathrm{T}}[k-d-1], \ldots, \boldsymbol{u}^{\mathrm{T}}[k-d-m], \boldsymbol{y}^{\mathrm{T}}[k-1], \ldots, \boldsymbol{y}^{\mathrm{T}}[k-n]\right] \\
\boldsymbol{f}^{\mathrm{T}}[\boldsymbol{u}, \boldsymbol{y}, \boldsymbol{\varsigma}, k] &= \left[\boldsymbol{u}^{\mathrm{T}}[k-d-1], \ldots, \boldsymbol{u}^{\mathrm{T}}[k-d-m], \boldsymbol{y}^{\mathrm{T}}[k-1], \ldots, \boldsymbol{y}^{\mathrm{T}}[k-n],\right. \\
&\qquad \left.\boldsymbol{\varsigma}^{\mathrm{T}}[k-1], \ldots, \boldsymbol{\varsigma}^{\mathrm{T}}[k-n]\right]
\end{aligned}
\tag{10.1.51}
$$

From Equations (10.1.47) and (10.1.49), we obtain an uncorrelated equation error term under the condition $\mathcal{C}(z^{-1}) = \mathbf{I}$, so this affords the possibility of using the simple *LS* method.

If the parameter estimation is based on N samples of the value pairs $\boldsymbol{u}[k]$, $\boldsymbol{y}[k]$ ($k = 1, 2, \ldots, N$) then here it is also reasonable to use the joint system equation applied for *SISO* systems referring to N samples. Introduce the following notations

$$
\begin{aligned}
\boldsymbol{U} &= [\boldsymbol{u}[1], \ldots \boldsymbol{u}[N]] \\
\boldsymbol{Y} &= [\boldsymbol{y}[1], \ldots, \boldsymbol{y}[N]] \\
\boldsymbol{E} &= [\boldsymbol{e}[1], \ldots, \boldsymbol{e}[N]] \\
\boldsymbol{F} &= [\boldsymbol{f}[1], \ldots, \boldsymbol{f}[N]]
\end{aligned}
\tag{10.1.52}
$$

Applying the joint matrix equation we obtain

$$\boldsymbol{Y} = \boldsymbol{P}_{\mathrm{BA}}\boldsymbol{F} + \boldsymbol{E} \tag{10.1.53}$$

Equation (10.1.53) can be rearranged by means of operation vec(…), that is, vector operation (see Appendix A.1.3)

$$\begin{aligned}\operatorname{vec}(\boldsymbol{Y}) &= \operatorname{vec}(\boldsymbol{P}_{\mathrm{BA}}\boldsymbol{F} + \boldsymbol{E}) = \operatorname{vec}(\boldsymbol{P}_{\mathrm{BA}}\boldsymbol{F}) + \operatorname{vec}(\boldsymbol{E}) \\ &= \left(\boldsymbol{F}^{\mathrm{T}} \otimes \boldsymbol{I}_p\right)\operatorname{vec}(\boldsymbol{P}_{\mathrm{BA}}) + \operatorname{vec}(\boldsymbol{E})\end{aligned} \tag{10.1.54}$$

where the identity (A.1.40) is used and $\boldsymbol{I}_p$ is a $p \times p$ unity matrix, where p is the number of the outputs. Here $\otimes$ denotes the Kronecker matrix product.

The value pairs $\boldsymbol{u}[k]$, $\boldsymbol{y}[k]$ can be stored in other forms. Introduce the vectors

$$\begin{aligned}\boldsymbol{u}_{\mathrm{M}} &= \left[\boldsymbol{u}^{\mathrm{T}}[1], \ldots, \boldsymbol{u}^{\mathrm{T}}[N]\right] \\ \boldsymbol{y}_{\mathrm{M}} &= \left[\boldsymbol{y}^{\mathrm{T}}[1], \ldots, \boldsymbol{y}^{\mathrm{T}}[N]\right] \\ \boldsymbol{\varsigma}_{\mathrm{M}} &= \left[\boldsymbol{\varsigma}^{\mathrm{T}}[1], \ldots, \boldsymbol{\varsigma}^{\mathrm{T}}[N]\right]\end{aligned} \tag{10.1.55}$$

By taking the identity of the Kronecker products and the meaning of the shift matrices into account we see that the form of the *MIMO* system Equation (10.1.47) analogous with (10.1.19) for N samples is

$$\begin{aligned}\boldsymbol{y}_{\mathrm{M}} &= \left[\sum_{i=0}^{n} \boldsymbol{S}_N^i(1) \otimes \mathbf{A}_i\right]^{-1} \left[\sum_{i=1}^{m} \boldsymbol{S}_N^{d+i}(1) \otimes \mathbf{B}_i\right] \boldsymbol{u}_{\mathrm{M}} \\ &\quad + \left[\sum_{i=0}^{n} \boldsymbol{S}_N^i(1) \otimes \mathbf{A}_i\right]^{-1} \left[\sum_{i=0}^{n} \boldsymbol{S}_N^i(1) \otimes \mathbf{C}_i\right] \boldsymbol{e}_{\mathrm{M}}\end{aligned} \tag{10.1.56}$$

where $\boldsymbol{S}_N^i(1)$ is an $N \times N$ step (shift) matrix and obviously $\mathbf{C}_0 = \mathbf{I}$. For simplicity the initial conditions are neglected here.

Comparing the structures (10.1.52) and (10.1.54) we obtain

$$\begin{aligned}\boldsymbol{u}_{\mathrm{M}} &= \operatorname{vec}[\boldsymbol{U}] \\ \boldsymbol{y}_{\mathrm{M}} &= \operatorname{vec}[\boldsymbol{Y}] \\ \boldsymbol{\varsigma}_{\mathrm{M}} &= \operatorname{vec}[\boldsymbol{E}]\end{aligned} \tag{10.1.57}$$

and with these notations Equation (10.1.54) becomes

$$\boldsymbol{y}_{\mathrm{M}} = \mathbf{G}(\boldsymbol{u}_{\mathrm{M}}, \boldsymbol{y}_{\mathrm{M}})\boldsymbol{p}_{\mathrm{BA}} + \boldsymbol{\varsigma}_{\mathrm{M}} = \left(\boldsymbol{F}^{\mathrm{T}} \otimes \boldsymbol{I}_p\right)\boldsymbol{p}_{\mathrm{BA}} + \boldsymbol{\varsigma}_{\mathrm{M}} \tag{10.1.58}$$

where

$$\boldsymbol{p}_{\mathrm{BA}} = \operatorname{vec}(\boldsymbol{P}_{\mathrm{BA}}) \tag{10.1.59}$$

It can be checked that we obtain also the system Equation (10.1.58) by detailed analysis of (10.1.56) under the assumption $\boldsymbol{C}(z^{-1}) = \mathbf{I}$, since because of the above

$$\begin{aligned} \boldsymbol{y}_{\mathrm{M}} &= \left[\sum_{i=1}^{m} \boldsymbol{S}_N^{d+i}(1) \otimes \mathbf{B}_i\right] \boldsymbol{u}_{\mathrm{M}} - \left[\sum_{i=0}^{n} \boldsymbol{S}_N^{i}(1) \otimes \mathbf{A}_i\right] + \boldsymbol{\varsigma}_{\mathrm{M}} \\ &= \mathbf{G}\left(\boldsymbol{u}_{\mathrm{M}}, \boldsymbol{y}_{\mathrm{M}}\right) \boldsymbol{p}_{\mathrm{BA}} + \boldsymbol{\varsigma}_{\mathrm{M}} \end{aligned} \tag{10.1.60}$$

On the basis of these representations and starting from the different approaches in A.6.2, the *LS* estimation of the parameter matrices of the system can be obtained in the form

$$\hat{\boldsymbol{P}}_{\mathrm{BA}} = \boldsymbol{Y}\boldsymbol{F}^{\mathrm{T}}\left(\boldsymbol{F}\boldsymbol{F}^{\mathrm{T}}\right)^{-1} \tag{10.1.61}$$

or

$$\hat{\boldsymbol{p}}_{\mathrm{BA}} = \left[\left(\boldsymbol{F}\boldsymbol{F}^{\mathrm{T}}\right)^{-1}\boldsymbol{F} \otimes \boldsymbol{I}_p\right]\boldsymbol{y}_{\mathrm{M}} \tag{10.1.62}$$

Ljung's Prediction Error Method

The theoretical foundation of modern identification methods is the prediction error method of Ljung [123]. This approach estimates the output of the process models (10.1.15) and (10.1.34) in the knowledge of the parameter $\hat{\boldsymbol{p}}$ of the model $(\hat{G}, \hat{H}_{\mathrm{n}})$ according to the relationship

$$\hat{y}_{\mathrm{pr}}[k] = W_{\mathrm{y}}\left(z^{-1}, \hat{\boldsymbol{p}}\right) y[k] + W_{\mathrm{u}}\left(z^{-1}, \hat{\boldsymbol{p}}\right) u[k] \tag{10.1.63}$$

where the components of the predictor are

$$\begin{aligned} W_{\mathrm{y}}\left(z^{-1}, \hat{G}\right) &= \left[1 - H_{\mathrm{n}}^{-1}\left(z^{-1}, \hat{\boldsymbol{p}}\right)\right] \quad \text{and} \\ W_{\mathrm{u}}\left(z^{-1}, \hat{\boldsymbol{p}}\right) &= H_{\mathrm{n}}^{-1}\left(z^{-1}, \hat{\boldsymbol{p}}\right) G\left(z^{-1}, \hat{\boldsymbol{p}}\right) \end{aligned} \tag{10.1.64}$$

These components can be easily determined for any of the above-discussed dynamic process models in time-delay-free $(d = \hat{d} = 0)$ cases. For time-delay cases predictors of type $\hat{y}_{\mathrm{pr}}[k + d|k]$ need to be computed, got acquainted with earlier at the predictive regulators. Observe that in these cases it is also possible to use the predictor (10.1.63), but the computation of W_{y} and W_{u} becomes more complicated.

Compute the error of the predictor (10.1.63)

$$\varepsilon_{\mathrm{pr}}[k] = y[k] - \hat{y}_{\mathrm{pr}}[k] = \frac{y[k] - G(z^{-1}, \hat{\boldsymbol{p}})u[k]}{H_{\mathrm{n}}(z^{-1}, \hat{\boldsymbol{p}})} = \frac{1}{\hat{H}_{\mathrm{n}}}\left[\Delta u[k] + \Delta H_{\mathrm{n}}\varsigma[k]\right]$$
$$= \frac{1}{\hat{H}_{\mathrm{n}}}\left[\left(G - \hat{G}\right)u[k] - \left(H_{\mathrm{n}} - \hat{H}_{\mathrm{n}}\right)\varsigma[k]\right] + \varsigma[k] = \hat{\varsigma}[k] \tag{10.1.65}$$

which is clearly the estimation $\hat{\varsigma}[k]$ of the source noise $\varsigma[k]$, and thus $\varepsilon_{\mathrm{pr}}[k]_{\Delta \to 0} = \varsigma[k]$. Here the notations

$$G = G(z^{-1}, \boldsymbol{p}); \quad \hat{G} = G(z^{-1}, \hat{\boldsymbol{p}}); \quad H_{\mathrm{n}} = H_{\mathrm{n}}(z^{-1}, \boldsymbol{p});$$
$$\hat{H}_{\mathrm{n}} = H_{\mathrm{n}}(z^{-1}, \hat{\boldsymbol{p}}) \tag{10.1.66}$$

and

$$\Delta = G - \hat{G} = G(z^{-1}, \boldsymbol{p}) - G(z^{-1}, \hat{\boldsymbol{p}});$$
$$\Delta H_{\mathrm{n}} = H_{\mathrm{n}} - \hat{H}_{\mathrm{n}} = H_{\mathrm{n}}(z^{-1}, \boldsymbol{p}) - H_{\mathrm{n}}(z^{-1}, \hat{\boldsymbol{p}}) \tag{10.1.67}$$

are used. The expression (10.1.65) clearly shows that the identification method, minimizing the variance of $\varepsilon_{\mathrm{pr}}$, converges to the points $\Delta = 0$ and $\Delta H_{\mathrm{n}} = 0$ to provide unbiased estimation.

If the first term of (10.1.65) and $\varsigma[k]$ are independent, then the square of the spectrum of the signal $\varepsilon_{\mathrm{pr}}[k]$ is obtained as

$$\Phi_{\varepsilon\varepsilon}(\omega, \hat{\boldsymbol{p}}) = \frac{\left|G + B(\hat{\boldsymbol{p}}) - \hat{G}\right|^2 \Phi_{\mathrm{uu}}}{\left|\hat{H}_{\mathrm{n}}\right|} + \frac{\left|H - \hat{H}_{\mathrm{n}}\right|^2 \left(\sigma_{\varsigma}^2 - \frac{\left|\Phi_{u\varsigma}\right|^2}{\Phi_{\mathrm{uu}}}\right)}{\left|\hat{H}_{\mathrm{n}}\right|} + \sigma_{\varsigma}^2 \tag{10.1.68}$$

It is clear from the expression that the seeking of the minimum means the convergence of $\hat{G}$ to $G + B(\hat{\boldsymbol{p}})$. Therefore, the term $B(\hat{\boldsymbol{p}})$ introduced here is called bias (10.1), that is,

$$B(\hat{\boldsymbol{p}}) = B(e^{j\omega}, \hat{\boldsymbol{p}}) = \frac{[H(e^{j\omega}, \boldsymbol{p}) - H(e^{j\omega}, \hat{\boldsymbol{p}})]\Phi_{u\varsigma}(\omega)}{\left|H(e^{j\omega}, \hat{\boldsymbol{p}})\right|^2} = \frac{\left[H - \hat{H}\right]\Phi_{u\varsigma}}{\left|\hat{H}\right|^2} \tag{10.1.69}$$

The bias $B(e^{j\omega}, \hat{\boldsymbol{p}})$ is a complex frequency function which can be zero in the case of $\hat{H} \to H$, that is, when $\Delta H = 0$. Consequently, any identification method, which does not estimate the noise model without bias will also cause bias in the estimation of the process parameters.

The covariance matrix of the conventional *LS* estimations can be computed by (10.1.24). This method, however, does not give any information on the behavior of the estimated model in the frequency region, though this could obviously have been computed for both $\hat{G}$ and $\hat{H}_{\mathrm{n}}$ in the knowledge of $\hat{\boldsymbol{p}}$. This relationship is given by the asymptotic covariance theorem of Ljung [123], according to which the covariance matrix of the n-order model $(\hat{G}, \hat{H}_{\mathrm{n}})$ obtained from N samples is

$$\mathrm{Cov}\begin{bmatrix}\hat{G}_N\left(e^{j\omega}\right)\\ \hat{H}_{\mathrm{n}}^N\left(e^{j\omega}\right)\end{bmatrix} \approx \frac{n}{N}\Phi_{\mathrm{vv}}(\omega)\begin{bmatrix}\Phi_{\mathrm{uu}}(\omega) & \Phi_{\mathrm{u}\varsigma}(-\omega)\\ \Phi_{\varsigma\mathrm{u}}(\omega) & \sigma_\varsigma^2\end{bmatrix}^{-1}$$
$$= \frac{n}{N}\frac{\Phi_{\mathrm{vv}}(\omega)}{\sigma_\varsigma^2\Phi_{\mathrm{uu}}(\omega)}\begin{bmatrix}\sigma_\varsigma^2 & 0\\ 0 & \Phi_{\mathrm{uu}}(\omega)\end{bmatrix} \tag{10.1.70}$$

Here Φ_{vv} and Φ_{uu} are the squares of the spectra of the noise-free output and the input, $\Phi_{\mathrm{u}\varsigma}$ and $\Phi_{\varsigma\mathrm{u}}$ are the cross-spectra of the input and the source noise, respectively. In typical open-loop identification, $u[k]$ and $\varsigma[k]$ are uncorrelated, and therefore $\Phi_{\mathrm{u}\varsigma} = \Phi_{\varsigma\mathrm{u}} \equiv 0$. Since $\varsigma[k]$ is assumed to be white noise, $\Phi_{\varsigma\varsigma}(\omega) = \sigma_\varsigma^2$. Observe that the result depends on the order of the model not on the number of parameters!

In this case, $\hat{G}_N$ and $\hat{H}_{\mathrm{n}}^N$ are asymptotically uncorrelated even when the predictor terms have common parameters. The independent covariance is

$$\mathrm{Cov}\left[\hat{G}_N\left(e^{j\omega}\right)\right] = \frac{n}{N}\frac{\Phi_{\mathrm{vv}}(\omega)}{\Phi_{\mathrm{uu}}(\omega)} \quad \text{and}$$
$$\mathrm{Cov}\left[\hat{H}_{\mathrm{n}}^N\left(e^{j\omega}\right)\right] = \frac{n}{N}\frac{\Phi_{\mathrm{vv}}(\omega)}{\sigma_\varsigma^2} = \frac{n}{N}\left|H_{\mathrm{n}}\left(e^{j\omega}\right)\right|^2 \tag{10.1.71}$$

The identity of primary importance, valid for stationary signals,

$$E\left\{\frac{1}{2}\boldsymbol{\varepsilon}_{\mathrm{pr}}^2[k,\hat{\boldsymbol{p}}]\right\} = \frac{1}{4\pi}\int_{-\pi}^{\pi}\Phi_{\varepsilon\varepsilon}(\omega,\hat{\boldsymbol{p}})\mathrm{d}\omega \tag{10.1.72}$$

is applied in the derivation of both expressions (10.1.68) and (10.1.70), which shows how to compute the square mean value in the time and frequency domain.

10.2 RECURSIVE PROCESS IDENTIFICATION METHODS

Those identification methods, which use N data pairs of the collected input and output signals, are called "off-line" or "batch" methods. The methods discussed above belong to this class.

There are also measurement situations, however, when the model obtained from the above methods is modified after obtaining a new data pair. These methods are called "online" or "recursive" identification methods [63,122].

Consider first the recursive version of the LS method. To do this, assume that N data pairs have been processed and the following LS estimation is available

$$\hat{\boldsymbol{p}}[N] = \left\{\boldsymbol{F}^{\mathrm{T}}[N]\boldsymbol{F}[N]\right\}^{-1}\boldsymbol{F}^{\mathrm{T}}[N]\boldsymbol{y}_N \tag{10.2.1}$$

If we want to modify the estimation (10.2.1) by the data pair $u[N+1]$ and $y[N+1]$ measured in the $[N+1]$-th time point then the new estimation can be computed by the following recursive expressions (see A.6.4 in Appendix A.6.)

$$\hat{\boldsymbol{p}}[N+1] = \hat{\boldsymbol{p}}[N] + \boldsymbol{R}[N+1]\boldsymbol{f}[N+1]\left\{y[N+1] - \boldsymbol{f}^{\mathrm{T}}[N+1]\hat{\boldsymbol{p}}[N]\right\} \tag{10.2.2}$$

and

$$\boldsymbol{R}[N+1] = \boldsymbol{R}[N] - \frac{\boldsymbol{R}[N]\boldsymbol{f}[N+1]\boldsymbol{f}^{\mathrm{T}}[N+1]\boldsymbol{R}[N]}{1+\boldsymbol{f}^{\mathrm{T}}[N+1]\boldsymbol{R}[N]\boldsymbol{f}[N+1]} \tag{10.2.3}$$

Here $\boldsymbol{f}[N+1]$ is general and independent of the model type, that is, regardless of whether the method is applied in the static or dynamic process model. $\boldsymbol{R}[N]$ is the so-called convergence matrix

$$\boldsymbol{R}[N] = \left\{\sum_{j=1}^{N}\boldsymbol{f}[j]\boldsymbol{f}^{\mathrm{T}}[j]\right\}^{-1} = \left\{\boldsymbol{F}^{\mathrm{T}}[N]\boldsymbol{F}[N]\right\}^{-1} \tag{10.2.4}$$

It is clear that Equation (10.2.2) of the recursive parameter estimation can also be written in the form

$$\hat{\boldsymbol{p}}[N+1] = \hat{\boldsymbol{p}}[N] + \frac{\boldsymbol{R}[N]\boldsymbol{f}[N+1]}{1+\boldsymbol{f}^{\mathrm{T}}[N+1]\boldsymbol{R}[N]\boldsymbol{f}[N+1]} \times\left\{y[N+1] - \boldsymbol{f}^{\mathrm{T}}[N+1]\hat{\boldsymbol{p}}[N]\right\} \tag{10.2.5}$$

(see the derivation of (A.6.31)).

The equation pair (10.2.2) and (10.2.3) belongs to the family of the so-called learning, adaptive estimation algorithms, which are included in the canonical equation of the general *stochastic approximation* (*SA*) [150,151]:

$$\hat{\boldsymbol{p}}[k+1] = \hat{\boldsymbol{p}}[k] + \boldsymbol{R}[k+1]\frac{\mathrm{d}V(\hat{\boldsymbol{p}},k)}{\mathrm{d}\hat{\boldsymbol{p}}} \tag{10.2.6}$$

These *SA* algorithms differ in how they choose the convergence matrix $\boldsymbol{R}[k]$ and the gradient. Here the most well-known method is discussed.

If the parameters of the process are changing in time then it might be necessary to forget the validity of the former model and to take the new measurements into account with emphasized significance. To solve this "forgetting" problem, it is assumed that the past is forgotten by applying the following matrix

$$\boldsymbol{F}[N+1] = \begin{bmatrix} \lambda\boldsymbol{F}[N] \\ \boldsymbol{f}^{\mathrm{T}}[N+1] \end{bmatrix} \quad \text{instead of} \quad \boldsymbol{F}[N+1] = \begin{bmatrix} \boldsymbol{F}[N] \\ \boldsymbol{f}^{\mathrm{T}}[N+1] \end{bmatrix} \tag{10.2.7}$$

where $0 \le \lambda \le 1$ is the forgetting factor [122]. If $\lambda =$ constant, the convergence matrix can be computed by the equation

$$\boldsymbol{R}[N+1] = \frac{1}{\lambda^2}\left\{\boldsymbol{R}[N] - \frac{\boldsymbol{R}[N]\boldsymbol{f}[N+1]\boldsymbol{f}^{\mathrm{T}}[N+1]\boldsymbol{R}[N]}{\lambda^2 + \boldsymbol{f}^{\mathrm{T}}[N+1]\boldsymbol{R}[N]\boldsymbol{f}[N+1]}\right\} \tag{10.2.8}$$

instead of (10.2.3) (see (A.6.34)).

The constant forgetting factor, however, can cause trouble, if the new measurement does not contain sufficient new information since this algorithm forgets the old information exponentially, and $\boldsymbol{R}[N]$ can become singular. Thus an adequate forgetting strategy is the most critical part of the adaptive estimation method.

Note that Equations (10.2.2), (10.2.3), and (10.2.8) represent the so-called naive programming forms which illustrate the method very simply albeit their numerical behavior is not satisfactory. They are only for demonstration and simulation purposes. In practice, the canonical, diagonal form of $\boldsymbol{R}[N]$ and its recursive forms are used (as *UD* or *UDV* factorization—the sequential algorithms of the Gram–Schmidt matrix factorization [22,23]), and this solution is the best from the numerical perspective. One of the best methods in practice is the Givens transformation [47].

It was shown in Section 10.1 that the *LS* estimation of the parameter matrix $\boldsymbol{P}_{\mathrm{BA}}$ of the *MIMO* system equation according to (10.1.60) is

$$\hat{\boldsymbol{P}} = \boldsymbol{Y}\boldsymbol{F}^{\mathrm{T}}\left(\boldsymbol{F}\boldsymbol{F}^{\mathrm{T}}\right)^{-1} \tag{10.2.9}$$

The recursive relationships can be obtained in the simplest way by using Equation (10.2.9) written for N-th and $(N+1)$-th samples like

$$\begin{aligned} \hat{\boldsymbol{P}}[N] &= \boldsymbol{Y}_N\boldsymbol{F}^{\mathrm{T}}[N]\left\{\boldsymbol{F}^{\mathrm{T}}[N]\boldsymbol{F}[N]\right\}^{-1} \\ \hat{\boldsymbol{P}}[N+1] &= \boldsymbol{Y}_{N+1}\boldsymbol{F}^{\mathrm{T}}[N+1]\left\{\boldsymbol{F}^{\mathrm{T}}[N+1]\boldsymbol{F}[N+1]\right\}^{-1} \end{aligned} \tag{10.2.10}$$

Furthermore, take into account that now

$$\boldsymbol{F}[N+1] = \{\boldsymbol{F}[N], \boldsymbol{f}[N+1]\} \tag{10.2.11}$$

and

$$\boldsymbol{Y}[N+1] = \{\boldsymbol{Y}[N], \boldsymbol{y}[N+1]\} \tag{10.2.12}$$

According to (10.2.4), introduce the dyadic equation

$$\begin{aligned} \{\boldsymbol{R}[N+1]\}^{-1} &= \boldsymbol{F}[N+1]\boldsymbol{F}^{\mathrm{T}}[N+1] = \boldsymbol{F}[N]\boldsymbol{F}^{\mathrm{T}}[N] + \boldsymbol{f}[N+1]\boldsymbol{f}^{\mathrm{T}}[N+1] \\ &= \{\boldsymbol{R}[N]\}^{-1} + \boldsymbol{f}[N+1]\boldsymbol{f}^{\mathrm{T}}[N+1] \end{aligned} \tag{10.2.13}$$

and so the recursive equation of $\boldsymbol{R}[N]$ is completely the same as (10.2.3) derived in A.6.4 of Appendix A.6.

Based on (10.2.12), the expression can be written as

$$\boldsymbol{Y}[N+1]\boldsymbol{F}^{\mathrm{T}}[N+1] = \boldsymbol{Y}[N]\boldsymbol{F}^{\mathrm{T}}[N] + \boldsymbol{y}[N+1]\boldsymbol{f}^{\mathrm{T}}[N+1] \tag{10.2.14}$$

and the recursive estimation equation of the parameter matrix can be obtained using (10.2.10)

$$\hat{\boldsymbol{P}}[N+1] = \hat{\boldsymbol{P}}[N] - \left\{\boldsymbol{y}[N+1] - \hat{\boldsymbol{P}}[N]\boldsymbol{f}[N+1]\right\}\boldsymbol{f}^{\mathrm{T}}[N+1]\boldsymbol{R}[N+1] \tag{10.2.15}$$

Great formal similarity appears in the recursive methods discussed for *SISO* systems since the renewal of the convergence matrix is performed according to the same equation. The only difference comes from the dimension owed to the *MIMO* character. The application of the forgetting factor completely corresponds with the recursive Equation (10.2.8).

The recursive *LS* algorithm shown above for *MIMO* systems can also be applied without any change in the *ELS* method.

Forgetting Strategies

In form (10.2.8) of the convergence matrix the forgetting factor has been introduced which means that the former measurements are weighted by λ. If $\lambda = 1$, then there is no forgetting, and the change in the parameters appears very slowly in the recursive estimation. If $\lambda < 1$, then it can happen that the estimation forgets the model, and the matrix becomes singular if the new measurement does not have sufficient new information. These two requirements are difficult to be fulfilled, and therefore different forgetting strategies have been developed to handle this problem [34,46,117,152,164].

The constant forgetting factor $\lambda < 1$ provides exponential forgetting, and the speed of the adaptation is determined by the asymptotical length of the memory $N_a = (1 - \lambda)^{-1}$. This means the virtual length of the measurement data the estimation made from.

Because of the expressions (A.6.35) and (A.6.36) in A.6.4 of Appendix A.6, in the case of $\lambda = 1$ the convergence matrix is positive and semi-definite, and thus the quadratic terms $\boldsymbol{f}^{\mathrm{T}}[N+1]\ \boldsymbol{R}[N]\,\boldsymbol{f}[N+1]$ are positive, that is, both the determinant and trace of the matrix are monotonously decreasing. In the parameter and function components domain ellipses correspond to the equations $\boldsymbol{f}^{\mathrm{T}}[N+1]\ \boldsymbol{R}[N]\,\boldsymbol{f}[N+1] = \text{constant}$, known as information ellipses. During the recursive estimation both the volume of the ellipses and the sum of the main diagonals are monotonously decreasing, and thus the quality of the estimation is successively improving. Unfortunately, this useful phenomenon exists only until $\boldsymbol{R}[N]$ becomes singular or close to it. The most obvious reason is that the vector $\boldsymbol{f}[N]$ does not provide sufficient excitation to the parameter estimation and the forgetting factor $\lambda < 1$ is applied. Consider the very special case when $\boldsymbol{f}[N]$ becomes constant from the time point N_u, and investigate the recursive estimation (10.2.8) [34]. The expression (A.6.33) is now

$$\{\boldsymbol{R}[N+1]\}^{-1} = \lambda^{N+1-N_u}\{\boldsymbol{R}[N_u]\}^{-1} + \frac{1-\lambda^{N+1-N_u}}{1-\lambda^2}\boldsymbol{f}[N_u]\,\boldsymbol{f}^{\mathrm{T}}[N_u] \tag{10.2.16}$$

hence

$$\boldsymbol{R}[N+1] = \frac{1}{\lambda^{2(N+1-N_u)}}\left\{\boldsymbol{R}[N_u] - \frac{\boldsymbol{R}[N_u]\boldsymbol{f}[N_u]\boldsymbol{f}^{\mathrm{T}}[N_u]\boldsymbol{R}[N_u]}{\frac{\left(1-\lambda^2\right)\lambda^{2(N+1-N_u)}}{1-\lambda^{2(N+1-N_u)}} + \boldsymbol{f}^{\mathrm{T}}[N_u]\boldsymbol{R}[N_u]\boldsymbol{f}[N_u]}\right\} \tag{10.2.17}$$

If $N+1-N_u$ is large, then after bringing to common denominator, the factor $\lambda^{2(N+1-N_u)}$ in the denominator of the second term can be neglected and so

$$\begin{aligned}\boldsymbol{R}[N+1] &= \frac{1-\lambda^2}{\boldsymbol{f}^{\mathrm{T}}[N_u]\boldsymbol{R}[N_u]\boldsymbol{f}[N_u]}\boldsymbol{R}[N_u] \\ &\quad + \frac{1}{\lambda^{2(N+1-N_u)}}\left\{\boldsymbol{R}[N_u] - \frac{\boldsymbol{R}[N_u]\boldsymbol{f}[N_u]\boldsymbol{f}^{\mathrm{T}}[N_u]\boldsymbol{R}[N_u]}{\boldsymbol{f}^{\mathrm{T}}[N_u]\boldsymbol{R}[N_u]\boldsymbol{f}[N_u]}\right\} \\ &\triangleq \frac{1-\lambda^2}{\boldsymbol{f}^{\mathrm{T}}[N_u]\boldsymbol{R}[N_u]\boldsymbol{f}[N_u]}\boldsymbol{R}[N_u] + \frac{1}{\lambda^{2(N+1-N_u)}}\boldsymbol{R}^*\end{aligned} \tag{10.2.18}$$

The matrix $\boldsymbol{R}^*$ introduced here is singular and $\boldsymbol{f}^{\mathrm{T}}[N_u]\ \boldsymbol{R}[N_u]\boldsymbol{f}[N_u]=0$ in the direction of $\boldsymbol{f}[N_u]$. At the same time $\boldsymbol{f}^{\mathrm{T}}[N_u]\ \boldsymbol{R}[N_u]\ \boldsymbol{f}[N_u]=1-\lambda^2$ in the direction of $\boldsymbol{R}[N_u]\boldsymbol{f}[N_u]$. Because of the above relationships, certain elements of $\boldsymbol{R}[N]$ tend to infinity if the number of the parameters is greater than one. Based on (10.2.16) $\det(\boldsymbol{R}^{-1}[N])$ tends to zero and $\det(\boldsymbol{R}[N])$ tends to infinity. Similarly, it is true that $\{\dim(\boldsymbol{R}[N])-1\}$ eigenvalues of the matrix $\boldsymbol{R}^{-1}[N]$ tend to zero, and thus $\mathrm{tr}(\boldsymbol{R}[N])$ tends to infinity. Given the above, the information ellipse does not decrease in the direction of $\boldsymbol{f}[N_u]$, and so the estimation is not deteriorated in this direction but in the perpendicular direction, and the deterioration of the estimation can be expected [152].

Several heuristic methods for handling this problem have been published. Their common feature is that they also use varying forgetting factor. The weighting factor of the exponential forgetting can be set on the basis of the information provided by the measurements and parameter estimation.

In expression (10.2.2) of the recursive estimation the so-called a priori error

$$\hat{e}[N+1] = y[N+1] - \boldsymbol{f}^{\mathrm{T}}[N+1]\hat{\boldsymbol{p}}[N] \tag{10.2.19}$$

is of vital importance and can always be corrected by the new estimated parameter vector to obtain the a posteriori error

$$\hat{\varepsilon}[N+1] = y[N+1] - \boldsymbol{f}^{\mathrm{T}}[N+1]\hat{\boldsymbol{p}}[N+1] \tag{10.2.20}$$

At the systems close to be deterministic the a posteriori estimation error assesses the merits of the estimation. Slight differences can appear if the process is not sufficiently excited or if the estimation is fairly good. In both cases it is advisable to set the forgetting value close to one ($\lambda \cong 1$).

In the case of large a posteriori error it is reasonable to adopt the parameters faster, that is, to set $\lambda < 1$, in order to decrease the estimation error. The above consideration can be helped by the knowledge or good estimation of the variance σ_ς^2 of the measurement error. Do not forget that in the constant parameter case, the best estimation of the variance of the measurement error is the variance of the a posteriori error. (Note that the a posteriori error can be computed by recursive expressions with the recursive *LS* method.)

Thus the different adaptive forgetting strategies permanently observe and average the a posteriori error. If the variance of the a priori error is close to this value, the forgetting factor $\lambda \cong 1$ is required; if it is larger than this value, then $\lambda < 1$ needs to be applied.

Directional Forgetting Method

Vajk elaborated a method which allows forgetting only in the direction where the new excitation has effect [152]. If there is no new excitation, then the forgetting is stopped and the estimation is not deteriorated and keeps the learned model. Due to the definition (10.1.7) of the covariance matrix and the dyadic Equation (A.6.32) we can write a recursive form (10.2.13) for the covariance matrix as

$$\boldsymbol{Z}^{-1}[N+1] = \boldsymbol{Z}^{-1}[N] + \frac{1}{\sigma_\varsigma^2}\boldsymbol{f}[N]\boldsymbol{f}^{\mathrm{T}}[N] \tag{10.2.21}$$

then follow the decomposition principle introduced in (10.2.18). To do this, decompose the matrix $\boldsymbol{Z}^{-1}$ in two terms. Let the first term be parallel to the direction of the new excitation and the second one perpendicular to this

$$\boldsymbol{Z}^{-1}[N] = \boldsymbol{S}[N] + \gamma\boldsymbol{f}[N]\boldsymbol{f}^{\mathrm{T}}[N] \tag{10.2.22}$$

where $\boldsymbol{S}[N]$ is a singular matrix. (Here the concept of parallel and perpendicular does not correspond completely with the exact geometrical concepts.) The consequence of the singularity of $\boldsymbol{S}[N]$ is

$$\det\left\{\boldsymbol{Z}^{-1}[N] - \gamma\boldsymbol{f}[N]\boldsymbol{f}^{\mathrm{T}}[N]\right\} = \det\left\{\boldsymbol{Z}^{-1}[N]\right\}\left\{1 - \gamma\boldsymbol{f}[N]^{\mathrm{T}}\boldsymbol{Z}[N]\boldsymbol{f}[N]\right\} = 0 \tag{10.2.23}$$

where the identity (A.1.18) is used. In the case of good estimation the determinant of $\boldsymbol{Z}[N]$ is not zero, and therefore the necessary condition of the identity is

$$1 - \gamma\boldsymbol{f}^{\mathrm{T}}[N]\boldsymbol{Z}[N]\boldsymbol{f}[N] = 0 \tag{10.2.24}$$

and hence

$$\gamma = \frac{1}{\boldsymbol{f}^{\mathrm{T}}[N]\boldsymbol{Z}[N]\boldsymbol{f}[N]} \tag{10.2.25}$$

Given the above result the accumulated information in the direction of the excitation can be written as

$$\boldsymbol{\Gamma} = \frac{1}{\boldsymbol{f}^{\mathrm{T}}[N]\boldsymbol{Z}[N]\boldsymbol{f}[N]}\boldsymbol{f}[N]\boldsymbol{f}^{\mathrm{T}}[N] \tag{10.2.26}$$

Since during the estimation the goal is to perform the forgetting only in terms of the information matrix, where the new information modifies the matrix $\boldsymbol{Z}^{-1}$, the special decomposition of the expression (10.2.21) needs to be done. Introduce the following decomposition

$$\boldsymbol{Z}^{-1}[N+1] = \left\{\boldsymbol{Z}^{-1}[N] - \boldsymbol{\Gamma}\right\} + \delta\boldsymbol{\Gamma} + \frac{1}{\sigma_\varsigma^2}\boldsymbol{f}[N]\boldsymbol{f}^{\mathrm{T}}[N] \tag{10.2.27}$$

where the first term is the information perpendicular to the excitation, the second is weighted in the direction of the excitation, and the third contains the new information. Using the method applied for the derivation of (A.6.28) and (A.6.34) and the matrix inversion lemma (A.1.17) we obtain the covariance matrix of the recursive estimation as

$$\boldsymbol{Z}[N+1] = \boldsymbol{Z}[N] - \frac{\boldsymbol{Z}[N]\boldsymbol{f}[N]\boldsymbol{f}^{\mathrm{T}}[N]\boldsymbol{Z}[N]}{\boldsymbol{f}^{\mathrm{T}}[N]\boldsymbol{Z}[N]\boldsymbol{f}[N]}\left\{1 - \frac{1}{\delta + \frac{1}{\sigma_\varsigma^2}\boldsymbol{f}^{\mathrm{T}}[N]\boldsymbol{Z}[N]\boldsymbol{f}[N]}\right\} \tag{10.2.28}$$

and the equation of the parameter estimation is

$$\hat{\boldsymbol{p}}[N+1] = \hat{\boldsymbol{p}}[N] + \frac{\boldsymbol{Z}[N]\boldsymbol{f}[N+1]}{\delta\sigma_\varsigma^2 + \boldsymbol{f}^{\mathrm{T}}[N+1]\boldsymbol{Z}[N]\boldsymbol{f}[N+1]} \times \left\{y[N+1] - \boldsymbol{f}^{\mathrm{T}}[N+1]\hat{\boldsymbol{p}}[N]\right\} \tag{10.2.29}$$

which can be considered as the generalization of (10.2.5). The algorithm elaborated by Vajk is very similar to the method published in [117] with the difference that it does not forget the actual measurement and thus uses the maximum information in the estimation.

In the case of directional forgetting, the weight δ has to fulfill the following inequality

$$\frac{1}{\sigma_\varsigma^2}\boldsymbol{f}^{\mathrm{T}}[N]\boldsymbol{Z}[N]\boldsymbol{f}[N] < \delta \le 1 \tag{10.2.30}$$

The value $\delta > 1$ cannot be chosen since this would mean the enhancement of the past. The choice of $\delta = 0$ means the accumulated information would be forgotten completely. If the left side of the inequality is not fulfilled then the information matrix is not positively definite. If $\delta = 1$, the parameter tracking becomes impossible, the algorithm averages the measurements weighted with the reciprocal of the noise. The choice of $\delta = 0$, which is theoretically possible but not reasonable, ensures the fastest parameter tracking. In this case the old information in the direction of the actual excitation is replaced by new. In the case of the fastest tracking, the renewal equations are

$$\hat{\boldsymbol{p}}[N+1] = \hat{\boldsymbol{p}}[N] + \frac{\boldsymbol{Z}[N]\boldsymbol{f}[N+1]}{\boldsymbol{f}^{\mathrm{T}}[N+1]\boldsymbol{Z}[N]\boldsymbol{f}[N+1]}\left\{y[N+1] - \boldsymbol{f}^{\mathrm{T}}[N+1]\hat{\boldsymbol{p}}[N]\right\}$$

$$\boldsymbol{Z}[N+1] = \boldsymbol{Z}[N] - \frac{\boldsymbol{Z}[N]\boldsymbol{f}[N]\boldsymbol{f}^{\mathrm{T}}[N]\boldsymbol{Z}[N]}{\boldsymbol{f}^{\mathrm{T}}[N]\boldsymbol{Z}[N]\boldsymbol{f}[N]}\left\{1 - \frac{\sigma_e^2}{\boldsymbol{f}^{\mathrm{T}}[N]\boldsymbol{Z}[N]\boldsymbol{f}[N]}\right\} \tag{10.2.31}$$

Here in all cases the a posteriori error is zero, that is,

$$y[N+1] - \boldsymbol{f}^{\mathrm{T}}[N+1]\hat{\boldsymbol{p}}[N+1] \equiv 0 \tag{10.2.32}$$

Obviously, several methods can be elaborated for the choice of δ or for the directional forgetting discussed above. It is possible to forget as much collected information as new information is obtained; in this way constant adaptation speed can be reached, etc.

Several heuristic algorithms have been published in the literature for the adaptive setting of the forgetting factor. Their common feature is that they try to obtain sufficient estimation for the measurement noise variance in the case of good models and set the forgetting factor close to one. The variances of the a priori and a posteriori errors are continuously examined. If there is a substantial step in these values it refers to the change of the process parameters, and therefore the forgetting factor needs to be suitably decreased.

10.3 PROCESS IDENTIFICATION IN CLOSED-LOOP CONTROL

The process identification methods discussed in the previous two sections basically perform parameter estimation in open loop, parallel to the process. Thus the model is accommodated between the input and output signal. The basic situation is shown in Figure 10.3.1.

This case is very rare in the design of control systems. In industry or other technologies usually a closed-loop control is in operation, and it is very rarely allowed to open them and to make the measurements according to Figure 10.3.1. So open-loop measurement is mostly possible when the system has started to operate and tried to control manually.

In most practical cases the process identification needs to be performed in closed loop as in Figure 10.3.2. In a suitable case an outer exciting signal u_g can be applied.

In this case the input u is not independent of the disturbance signal y_n or its source noise ς, and thus the condition of the unbiased estimation is not fulfilled. There are special methods which assume a model between the

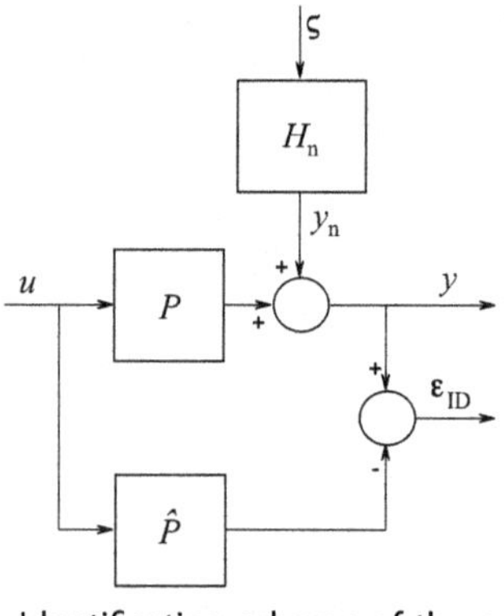

Figure 10.3.1 Identification scheme of the parallel model.

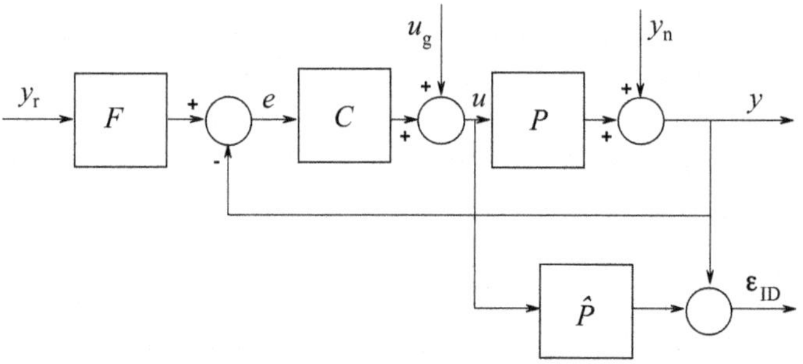

Figure 10.3.2 Parallel model in closed loop.

independent outer signal u_g and the process output y, but in this case the same reference signal y_r and the knowledge of the model C of the regulator are required. Söderström and Stoica elaborated several further process identification (ID) methods for closed loops [147].

Identification Methods Using *Youla–Kučera* (*Y–K*) Parameterization

A comprehensive literature deals with ID methods used in control loops when the parameter Q_P is estimated instead of the original process P. This method is based on the recognition that the *dual*-Y–K-parameterized closed loop (see Section 1.4), considering also the output disturbance, can be rearranged into the equivalent block scheme shown in Figure 1.4.9 [5,33,143,153].

Figure 10.3.3 contains an embedded open-loop subsystem (see Figure 10.3.4 drawn separately), whose input is

$$x = \mathcal{Y}r \tag{10.3.1}$$

and the output is

$$z = Q_P x + (\mathcal{A} - \mathcal{Y}Q_P)y_n = Q_P x + y_n' \tag{10.3.2}$$

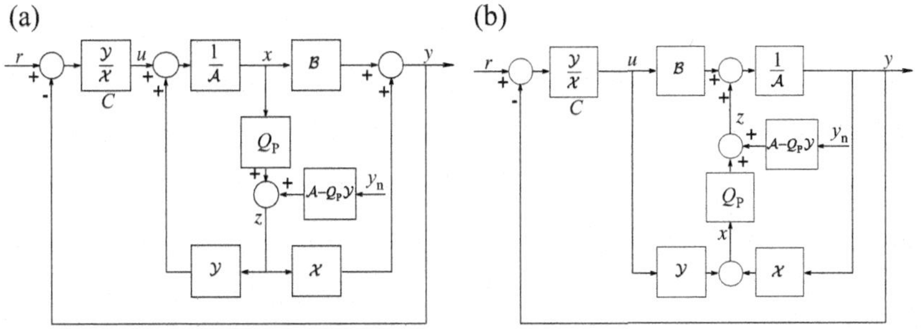

Figure 10.3.3 Equivalent block schemes of the *dual*-*Y*–*K*-parameterized closed-loop control assuming output noise.

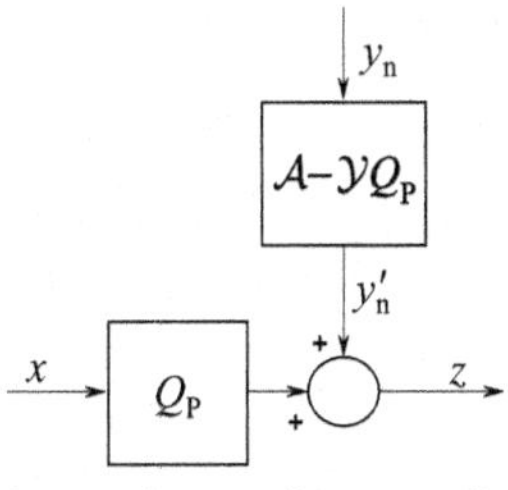

Figure 10.3.4 Embedded open-loop subsystem for identification purposes.

Here x and z can be determined from the measurements by known filters. It is important to note that r and y_n are independent, and then x and y'_n are also independent. Therefore, in an embedded system, the identification of Q_P can be performed by a conventional open-loop ID method.

Let us express the *dual-Y–K* parameter Q_P from (1.4.34)

$$Q_P = \frac{\mathcal{A}}{\mathcal{X}(1+CP)}(P-P^*) = \frac{\mathcal{A}(1-QP)}{\mathcal{X}}(P-P^*) = \frac{\mathcal{A}(1-HC)}{\mathcal{X}}(P-P^*) \tag{10.3.3}$$

This expression unambiguously shows that $Q_P \to 0$ if $P^* \to P$. If the estimation $\hat{Q}_P$ is available by means of an iterative ID method then the estimation of the process transfer function can be obtained from the following equation

$$\hat{P} = \left(\mathcal{B} + \mathcal{X}\hat{Q}_P\right) \Big/ \left(\mathcal{A} - \mathcal{Y}\hat{Q}_P\right) \tag{10.3.4}$$

which is the consequence of the definition (1.4.34). The next iteration step starts with the assumption $P^* = \hat{P}$ and the new regulator $C = C(\hat{P})$ needs to be set. If the iterative method works, then smaller and smaller $\hat{Q}_P$ needs to be estimated to provide worse and worse conditions concerning the numerical behavior of the identification task. This paradigm is well known from the literature. This solution does not provide the best result in terms of closed-loop identification.

Parallel Model-Based Closed-Loop Identification Methods

It was shown in Chapter 4 (Section 4.3) that the modeling loss in the sensitivity function of the closed loop is

$$S_{\text{mod}} = S - \hat{S} = \hat{T} - T = -\frac{\hat{T}\hat{S}\ell}{1+\hat{T}\ell} = -\hat{T}S\ell\Big|_{\ell \to 0} \approx -\hat{T}\hat{S}\ell \tag{10.3.5}$$

That process identification method fits to this observation what approaches the original complementary sensitivity function T with the model-based complementary sensitivity function $\hat{T}$ and does not approach the true process P with the model $\hat{P}$. Closed-loop identification cannot be substituted by the open loop one.

Irrespective of this observation, in the early days of the identification techniques Young suggested a new direction [163], which assumes a complete model-based closed loop parallel to the closed-loop control. The identification is performed according to Figure 10.3.5.

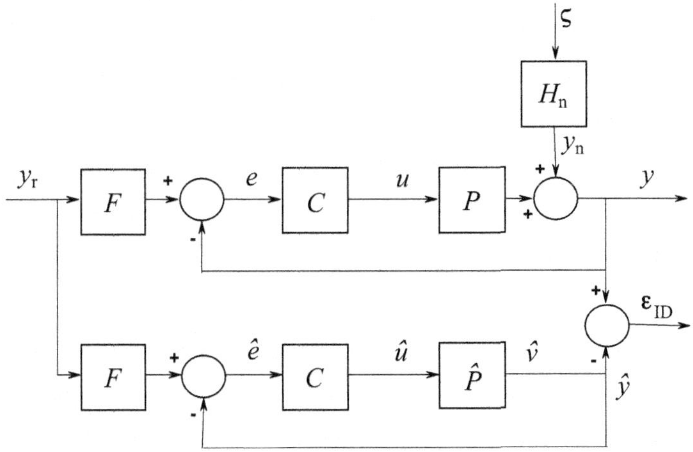

Figure 10.3.5 Identification scheme for modeling the closed-loop control.

Here the question of the independence of the inner signals of the model-based closed loop of the disturbance signal y_n or its source noise ς is obviously not raised. The method best fitting this scheme is the *IV* method shown in Section 10.1, according to which the estimation of the process parameters can be given by the following equation

$$\hat{\boldsymbol{p}}_{ba} = \left[\mathbf{G}_{uv}^{T}\boldsymbol{F}_{uy}\right]^{-1}\mathbf{G}_{uv}^{T}\boldsymbol{y} = \left[\mathbf{G}^{T}(\boldsymbol{u},\boldsymbol{v})\boldsymbol{F}(\boldsymbol{u},\boldsymbol{y})\right]^{-1}\mathbf{G}^{T}(\boldsymbol{u},\boldsymbol{v})\boldsymbol{y} \qquad (10.3.6)$$

where $\boldsymbol{F}_{uy}$ is created according to the classical methods (10.1.18). The matrix $\mathbf{G}(\hat{\boldsymbol{u}},\hat{\boldsymbol{v}})$ of the *IV*s is completely composed of the signals $\hat{u}$ and $\hat{v}$ of the computed parallel closed loop.

Process Identification Methods Based on *Keviczky–Bányász* (*KB*) Parameterization

Apply the Young method for the scheme of the *YP* control loop shown in Figure 2.1.7 and according to Figure 10.3.6.

The figure can be further simplified by block manipulations and the resulting special block scheme is presented in Figure 10.3.7. The figure shows a very important behavior, namely that in the *KB*-parameterized closed loop the error e is equal to the identification error ε_{ID} but with a negative sign. Thus the satisfactory model leads to small control error. This feature exists only in *KB*-parameterized closed-loop controls. Now the reasonable realization of the process identification method is performed between the generated (computed) signal $\hat{u}$ and the output of the closed loop. The advantage of the method is that the closed loop does not have to be opened to the measurements necessary for the identification.

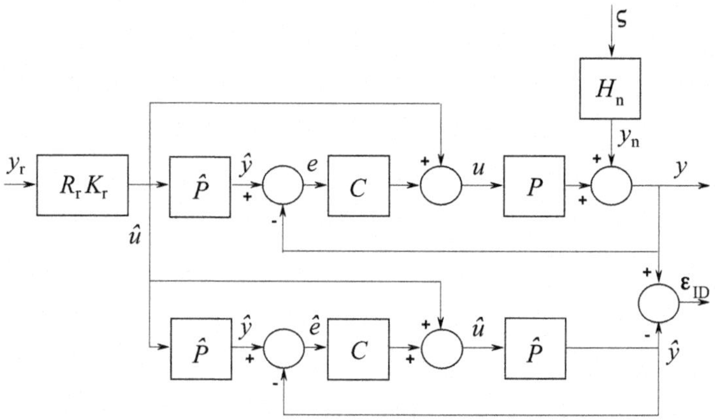

Figure 10.3.6 Identification scheme for modeling the *YP* closed-loop control.

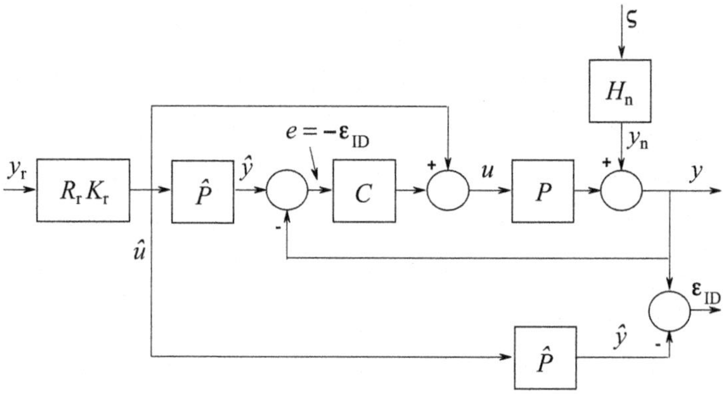

Figure 10.3.7 Equivalent identification scheme for modeling the *KB*-parameterized closed control loop.

As mentioned, the most important condition in closed-loop identification is that the model input signal must be uncorrelated with the source noise. This condition is obviously fulfilled in Figure 10.3.7. It is also, however, fulfilled in the original method of Young. The advantage of *KB* parameterization is that it virtually opens the closed loop, and thus identification methods have to be handled like classical open-loop methods. Let us see first what the asymptotic forms of the equivalent process and noise models look like.

Since the source noise ς has an effect on the output signal via the closed system, the original noise model H_n of the open loop shown in Figure 10.3.7 virtually becomes higher order in the identification scheme of Figure 10.3.7. Consider the *KB*-parameterized closed loop with input signal $r = \hat{u}$ for the ideal case $\hat{P} = P$, $C_o = C(P)$ in Figure 10.3.8.

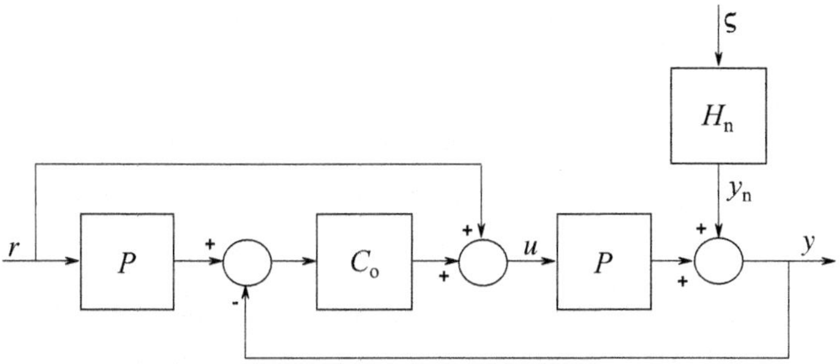

Figure 10.3.8 *KB*-parameterized closed-loop control for the ideal case $\hat{P} = P$.

This scheme is equivalent to that of Figure 10.3.9, which clearly shows that the noise model H_n changes to H'_n.

The virtual noise model derived from the closed loop has the form

$$H'_n = H_n S = \frac{H_n}{1 + C_o P} = H_n(1 - Q_o P) = H_n\left(1 - R_n G_n P_- z^{-d}\right) \tag{10.3.7}$$

Next investigate the effect of model-based *KB* parameterization on the above approach, as shown in Figure 10.3.10.

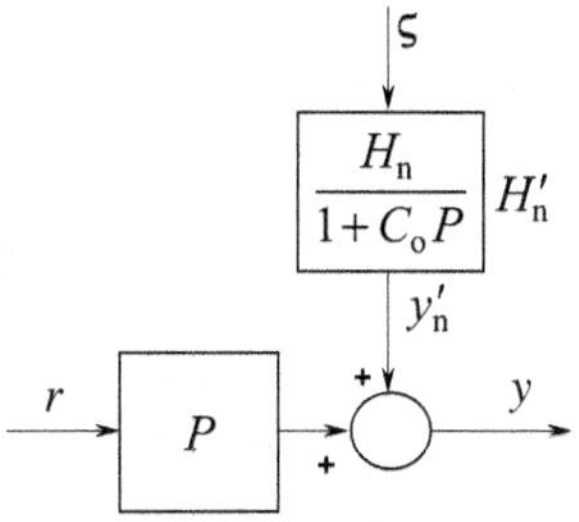

Figure 10.3.9 Equivalent form of the *KB*-parameterized closed-loop control for the ideal case $\hat{P} = P$.

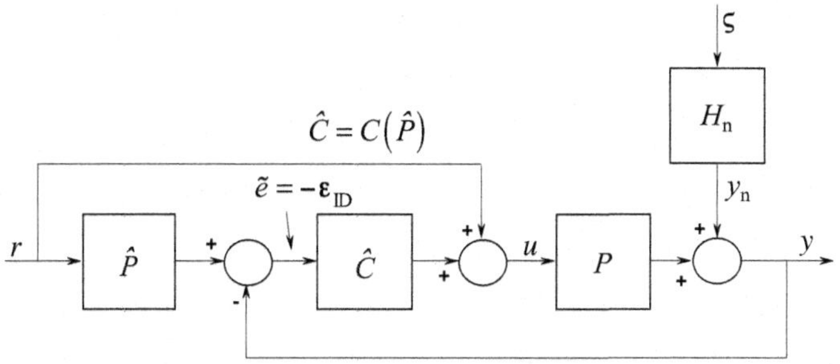

Figure 10.3.10 *KB*-parameterized model-based closed-loop control.

Considering the sensitivity investigations of Section 9, the resulting relationships in Figure 10.3.10 can be given as

$$\begin{aligned} y &= \frac{1+\hat{C}\hat{P}}{1+\hat{C}P}Pr + \frac{1}{1+\hat{C}P}y_{\mathrm{n}} = \frac{1+\hat{C}\hat{P}}{1+\hat{C}P}Pr + \frac{H_{\mathrm{n}}}{1+\hat{C}P}\varsigma \\ &= P_{\Delta}r + H'_{\Delta}\varsigma = P_{\Delta}\left(P,\hat{P},\hat{C}\right)r + H'_{\Delta}\left(H_{\mathrm{n}},P,\hat{C}\right)\varsigma \end{aligned} \tag{10.3.8}$$

where $P = P_{\Delta}(\Delta = 0)$ and $H' = H_{\mathrm{n}}(1 - Q_{\mathrm{o}}P) = H'_{\Delta}(\Delta = 0)$. The transfer functions to be identified are

$$P_{\Delta} = P - \frac{\hat{C}P\Delta}{1+\hat{C}P} = P - \tilde{Q}P\Delta = P - \frac{\hat{Q}P\Delta}{1+\hat{Q}\Delta} = \frac{P}{1+\hat{Q}\Delta} \tag{10.3.9}$$

$$H'_{\Delta} = H_{\mathrm{n}}\tilde{S} = \frac{H_{\mathrm{n}}}{1+\hat{C}P} = \frac{H_{\mathrm{n}}}{1+\frac{\hat{Q}P}{1-\hat{Q}\hat{P}}} = H_{\mathrm{n}}\left(1 - \frac{\hat{Q}P}{1+\hat{Q}\Delta}\right) \tag{10.3.10}$$

The simpler version of the real closed system of (10.3.8) is

$$y = \frac{P}{1+\hat{Q}\Delta}r + H_{\mathrm{n}}\left(1 - \frac{\hat{Q}P}{1+\hat{Q}\Delta}\right)\varsigma = P_{\Delta}r + H'_{\Delta}\varsigma \tag{10.3.11}$$

Consequently, the *KB*-parameterized model-based regulator presented in Figure 10.3.11 makes possible parallel ID. Since P_{Δ} depends on $\hat{P}$, only iterative ID method can be applied which tunes the model step by step.

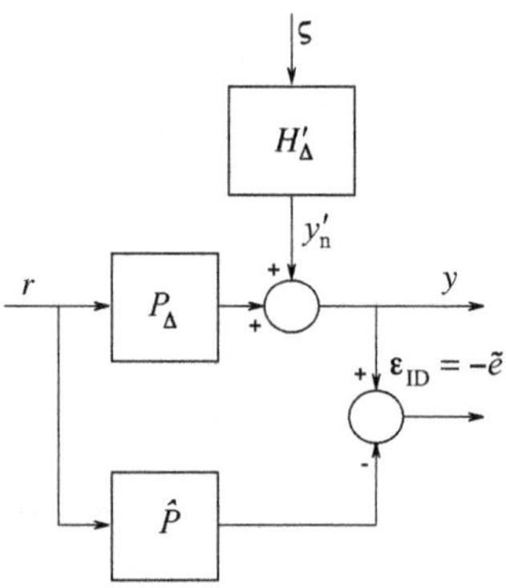

Figure 10.3.11 Parallel identification in the *KB*-parameterized model-based closed-loop control.

Investigate the feasible convergence points where $\hat{P}$ and $\hat{H}'_n$ converge during the iteration. Now the model errors have the forms

$$P_\Delta - \hat{P} = \frac{P}{1+\hat{Q}\Delta} - \hat{P} = \frac{1-\hat{Q}\hat{P}}{1+\hat{Q}\Delta}\Delta \tag{10.3.12}$$

and

$$\begin{aligned} H'_\Delta - \hat{H}'_n &= H_n\left(1 - \frac{\hat{Q}P}{1+\hat{Q}\Delta}\right) - \hat{H}'_n = H_n\left(1-\hat{Q}P\right)\frac{1+\tilde{C}\Delta}{1+\hat{Q}\Delta} - \hat{H}'_n \\ &= H'_n\frac{1+\hat{C}\Delta}{1+Q\Delta} - \hat{H}'_n \end{aligned} \tag{10.3.13}$$

Thus the convergence points are $P_\Delta \to P_\Delta(\Delta = 0) = P$ for the process and $H'_\Delta \to H'_\Delta(\Delta = 0) = H'_n$ for the noise model in the case of $\Delta \to 0$, if $\hat{P}$ and P belong to the same model class and H'_Δ and H'_n belong to the same model class of increased order.

Asymptotic Variances of Process Identification

Using form (10.3.11) of the model-based *KB*-parameterized closed system, we can write the fundamental equation of the Ljung [123] prediction error method (10.1.65) as

$$\varepsilon_{pr}[k] = \frac{1}{\hat{H}'_n}\left\{\frac{1-\hat{Q}\hat{P}}{1+\hat{Q}\Delta}\Delta r + \left[H_n\left(1-\hat{Q}P\right)\frac{1+\tilde{C}\Delta}{1+\hat{Q}\Delta} - \hat{H}'_n\right]\varsigma\right\} + \varsigma \tag{10.3.14}$$

In the case of $\Delta \to 0$ the prediction error is

$$\varepsilon_{pr}(\Delta = 0) = \frac{1}{\hat{H}'_n}\left\{S\Delta r + \left[H'_n - \hat{H}'_n\right]\varsigma\right\} + \varsigma = \frac{1}{\hat{H}'_n}\left\{S\Delta r + \tilde{H}'_n\varsigma\right\} + \varsigma \tag{10.3.15}$$

in the convergence point, where

$$\tilde{H}'_n = H'_n - \hat{H}'_n \quad \text{and} \quad S = \frac{1}{1+C_o P} = 1 - Q_o P \tag{10.3.16}$$

The frequency spectrum of the prediction error is

$$\Phi_{\varepsilon\varepsilon}(\omega, \hat{\boldsymbol{p}}) = \frac{1}{\left|\hat{H}'_n\right|^2}\left\{|S|^2|\Delta|^2\phi_{rr} + \left|\tilde{H}'\right|^2\sigma_\varsigma^2\right\} + \sigma_\varsigma^2; \quad E\{\varsigma^2\} = \Phi_{\varsigma\varsigma} = \sigma_\varsigma^2 \tag{10.3.17}$$

Because of this latter expression, the virtual models $\Phi'_{\mathrm{uu}} = |S|^2\Phi_{\mathrm{rr}}$ and $\Phi'_{\mathrm{vv}} = |S|^2|\boldsymbol{H}_{\mathrm{n}}|^2\sigma_{\varsigma}^2$ contain the spectrum of the virtual input signal u' and the virtual output noise, respectively. Because of Ljung's [123] asymptotic covariance theorem and in the case of the $(\hat{G}, \hat{H}'_{\mathrm{n}})$ model-based *KB*-parameterized closed-loop identification, the (10.1.70) covariance matrix becomes

$$\mathrm{Cov}\begin{bmatrix} \hat{G}_N\left(e^{j\omega}\right) \\ \hat{H}_{\mathrm{n}}^{\prime N}\left(e^{j\omega}\right) \end{bmatrix} \approx \frac{n}{N}\Phi'_{\mathrm{vv}}(\omega)\begin{bmatrix} \Phi'_{uu}(\omega) & \Phi_{u'\varsigma}(-\omega) \\ \Phi_{\varsigma u'}(\omega) & \sigma_{\varsigma}^2 \end{bmatrix}^{-1}$$
$$= \frac{n}{N}\Phi'_{\mathrm{vv}}(\omega)\begin{bmatrix} \sigma_{\varsigma}^2 & -\Phi_{\varsigma u'}(\omega) \\ -\Phi_{u'\varsigma}(\omega) & \Phi'_{uu}(\omega) \end{bmatrix}\frac{1}{\det} \tag{10.3.18}$$

where

$$\Phi'_{\mathrm{vv}} = |S|^2\Phi_{\mathrm{vv}} = |S|^2|H_{\mathrm{n}}|^2\sigma_{\varsigma}^2 \quad \text{and} \quad \det = \sigma_{\varsigma}^2\Phi'_{\mathrm{uu}} - |\Phi_{u'\varsigma}|^2 = \sigma_{\varsigma}^2\Phi'_{\mathrm{uu}} \tag{10.3.19}$$

since $\Phi_{u'\varsigma} = \Phi_{\varsigma u'} = 0$ and $\Phi'_{uu}(\omega) = \Phi_{u'u'}$. Finally, the partial covariance of the process and noise model is

$$\mathrm{Cov}\left[\hat{G}_N\left(e^{j\omega}\right)\right] = \frac{n}{N}\frac{\Phi'_{\mathrm{vv}}(\omega)}{\Phi'_{\mathrm{uu}}(\omega)} = \frac{n}{N}\frac{|S|^2\Phi_{\mathrm{vv}}(\omega)}{|S|^2\Phi_{\mathrm{rr}}(\omega)} = \frac{n}{N}\frac{\Phi_{\mathrm{vv}}(\omega)}{\Phi_{\mathrm{uu}}(\omega)} \tag{10.3.20}$$

and

$$\mathrm{Cov}\left[\hat{H}_{\mathrm{n}}^{\prime N}\left(e^{j\omega}\right)\right] = \frac{n}{N}\frac{\Phi'_{\mathrm{vv}}(\omega)}{\sigma_{\varsigma}^2} = \frac{n}{N}\frac{\Phi_{\mathrm{vv}}(\omega)}{\sigma_{\varsigma}^2}|S|^2 = \frac{n}{N}\left|H_{\mathrm{n}}\left(e^{j\omega}\right)\right|^2|S|^2 \tag{10.3.21}$$

Practically the same result is obtained as in the classical Ljung method for open-loop identification (see formulas (10.1.70) and (10.1.71)). The only difference is that the covariance matrix of the noise has the weight $|S|^2$. This weight is generally less than one, except in the medium frequency region, as shown for a typical case in Figure 10.3.12.

The result obtained for the noise model is completely understandable since the model-based *KB* parameterization virtually increases the dimension of the noise model according to $H'_{\mathrm{n}} = SH_{\mathrm{n}}$, so it can be retrieved from the estimation as $\hat{H}_{\mathrm{n}} = \hat{H}'_{\mathrm{n}}/\hat{S}$. This means that in the partial covariance of $\hat{H}_{\mathrm{n}}$ the weighting by $|S|^2$ no longer appears.

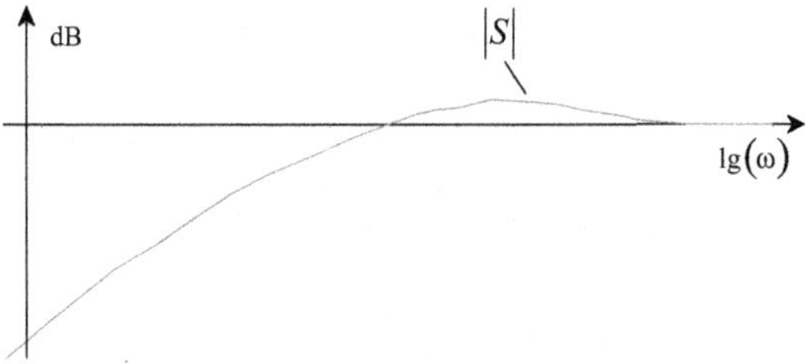

Figure 10.3.12 Typical medium frequency weighting in the covariance matrix of the noise model by the absolute value $|S|$.

Table 10.3.1 Partial covariance matrices of the process and noise model for different identification methods

	Open loop	*Keviczky–Bányász* parameterization	Closed system
$\hat{G}_N$	$\frac{n}{N}\frac{\Phi_{vv}}{\Phi_{uu}}$	$\frac{n}{N}\frac{\Phi_{vv}}{\Phi_{rr}}$	$\frac{n}{N}\frac{\Phi_{vv}}{\Phi_{uu}}\frac{1}{1-\frac{\Phi^{\varsigma}_{uu}}{\Phi_{uu}}}$
$\hat{H}^N_n$	$\frac{n}{N}\frac{\Phi_{vv}}{\sigma^2_{\varsigma}}$	$\frac{n}{N}\frac{\Phi'_{vv}}{\sigma^2_{\varsigma}}=\frac{n}{N}\frac{\Phi_{vv}}{\sigma^2_{\varsigma}}\lvert S\rvert^2$	$\frac{n}{N}\frac{\Phi_{vv}}{\sigma^2_{\varsigma}}\left(1+\frac{\Phi^{\varsigma}_{uu}}{\Phi^{r}_{uu}}\right)$

Gevers [48] demonstrated that the covariance of closed-loop identification is always higher than that obtained in open-loop identification. In Table 10.3.1 the partial covariance matrices of the process and noise model are summarized for the case of open loop, *KB* parameterization-based and closed-loop identification. The expressions in the last column are given without detailed derivation from [48–50].

It is assumed in the computation of the partial covariance of closed-loop identification that the spectrum of the input signal derives from two effects

$$\Phi_{uu} = \Phi^{r}_{uu} + \Phi^{\varsigma}_{uu} \tag{10.3.22}$$

where Φ^{r}_{uu} and Φ^{ς}_{uu} come from the reference signal and the source noise, respectively. Obviously

$$\frac{1}{1-\frac{\Phi^{\varsigma}_{uu}}{\Phi^{r}_{uu}}} > 1 \tag{10.3.23}$$

exists, since

$$\Phi^{\varsigma}_{uu} < \Phi_{uu} \quad \text{and} \quad 1+\frac{\Phi^{\varsigma}_{uu}}{\Phi^{r}_{uu}} > 1 \tag{10.3.24}$$

It is clear from Table 10.3.1 that the covariance of the *KB* parameterization-based identification is the same as in the open loop, and the covariance in the closed-loop identification is always greater. It can be easily checked that

$$\frac{\Phi^{\mathrm{r}}_{\mathrm{uu}}+\Phi^{\varsigma}_{\mathrm{uu}}}{\Phi^{\mathrm{r}}_{\mathrm{uu}}}=\frac{|S|^2\Phi_{\mathrm{rr}}+|C|^2|S|^2|S|^2\sigma_{\varsigma}^2}{|S|^2\Phi_{\mathrm{rr}}}=1+\frac{|C|^2|S|^2\sigma_{\varsigma}^2}{\Phi_{\mathrm{rr}}}=1+\frac{|T|^2\sigma_{\varsigma}^2}{|S|^2\Phi_{\mathrm{rr}}}\geq 1 \tag{10.3.25}$$

where the second term is a special noise/signal ratio. Here T and S refer to the ideal case $\Delta = 0$.

Error Properties of *KB* Parameterization-Based Identification Methods

Based on Figure 10.3.7 the equation of the control and identification error can be easily written for the equivalent scheme modeling the *KB*-parameterized closed-loop control

$$\varepsilon_{\mathrm{ID}}=-\tilde{e}=y-\hat{y}=\frac{R_{\mathrm{r}}\hat{G}_{\mathrm{r}}\hat{\overline{P}}_{-}}{1+\hat{C}P}\ell y_{\mathrm{r}}+\frac{1}{1+\hat{C}P}y_{\mathrm{n}};\quad y_{\mathrm{n}}=H_{\mathrm{n}}\varsigma \tag{10.3.26}$$

Consequently the uncertainty scheme of the closed system can be drawn as shown in Figure 10.3.13.

Equation (10.3.26) of the identification error can be further modified

$$\varepsilon_{\mathrm{ID}}=\underbrace{\frac{\hat{G}_{\mathrm{r}}\hat{\overline{P}}_{-}}{1+\hat{C}P}R_{\mathrm{r}}\ell y_{\mathrm{r}}}_{\tilde{\varepsilon}_{\mathrm{r}}}+\underbrace{\frac{\hat{G}_{\mathrm{n}}\hat{\overline{P}}_{-}}{1+\hat{C}P}R_{\mathrm{n}}\ell y_{\mathrm{n}}}_{\tilde{\varepsilon}_{\mathrm{n}}}+\underbrace{\left(1-R_{\mathrm{n}}\hat{G}_{\mathrm{n}}\hat{\overline{P}}_{-}\right)y_{\mathrm{n}}}_{\tilde{\varepsilon}_{\mathrm{o}}} \tag{10.3.27}$$

where $\tilde{\varepsilon}_{\mathrm{r}}$ and $\tilde{\varepsilon}_{\mathrm{n}}$ are the contributions of the reference signal and the output noise, respectively, to the identification error. The last term $\tilde{\varepsilon}_{\mathrm{o}}$ is the residual uncorrelated error. It has been seen in Sections 4.2 and 6 that the

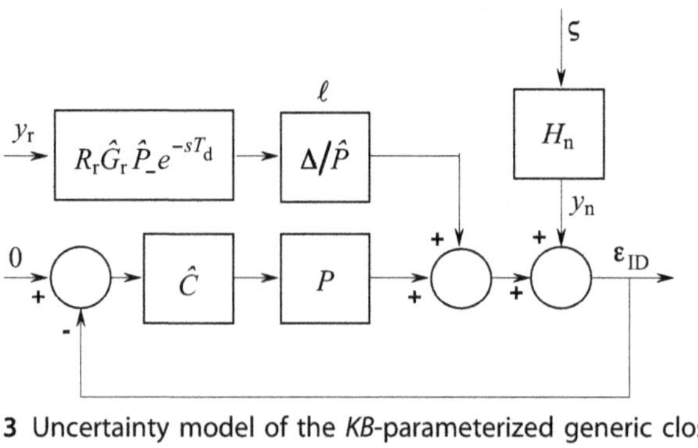

Figure 10.3.13 Uncertainty model of the *KB*-parameterized generic closed system.

optimal predictor of the disturbance signal can be used in the optimization of the regulator. The optimal predictor that can be used in the closed loop depends also on the parameters of the process and the noise model. In the *YP*-regulator the reference model R_n can be chosen arbitrarily and it is not the same as the respective optimal predictor. Mark the optimal predictor by R_n^o and the relative error between the applied predictor and the respective optimal predictor as

$$\ell_n = \frac{R_n^o - R_n}{R_n} = \frac{\Delta R_n}{R_n} \tag{10.3.28}$$

Using the above expression we obtain

$$\begin{aligned}
\boldsymbol{\varepsilon}_{ID} &= \underbrace{\frac{\hat{G}_r \hat{\tilde{P}}_-}{1+\hat{C}P} R_r \ell y_r}_{\tilde{\varepsilon}_r} + \underbrace{\frac{\hat{G}_n \hat{\tilde{P}}_-}{1+\hat{C}P} R_n \ell y_n}_{\tilde{\varepsilon}_n} + \underbrace{\tilde{G}_n \hat{\tilde{P}}_- R_n \ell_n y_n}_{\tilde{\varepsilon}_{\Delta R_n}} + \underbrace{\left(1 - R_n^o \tilde{G}_n \hat{\tilde{P}}_-\right) y_n}_{\tilde{\varepsilon}_o^o} \\
&= \underbrace{\frac{\hat{\tilde{P}}_-}{1+\hat{C}P} \ell \hat{y}_r}_{\tilde{\varepsilon}_r} + \underbrace{\frac{\hat{\tilde{P}}_-}{1+\hat{C}P} \ell \hat{y}_n}_{\tilde{\varepsilon}_n} + \underbrace{\tilde{G}_n \hat{\tilde{P}}_- R_n \ell_n y_n}_{\tilde{\varepsilon}_{\Delta R_n}} + \underbrace{\left(1 - R_n^o \tilde{G}_n \hat{\tilde{P}}_-\right) y_n}_{\tilde{\varepsilon}_o^o} \\
&= \frac{\hat{\tilde{P}}_-}{1+\hat{C}P} \ell \left(\hat{y}_r - \hat{y}_n\right) + \hat{\tilde{P}}_- \ell_n \hat{y}_n + \tilde{\boldsymbol{\varepsilon}}_o^o
\end{aligned} \tag{10.3.29}$$

Here $\tilde{\boldsymbol{\varepsilon}}_o^o$ is the error when the optimal predictor is applied and $\ell_n = 0$, and the error owed to the nonideal predictor is

$$\tilde{\boldsymbol{\varepsilon}}_{\Delta R_n} = \tilde{G}_n \hat{\tilde{P}}_- R_n \ell_n y_n = \hat{\tilde{P}}_- \ell_n y_n \tag{10.3.30}$$

$\hat{y}_r$ and $\hat{y}_n$ are filter variables computed according to

$$\hat{y}_r = R_r \hat{G}_r y_r; \quad \hat{y}_n = R_n \hat{G}_n y_n \tag{10.3.31}$$

More interesting results can be obtained by investigating the error signal in the frequency domain. Four methods will be compared:

1. The first method is the open loop, parallel model-based identification method shown in Figure 10.3.1.
2. The second one is the *KB*-parameterized identification method shown in Figure 10.3.7.

3. The third identification method is modeling the parallel closed control loop.
4. The fourth method is presented in Figure 10.3.14, where the parallel model is in the closed loop.

In the first method the following frequency-dependent weighting function can be recognized in the identification error

$$\varepsilon_{\mathrm{ID}} = \varepsilon_{\mathrm{ID}}^{1} = \left(P - \hat{P}\right)u = \hat{P}\frac{\Delta}{\hat{P}}u = W_1 \ell u; \quad W_1^{\mathrm{o}} = W_1\big|_{\ell=0} = \hat{P}\big|_{\ell=0} = P \tag{10.3.32}$$

In the second method, using (10.3.26) we obtain

$$\varepsilon_{\mathrm{ID}} = \varepsilon_{\mathrm{ID}}^{2} = \frac{1}{1+\hat{C}P}\ell y_{\mathrm{r}} + \frac{1}{1+\hat{C}P}y_{\mathrm{n}} = W_2 \ell y_{\mathrm{r}} + \tilde{S} y_{\mathrm{n}} \tag{10.3.33}$$

where

$$W_2^{\mathrm{o}} = W_2\big|_{\ell=0} = \frac{1}{1+\hat{C}P}\bigg|_{\ell=0} = \frac{1}{1+CP} = S \tag{10.3.34}$$

This is the only method where the equation $e = e_2 = -\varepsilon_{\mathrm{ID}}^{2}$ exists for the control error.

In the third method the identification error is

$$\varepsilon_{\mathrm{ID}} = \varepsilon_{\mathrm{ID}}^{3} = \frac{\hat{C}\hat{P}}{\left(1+\hat{C}P\right)\left(1+\hat{C}\hat{P}\right)}\ell y_{\mathrm{r}} + \frac{1}{1+\hat{C}P}y_{\mathrm{n}} = W_3 \ell y_{\mathrm{r}} + \tilde{S} y_{\mathrm{n}} \tag{10.3.35}$$

where

$$W_3^{\mathrm{o}} = W_3\big|_{\ell=0} = \frac{\hat{C}\hat{P}}{\left(1+\hat{C}P\right)\left(1+\hat{C}\hat{P}\right)}\Bigg|_{\ell=0} = \frac{CP}{(1+CP)^2} = \frac{L}{(1+L)^2} \tag{10.3.36}$$

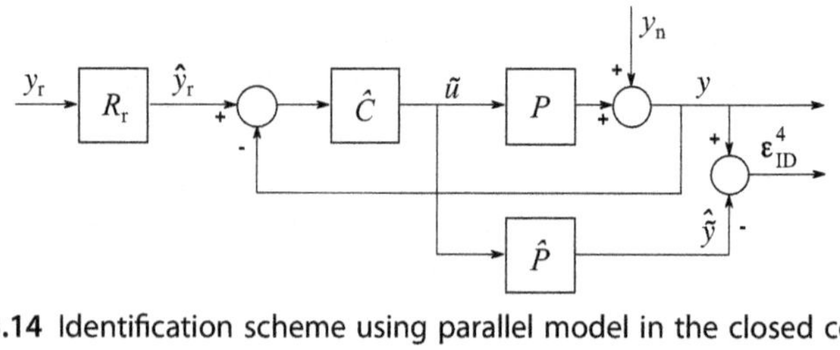

Figure 10.3.14 Identification scheme using parallel model in the closed control loop.

This is the most frequently used and investigated method in the practice of iterative identification/control tasks. Note that the maximum of the weight $|W_3^{\rm o}|$ is at $|L| = 1$, and this condition corresponds to the cutoff frequency $\omega_{\rm c}$. In the two loops the control errors are different

$$\tilde{e} = \tilde{e}_3 = \frac{1}{1+\hat{C}P}\hat{y}_{\rm r} + \frac{1}{1+\hat{C}P}y_{\rm n} \quad \text{and} \quad \hat{e} = \hat{e}_3 = \frac{1}{1+\hat{C}\hat{P}}\hat{y}_{\rm r} \tag{10.3.37}$$

The fourth method more or less corresponds to the so-called *self-tuning* (ST) regulator executing the adaptive version of the MV regulator, if the identification is based on the optimal predictor providing independent error. In other cases the method is unfavorable from an identification perspective, since it does not ensure the separateness of the input and the source noise. Now the identification error is

$$\varepsilon_{\rm ID} = \varepsilon_{\rm ID}^4 = \frac{\hat{C}\hat{P}}{1+\hat{C}P}\ell y_{\rm r} + \left(1 - \frac{\hat{C}\Delta}{1+\hat{C}P}\right)y_{\rm n} = W_4\ell\hat{y}_{\rm r} + \tilde{S}_4(\Delta)y_{\rm n} \tag{10.3.38}$$

where

$$W_4^{\rm o} = W_4\big|_{\ell=0} = \left.\frac{\hat{C}P}{1+\hat{C}P}\right|_{\ell=0} = \tilde{T}\big|_{\ell=0} = \frac{CP}{1+CP} \tag{10.3.39}$$

and the control error becomes

$$\tilde{e} = \tilde{e}_4 = \frac{1}{1+\hat{C}P}R_{\rm r}y_{\rm r} + \frac{1}{1+\hat{C}P}y_{\rm n} = \frac{1}{1+\hat{C}P}\hat{y}_{\rm r} + \frac{1}{1+\hat{C}P}y_{\rm n} \tag{10.3.40}$$

Observe that $\tilde{S}_4(\Delta)\big|_{\Delta=0} = 1$, which means that if the algorithm converges, that is, $\Delta \rightarrow 0$, then the additive noise becomes independent (special feature of the ST regulators).

Table 10.3.2 summarizes the weighting function $W_{\rm x}^{\rm o}$ obtained in the four cases. The weights are computed in all cases for the ideal $\Delta = 0$.

Table 10.3.2 Weighting functions in the different identification methods

$W_1^{\rm o}$	$W_2^{\rm o}$	$W_3^{\rm o}$	$W_4^{\rm o}$
P	$\frac{1}{1+CP}$	$\frac{CP}{(1+CP)^2}$	$\frac{CP}{1+CP}$

The theoretical and experimental result is that the modeling error is smallest in the frequency region where the weights are the greatest. The weights W_1^o and W_4^o emphasize the low frequency regions, and W_2^o emphasizes the high frequencies. The weight W_3^o, however, emphasizes the medium frequency region in the vicinity of the cutoff frequency. Given the above it can be stated that W_1^o and W_4^o are not perfect in iterative design schemes because they provide the worst model in those regions where the features of the regulator are intended to be improved. The weight W_3^o can be applied in iterative methods, since it provides the smallest identification error in the vicinity of ω_c, but a cautious ("windsurfing" [1]) procedure is required because the model is very false in the other frequencies. Therefore, the weight W_2^o is considered the best because it emphasizes the high frequencies, and its precise model makes it possible to increase the bandwidth of the control loop via the small identification error. As a compromise it might be the best, if the combined identification and control design is started by W_2^o, then in the vicinity of the optimum the iteration is stopped by W_3^o.

Typical shapes of the above weighting functions penalizing the modeling error in the different identification methods are shown in Figure 10.3.15.

It is very interesting to compare the identification and control errors. Consider the simple case without output disturbance, that is, $y_n = 0$, where

$$\boldsymbol{\varepsilon}_{\mathrm{ID}} = \boldsymbol{\varepsilon}_{\mathrm{ID}}^2 = -\tilde{e} = -\tilde{e}_2 \tag{10.3.41}$$

$$\boldsymbol{\varepsilon}_{\mathrm{ID}} = \boldsymbol{\varepsilon}_{\mathrm{ID}}^3 = \boldsymbol{\Delta}\frac{\hat{C}}{1+\hat{C}\hat{P}}\tilde{e} = \boldsymbol{\Delta}\frac{\hat{C}}{1+\hat{C}\hat{P}}\tilde{e}_3 = \boldsymbol{\Delta}\hat{S}\hat{C}\tilde{e}_3 = \boldsymbol{\Delta}\hat{Q}\tilde{e}_3 \tag{10.3.42}$$

$$\boldsymbol{\varepsilon}_{\mathrm{ID}} = \boldsymbol{\varepsilon}_{\mathrm{ID}}^4 = \boldsymbol{\Delta}\hat{C}\tilde{e} = \boldsymbol{\Delta}\hat{C}\tilde{e}_4 \tag{10.3.43}$$

Equations (10.3.42) and (10.3.43) show phenomena which are very uncertain and not much discussed in the literature. It can cause fundamental problems when these identification methods are applied in adaptive and iterative control schemes. The trouble is that $\boldsymbol{\varepsilon}_{\mathrm{ID}}^3$ and $\boldsymbol{\varepsilon}_{\mathrm{ID}}^4$ can become zero even if the model error $\boldsymbol{\Delta}$ is not zero, if the relating control errors $\tilde{e}_3$ and $\tilde{e}_4$ are zero. The simplest case when this problem appears is when the closed-loop control gets excitation causing zero control error. In these methods, persistent excitation is extremely important.

Considering the different gap metrics discussed in Section 9.3 it is easy to find which gap metric is the best to compare identified models at the

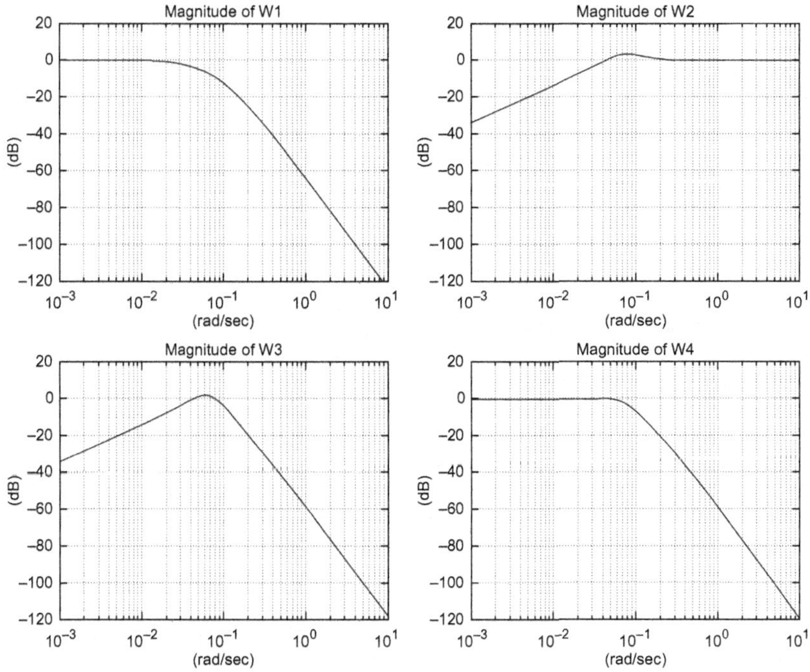

Figure 10.3.15 Typical forms of the weighting functions penalizing the modeling error in the different identification methods.

above identification methods. The classical Vinnicombe metric is best for the weight W_3^o, which has a maximum around the cutoff frequency ω_c. From the newly introduced metrics the $K-$ sphere-based metric fits to the weight W_2^o corresponds to the KB parameterization-based identification. The $KM-$ sphere-based metric should be used to the weight W_4^o corresponds to the identification method using parallel model in the closed loop.

CHAPTER 11

Adaptive Regulators and Iterative Tuning

The preferred topic in the control engineering speaking about the model-based control methods has no vital importance. Namely, the control design, except in some special cases, is always based on models. The main question is how the model determination (generally identification from experiments, measurements) relates to the control design. The well-known classical case is shown in Figure 11.1 where the identification is performed independently of the closed-loop control and then, in the next step, the regulator is designed and the resulting regulator is applied in the closed loop. The result of the design determines the regulator $\hat{C}(\hat{G})$ belonging to the given model $\hat{G}$ in both structure and parameters. (Since the realization is always discrete-time (DT), the process is denoted shortly by G and its model by $\hat{G}$.)

In the technological environment open-loop measurements are generally not welcome, so the above classical method can be applied very rarely. It is mostly used in laboratory tests or processes when improper operation has no high risk or costs.

It can be generally stated that the better the process model, the better the control design. Therefore it is a continuous requirement to improve the

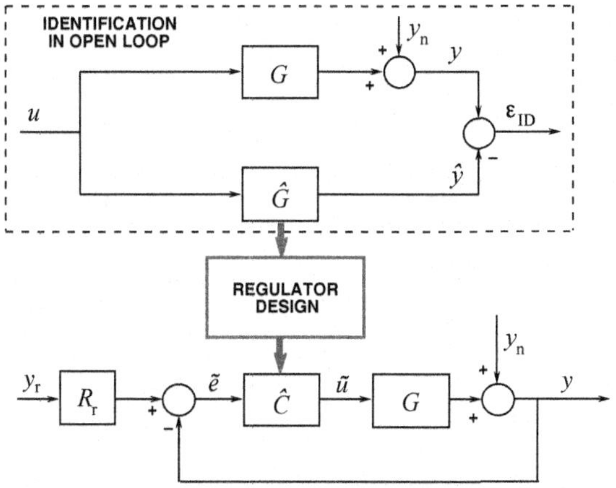

Figure 11.1 Identification in open loop and control design.

quality of the model obtained from the *identification* (ID). It can be done by repeated measurement (experiment). Techniques where the new measurements are used to renew the model parameter estimation in each sampling are called learning or adaptive methods referring both to the renewal of the process model and to the redesign of the regulator. Thus learning adaptive methods apply the recursive ID methods discussed in Section 10.2. These algorithms are usually called online ones, too. Recursive ID methods are suitable also for tracking the changing parameters of the process, and that is why they are called learning (*online*) algorithms. This procedure is also called adaptive control, in some publications dual control.

The formation of an adequate learning algorithm (or forgetting algorithm) is not a simple task and requires deep knowledge. This explains why it is widely used when longer measurement records (with N samples) are processed by usually off-line (*batch*) identification techniques and the control design is based on the result of this procedure. In this measurement situation, the closed system is not opened and the data are collected as normal. This procedure is known as *simultaneous identification and control* iterative strategy. In the next off-line experiment, the formerly obtained best model is obviously used. By this technique the optimality of the regulator can be gradually improved as the model becomes more and more accurate, while the operation of the process is only slightly disturbed.

After selection of the estimation criterion and model structure, the accuracy of the estimation cannot be improved unless the excitation signal is optimized. This is valid for both open-loop and closed-loop ID. The most reasonable option in the closed-loop case is to choose a proper reference signal. The optimization of the input signal is known as *optimal input design*, and its methods were discussed in Section 4.3. It is therefore possible to elaborate adaptive learning control where the exciting signal is also optimized, a technique called *triple control*.

Modern microprocessor-based compact regulators already have certain embedded facilities for automatic tuning. Such commercial regulators usually apply the tuning method of Åström [9], which generates oscillations of small amplitude by substituting the classical regulator with a relay of hysteresis characteristic of a certain tuning period. The *proportional-integral-derivative* (*PID*) regulator parameters are computed from this transient. The method does not use the true model.

Today adaptive control is a separate discipline. This book does not give a complete survey of the subject but concentrates only on the control design methods discussed above, especially on closed-loop identification using the *Keviczky–Bányász* (*KB*) parameterization and *Youla parameterization*-based (*YP*) control design techniques.

In open-loop identification, we practically always have the chance of iterative adaptation. First, a model class $\mathcal{M}$ needs to be chosen (e.g., linear dynamic n order processes) and an identification criterion $J_{\mathrm{ID}}(\boldsymbol{\varepsilon}_{\mathrm{ID}})$ (e.g., least squares (*LS*) method). Here $\boldsymbol{\varepsilon}_{\mathrm{ID}}$ is the modeling error. Similarly a control class (type) C needs to be selected and a control criterion $J_{\mathrm{C}}(e_{\mathrm{C}})$ (e.g., optimization according to the $\mathcal{H}_2$ norm). e_{C} is the control error. The process is as follows:

1. Determine the input signal series and apply to the input of the process. At the same time, start the collection of the N data pairs $\{u[k],\ y[k];\ k=1,\ldots,N\}$. (In the *multiple input multiple output* (*MIMO*) case the input and output signal are vectors, and thus $\{\boldsymbol{u}[k],\ \boldsymbol{y}[k];\ k=1,\ldots,N\}$.)
2. Then the estimation of the process parameters, that is, the ID of the process needs to be performed. This step can be formulated by the following expression

$$\begin{aligned}\hat{P} &= \arg\min_{\hat{P}\in\mathcal{M}} J_{\mathrm{ID}}(\boldsymbol{\varepsilon}_{\mathrm{ID}}) = \arg\min_{\hat{P}\in\mathcal{M}} J_{\mathrm{ID}}(\boldsymbol{\varepsilon}_{\mathrm{ID}})\Big|_{\{u[k],\ y[k];\ k=1,\ldots,N\}} \\ &= \hat{P}\big(\{u[k]; k=1,\ldots,N\}, \mathcal{M}, P\big)\end{aligned} \tag{11.1}$$

3. Formally, the regulator can be obtained by the solution of the following optimization task

$$\hat{C} = \arg\min_{\hat{C}\in C} J_{\mathrm{ID}}(e_{\mathrm{C}}) = \arg\min_{\hat{C}\in C} J_{\mathrm{ID}}\left(e_{\mathrm{C}}, \hat{G}\right) = \hat{C}_{\mathrm{opt}}\left(C, \hat{G}\right) \tag{11.2}$$

4. Investigate the accuracy of the obtained model. If it is acceptable then the iteration is stopped, if not, then the successive improvement of the model and the regulator is continued.
5. Based on the results, the input signal series can be optimized by the methods discussed in Sections 11.3 and 4.3. Choose the $\mathcal{D}$ design method according to which

$$\{u[k]; k=1,\ldots,N\} = \mathcal{D}\left\{\mathcal{M}, \hat{G}, U_{\max}\right\} \tag{11.3}$$

Then using the generated optimal input signal, turn back to the first point and continue the method. Here the input signal has the amplitude constraint $U_{\max}$.

11.1 ALGORITHMS OF ADAPTIVE LEARNING METHODS

The recursive parameter estimation technique of adaptive learning control improves the parameter estimation of the process in each sampling point, and the output of the resulting regulator is put to the process with the delay caused by the computation of the optimal regulator. By applying modern, fast, computational devices to slower processes, the delay can be almost eliminated, that is, the identification and control can be considered as a simultaneous step. This strategy is called adaptive control. The elaboration of a reliably working adaptive regulator is not an easy task. A recursive parameter estimation algorithm is required, which does not forget the process model learned if the measurements do not have a large amount of new information. If the quantity of the new information is, however, considerable, then the method can follow the slow parameter changes with a suitable forgetting strategy.

In the case of certain so-called predictive regulators, the process model is not directly identified but the determination of the process model is always hidden somehow in these algorithms.

In adaptive control, the control design algorithm usually takes over the parameters renewed in the identification step and does not investigate the fact that they derive from a statistical estimation. This approach is known as the *certainty equivalence principle*. There is a way to elaborate methods by taking the statistical behavior of the parameter estimations into account. This approach leads to the so-called *cautious control* regarding the modification of the control parameters. The canonical forms of these methods were developed by Tsypkin [149,150], who proved that methods developed earlier are based practically on the same principle and the different approaches can be formally arranged into identical schemes.

The following ID and control target functions need to be minimized

$$V_{\mathrm{ID}}\left(\hat{\boldsymbol{p}}_{\mathrm{ID}}[k], \hat{\boldsymbol{p}}_{\mathrm{C}}[k] \,\middle|\, k-1, \mathcal{U}[k-1], \mathcal{Y}[k-1]\right) \tag{11.1.1}$$

and

$$V_{\mathrm{C}}\left(\hat{\boldsymbol{p}}_{\mathrm{C}}[k], \hat{\boldsymbol{p}}_{\mathrm{ID}}[k] \,\middle|\, k-1, \mathcal{U}[k-1], \mathcal{Y}[k-1]\right) \tag{11.1.2}$$

Here $\hat{\boldsymbol{p}}_{\mathrm{ID}}[k]$ and $\hat{\boldsymbol{p}}_{\mathrm{C}}[k]$ are the parameter vectors of the process model and the regulator, respectively; $\mathcal{U}[k-1]$ and $\mathcal{Y}[k-1]$ are the data set of the input and output signals available until the sampling point $[k-1]$. After computing $\hat{\boldsymbol{p}}_{\mathrm{C}}[k]$ the control algorithm determines $u[k]$ and putting it to process the answer $y[k]$ of the process is obtained. The target functions usually have the forms

$$\begin{aligned} & V_{\mathrm{ID}}\left(\hat{\boldsymbol{p}}_{\mathrm{ID}}[k], \hat{\boldsymbol{p}}_{\mathrm{C}}[k] \middle| k-1, \mathcal{U}[k-1], \mathcal{Y}[k-1]\right) \\ & \quad = E\left\{Q_{\mathrm{ID}}\left(\hat{\boldsymbol{p}}_{\mathrm{ID}}[k], \hat{\boldsymbol{p}}_{\mathrm{C}}[k] \middle| k-1, \mathcal{U}[k-1], \mathcal{Y}[k-1]\right)\right\} \end{aligned} \tag{11.1.3}$$

$$\begin{aligned} & V_{\mathrm{C}}\left(\hat{\boldsymbol{p}}_{\mathrm{C}}[k], \hat{\boldsymbol{p}}_{\mathrm{ID}}[k] \middle| k-1, \mathcal{U}[k-1], \mathcal{Y}[k-1]\right) \\ & \quad = E\left\{Q_{\mathrm{C}}\left(\hat{\boldsymbol{p}}_{\mathrm{C}}[k], \hat{\boldsymbol{p}}_{\mathrm{ID}}[k] \middle| k-1, \mathcal{U}[k-1], \mathcal{Y}[k-1]\right)\right\} \end{aligned} \tag{11.1.4}$$

and

$$\hat{\boldsymbol{p}}_{\mathrm{ID}}[k] = \arg\min_{\hat{\boldsymbol{p}}_{\mathrm{ID}}[k]} E\left\{Q_{\mathrm{ID}}\left(\hat{\boldsymbol{p}}_{\mathrm{ID}}[k], \hat{\boldsymbol{p}}_{\mathrm{C}}[k] \middle| k-1, \mathcal{U}[k-1], \mathcal{Y}[k-1]\right)\right\} \tag{11.1.5}$$

$$\hat{\boldsymbol{p}}_{\mathrm{C}}[k] = \arg\min_{\hat{\boldsymbol{p}}_{\mathrm{C}}[k]} E\left\{Q_{\mathrm{C}}\left(\hat{\boldsymbol{p}}_{\mathrm{ID}}[k], \hat{\boldsymbol{p}}_{\mathrm{C}}[k] \middle| k-1, \mathcal{U}[k-1], \mathcal{Y}[k-1]\right)\right\} \tag{11.1.6}$$

Tsypkin recommended looking for a solution among the following algorithm family

$$\begin{aligned} \hat{\boldsymbol{p}}_{\mathrm{ID}}[k] = {} & \hat{\boldsymbol{p}}_{\mathrm{ID}}[k-1] \\ & - \boldsymbol{R}_{\mathrm{ID}}[k]\left\{\boldsymbol{I} + \boldsymbol{S}_{\mathrm{C}}^{\mathrm{ID}}[k]\right\} \frac{\mathrm{d}Q_{\mathrm{ID}}\left(\hat{\boldsymbol{p}}_{\mathrm{ID}}[k-1], \hat{\boldsymbol{p}}_{\mathrm{C}}[k-1] \middle| k-1, \mathcal{U}[k-1], \mathcal{Y}[k-1]\right)}{\mathrm{d}\hat{\boldsymbol{p}}_{\mathrm{ID}}[k-1]} \end{aligned} \tag{11.1.7}$$

$$\begin{aligned} \hat{\boldsymbol{p}}_{\mathrm{C}}[k] = {} & \hat{\boldsymbol{p}}_{\mathrm{C}}[k-1] \\ & - \boldsymbol{R}_{\mathrm{C}}[k]\left\{\boldsymbol{I} + \boldsymbol{S}_{\mathrm{ID}}^{\mathrm{C}}[k]\right\} \frac{\mathrm{d}Q_{\mathrm{C}}\left(\hat{\boldsymbol{p}}_{\mathrm{C}}[k-1], \hat{\boldsymbol{p}}_{\mathrm{ID}}[k-1] \middle| k-1, \mathcal{U}[k-1], \mathcal{Y}[k-1]\right)}{\mathrm{d}\hat{\boldsymbol{p}}_{\mathrm{C}}[k-1]} \end{aligned} \tag{11.1.8}$$

where

$$\boldsymbol{S}_{\mathrm{C}}^{\mathrm{ID}}[k] = \boldsymbol{S}_{\mathrm{C}}^{\mathrm{ID}}[k-1] + \boldsymbol{R}_{\mathrm{C}}^{\mathrm{S}}[k] \frac{\mathrm{d}\hat{\boldsymbol{p}}_{\mathrm{ID}}[k]}{\mathrm{d}\hat{\boldsymbol{p}}_{\mathrm{C}}[k]} \tag{11.1.9}$$

$$\boldsymbol{S}_{\mathrm{ID}}^{\mathrm{C}}[k] = \boldsymbol{S}_{\mathrm{ID}}^{\mathrm{C}}[k-1] + \boldsymbol{R}_{\mathrm{ID}}^{\mathrm{S}}[k] \frac{\mathrm{d}\hat{\boldsymbol{p}}_{\mathrm{C}}[k]}{\mathrm{d}\hat{\boldsymbol{p}}_{\mathrm{ID}}[k]} \tag{11.1.10}$$

Here $R_{\mathrm{ID}}[k]$, $R_{\mathrm{C}}[k]$, $R_{\mathrm{ID}}^{\mathrm{S}}[k]$, $R_{\mathrm{C}}^{\mathrm{S}}[k]$ are the convergence matrices of the algorithms, $\boldsymbol{S}_{\mathrm{ID}}^{\mathrm{C}}[k]$ and $\boldsymbol{S}_{\mathrm{C}}^{\mathrm{ID}}[k]$ are the so-called sensitivity matrices. These latter matrices represent the following Jacobi matrices of the interactions

$$\boldsymbol{J}_{\mathrm{ID}}^{\mathrm{C}}[k] = \frac{\mathrm{d}\hat{\boldsymbol{p}}_{\mathrm{C}}[k]}{\mathrm{d}\hat{\boldsymbol{p}}_{\mathrm{ID}}[k]} \tag{11.1.11}$$

$$\boldsymbol{J}_{\mathrm{C}}^{\mathrm{ID}}[k] = \frac{\mathrm{d}\hat{\boldsymbol{p}}_{\mathrm{ID}}[k]}{\mathrm{d}\hat{\boldsymbol{p}}_{\mathrm{C}}[k]} \tag{11.1.12}$$

The collection of the algorithms (11.1.7)–(11.1.10) is called dual control by Tsypkin. The name refers to the canonical dual relation between identification and control. (In the literature, other algorithms of adaptive control are also called dual control.)

In most cases $\hat{\boldsymbol{p}}_{\mathrm{ID}}[k]/\hat{\boldsymbol{p}}_{\mathrm{C}}[k] \equiv 0$, and therefore the canonical adaptive process control algorithm (11.1.7) becomes

$$\hat{\boldsymbol{p}}_{\mathrm{ID}}[k] = \hat{\boldsymbol{p}}_{\mathrm{ID}}[k-1] - \boldsymbol{R}_{\mathrm{ID}}[k] \frac{\mathrm{d}Q_{\mathrm{ID}}\left(\hat{\boldsymbol{p}}_{\mathrm{ID}}[k-1], \hat{\boldsymbol{p}}_{\mathrm{C}}[k-1] \middle| k-1, \mathcal{U}[k-1], \mathcal{Y}[k-1]\right)}{\mathrm{d}\hat{\boldsymbol{p}}_{\mathrm{ID}}[k-1]} \tag{11.1.13}$$

(In closed-loop ID, the accuracy of the process parameter estimations obviously depends on the control parameters. This second-order dependence, however, cannot be expressed in analytical (explicit) form and therefore it is usually neglected.)

Similarly, it is exceptional that adaptive estimation of the control parameters is done according to (11.1.8). In practical applications, on the one hand, more steps are wanted until the model parameters become accurate, and minimization (11.1.6), on the other hand, is performed in one step according to the following expression

$$\hat{\boldsymbol{p}}_{\mathrm{C}}[k] = \arg \min_{\hat{\boldsymbol{p}}_{\mathrm{C}}[k]} Q_{\mathrm{C}}\left(\hat{\boldsymbol{p}}_{\mathrm{ID}}[k], \hat{\boldsymbol{p}}_{\mathrm{C}}[k]\right) \tag{11.1.14}$$

and the computation of the sensitivity model or (11.1.8) and (11.1.10) in the complete recursive procedure can be omitted.

Investigate the adaptive process identification of the dynamic linear systems, where the reasonable choice of the target function (11.1.3) leads to a model linear in parameters

$$Q_{\mathrm{ID}}\left(\hat{\boldsymbol{p}}_{\mathrm{ID}}[k] \middle| k-1, \mathcal{U}[k-1], \mathcal{Y}[k-1]\right) = \frac{1}{2}\left\{y[k] - \boldsymbol{f}^{\mathrm{T}}[u, y, k-1]\hat{\boldsymbol{p}}_{\mathrm{ID}}\right\}^2 \tag{11.1.15}$$

This choice suits the *LS* method discussed in Sections 10.1 and 10.2. With this method the adaptive algorithm has the form

$$\hat{\boldsymbol{p}}_{\mathrm{ID}}[k] = \hat{\boldsymbol{p}}_{\mathrm{ID}}[k-1] - \boldsymbol{R}_{\mathrm{ID}}[k]\,\boldsymbol{f}[u,y,k-1]\{y[k] - \boldsymbol{f}^{\mathrm{T}}[u,y,k-1]\hat{\boldsymbol{p}}_{\mathrm{ID}}\} \tag{11.1.16}$$

The canonical forms belong to the family of stochastic approximation algorithms. The optimal choice of $\boldsymbol{R}_{\mathrm{ID}}[k]$ in (11.1.13) is obtained in the form given in (11.1.18), if the Hesse matrix of the second-order partial derivates of the target function is

$$\boldsymbol{H}[k] = \left[\frac{\partial^2 Q_{\mathrm{ID}}(\hat{\boldsymbol{p}}_{\mathrm{ID}}[k]\,|\,k-1, \mathcal{U}[k-1], \mathcal{Y}[k-1])}{\partial\hat{p}^{i}_{\mathrm{ID}}\partial\hat{p}^{j}_{\mathrm{ID}}}\right] = \boldsymbol{f}[u,y,k]\,\boldsymbol{f}^{\mathrm{T}}[u,y,k] \tag{11.1.17}$$

$$\boldsymbol{R}_{\mathrm{ID}}[k] = \left\{\sum_{j=1}^{k}\boldsymbol{H}[k]\right\}^{-1} = \left\{\sum_{j=1}^{k}\boldsymbol{f}[u,y,k]\boldsymbol{f}^{\mathrm{T}}[u,y,k]\right\}^{-1} \tag{11.1.18}$$

Since $\boldsymbol{R}_{\mathrm{ID}}[k]$ is a dyadic sum, the derivations (A.6.26)–(A.6.28) can also be applied, according to which

$$\boldsymbol{R}_{\mathrm{ID}}[k] = \boldsymbol{R}_{\mathrm{ID}}[k-1] - \frac{\boldsymbol{R}_{\mathrm{ID}}[k-1]\boldsymbol{f}[u,y,k]\boldsymbol{f}^{\mathrm{T}}[u,y,k]\boldsymbol{R}_{\mathrm{ID}}[k-1]}{1 + \boldsymbol{f}^{\mathrm{T}}[u,y,k]\boldsymbol{R}_{\mathrm{ID}}[k-1]\boldsymbol{f}[u,y,k]} \tag{11.1.19}$$

Thus the same algorithm is obtained which was investigated in the recursive *LS* method. The optimal stochastic approximation algorithm of the adaptive ID is equal to the *LS* method in the case of $\lambda = 1$. It follows from the complete conformity that the statements made for the forgetting strategies of the recursive *LS* method can also be applied here.

Thus the following steps have to be performed at the adaptive dual control in a sampling cycle.

1. Using the a priori process model $\hat{G}[k-1]$, the optimal regulator can be formally computed by solving the optimization task

$$\begin{aligned}\hat{C}[k] &= \arg\min_{\hat{C}[k]\in C} J_{\mathrm{ID}}(e_{\mathrm{C}}[k]) = \arg\min_{\hat{C}[k]\in C} J_{\mathrm{ID}}\left[\tilde{e}[k], \hat{G}[k-1]\right] \\ &= \hat{C}_{\mathrm{opt}}\left(C, \hat{G}[k-1]\right)\end{aligned} \tag{11.1.20}$$

2. Determine the actual value of the reference signal $y_{\mathrm{r}}[k]$ and then add it to the input of the closed system.

3. In the knowledge of the new regulator $\hat{C}[k]$ and the reference signal, determine the actual value of the actuating signal and put it to the input of the process. The ID algorithm works from $\hat{\tilde{u}}[k]$ and the obtained process output signal $y[k]$. (In multivariable systems these signals are vectors, i.e., $\boldsymbol{y}[k]$ and $\hat{\tilde{\boldsymbol{u}}}[k]$.)
4. Then the recursive estimation of the process parameters is performed from the measured data. This step means the updating equations of the convergence matrix $\boldsymbol{R}[k] = \boldsymbol{R}_{\mathrm{ID}}[k]$ and parameter vector $\hat{\boldsymbol{p}}_{\mathrm{ID}}[k]$. At the renewing of $\boldsymbol{R}[k] = \boldsymbol{R}_{\mathrm{ID}}[k]$ the forgetting strategy discussed in Section 10.2 needs to be reasonably applied. The updating of $\hat{\boldsymbol{p}}_{\mathrm{ID}}[k]$ is made according to (10.1.13). (In multivariable case, the equation of the recursive parameter estimation is the same as (10.2.15).)
5. In each step, the merit of the obtained model and the applicability of the regulator resulting from the optimization are investigated. If necessary, the forgetting strategy is changed.
6. Because of basic character of the recursive operation, the memories of the DT filters of both the process model and the regulator are prepared for the next sampling cycle (shifting or stack operation).
7. In the next cycle, the operation is continued from the first point.

The block scheme of the adaptive dual control discussed in the above steps is presented in Figure 11.1.1 in the $[k]$-th sampling cycle.

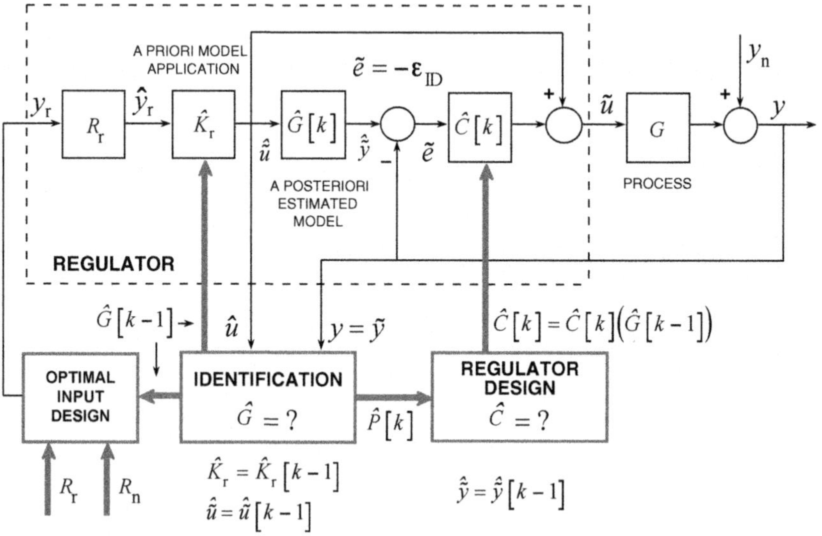

Figure 11.1.1 Block scheme of the adaptive dual control in closed loop.

The above adaptive dual algorithm, which actually consists of a recursive identification step and the computation of the regulator based on the obtained parameter estimation, is usually called indirect adaptive control.

Simulation Examples

Example 11.1.1

Let the process be a second-order time-delay system whose pulse transfer function is

$$G = \frac{0.001(1.1z^{-1} + 1z^{-2})}{1 - 1.6693z^{-1} + 0.7788z^{-2}} z^{-1} = \frac{0.001(1.1 + 1z^{-1})}{1 - 1.6693z^{-1} + 0.7788z^{-2}} z^{-2} \tag{11.1.21}$$

This equation describes the second-order effect of the rod control of a helicopter on the sideslip angle. Here the sampling time $T_s = 0.05$ [sec] is applied and the time delay of the process is $d = 1$.

The following reference models with unity gain are used in the design

$$R_r = \frac{0.5z^{-1}}{1 - 0.5z^{-1}} \quad \text{and} \quad R_n = \frac{0.2z^{-1}}{1 - 0.8z^{-1}} \tag{11.1.22}$$

To start the iteration the following initial model

$$\hat{G}_o = \frac{0.001(20 + 0.5z^{-1})}{1 - 1.5z^{-1} + 0.8z^{-2}} z^{-2} \tag{11.1.23}$$

is assumed and the process time delay is assumed to be known. A square signal with periodic time of 40 samples is applied as reference signal. In the simulation it is assumed that the additive noise y_n is white noise, whose variance is small, that is, $\sigma_{y_n} = 0.01$. The regulator is designed by the *YP* method, assuming an *inverse stable* (*IS*) process. Adaptive dual control is applied.

The reference model R_r and the process output, the control and identification errors are shown in Figure 11.1.2 during adaptive dual control using the *LS* method. The initial value of the convergence matrix was $\boldsymbol{R}[0] = 100\boldsymbol{I}$ and the forgetting factor had the value $\lambda^2 = 0.95$.

Example 11.1.2

Let the process be a first-order time-delay system whose pulse transfer function is

$$G = \frac{0.007869z^{-1}}{1 - 0.606531z^{-1}} z^{-3} = \frac{0.007869}{1 - 0.606531z^{-1}} z^{-4} \tag{11.1.24}$$

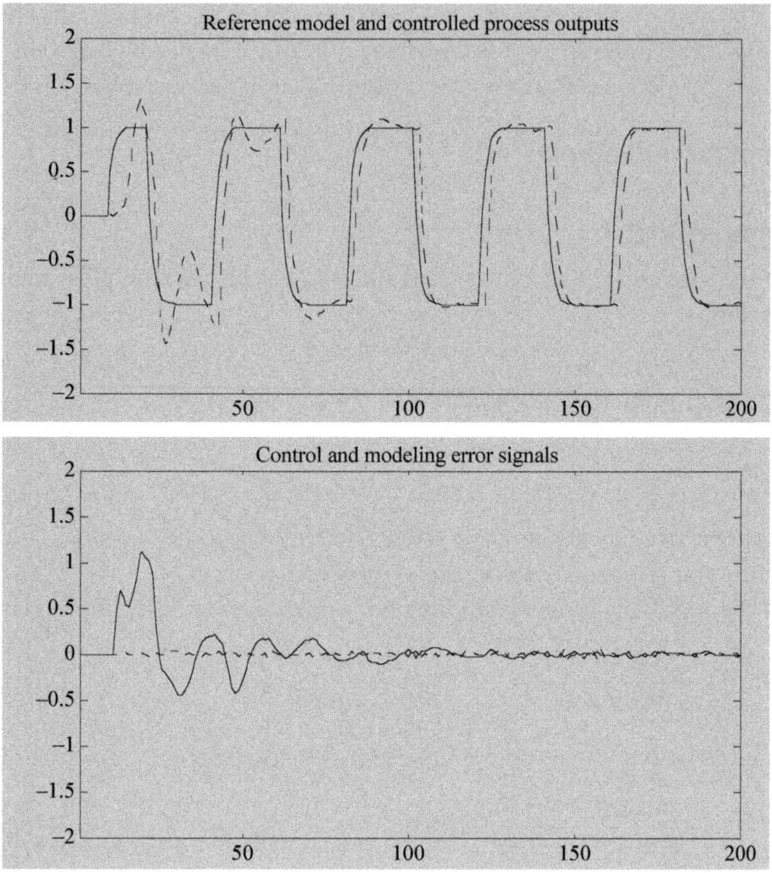

Figure 11.1.2 Plot of the reference model R_r and the process outputs, the control and modelling errors during the adaptive dual control (second-order example).

This form is the simpler, first-order effect of the rod control of the helicopter on the sideslip angle. The applied sampling time is $T_s = 0.05$ [sec] and the time delay of the process is $d = 3$. The same reference models, output noise, and reference excitation are used as in the previous example. Let the initial model be

$$\hat{G}_o = \frac{0.01z^{-1}}{1 - 0.4z^{-1}} \tag{11.1.25}$$

The adaptive dual control is performed with a *YP* regulator and an *IS* process is assumed. The reference model R_r and the process outputs, the control and identification errors are shown in Figure 11.1.3 during

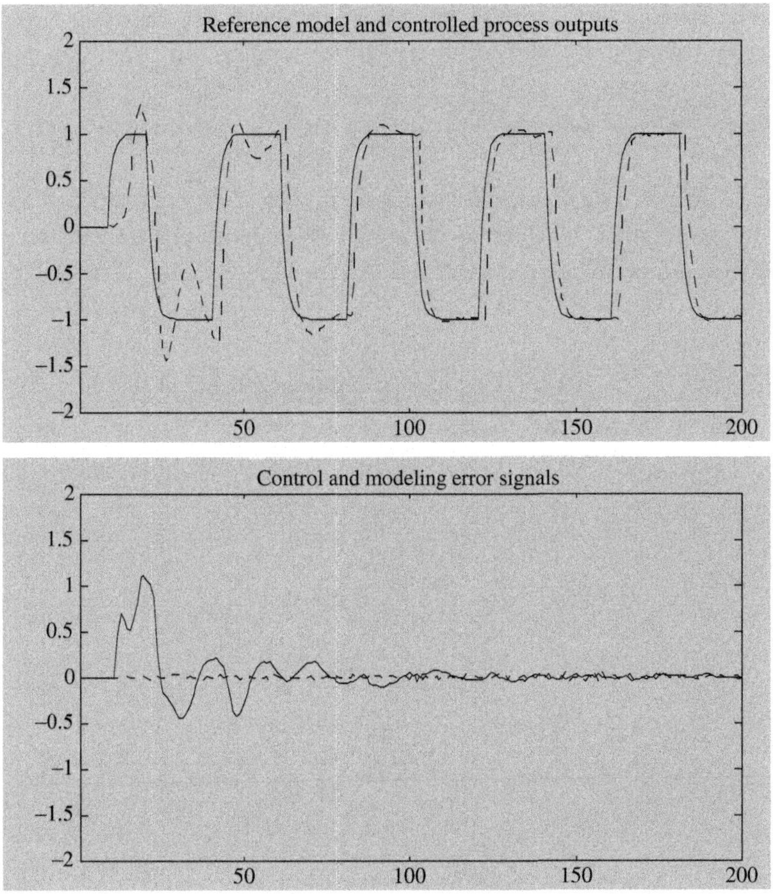

Figure 11.1.3 Plot of the reference model R_r and the process outputs, the control and modelling errors during adaptive dual control (first-order example).

adaptive dual control using the *LS* method. The initial value of the covariance matrix was $\boldsymbol{R}[0] = 100\boldsymbol{I}$ and the forgetting factor had the value $\lambda^2 = 0.95$.

Example 11.1.3

Let the system be a process with unstable zero, whose pulse transfer function is

$$G = \frac{0.0364z^{-1}\left(1 + 1.2z^{-1}\right)}{1 - 1.6z^{-1} + 0.68z^{-2}} \tag{11.1.26}$$

The same reference models and output noise are applied as in the previous example. Let the initial model be

$$\hat{G}_o = \frac{0.04z^{-1}\left(1 + 1.0z^{-1}\right)}{1 - 1.4z^{-1} + 0.5z^{-2}} \tag{11.1.27}$$

The adaptive dual control is performed with a *YP* regulator and an *IU* process is assumed. The reference model R_r and the process outputs, the control and identification errors are shown in Figure 11.1.4 during adaptive

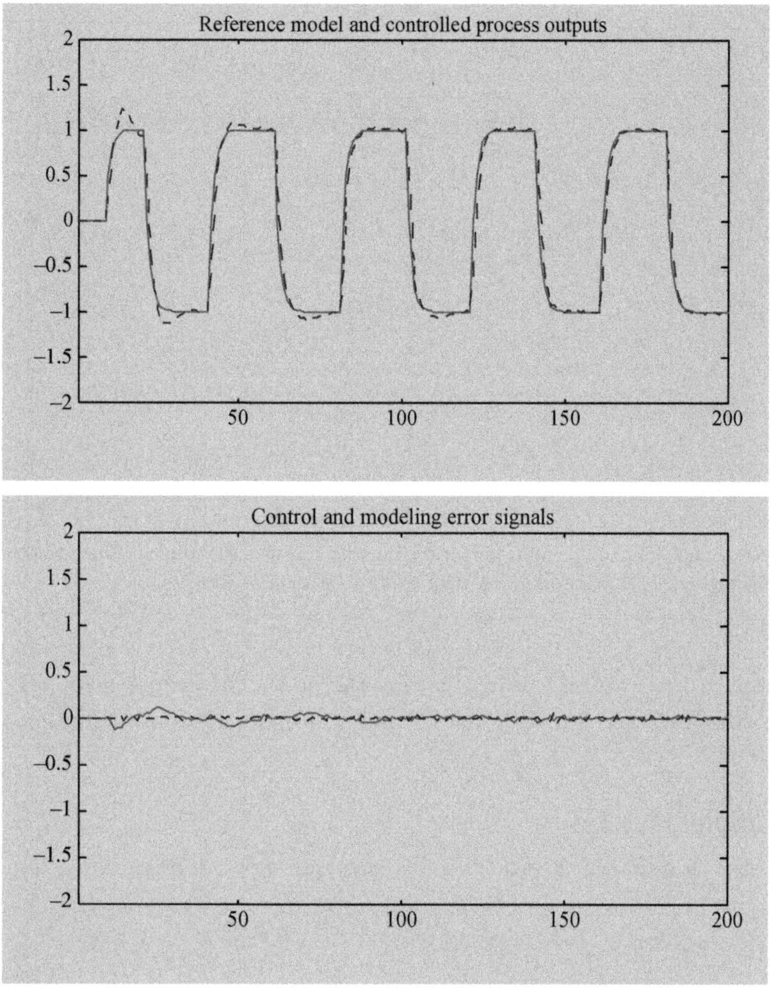

Figure 11.1.4 Plot of the reference model R_r and the process outputs, the control and modelling errors during adaptive dual control (second-order example).

dual control using the *LS* method. The initial value of the covariance matrix was $\boldsymbol{R}[0] = 100\boldsymbol{I}$ and the forgetting factor had the value $\lambda^2 = 0.95$.

Self-Tuning Regulators

The special solution of adaptive dual control, when the regulator can be determined via a simple (trivial) computation, is called the *self-tuning* control [11]. This method is presented via the predictive regulators discussed in Section 2.4. The *d*-step predictor of the DT process (2.4.1) was given by Equation (2.4.5). Let us write this in the form

$$y[k+d] = \mathcal{GT}\, u[k] + \mathcal{P}y[k] = \mathcal{Q}u[k] + \mathcal{P}y[k] = y[k+d|k]; \quad \mathcal{Q} = \mathcal{GT} \tag{11.1.28}$$

which is an expression linear in parameters and is identical with

$$y[k+d] = \boldsymbol{f}^{\mathrm{T}}[u, y, k]\boldsymbol{p}_{\mathrm{qp}} \tag{11.1.29}$$

Here

$$\begin{aligned}\boldsymbol{f}^{\mathrm{T}}[u, y, k] &= \begin{bmatrix} u[k] & u[k-1] & \dots & u\left[k-n_{\mathrm{q}}\right] & y[k] & \dots & -y\left[k-n_{\mathrm{p}}\right]\end{bmatrix} \\ &= \begin{bmatrix} u[k] & \tilde{\boldsymbol{f}}^{\mathrm{T}}\end{bmatrix}^{\mathrm{T}} \\ \boldsymbol{p}_{\mathrm{qp}} &= \begin{bmatrix} q_{\mathrm{o}} & q_1 & \dots & q_{n_{\mathrm{q}}} & p_{\mathrm{o}} & \dots & p_{n_{\mathrm{p}}}\end{bmatrix}^{\mathrm{T}} = \begin{bmatrix} q_{\mathrm{o}} & \tilde{\boldsymbol{p}}^{\mathrm{T}}_{\mathrm{qp}}\end{bmatrix}^{\mathrm{T}}\end{aligned} \tag{11.1.30}$$

For simplicity, choose a zero reference signal, that is, $y_{\mathrm{r}}[k] \equiv 0$. Equation (2.4.7) of the predictive regulator becomes

$$u[k] = -\frac{\mathcal{P}}{\mathcal{GT}}y[k] = \frac{1}{q_{\mathrm{o}}}\tilde{\boldsymbol{f}}^{\mathrm{T}}\tilde{\boldsymbol{p}}_{\mathrm{qp}} \tag{11.1.31}$$

Given equation (11.1.29) it is easy to construct a recursive identification method and the predictive regulator can be obtained from the resulting estimated parameters $\hat{\boldsymbol{p}}_{\mathrm{qp}} = \begin{bmatrix} \hat{q}_{\mathrm{o}} & \hat{\tilde{\boldsymbol{p}}}^{\mathrm{T}}_{\mathrm{qp}}\end{bmatrix}^{\mathrm{T}}$ in the form

$$u[k] = \frac{1}{\hat{q}_{\mathrm{o}}}\tilde{\boldsymbol{f}}^{\mathrm{T}}\hat{\tilde{\boldsymbol{p}}}_{\mathrm{qp}} \tag{11.1.32}$$

The computation of the regulator is really trivial since it requires only division by $\hat{q}_{\mathrm{o}}$.

Based on a similar principle, but for a slightly complicated stochastic case, Åström and Wittenmark [11] elaborated the first self-tuning regulator. Such regulators are usually known as direct adaptive regulators, too.

Keviczky and his coworkers played a key role in the first applications of the self-tuning regulator in the cement, glass, and energy industries [16,70,75–78].

Investigate the derivation of the *MIMO* self-tuning regulator. The d-step predictor of the DT process (7.103) was given by Equation (7.111). Let us write this in the following form

$$\boldsymbol{y}[k+d] = \mathcal{T}_{\mathrm{L}}\mathcal{N}_{\mathrm{L}}\boldsymbol{u}[k] + \mathcal{P}_{\mathrm{L}}\,\boldsymbol{y}[k] = \mathcal{Q}_{\mathrm{L}}\,\boldsymbol{u}[k] + \mathcal{P}_{\mathrm{L}}\,\boldsymbol{y}[k] = \boldsymbol{y}[k+d|k] \tag{11.1.33}$$

which is finally linear in parameter matrices and identical to the form

$$\boldsymbol{y}[k+d] = \boldsymbol{P}_{\mathrm{QP}}\boldsymbol{f}[\boldsymbol{u},\boldsymbol{y},k] \tag{11.1.34}$$

Here

$$\begin{aligned} \boldsymbol{P}_{\mathrm{QP}} &= \left[\mathbf{Q}_{\mathrm{o}}, \mathbf{Q}_2, \ldots, \mathbf{Q}_{n_{\mathrm{q}}}, \mathbf{P}_1, \ldots, \mathbf{P}_{n_{\mathrm{p}}}\right] = \left[\mathbf{Q}_{\mathrm{o}}, \tilde{\boldsymbol{P}}_{\mathrm{QP}}\right] \\ \boldsymbol{f}[\boldsymbol{u},\boldsymbol{y},k] &= \left[\boldsymbol{u}^{\mathrm{T}}[k], \ldots, \boldsymbol{u}^{\mathrm{T}}\left[k-n_{\mathrm{q}}\right], \boldsymbol{y}^{\mathrm{T}}[k-1], \ldots, \boldsymbol{y}^{\mathrm{T}}\left[k-n_{\mathrm{p}}\right]\right]^{\mathrm{T}} \\ &= \left[\boldsymbol{u}^{\mathrm{T}}[k], \tilde{\boldsymbol{f}}^{\mathrm{T}}[\boldsymbol{u},\boldsymbol{y},k]\right]^{\mathrm{T}} \end{aligned} \tag{11.1.35}$$

For simplicity, choose a zero reference signal, that is, $\boldsymbol{y}_{\mathrm{r}}[k] \equiv \mathbf{0}$. Equation (2.4.5) of the predictive regulator becomes

$$\boldsymbol{y}[k+d|k] = \mathcal{T}_{\mathrm{L}}\mathcal{N}_{\mathrm{L}}\,\boldsymbol{u}[k] + \mathcal{P}_{\mathrm{L}}\,\boldsymbol{y}[k] = \boldsymbol{y}[k+d] = \boldsymbol{P}_{\mathrm{QP}}\boldsymbol{f}[\boldsymbol{u},\boldsymbol{y},k] = \mathbf{0} \tag{11.1.36}$$

whence

$$\boldsymbol{u}[k] = \mathbf{Q}_{\mathrm{o}}^{-1}\tilde{\boldsymbol{P}}_{\mathrm{QP}}\tilde{\boldsymbol{f}}^{\mathrm{T}}[\boldsymbol{u},\boldsymbol{y},k] \tag{11.1.37}$$

Given equation (11.1.34) it is easy to construct a recursive identification method and the predictive *MIMO* regulator can be obtained from the resulted estimated parameter matrices $\hat{\boldsymbol{P}}_{\mathrm{QP}} = \left[\hat{\mathbf{Q}}_{\mathrm{o}}, \hat{\tilde{\boldsymbol{P}}}_{\mathrm{QP}}\right]$ in the form

$$\boldsymbol{u}[k] = \hat{\mathbf{Q}}_{\mathrm{o}}^{-1}\hat{\tilde{\boldsymbol{P}}}_{\mathrm{QP}}\tilde{\boldsymbol{f}}^{\mathrm{T}}[\boldsymbol{u},\boldsymbol{y},k] \tag{11.1.38}$$

The computation of the regulator is really trivial since it requires only multiplication by a matrix coefficient, that is, by the inverse of $\hat{\mathbf{Q}}_{\mathrm{o}}$.

The extension of the self-tuning regulators to the multivariable stochastic systems was elaborated by Åström and his coworkers [8,26,27] and Keviczky and Hetthéssy [58,69]. The essence of the method was briefly discussed in Section 7.

11.2 ITERATIVE METHODS: SIMULTANEOUS IDENTIFICATION AND CONTROL

The *simultaneous identification and control* strategy of the iterative methods processes long measurement records (N samples) usually by off-line (*batch*) identification technique, and the control is based on the result of the ID [33,48]. In the next, off-line, experiment the most accurate process model previously obtained is obviously used. With this technique, the optimality of the regulator can be gradually improved as the model becomes more and more accurate, while the normal operation of the system is only slightly disturbed. The algorithms of the simultaneous identification and control apply the *KB* parameterization-based closed-loop ID.

Before the iteration the model class $\mathcal{M}$ (e.g., linear dynamic n-order processes), and the ID criterion $J_{\text{ID}}(\varepsilon_{\text{ID}})$ (e.g., *LS* method) is chosen. Here ε_{ID} is the modeling error. Similarly, the control class (type) C and the control criterion $J_{\text{C}}(e_{\text{C}})$ (e.g., optimization by the $\mathcal{H}_2$ norm) are also chosen. Here e_{C} is the control error. The coherent order of the steps to be taken in the i-th iteration is as follows:

1. The a priori process model $\hat{G}_{i-1}$ is used, and the optimal regulator is computed by solving the optimality task

$$\hat{C}_i = \arg\min_{\hat{C}_i \in C} J_{\text{ID}}\left(e_{\text{C}}^i\right) = \arg\min_{\hat{C}_i \in C} J_{\text{ID}}\left[\tilde{e}\left(e_{\text{C}}^i\right), \hat{G}_{i-1}\right] = \hat{C}_{\text{opt}}\left(C, \hat{G}_{i-1}\right) \tag{11.2.1}$$

2. Then the input signal series is determined, that is, the series $\mathcal{Y}_r^i = \left\{y_r^i[k]; k = 1, \ldots, N\right\}$ of the reference signal y_{r} and it is applied to the input of the closed system.
3. With knowledge of the new regulator $\hat{C}_i$ and the reference signal, the actual value of the actuating signal series $\hat{\tilde{u}}_i$ is determined and N data pairs $\mathcal{U}^i = \left\{\hat{u}^i[k], y^i[k]; k = 1, \ldots, N\right\}$ collected. (In multivariable systems $\mathcal{U}^i = \left\{\hat{\boldsymbol{u}}^i[k], \boldsymbol{y}^i[k]; k = 1, \ldots, N\right\}$, so these signals are vectors.)
4. Then the estimation of the process parameters, that is, the ID of the process is performed from the measured data. This step can be formally described by the following expression

$$\hat{G}_i = \arg\min_{\hat{G}_i \in \mathcal{M}} J_{\text{ID}}\left(\varepsilon^i_{\text{ID}}\right) = \arg\min_{\hat{G}_i \in \mathcal{M}} J_{\text{ID}}\left(\varepsilon^i_{\text{ID}}\right)\bigg|_{\mathcal{U}^i} = \hat{G}\left(\mathcal{U}^i, \mathcal{M}, G\right) \tag{11.2.2}$$

5. In this step, the merit of the obtained model and its distance from the model obtained in the previous iteration is investigated: if the model is acceptable or the distance is correspondingly small, then the iteration is stopped; otherwise the successive improvement of the model and regulator is continued.
6. Based on the obtained results, the input series can be optimized by the methods discussed in Sections 11.3 and 4.3. Choose the design method $\mathcal{D}$, according to which

$$\mathcal{Y}^{i+1}_r = \mathcal{D}\left\{\mathcal{M}, \hat{G}_i, Y^{\text{Max}}_{\text{r}}\right\} \tag{11.2.3}$$

Here $Y^{\text{Max}}_{\text{r}}$ is the biggest amplitude admissible for the reference signal. Given knowledge of the generated optimal reference signal, the method is continued from the first point.

The block scheme of the iterative simultaneous identification and control discussed in the above steps is presented in Figure 11.2.1 in the i-th iteration step. It is important to note that the signal $\hat{\hat{u}}_i$ needs to be determined by using the a priori model $\hat{G}_{i-1}$ and the computation of $\hat{K}^{i-1}_{\text{r}}$.

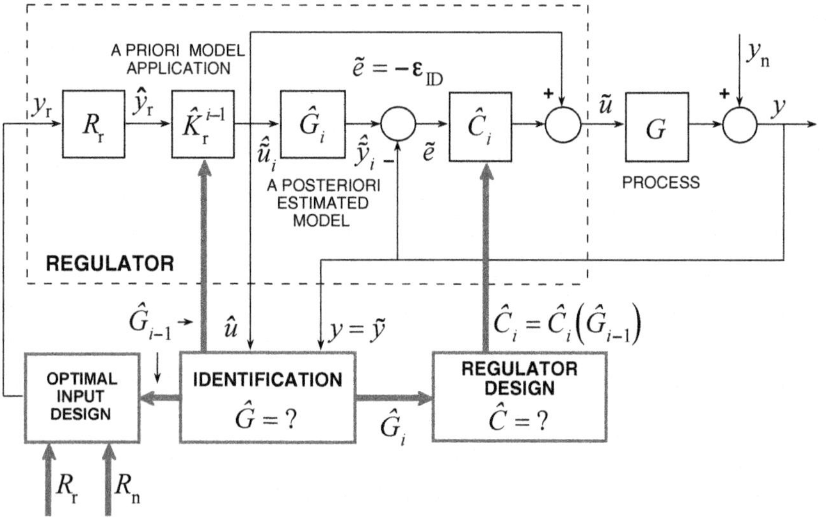

Figure 11.2.1 Block scheme of iterative joint identification and control in closed loop.

Simulation Examples

The algorithms of the simultaneous identification and control have been tested via several simulation examples. Some of these are shown below.

Example 11.2.1

Let the process be a first-order time-delay system, whose pulse transfer function is

$$G = \frac{0.007869z^{-1}}{1 - 0.606531z^{-1}} z^{-3} = \frac{0.007869}{1 - 0.606531z^{-1}} z^{-4} \tag{11.2.4}$$

This form represents the effect of the rod control of a helicopter on the sideslip angle. Here the sampling time $T_s = 0.05$ [sec] is applied and the time delay of the process is $d = 3$. Combined iterative identification and control tests are performed. The reference models with unity gain

$$R_r = \frac{0.5z^{-1}}{1 - 0.5z^{-1}} \quad \text{and} \quad R_n = \frac{0.2z^{-1}}{1 - 0.8z^{-1}} \tag{11.2.5}$$

are used in the design. The iteration starts with the model

$$\hat{G}_o = \frac{0.01z^{-1}}{1 - 0.4z^{-1}} \tag{11.2.6}$$

A square signal with periodic time of 40 samples times is applied as reference signal. In the simulation it is assumed that the additive noise y_n is white noise, whose variance is very small, that is, $\sigma_{y_n} = 0.01$. In each step $N = 100$ samples are processed. Because of the small output noise, a simple off-line *LS* method is used for the identification. The regulator is designed by the *YP* method assuming an *IS* process.

The shape of the identification and control loss functions (variances) during the iteration is shown in Figure 11.2.2. It is evident that the iteration is very fast and reaches the optimal value in four steps. Figure 11.2.3 shows the shape of the reference model R_r (continuous line) and the process outputs (dashed line) at the beginning and end of the iteration.

Example 11.2.2

Let the process have an unstable zero whose pulse transfer function is

$$G = \frac{0.0364z^{-1}(1 + 1.2z^{-1})}{1 - 1.6z^{-1} + 0.68z^{-2}} \tag{11.2.7}$$

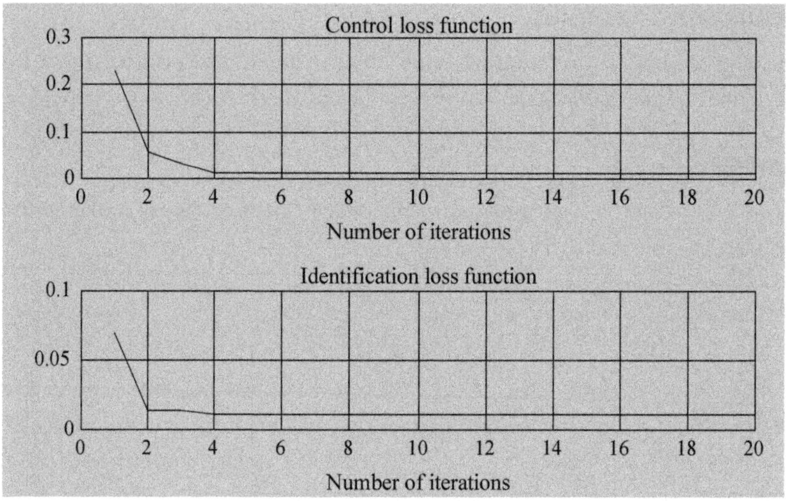

Figure 11.2.2 Plot of the loss functions during the iteration (first-order example).

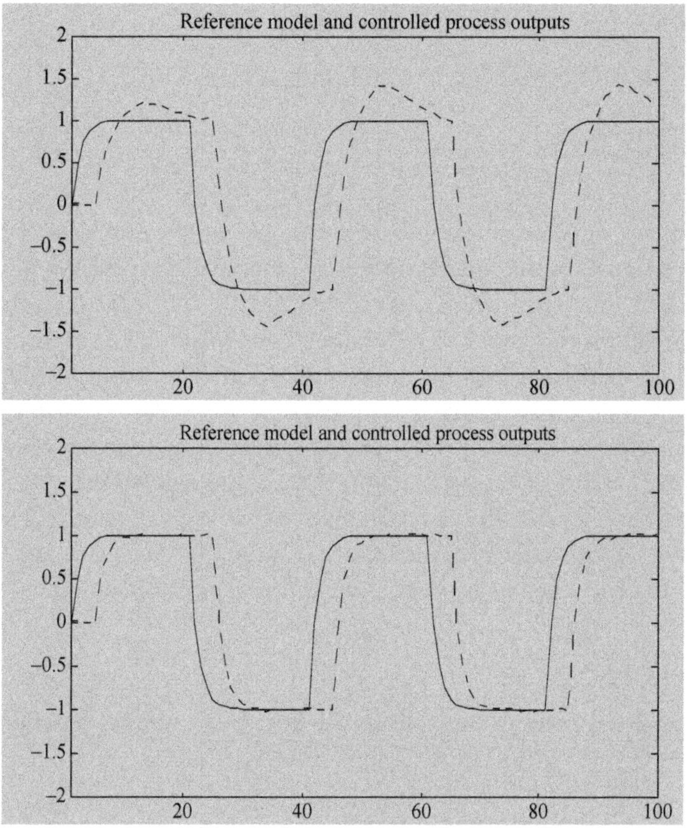

Figure 11.2.3 Plot of the reference model R_r and the process outputs at the beginning and the end of the iteration (second-order example).

The same reference models and output noise are applied as in the previous example. The number of the processed samples is also the same. Let the initial model be

$$\hat{G}_{\mathrm{o}} = \frac{0.04z^{-1}\left(1+1.0z^{-1}\right)}{1-1.4z^{-1}+0.5z^{-2}} \tag{11.2.8}$$

The control and ID loss functions (variances) can be seen in Figure 11.2.4 during the iteration. Figure 11.2.5 shows the shape of the reference model R_{r} (continuous line) and the process outputs (dashed line) at the beginning and end of the iteration.

The output of the closed system (continuous line) and the identified process model output (dashed line) are shown in Figure 11.2.6. It is clear that these are almost completely the same at the start and end of the iteration. This example clearly illustrates the necessity of the iterative strategy since in spite of the very good model fitting, the control error is very high at the beginning of the iteration and has almost completely vanished by the end.

In this example, the off-line *LS* method is also used for process identification. The regulator is designed by *YP* method, assuming an *IU* process.

Example 11.2.3

Let the process again be a second-order one whose pulse transfer function is

$$G = \frac{0.125z^{-1}\left(1+0.6z^{-1}\right)}{\left(1-0.5z^{-1}\right)\left(1-0.8z^{-1}\right)} \tag{11.2.9}$$

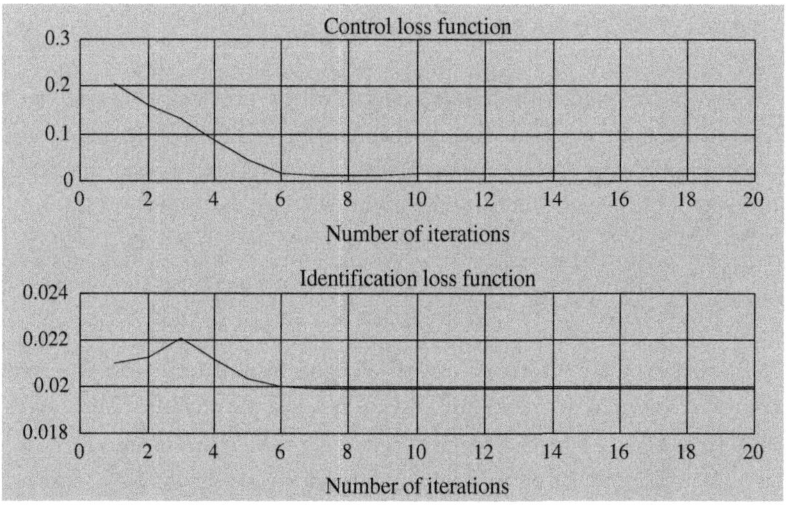

Figure 11.2.4 Plot of the loss functions during the iterations (second-order example).

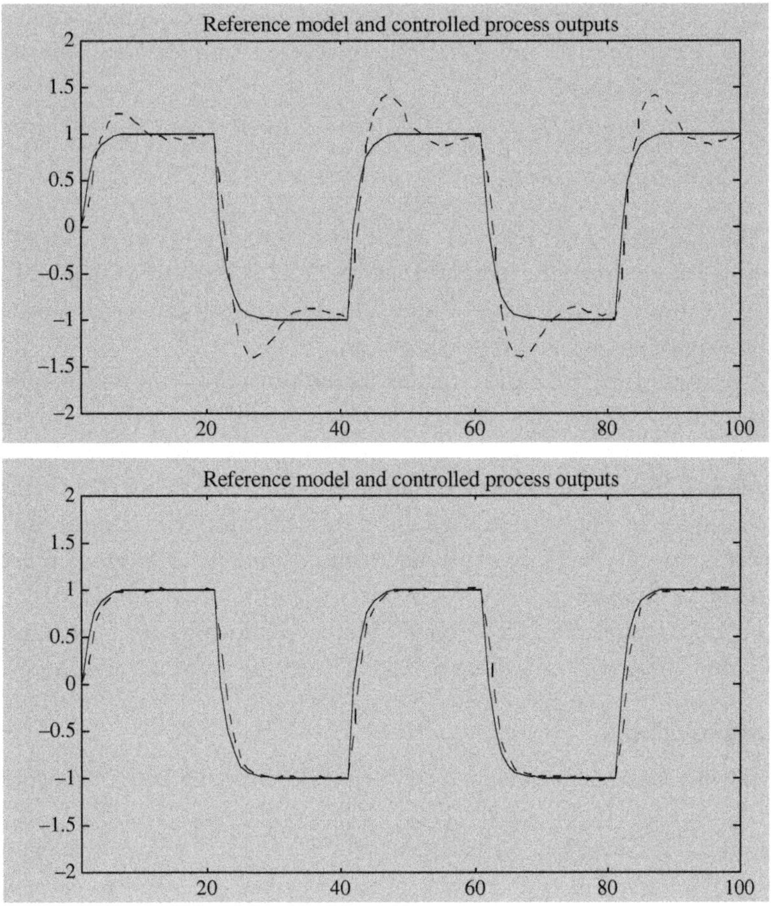

Figure 11.2.5 Plot of the reference model R_r and the process outputs at the beginning and the end of the iteration (second-order example).

The same reference models are used as in the previous example but the output noise is now much larger, that is, $\sigma_{y_n} = 0.1$. The number of the processed samples is also the same. Because of the larger noise, the *extended least squares* (*ELS*) method is used for the ID. Let the initial model be

$$\hat{G}_o = \frac{0.1z^{-1}\left(1 - 0.1z^{-1}\right)}{\left(1 - 0.2z^{-1}\right)\left(1 - 0.9z^{-1}\right)} \tag{11.2.10}$$

The shape of the reference model R_r (continuous line) and the process output (dashed line) at the beginning and end of the iteration are shown in Figure 11.2.7.

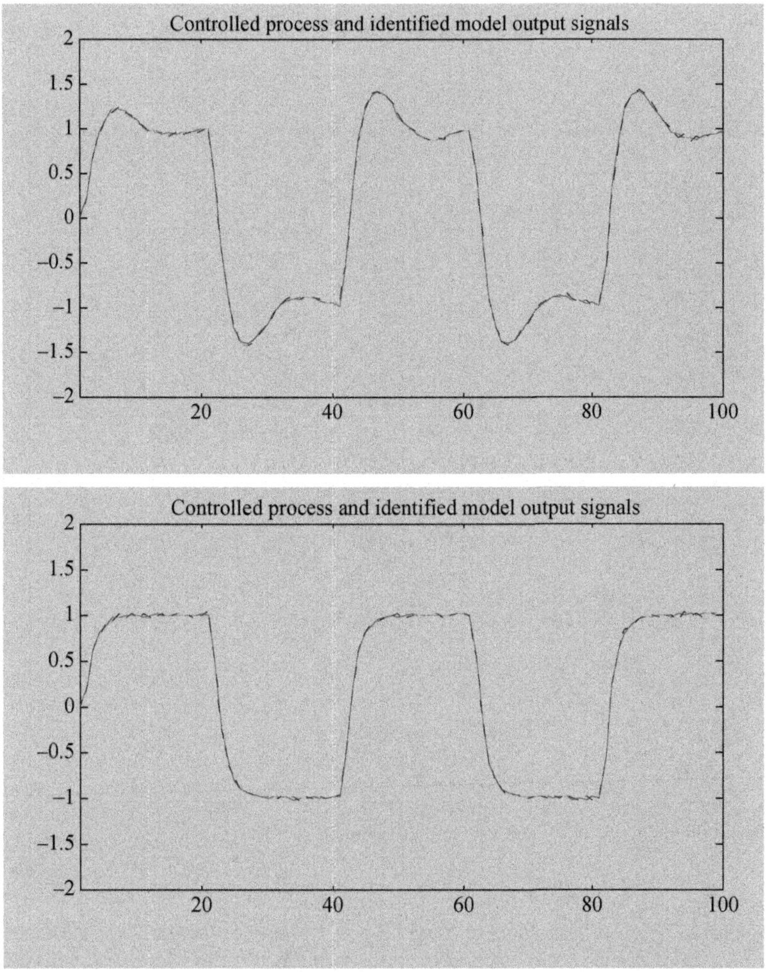

Figure 11.2.6 Plot of the reference model R_r and the process outputs at the beginning and the end of the iteration (second-order example).

It can be observed in Figure 11.2.7 that the gradual improvement of the control by iterations is working well and the desired transient process, tracking the reference signal, is obtained at the end of the iteration. The *ELS* method is suitable even in a very noisy situation.

Figure 11.2.8 presents the shape of the reference model R_r (continuous line) and the process outputs (dashed line) at the beginning and the end of the iteration using the *LS* method. The figure clearly illustrates that the *LS* method is not suitable in the case of large noise. Though the iterative control design works even in this case, the noise cannot be filtered.

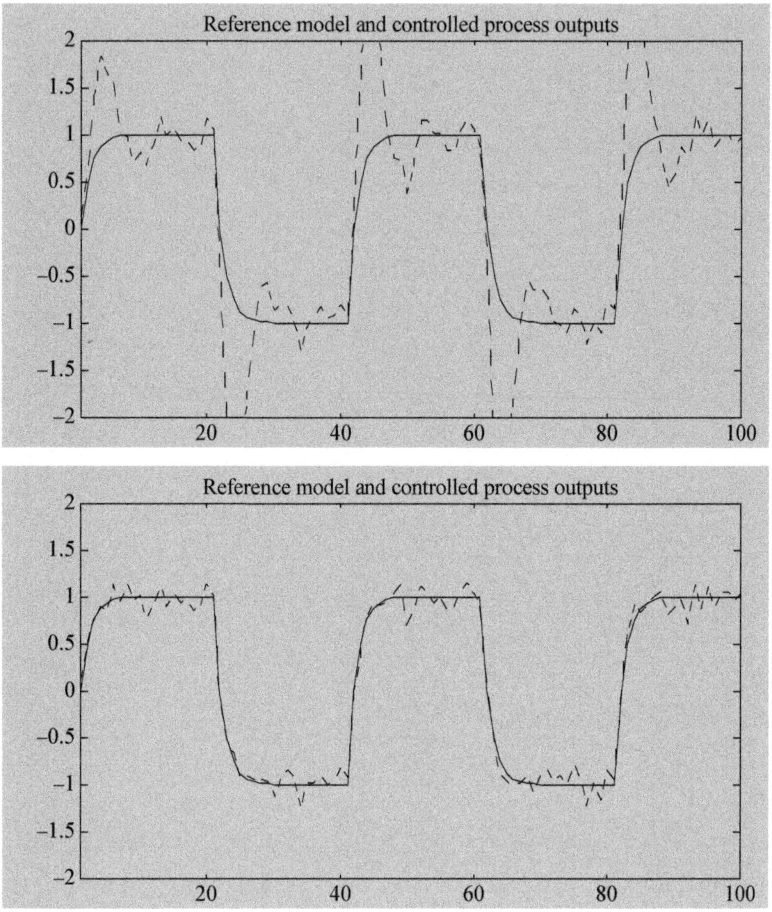

Figure 11.2.7 Plot of the reference model R_r and the process outputs at the beginning and the end of the iteration (second-order example with noise) for the *extended least squares* method.

Optimization of the Reference Signal

Example 11.2.4

Let the process be a second-order one whose pulse transfer function is

$$G = \frac{0.125z^{-1}\left(1 + 1.6z^{-1}\right)}{\left(1 - 0.5z^{-1}\right)\left(1 - 0.8z^{-1}\right)} \tag{11.2.11}$$

The same reference models are applied as in the previous examples. Let the output noise be $\sigma_{y_n} = 0.01$. The ID is performed by an off-line *LS* method.

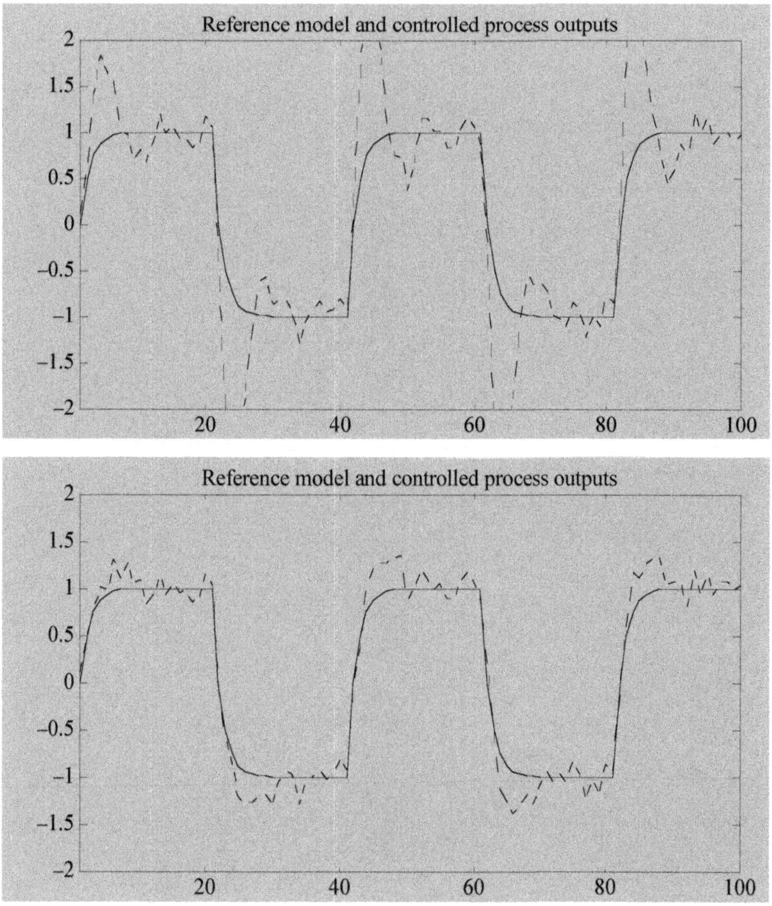

Figure 11.2.8 Plot of the reference model R_r and the process outputs at the beginning and the end of the iteration (second-order example with noise) for the *least squares* method.

The number of the processed samples is also the same. The regulator is designed by the *YP* method, assuming an *IU* process. Let the model be

$$\hat{G}_o = \frac{0.1z^{-1}\left(1 + 4.0z^{-1}\right)}{\left(1 - 0.2z^{-1}\right)\left(1 - 0.9z^{-1}\right)} \tag{11.2.12}$$

The optimal reference signal series $\{y_r^{\circledast}[k]; k = 1, \ldots, N\}$ is generated by (4.3.21). The reference model R_r (continuous line) and the process outputs (dashed line) are shown in Figure 11.2.9 at the beginning and end of the iteration. Figure 11.2.10 shows the amplitude characteristic of the input-generating filter $H_{KB}(j\omega)$ (continuous line) and the frequency characteristic

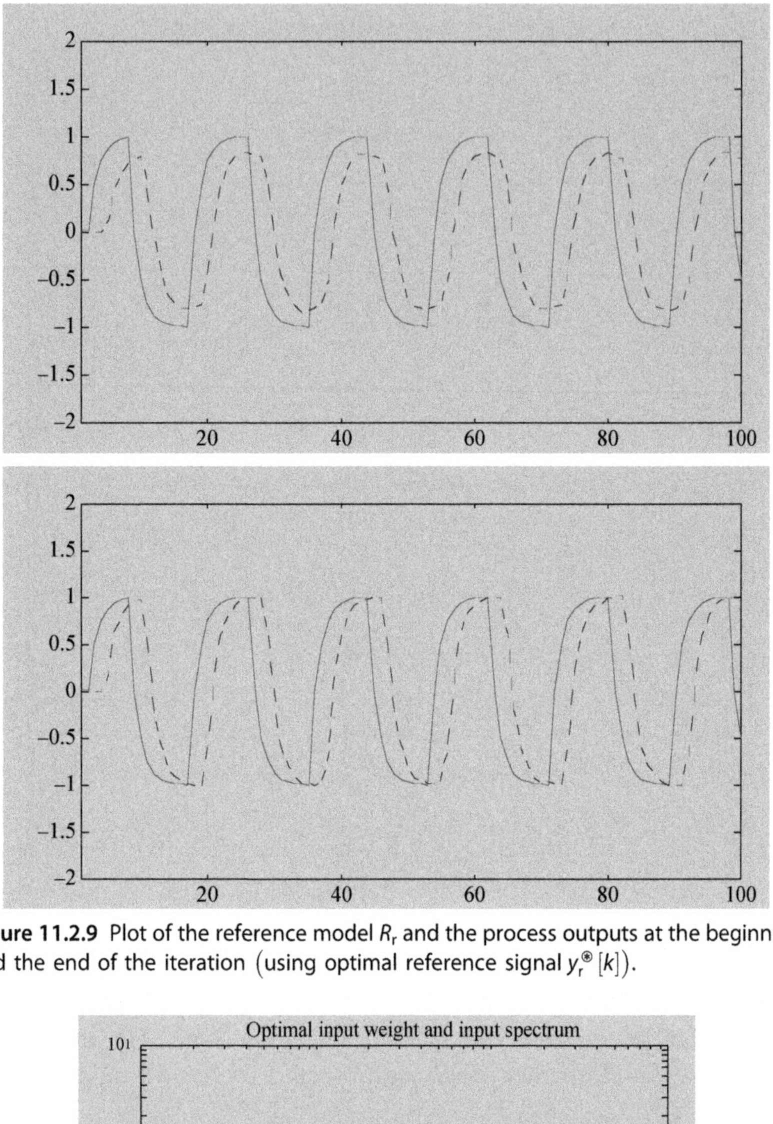

Figure 11.2.9 Plot of the reference model R_r and the process outputs at the beginning and the end of the iteration (using optimal reference signal $y_r^{\circledast}[k]$).

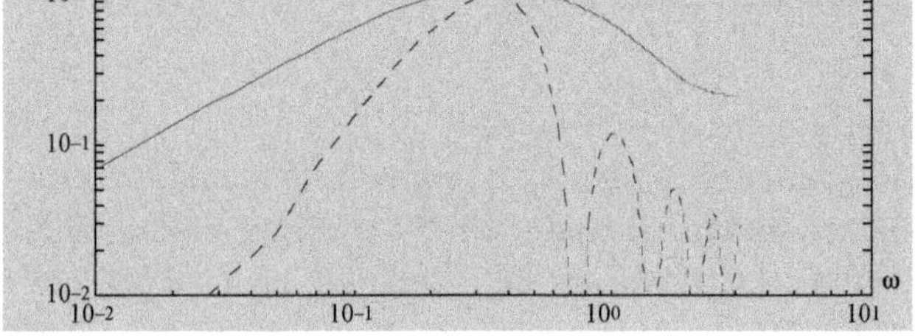

Figure 11.2.10 Spectra of the frequency function $|H_{KB}(j\omega)|$ (solid line) and the obtained optimal input (dotted line).

of the optimal signal $y_r^{\circledast}[k]$ at the end of the iteration. The maximum of $H_{KB}(j\omega)$ is in the vicinity of the cutoff frequency and it is evident that the signal $y_r^{\circledast}[k]$ concentrates the maximal components in this region.

Figure 11.2.11 shows the effect of the optimization of the exciting signal, where three input signals are compared: a unit step (dotted), an arbitrary chosen square *pseudo random binary signal* (PRBS) (dashed), and a signal (continuous) optimized by (4.3.21). The control loss function is also presented in the figure. It is clear that none of the signals can compete with the optimal excitation in terms of the speed of the iteration. The reason is that the optimal signal provides increasingly accurate model in each step in the vicinity of the frequency region, which is important from the regulator design perspective.

Redesign of the Reference Model

This is perhaps the best place to discuss the redesign algorithm of the reference model. The redesign requires that in the i-th step the coefficient $\hat{f}_{r,1}^{i}$ of the reference model $\hat{R}_r^i$ is computed by the coefficient $\hat{g}_1^{i-1}$ of the model $\hat{G}_{i-1}$ obtained by identification in the $[i-1]$-th iteration step, that is,

$$\hat{f}_{r,1}^{i} \le \hat{g}_1^{i-1} U_{max} - 1 \tag{11.2.13}$$

The new reference model is obtained as

$$\hat{R}_r^i = \frac{\left(1 + \hat{f}_{r,1}^{i}\right) z^{-1}}{1 + \hat{f}_{r,1}^{i} z^{-1}} \tag{11.2.14}$$

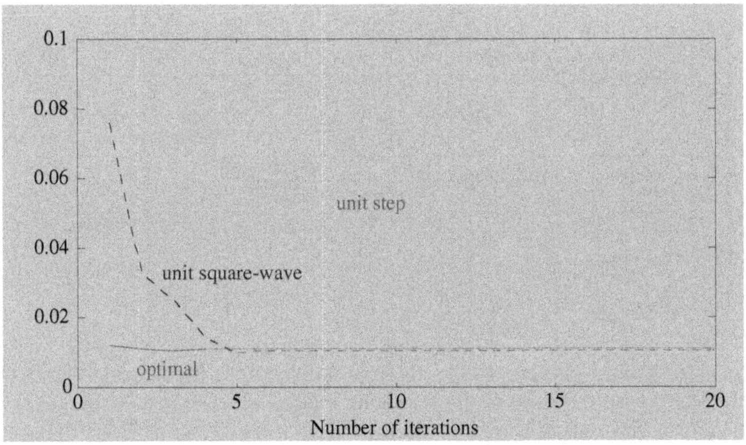

Figure 11.2.11 Loss functions of the control for different exciting signals.

Then the strategy of the iterative methods detailed above is continued in the closed loop.

The block scheme of the iterative simultaneous ID and control discussed above is shown in Figure 11.2.12 with the redesign of the reference model in the *i*-th iteration step.

The typical response of the classical *PID* regulator can be seen in Figure 11.2.13 for square-wave reference signal disturbance.

The same kind of transient can be expected at the output of the *YP* regulator if $y_r[k]$ has a square signal form. Using the notations of the figure, the overexcitation (4.1.6), (4.1.14) is obtained as

$$u_t = \frac{\Delta u^+(0)}{\Delta u^+(\infty) - \Delta u^-(\infty)} = \frac{\Delta u^+(\infty)}{2\Delta u^+(\infty)} = \frac{u^+_{\max}(0) + \Delta u^+(\infty)}{2\Delta u^+(\infty)} \tag{11.2.15}$$

Example 11.2.5

Let the process be given by the pulse transfer function used in Example 11.2.4. Combined iterative identification and control tests are performed. The following reference models of unity gain are used for the design

$$R_r = \frac{0.9z^{-1}}{1 - 0.1z^{-1}} \quad \text{and} \quad R_n = \frac{0.2z^{-1}}{1 - 0.8z^{-1}} \tag{11.2.16}$$

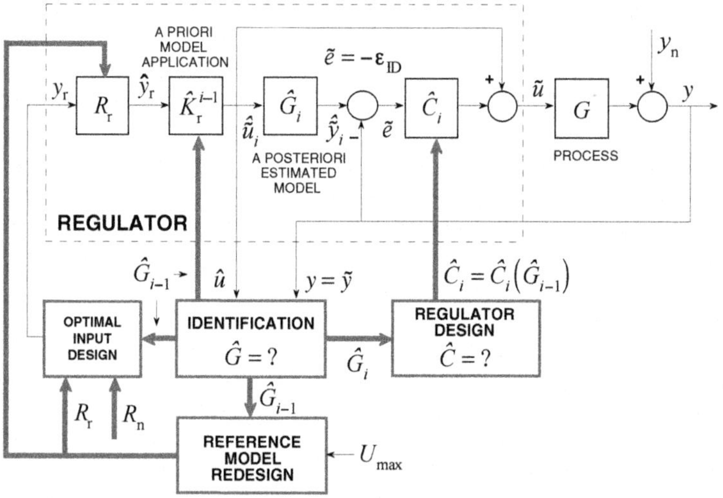

Figure 11.2.12 Block scheme of the iterative simultaneous identification and control in closed loop with redesign of the reference model.

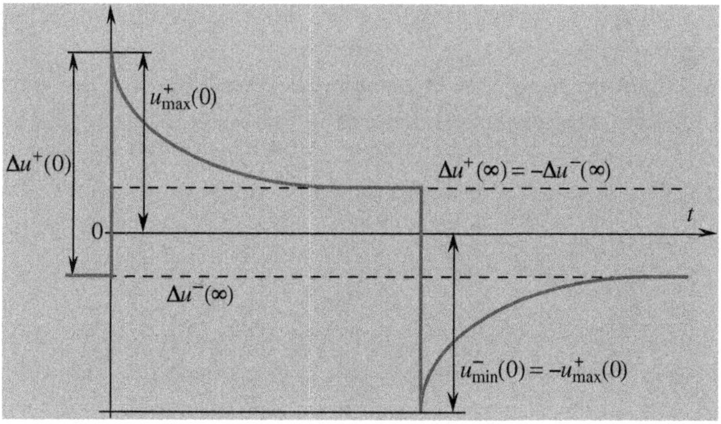

Figure 11.2.13 Typical response of the classical *proportional-integral-derivative* regulator in the case of a square-wave disturbance signal.

At the start of the iteration, the model

$$\hat{G}_o = \frac{0.1z^{-1}\left(1 + 4.0z^{-1}\right)}{\left(1 - 0.2z^{-1}\right)\left(1 - 0.9z^{-1}\right)} \tag{11.2.17}$$

is assumed. A square signal with periodic time of 40 samples is applied as reference signal. In the simulation it is assumed that the additive noise y_n is white noise, whose variance is $\sigma_{y_n} = 0.01$. The number of the processed samples is $N = 100$. Because of the small output noise, the identification is performed by a simple off-line *LS* method. The regulator is designed by the *YP* method, assuming an *IU* process.

The output of the regulator is presented in Figure 11.2.14 where it is seen that the overexcitation is very high at 900%, that is, $u_t = 9$. Assume

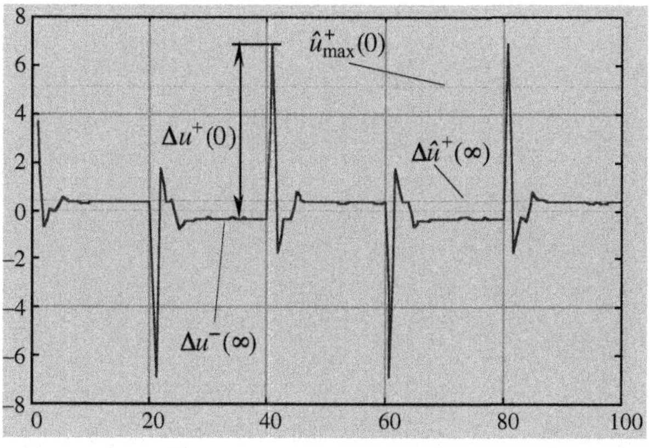

Figure 11.2.14 Response of the *YP* regulator before the iteration.

that the actuator can realize only $\overline{u}_t = 5$. This requires the redesign of the reference model R_r.

The condition $u_t \leq \overline{u}_t = 5$ needs to be prescribed for the reference model redesign iteration. It can be seen in Figure 11.2.15 that the control output is according to the prescribed over actuation by the end of the iteration. The obtained redesigned reference model is

$$\overline{R}_r = \frac{0.5022z^{-1}}{1 - 0.4978z^{-1}} \tag{11.2.18}$$

Figure 11.2.16 shows the time functions of the output of the reference model (continuous line) and the closed system (dashed line). Figure 11.2.17 presents the Bode diagram of the original (continuous line) and the

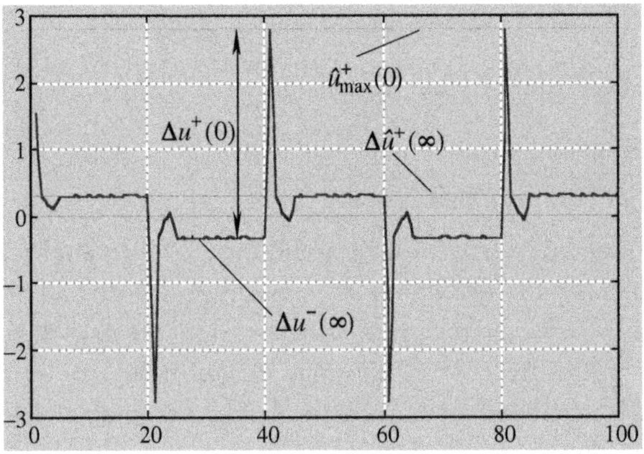

Figure 11.2.15 Response of the *YP* regulator after the iteration.

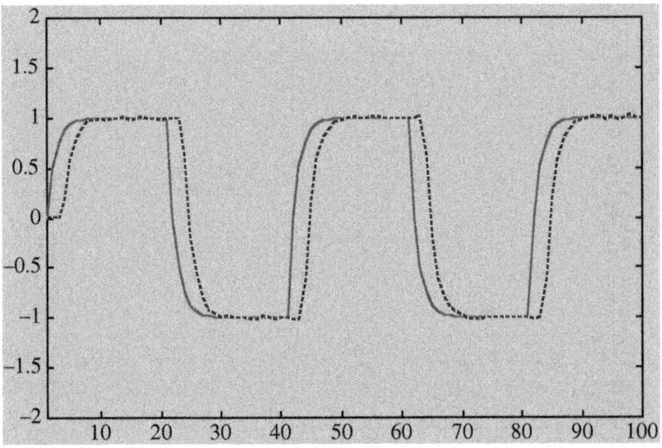

Figure 11.2.16 Outputs of the reference model and the process.

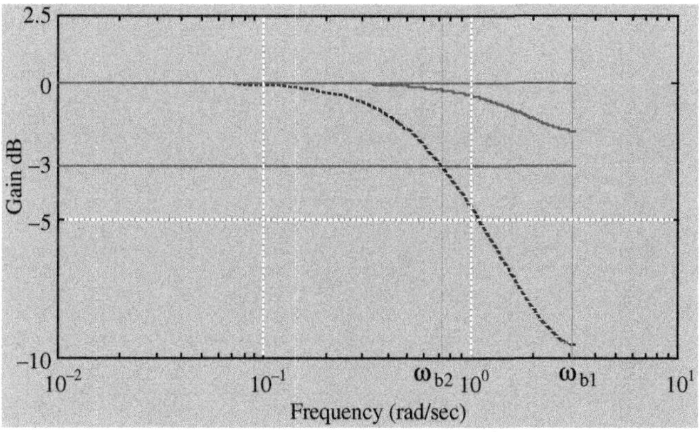

Figure 11.2.17 Bode diagram of the original and the new reference model

redesigned (dashed line) reference model. The prescribed overexcitation could be reached only by substantially slowing down the closed system.

Example 11.2.6

Solve the previous task allowing less overexcitation $u_t \leq \overline{u}_t = 3$. The iteration strategy is operated in the same circumstances. The output of the regulator is seen in Figure 11.2.18 at the start of the iteration and in Figure 11.2.19 at the end of the iteration.

Figure 11.2.20 contains the time function of the reference model (continuous line) and of the closed system (dashed line). Figure 11.2.21 shows the Bode diagram of the original (continuous line) and the redesigned (dashed line) reference model. The prescribed overexcitation could

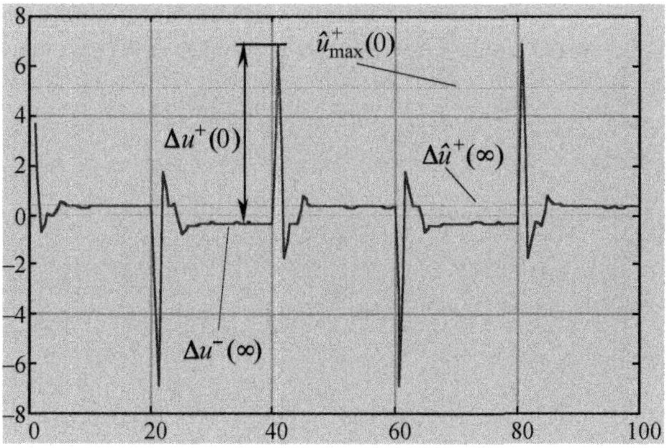

Figure 11.2.18 Actuating signal before the iteration.

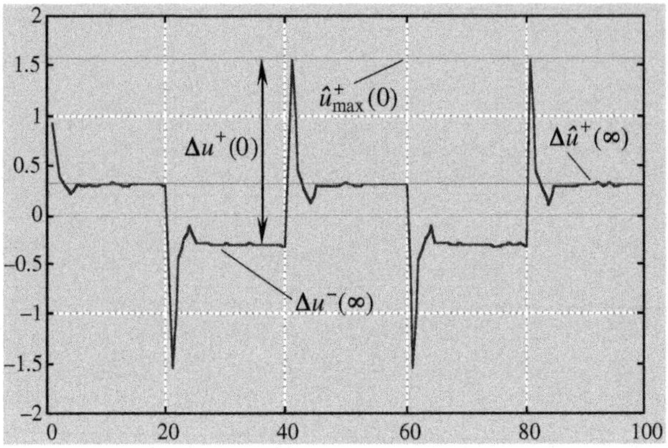

Figure 11.2.19 Actuating signal after the iteration.

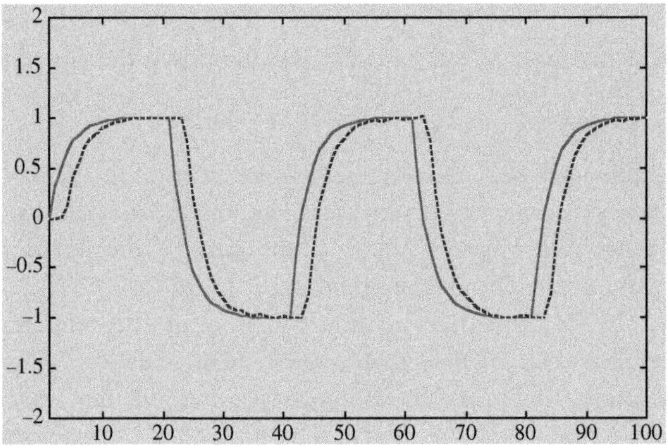

Figure 11.2.20 Outputs of the reference model and the process.

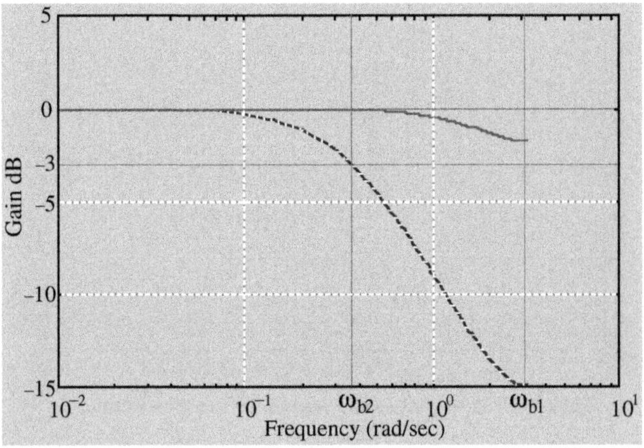

Figure 11.2.21 Bode diagram of the original and the new reference model.

be reached only by further slowing down the closed system. The resulting redesigned reference model is

$$\overline{R}_{\mathrm{r}} = \frac{0.3051z^{-1}}{1 - 0.6949z^{-1}} \tag{11.2.19}$$

11.3 TRIPLE CONTROL

The accuracy of the model provided by the ID significantly depends on the applied excitation either in open or in closed loop. In the optimization of the dynamic model identification in Section 4.3, it was clear that the optimal input signal depends on the process parameters. This paradigm (*catch-22*) can only be solved by applying an iterative method where in each step, using the best parameters, the excitation of the process is optimized, and then this signal is used during the next experiment. The iterative procedure means learning the process model in the fastest way.

In Section 4.3 the optimization possibilities of the modeling loss were discussed in detail. The optimization of the exciting input signal for the purpose of the ID is called input design [51]. In the case of static models this optimization is called experiment design, which has a somewhat longer history in the literature than the input design of dynamic systems. The first method was published by Keviczky [67,68] who recommended the sequential optimization of the determinant of the information matrix. This input signal, because of the conventional notations in the experiment design, is known as the *D*-optimal input signal. The method applies the simple relationship whereby the information matrix of the *LS* estimation, which is linear in parameters, is the inverse of the covariance matrix (10.1.24)

$$\boldsymbol{M}_{\mathrm{inf}} = \boldsymbol{K}_{\hat{p}}^{-1} = \frac{1}{\sigma_{\varsigma}^{2}}\left[\boldsymbol{F}_{\mathrm{uy}}^{\mathrm{T}}\boldsymbol{F}_{\mathrm{uy}}\right] \tag{11.3.1}$$

The renewing equations of the convergence matrices have been seen in recursive estimations (10.2.2), (10.2.3) and adaptive identification methods (10.1.18), (10.1.21). On the basis of these equations we can write

$$\boldsymbol{M}_{\mathrm{inf}}[k] = \boldsymbol{M}_{\mathrm{inf}}[k-1] + \boldsymbol{f}[u, y, k]\boldsymbol{f}^{\mathrm{T}}[u, y, k] \tag{11.3.2}$$

According to (A.1.18) the sequential form of the determinant of the information matrix is

$$\det\{\boldsymbol{M}_{\mathrm{inf}}[k]\} = \det\{\boldsymbol{M}_{\mathrm{inf}}[k-1]\}\left\{1 + \boldsymbol{f}^{\mathrm{T}}[u, y, k]\boldsymbol{M}_{\mathrm{inf}}^{-1}[k-1]\boldsymbol{f}[u, y, k]\right\} \tag{11.3.3}$$

Thus the next input signal $u[k]$ in the situation (memory) vector $f[u,y,k]$ needs to be determined by maximizing the quadratic form $\boldsymbol{f}^{\mathrm{T}}[u,y,k]\boldsymbol{M}_{\mathrm{inf}}^{-1}[k-1]\boldsymbol{f}[u,y,k]$ under a given constraint. The input signal usually has an amplitude constraint.

Let the equation of the n-order process be

$$y[k] = G\left(z^{-1}\right)u[k] = \frac{\mathcal{B}(z^{-1})z^{-d}}{\mathcal{A}(z^{-1})}u[k] = \frac{b_o + b_1 z^{-1} + \ldots + b_n z^{-n}}{1 + a_1 z^{-1} + \ldots + a_n z^{-n}}u[k] \tag{11.3.4}$$

where again the following form which is linear in parameters can be used

$$y[k] = \boldsymbol{f}^{\mathrm{T}}[u,y,k]\boldsymbol{p}_{\mathrm{ba}} \tag{11.3.5}$$

where

$$\begin{aligned}\boldsymbol{f}^{\mathrm{T}}[u,y,k] &= \left[\, u[k] \quad u[k-1] \quad \ldots \quad u[k-n] \quad -y[k-1] \quad \ldots \quad -y[k-n] \,\right] \\ \boldsymbol{p}_{\mathrm{ba}} &= \left[\, b_o \quad b_1 \quad b_2 \quad \ldots \quad b_n \quad a_1 \quad \ldots \quad a_n \,\right]^{\mathrm{T}}\end{aligned} \tag{11.3.6}$$

Let

$$\boldsymbol{f}[u,y,k] = \left\{u[k], \boldsymbol{g}^{\mathrm{T}}[k-1]\right\}^{\mathrm{T}} \tag{11.3.7}$$

and the partition of the covariance matrix (11.1.19) is

$$\boldsymbol{R}[k] = \begin{bmatrix} r[k] & \boldsymbol{d}^{\mathrm{T}}[k] \\ \boldsymbol{d}[k] & \mathbf{Q}[k] \end{bmatrix} \tag{11.3.8}$$

Based on (11.3.3) and the above notations, the increment of the determinant of the information matrix becomes

$$\begin{aligned}\frac{\det\{\boldsymbol{M}_{\mathrm{inf}}[k+1]\}}{\det\{\boldsymbol{M}_{\mathrm{inf}}[k]\}} &= 1 + \boldsymbol{f}^{\mathrm{T}}[u,y,k+1]\boldsymbol{R}[k]\boldsymbol{f}[u,y,k+1] \\ &= r[k]u^2[k+1] + 2u[k+1]\boldsymbol{g}^{\mathrm{T}}[k]\boldsymbol{d}[k] + \boldsymbol{g}^{\mathrm{T}}[k]\mathbf{Q}[k]\boldsymbol{d}[k] + 1\end{aligned} \tag{11.3.9}$$

This characteristic is a second-order parabola, whose peak is upside down since $r[k] > 0$. If the amplitude constraint for the input is $-U_{\max} \leq u[k] \leq U_{\max}$ then the algorithm generating the D-optimal input signal has the form

$$u_{\mathrm{Dopt}}[k+1] = U\mathrm{sign}\left[\boldsymbol{g}^{\mathrm{T}}[k]\boldsymbol{d}[k]\right] = \begin{cases} +U_{\max}; & \text{if } \boldsymbol{g}^{\mathrm{T}}[k]\boldsymbol{d}[k] \geq 0 \\ -U_{\max}; & \text{if } \boldsymbol{g}^{\mathrm{T}}[k]\boldsymbol{d}[k] < 0 \end{cases} \tag{11.3.10}$$

It can be checked that the algorithm provides an output signal of maximum variance under the given constraint.

This approach actually applies the steps of an adaptive control and works very well in practice. The method does not deal with joint parameter estimation and input design, so it can be considered only as a suboptimal solution.

For realization of *KB* parameterization-based closed-loop adaptive control, it is reasonable to use the optimal method shown in Section 4.3 for the input signal design. This means that besides the two subsystems of the dual control a third one appears, which, by gradually learning the process parameters, generates more and more optimal exciting input signals and provides the best parameter estimation. This strategy is known as triple control [83], a reference to the extension of the dual character.

Thus the following steps have to be performed in each sampling cycle of adaptive triple control. The assumption $-U \leq y_r[k] \leq U$ is made for the reference signal under constraint ($U = 1$).

1. Using the a priori process model $\hat{G}[k-1]$ the optimal regulator is computed by solving the following optimization task

$$\hat{C}[k] = \arg \min_{\hat{C}[k] \in C} J_{\mathrm{ID}}(e_{\mathrm{C}}[k]) = \arg \min_{\hat{C}[k] \in C} J_{\mathrm{ID}}\left[\tilde{e}[k], \hat{G}[k-1]\right]$$
$$= \hat{C}_{\mathrm{opt}}\left(C, \hat{G}[k-1]\right) \tag{11.3.11}$$

2. Then the optimal value $y_r^{\circledast}[k]$ of the reference signal is determined

$$y_r^{\circledast}[k] = -U\mathrm{sign}\left[y_r^*\right] = -U\mathrm{sign}\left[-\frac{\boldsymbol{g}^{\mathrm{T}}[k-1]\boldsymbol{q}}{g_{\hat{0}}}\right]; \quad y_r^* = -\frac{\boldsymbol{g}^{\mathrm{T}}[k-1]\boldsymbol{q}}{g_{\hat{0}}} \tag{11.3.12}$$

and applied to the input of the closed system (see (4.3.21))

3. In the knowledge of the new regulator $\hat{C}[k]$ and the reference signal, the actual value of the actuating signal is determined and added to the input of the process. The resulting output signal $y[k]$ and $\hat{\tilde{u}}[k]$ is processed by the ID algorithm. (In the case of multivariable processes these values are vectors, and thus $\boldsymbol{y}[k]$ and $\hat{\tilde{\boldsymbol{u}}}[k]$.)
4. The recursive estimation of the process parameters is obtained from the measured data. This step requires the renewing equation of the convergence matrix $\boldsymbol{R}[k]$ and the parameter vector $\hat{\boldsymbol{p}}_{\mathrm{ID}}[k]$. For updating of the

convergence matrix, a chosen forgetting strategy (Section 10.2) needs to be applied. The updating of $\hat{\boldsymbol{p}}_{\mathrm{ID}}[k]$ is done according to (10.1.15). (In the multivariable case, the equation of the recursive parameter estimation is (10.2.15).)

5. The merit of the obtained model and the applicability of the regulator obtained from the optimization are investigated step by step. If necessary the forgetting strategy is changed.
6. Because of the recursive operation, the DT memories of the filters of both the regulator and the process model are prepared for the next sampling cycle (shifting or stack operation).
7. In the next cycle, the operation is continued from the first point.

The block scheme of the adaptive triple control shown above is presented in Figure 11.3.1 in the $[k]$-th sampling cycle.

Simulation Examples

Example 11.3.1

Let the process be a second-order time-delay system as in Example 11.1.1. The same reference models are applied and the output noise is also the same, that is, $\sigma_{y_n} = 0.01$.

The shape of the reference model R_r, the process output, the control, and ID errors are presented in Figure 11.3.2 during adaptive triple control, whereby the *LS* method is applied. The regulator is designed by the *YP*

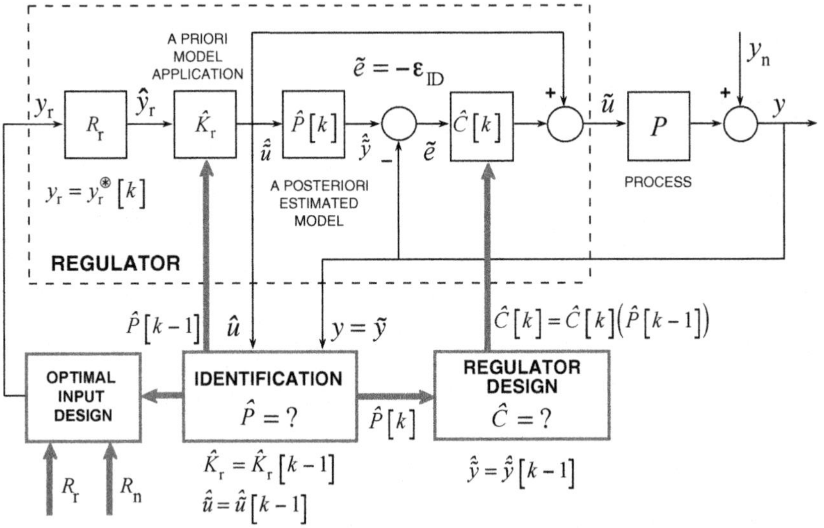

Figure 11.3.1 Block scheme of the adaptive triple control in closed loop.

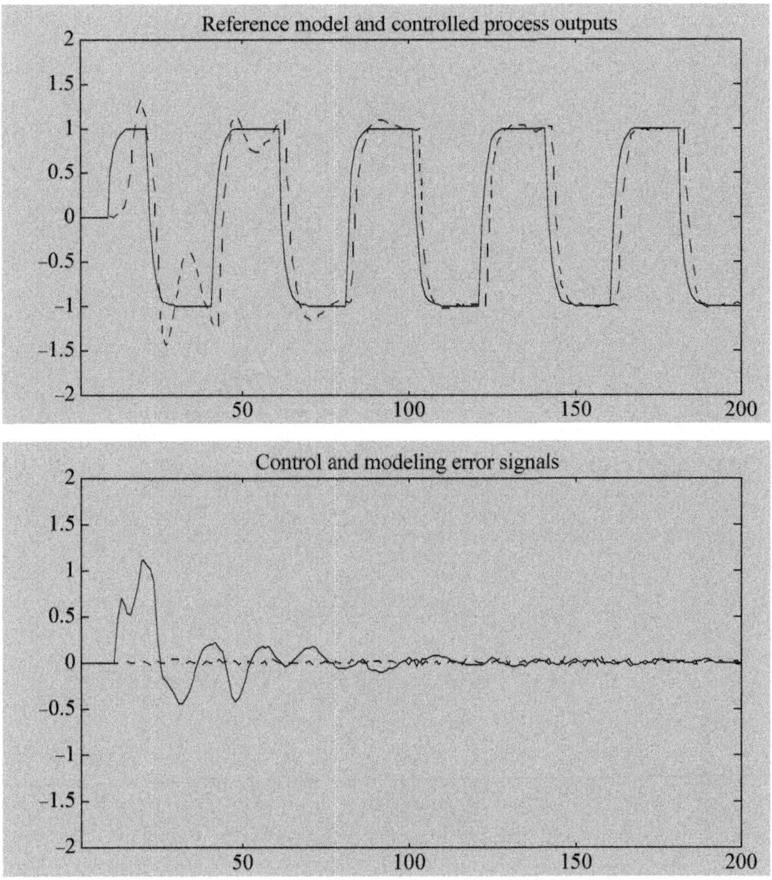

Figure 11.3.2 The shape of the reference model R_r, the process output, the control, and ID errors during adaptive triple control in the case of optimal input design.

method, assuming an *IS* process. The initial value of the covariance matrix is chosen as $\mathbf{R}[0]=100\mathbf{I}$, and the forgetting factor is $\lambda^2=0.95$. At the start of the iteration, the initial model is chosen according to (11.1.23). In each step, the optimal reference signal $y_r^{\circledast}[k]$ is generated by (4.3.21).

Example 11.3.2

Let the process be a first-order time-delay system as in Example 11.1.2. The same reference models are applied and the output noise is also unchanged, that is, $\sigma_{y_n} = 0.01$.

The shape of the reference model R_r, the process outputs, the control, and ID errors are seen in Figure 11.3.3 during adaptive triple control,

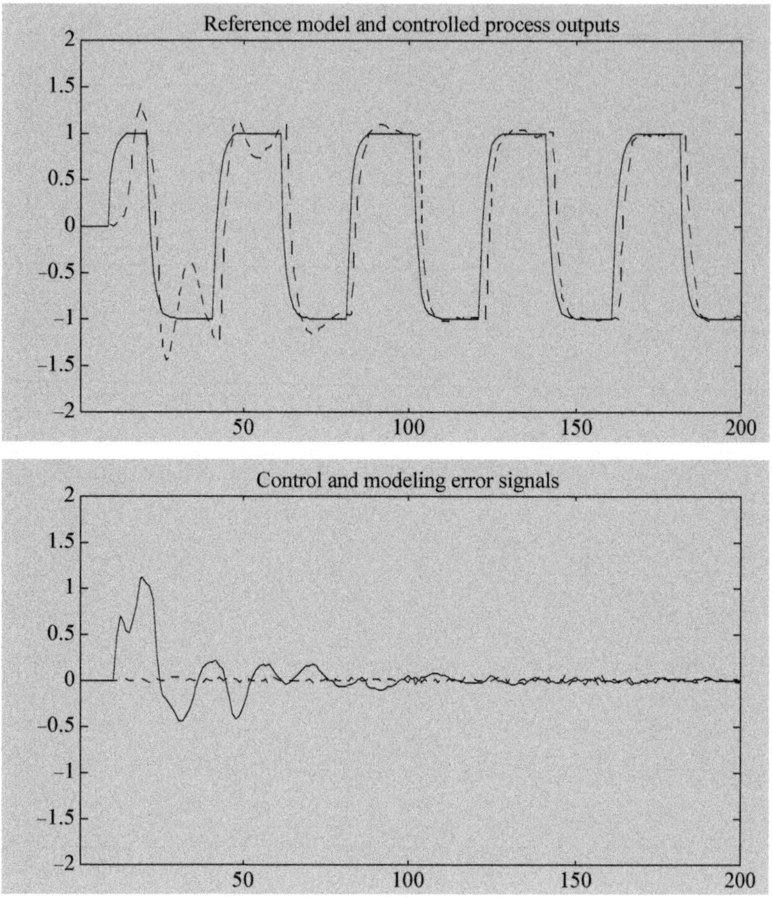

Figure 11.3.3 The shape of the reference model R_r, the process outputs, the control, and ID errors during adaptive triple control in the case of optimal input design.

whereby the *LS* method is applied. The regulator is designed by the *YP* method, assuming an *IS* process. The initial value of the covariance matrix is chosen as $\boldsymbol{R}[0]=100\boldsymbol{I}$, and the forgetting factor is $\lambda^2=0.95$. At the start of the iteration, the initial model is also chosen according to (11.1.25). In each step, the optimal reference signal $y_r^{\circledast}[k]$ is generated by (4.3.21).

Example 11.3.3

Let the process be an *IU* second-order time-delay system as used in Example 11.1.3. The same reference models are applied and the output noise is also unchanged, that is, $\sigma_{y_n}=0.01$.

The shape of the reference model R_r, the process outputs, the control, and ID errors are seen in Figure 11.3.4 during adaptive triple control whereby the recursive *LS* method is applied. The regulator is designed by the *YP* method, assuming an *IU* process. The initial value of the covariance matrix is chosen as $\mathbf{R}[0]=100\boldsymbol{I}$, and the forgetting factor is $\lambda^2 = 0.95$. At the start of the iteration, the initial model is also chosen according to (11.1.27). In each step, the optimal reference signal $y_r^{\circledast}[k]$ is generated by (4.3.21).

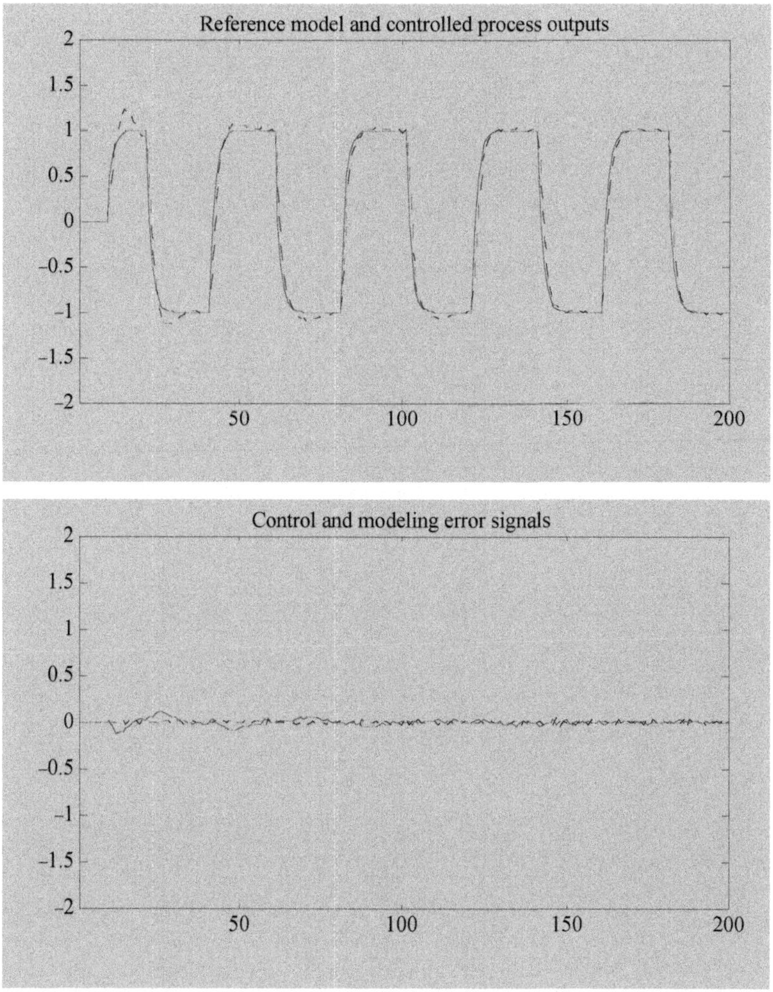

Figure 11.3.4 The shape of the reference model R_r, the process outputs, the control and ID errors during adaptive triple control in the case of optimal input design.

APPENDIX 1

Mathematical Summary

A.1.1 SOME BASIC THEOREMS OF MATRIX ALGEBRA

The tabular arrangement below is called a *matrix*

$$A = \begin{bmatrix} a_{11} & a_{12} & \cdots & a_{1n} \\ a_{21} & a_{22} & \cdots & a_{2n} \\ \vdots & \vdots & \ddots & \vdots \\ a_{m1} & a_{m2} & \cdots & a_{mn} \end{bmatrix} \tag{A.1.1}$$

The a_{ij} quantities are the elements of the matrix. If the elements are real, then the matrix is a *real* one; if they are complex, the matrix is *complex*. Matrix $\boldsymbol{A}$ has m rows and n columns, its dimension (size) is $m \times n$. The $m \times n$ matrix is *rectangular*, the $n \times n$ matrix is *square* (or *quadratic*), the size $m \times 1$ means a *column matrix* (*column vector*), the $1 \times n$ matrix is a *row matrix* (*row vector*), and the 1×1 matrix is *scalar*.

The matrices are usually noted by bold (fat) capital letters and the column and row vectors are marked by bold small letters. The determinant constructed from the elements of the quadratic matrix $\boldsymbol{A}$ is noted by $|\boldsymbol{A}|$ or, rarely, det($\boldsymbol{A}$).

The definition of the transpose of matrix $\boldsymbol{A}$ is that its elements are mirrored to the main diagonal and noted by $\boldsymbol{A}^{\mathrm{T}}$ as

$$\boldsymbol{A}^{\mathrm{T}} = \begin{bmatrix} a_{11} & a_{21} & \cdots & a_{m1} \\ a_{12} & a_{22} & \cdots & a_{m1} \\ \vdots & \vdots & \ddots & \vdots \\ a_{1n} & a_{2n} & \cdots & a_{nm} \end{bmatrix} \tag{A.1.2}$$

If $\boldsymbol{A}$ is an $m \times n$ matrix, then its mirrored form (*transposed* form) is an $n \times m$ matrix, that is, $(\boldsymbol{A}^{\mathrm{T}})^{\mathrm{T}} = \boldsymbol{A}$. If $\boldsymbol{A}^{\mathrm{T}} = \boldsymbol{A}$, then $\boldsymbol{A}$ can be considered a mirror matrix.

The vectors are usually represented as column matrices, the row matrices are usually noted as the transpose of the column matrices, for example,

$$\boldsymbol{x} = \begin{bmatrix} x_1 \\ \vdots \\ x_n \end{bmatrix} = [x_1 \quad , \ldots, \quad x_n]^{\mathrm{T}} = \left[\boldsymbol{x}^{\mathrm{T}}\right]^{\mathrm{T}} \tag{A.1.3}$$

All elements of the zero matrix **O** or zero vector **0** are zeros. The *diagonal* matrix has elements different from zero only in its main diagonal

$$\boldsymbol{D} = \mathbf{diag}[a_{11}, a_{22}, \ldots, a_{nn}] \tag{A.1.4}$$

It is an *identity* matrix $\boldsymbol{I}$, if all of its diagonal elements have value identity $\boldsymbol{I} = \mathbf{diag}[1,1,\ldots,1]$.

Two matrices are considered equal if their corresponding elements are equal. The sum of two or more matrices of the same type is composed of their corresponding elements. The multiplication of a matrix by scalar means to multiply each element of the matrix by the scalar value. The most characteristic rule is the multiplication of two matrices, that is, $m \times l$ matrix $\boldsymbol{A}$ and $l \times n$ matrix $\boldsymbol{B}$:

$$\boldsymbol{C} = \boldsymbol{AB} \tag{A.1.5}$$

where

$$c_{ij} = \sum_{k=1}^{l} a_{ik} b_{kj}; \quad \begin{cases} i = 1, 2, \ldots, m \\ j = 1, 2, \ldots, n \end{cases} \tag{A.1.6}$$

that is, the element in the ith row and in the jth column of the $m \times n$ matrix $\boldsymbol{C}$ is obtained such that the complete ith row of matrix $\boldsymbol{A}$ is composed (multiplied by) of the jth column of matrix $\boldsymbol{B}$. (The number l of the columns of matrix $\boldsymbol{A}$ must be equal to the number l of the rows of matrix $\boldsymbol{B}$.) The matrix multiplication is associative and distributive but generally it is not commutative: $\boldsymbol{AB} \neq \boldsymbol{BA}$. But if $\boldsymbol{AB} = \boldsymbol{BA}$, then matrices $\boldsymbol{A}$ and $\boldsymbol{B}$ are called interchangeable. Note that the determinant $|\boldsymbol{C}| = \det(\boldsymbol{C})$ of the quadratic product matrix $\boldsymbol{C}$ is equal to the product of the determinants $|\boldsymbol{A}|$ and $|\boldsymbol{B}|$ of the factor matrices: $|\boldsymbol{C}| = |\boldsymbol{A}||\boldsymbol{B}|$.

The scalar product of two vectors of same dimension can be expressed as

$$\boldsymbol{a} \cdot \boldsymbol{b} = \boldsymbol{a}^{\mathrm{T}} \boldsymbol{b} = \boldsymbol{b}^{\mathrm{T}} \boldsymbol{a} = \boldsymbol{b} \cdot \boldsymbol{a} \tag{A.1.7}$$

If the scalar product of two nonzero vectors is zero, then the two vectors are called *orthogonal* (perpendicular to each other) vectors.

Observe the following rule

$$[\boldsymbol{AB}]^{\mathrm{T}} = \boldsymbol{B}^{\mathrm{T}} \boldsymbol{A}^{\mathrm{T}} \tag{A.1.8}$$

The inverse $\boldsymbol{A}^{-1}$ of the quadratic, nonzero, that is, nonsingular, regular matrix $\boldsymbol{A}$ with a nonzero determinant can be formulated as

$$\boldsymbol{A}^{-1} \boldsymbol{A} = \boldsymbol{A} \boldsymbol{A}^{-1} = \boldsymbol{I} \tag{A.1.9}$$

The inverse of $\boldsymbol{A}$ can be determined by

$$\boldsymbol{A}^{-1} = \frac{\mathbf{adj}(\boldsymbol{A})}{|\boldsymbol{A}|} \tag{A.1.10}$$

Here $|\boldsymbol{A}|$ is the (nonzero) determinant of $\boldsymbol{A}$, the adjunct matrix $\mathbf{adj}(\boldsymbol{A})$ is obtained by mirroring the matrix constructed from the signed subdeterminants belonging to each element of matrix $\boldsymbol{A}$. Since the rule $|\boldsymbol{AB}| = |\boldsymbol{A}|\ |\boldsymbol{B}|$ exists, according to $1 = |\boldsymbol{I}| = |\boldsymbol{A}^{-1}\boldsymbol{A}| = |\boldsymbol{A}^{-1}|\ |\boldsymbol{A}|$, matrix $\boldsymbol{A}$ has unambiguously determined as the inverse of $\boldsymbol{A}^{-1}$ only if $|\boldsymbol{A}| \neq 0$, that is, matrix $\boldsymbol{A}$ is not singular. Obviously

$$\left[\boldsymbol{A}^{-1}\right]^{-1} = \boldsymbol{A} \quad \text{and} \quad \left[\boldsymbol{A}^{-1}\right]^{\mathrm{T}} = \left[\boldsymbol{A}^{\mathrm{T}}\right]^{-1} \tag{A.1.11}$$

Moreover, if $\boldsymbol{A}$ and $\boldsymbol{B}$ are regular quadratic matrices

$$[\boldsymbol{AB}]^{-1} = \boldsymbol{B}^{-1}\boldsymbol{A}^{-1} \tag{A.1.12}$$

The matrix $s\boldsymbol{I} - \boldsymbol{A}$ or $\boldsymbol{A} - s\boldsymbol{I}$ is called the characteristic matrix of the quadratic matrix $\boldsymbol{A}$, and the equation $\mathcal{A}(s) = |s\boldsymbol{I} - \boldsymbol{A}| = 0$ is called the characteristic equation. The roots $\lambda_i (i = 1,2,\ldots,n)$ of the characteristic equation are the eigenvalues of matrix $\boldsymbol{A}$. According to the main axes theorem the eigenvectors $\boldsymbol{v}_i (i = 1,2,\ldots n)$ of matrix $\boldsymbol{A}$ fulfill the vector equations

$$\boldsymbol{A}\boldsymbol{v}_i = \lambda_i \boldsymbol{v}_i (i = 1, 2, \ldots, n) \tag{A.1.13}$$

This is the practical definition of the eigenvectors. If the vectors $\boldsymbol{v}_i$ are linearly independent, matrix $\boldsymbol{A}$ is of simple structure; if they are not independent, then matrix $\boldsymbol{A}$ is degenerated.

For square matrices $\boldsymbol{A}$ and $\boldsymbol{B}$, the product matrices $\boldsymbol{AB}$ and $\boldsymbol{BA}$ have the same eigenvalues.

One of the basic theorems in the matrix theory is the Cayley–Hamilton theorem: any matrix $\boldsymbol{A}$ fulfills its characteristic equation, that is, $\mathcal{A}(\boldsymbol{A}) = 0$. (Here in the scalar polynomial equation $\mathcal{A}(s) = 0$, s^i must be substituted by $\boldsymbol{A}^i (i = 1,2,\ldots,n)$, whereas s^0 must be substituted by $\boldsymbol{A}^0 = \boldsymbol{I}$, and finally a matrix polynomial equation is obtained.)

It may be necessary to express the inner structure of matrices by the so-called block matrices, like

$$\boldsymbol{M} = \begin{bmatrix} \boldsymbol{A} & \boldsymbol{B} \\ \boldsymbol{C} & \boldsymbol{D} \end{bmatrix} \tag{A.1.14}$$

According to the matrix multiplication rule

$$\begin{bmatrix} A & B \\ C & D \end{bmatrix} \begin{bmatrix} E & F \\ G & H \end{bmatrix} = \begin{bmatrix} AE + BG & AF + BH \\ CE + DG & CF + DH \end{bmatrix} \tag{A.1.15}$$

The determinant of the quasi-diagonal matrix is

$$\begin{vmatrix} A & B \\ O & D \end{vmatrix} = \det \begin{bmatrix} A & B \\ O & D \end{bmatrix} = \det(\boldsymbol{A}) \det(\boldsymbol{D}) = |\boldsymbol{A}| \; |\boldsymbol{D}| \tag{A.1.16}$$

The product $\boldsymbol{ab}^{\mathrm{T}}$ is called the diadic product. The inverse of a matrix extended by a diadic product can be easily determined, if the inverse of $\boldsymbol{A}$ is known

$$\left(\boldsymbol{A} + \boldsymbol{ab}^{\mathrm{T}}\right)^{-1} = \boldsymbol{A}^{-1} - \frac{(\boldsymbol{A}^{-1}\boldsymbol{a})(\boldsymbol{b}^{\mathrm{T}}\boldsymbol{A}^{-1})}{1 + \boldsymbol{b}^{\mathrm{T}}\boldsymbol{A}^{-1}\boldsymbol{a}} \tag{A.1.17}$$

and for determinant of the extension the following expression exists

$$\left|\boldsymbol{A} + \boldsymbol{ab}^{\mathrm{T}}\right| = \left(1 + \boldsymbol{b}^{\mathrm{T}}\boldsymbol{A}^{-1}\boldsymbol{a}\right)|\boldsymbol{A}| \tag{A.1.18}$$

A.1.2 FOUNDATIONS OF VECTOR ANALYSIS

The vector analysis in the Euclidian sphere, besides the so-called scalar–scalar function

$$f = f(x) \tag{A.1.19}$$

uses the definition of the scalar–vector functions

$$f = f(\boldsymbol{x}) \tag{A.1.20}$$

and also the vector–vector functions

$$\boldsymbol{f} = \boldsymbol{f}(\boldsymbol{x}) \tag{A.1.21}$$

(These are the special cases of the most general but rare matrix–matrix functions $\boldsymbol{F} = \boldsymbol{F}(\boldsymbol{X})$.) The multivariable scalar–scalar, scalar–vector, or vector–vector functions are more frequent, for example,

$$f = f(x, u); \quad f = f(\boldsymbol{x}, \boldsymbol{u}); \quad \boldsymbol{f} = \boldsymbol{f}(\boldsymbol{x}, \boldsymbol{u}) \tag{A.1.22}$$

or similarly the functions with an independent variable (usually the time or parameter)

$$f = f(x, u, t); \quad f = f(\boldsymbol{x}, \boldsymbol{u}, t); \quad \boldsymbol{f} = \boldsymbol{f}(\boldsymbol{x}, \boldsymbol{u}, t) \tag{A.1.23}$$

It is extremely important to show some differentiating rules. The derivation by scalar is simple, for example,

$$\frac{\mathrm{d}\boldsymbol{x}(t)}{\mathrm{d}t} = \left[\frac{\mathrm{d}x_1}{\mathrm{d}t}, \frac{\mathrm{d}x_2}{\mathrm{d}t}, \ldots, \frac{\mathrm{d}x_n}{\mathrm{d}t}\right]^{\mathrm{T}} = [\dot{x}_1, \dot{x}_2, \ldots, \dot{x}_n] = \dot{\boldsymbol{x}} \tag{A.1.24}$$

$$\frac{\mathrm{d}\boldsymbol{A}(t)}{\mathrm{d}t} = \begin{bmatrix} \dot{a}_{11} & \dot{a}_{12} & \ldots & \dot{a}_{1n} \\ \dot{a}_{21} & \dot{a}_{22} & \ldots & \dot{a}_{2n} \\ \vdots & \vdots & \ddots & \vdots \\ \dot{a}_{m1} & \dot{a}_{m2} & \ldots & \dot{a}_{mn} \end{bmatrix} = \dot{\boldsymbol{A}} \tag{A.1.25}$$

The gradient of a scalar vector function is a column vector

$$\mathbf{grad}\,[f(\boldsymbol{x})] = \frac{\mathrm{d}f(\boldsymbol{x})}{\mathrm{d}\boldsymbol{x}} = \left[\frac{\mathrm{d}f(\boldsymbol{x})}{\mathrm{d}x_1} \quad \frac{\mathrm{d}f(\boldsymbol{x})}{\mathrm{d}x_2} \quad \ldots \quad \frac{\mathrm{d}f(\boldsymbol{x})}{\mathrm{d}x_n}\right]^{\mathrm{T}} \tag{A.1.26}$$

which means the application of a multivariable differential operator

$$\frac{\mathrm{d}}{\mathrm{d}\boldsymbol{x}} = \left[\frac{\mathrm{d}}{\mathrm{d}x_1} \quad \frac{\mathrm{d}}{\mathrm{d}x_2} \quad \ldots \quad \frac{\mathrm{d}}{\mathrm{d}x_n}\right]^{\mathrm{T}} \tag{A.1.27}$$

that is,

$$\mathbf{grad}\,[f(\boldsymbol{x})] = \frac{\mathrm{d}}{\mathrm{d}\boldsymbol{x}} f(\boldsymbol{x}) = \frac{\mathrm{d}f(\boldsymbol{x})}{\mathrm{d}\boldsymbol{x}} = \boldsymbol{J}(f, \boldsymbol{x}) \tag{A.1.28}$$

Here, the so-called Jacobi derivation matrix (or Jacobian) appears, which has the general form

$$\boldsymbol{J} = \boldsymbol{J}(\boldsymbol{f}, \boldsymbol{x}) = \begin{bmatrix} \frac{\mathrm{d}f_1}{\mathrm{d}x_1} & \frac{\mathrm{d}f_1}{\mathrm{d}x_2} & \ldots & \frac{\mathrm{d}f_1}{\mathrm{d}x_n} \\ \frac{\mathrm{d}f_2}{\mathrm{d}x_1} & \frac{\mathrm{d}f_2}{\mathrm{d}x_2} & \ldots & \frac{\mathrm{d}f_2}{\mathrm{d}x_n} \\ \vdots & \vdots & \ddots & \vdots \\ \frac{\mathrm{d}f_m}{\mathrm{d}x_1} & \frac{\mathrm{d}f_m}{\mathrm{d}x_2} & \ldots & \frac{\mathrm{d}f_m}{\mathrm{d}x_n} \end{bmatrix} \tag{A.1.29}$$

Avoiding complicated notation, the Jacobi matrix is used in its symbolic form

$$J(f,x)=\frac{\mathrm{d}f(x)}{\mathrm{d}x^{\mathrm{T}}},\ \text{and its transpose in the form}\ J^{\mathrm{T}}(f,x)=\frac{\mathrm{d}f^{\mathrm{T}}(x)}{\mathrm{d}x} \tag{A.1.30}$$

Thus the transpose of the gradient vector is

$$\mathbf{grad}^{\mathrm{T}}[f(x)]=\left[\frac{\mathrm{d}f(x)}{\mathrm{d}x}\right]^{\mathrm{T}}=\frac{\mathrm{d}f(x)}{\mathrm{d}x^{\mathrm{T}}}=J(f,x^{\mathrm{T}})$$

A.1.3 KRONECKER PRODUCT OF MATRICES

Let A and B be an $m \times n$ and $r \times s$ matrix. The Kronecker product of matrices A and B is defined as

$$A\otimes B=\begin{bmatrix} a_{11}B & a_{12}B & \dots & a_{1n}B \\ a_{21}B & a_{22}B & \dots & a_{2n}B \\ \vdots & \vdots & \ddots & \vdots \\ a_{m1}B & a_{m2}B & \dots & a_{mn}B \end{bmatrix} \tag{A.1.31}$$

where matrix $A\otimes B$ measures $m\,r \times n\,s$. Next, the most important identities regarding the Kronecker matrix product are summarized.

If the matrices A and C, B and D fulfill the condition of multiplication, then

$$(A\otimes B)\,(C\otimes D)=A\,C\otimes B\,D \tag{A.1.32}$$

If the inverse of matrices A and B exists, then

$$(A\otimes B)^{-1}=A^{-1}\otimes B^{-1} \tag{A.1.33}$$

The transposition rule for the Kronecker matrix product can be formulated as

$$(A\otimes B)^{\mathrm{T}}=A^{\mathrm{T}}\otimes B^{\mathrm{T}} \tag{A.1.34}$$

If the matrices fulfill the condition of summing up (regarding their dimensions), then

$$(A+B)\otimes(C+D)=A\otimes C+A\otimes D+B\otimes C+B\otimes D \tag{A.1.35}$$

The operations can be grouped as

$$A \otimes (B \otimes C) = (A \otimes B) \otimes C \tag{A.1.36}$$

The trace of the Kronecker products can be calculated as

$$\mathrm{tr}(A \otimes B) = \mathrm{tr}(A)\ \mathrm{tr}\ (B) \tag{A.1.37}$$

The above identities can be verified via simple examples.

If A is an $n \times n$ matrix with eigenvalues $\{\lambda_1, \lambda_2, \ldots, \lambda_n\}$, and B is an $m \times m$ matrix with eigenvalues $\{\mu_1, \mu_2, \ldots, \mu_m\}$, then the Kronecker matrix product $C = A \otimes B$ has eigenvalues of $\{\lambda_i \mu_j\}$. The determinant of C can be calculated by $\det(A \otimes B) = [\det(A)^n][\det(B)^m]$.

The determination of the vectors derived from matrices can be defined by the Kronecker products. Let A be an $m \times n$ matrix, and note the vectors derived from A by vec(A) as

$$\mathrm{vec}(A) = \begin{bmatrix} a_1 \\ a_2 \\ \vdots \\ a_n \end{bmatrix} = \begin{bmatrix} a_1^T & a_2^T & \ldots & a_n^T \end{bmatrix} \tag{A.1.38}$$

where $a_1, a_2, \ldots, a_n$ are the column vectors of A. The most important rules are summarized below.

If y is a column vector, then

$$\mathrm{vec}(y) = \mathrm{vec}(y^T) = y \tag{A.1.39}$$

The formulation of the product matrices by vectors can be defined by the Kronecker products as

$$\mathrm{vec}(ABC) = (C^T \otimes A)\mathrm{vec}(B) \tag{A.1.40}$$

If A and B are $m \times n$ and $n \times r$ matrices, respectively, then, based on (A.1.40), it is obvious that

$$\mathrm{vec}(AB) = (B^T \otimes I_m)\mathrm{vec}(A) = (B^T \otimes A)\mathrm{vec}(I_m) = (I_r \otimes A)\mathrm{vec}(B) \tag{A.1.41}$$

where the subscript of the identity matrices refers to their size. Here the trivial identity

$$\mathrm{vec}(AB) = \mathrm{vec}(I_m AB) = \mathrm{vec}(A I_n B) = \mathrm{vec}(ABI_r) \tag{A.1.42}$$

is applied.

The identities below refer to the calculation of the trace of the product matrices

$$\mathrm{tr}(\boldsymbol{ABC}) = \left[\mathrm{vec}\left(\boldsymbol{A}^{\mathrm{T}}\right)\right]^{\mathrm{T}}(\boldsymbol{I}\otimes\boldsymbol{B})\mathrm{vec}(\boldsymbol{C}) \tag{A.1.43}$$

$$\mathrm{tr}(\boldsymbol{AB}) = \left[\mathrm{vec}\left(\boldsymbol{A}^{\mathrm{T}}\right)\right]^{\mathrm{T}}\mathrm{vec}(\boldsymbol{B}) \tag{A.1.44}$$

$$\begin{aligned}\mathrm{tr}\left(\boldsymbol{AZ}^{\mathrm{T}}\boldsymbol{BZC}\right) &= [\mathrm{vec}(\boldsymbol{Z})]^{\mathrm{T}}\left(\boldsymbol{A}^{\mathrm{T}}\boldsymbol{C}^{\mathrm{T}}\otimes\boldsymbol{B}\right)\mathrm{vec}(\boldsymbol{Z})\\ &= [\mathrm{vec}(\boldsymbol{Z})]^{\mathrm{T}}\left(\boldsymbol{CA}\otimes\boldsymbol{B}^{\mathrm{T}}\right)\mathrm{vec}(\boldsymbol{Z})\end{aligned} \tag{A.1.45}$$

The identities can be proved by simple examples.

A.1.4 TOEPLITZ MATRICES

The following matrix operator is called a generalized Toeplitz matrix

$$\boldsymbol{S}_N^i(\boldsymbol{A}) = \begin{bmatrix} \boldsymbol{0} & & & & & & \\ \vdots & \ddots & & & & & \\ \boldsymbol{0} & \ddots & \ddots & & & & \\ \boldsymbol{A} & \boldsymbol{0} & \ddots & \ddots & & & \\ \boldsymbol{0} & \boldsymbol{A} & \ddots & \ddots & \ddots & & \\ \vdots & \vdots & \ddots & \ddots & \ddots & \ddots & \\ \boldsymbol{0} & \boldsymbol{0} & \ldots & \boldsymbol{A} & \boldsymbol{0} & \boldsymbol{0} & \boldsymbol{0} \end{bmatrix} \tag{A.1.46}$$

and $\boldsymbol{S}_N^i(\boldsymbol{A}) = \boldsymbol{0}$ if $i > N$. If $\boldsymbol{A}$ is an $m\times n$ matrix, then the size of $\boldsymbol{S}_N^i(\boldsymbol{A})$ is $Nm\times Nn$. The generalized Toeplitz matrix can be defined easily by the Kronecker matrix product, so the form (A.1.46) can also be written as

$$\boldsymbol{S}_N^i(\boldsymbol{A}) = \boldsymbol{S}_N^i(1)\otimes\boldsymbol{A} \tag{A.1.47}$$

Here $\boldsymbol{S}_N^i(1)$ measures $N\times N$ and is defined by (A.1.46), and thus it has values of one only in the ith row under the main diagonal. All its other elements are zero. Thus the Toeplitz matrices have a special structure, that is, they are lower triangle matrices.

Based on the simple example below, $\boldsymbol{S}_N^i(1)$ is a shift matrix and denoted by $\boldsymbol{S}^i$.

Let

$$\boldsymbol{S}_N^1(1) = \boldsymbol{S}^1 = \boldsymbol{S} = \begin{bmatrix} 0 & 0 & 0 \\ 1 & 0 & 0 \\ 0 & 1 & 0 \end{bmatrix} \tag{A.1.48}$$

and

$$\boldsymbol{y} = \begin{bmatrix} y_1 & y_2 & y_3 \end{bmatrix}^{\mathrm{T}} = \begin{bmatrix} y_1 \\ y_2 \\ y_3 \end{bmatrix} \tag{A.1.49}$$

Then

$$\boldsymbol{S}_N^1(1)\boldsymbol{y} = \boldsymbol{S}\boldsymbol{y} = \begin{bmatrix} 0 \\ y_1 \\ y_2 \end{bmatrix} \tag{A.1.50}$$

and the name is clearly appropriate. Thanks to the above features the matrix is used in *discrete-time* process identification methods and preparation of computer algorithms.

Next, the most important identities regarding the generalized Toeplitz matrices and their connection with the matrix polynomials discussed in Chapter 7 are briefly presented. Let

$$\boldsymbol{T}_{NL} = \sum_{i=0}^{n_L} \boldsymbol{S}_N^i(\mathbf{L}_i) \tag{A.1.51}$$

and

$$\boldsymbol{T}_{NM} = \sum_{i=0}^{n_M} \boldsymbol{S}_N^i(\mathbf{M}_i) \tag{A.1.52}$$

where $\mathbf{L}_i$ and $\mathbf{M}_i$ are parameter matrices

$$\mathcal{L}(z) = \sum_{i=0}^{n_L} \mathbf{L}_i z^i \tag{A.1.53}$$

and

$$\mathcal{M}(z) = \sum_{i=0}^{n_L} \mathbf{M}_i z^i \tag{A.1.54}$$

are the coefficients of the matrix polynomials. The following rules can be formulated:

1. If

$$\mathcal{L}(z) + \mathcal{M}(z) = \mathcal{K}(z) = \sum_{i=0}^{n_K} \mathbf{K}_i z^i \tag{A.1.55}$$

where

$$\mathbf{K}_i = \mathbf{L}_i + \mathbf{M}_i \quad \text{and} \quad n_K = \max\{n_L, n_M\} \tag{A.1.56}$$

then the relationship

$$\boldsymbol{T}_{NL} + \boldsymbol{T}_{NM} = \sum_{i=0}^{n_K} \boldsymbol{S}_N^i(\mathbf{K}_i) \tag{A.1.57}$$

exists.

2. If

$$\mathcal{L}(z)\mathcal{M}(z) = \sum_{i=0}^{n_K} \mathbf{K}_i z^i \tag{A.1.58}$$

where

$$\mathbf{K}_i = \sum_{j=0}^{i} \mathbf{L}_j \mathbf{M}_{i-j} \quad \text{and} \quad n_K = n_L + n_M \tag{A.1.59}$$

then the relationship

$$\boldsymbol{T}_{NL}\ \boldsymbol{T}_{NM} = \sum_{i=0}^{n_K} \boldsymbol{S}_N^i(\mathbf{K}_i) \tag{A.1.60}$$

exists.

3. Let

$$\mathcal{L}(z) = \sum_{i=0}^{n_L} \mathbf{L}_i z^i = \mathbf{L}_0 + \sum_{i=1}^{n_L} \mathbf{L}_i z^i \tag{A.1.61}$$

If $\mathbf{L}_i$ is quadratic and $\mathbf{L}_0$ is nonsingular, then

$$\boldsymbol{T}_{NL}^{-1} = \sum_{i=0}^{N-1} \boldsymbol{S}_N^i(\mathbf{K}_i) \tag{A.1.62}$$

where

$$\mathcal{K}(z) = \mathcal{L}^{-1}(z) = \sum_{i=0}^{\infty} \mathbf{K}_i z^i \tag{A.1.63}$$

and

$$
\begin{aligned}
\mathbf{K}_0 &= \mathbf{L}_0^{-1} \\
\mathbf{K}_1 &= -\mathbf{L}_0^{-1}\mathbf{L}_1\mathbf{K}_0 \\
\mathbf{K}_2 &= -\mathbf{L}_0^{-1}(\mathbf{L}_1\mathbf{K}_1 + \mathbf{L}_2\mathbf{K}_0) \\
&\vdots \\
\mathbf{K}_{N-1} &= -\mathbf{L}_0^{-1}(\mathbf{L}_1\mathbf{K}_{N-2} + \mathbf{L}_2\mathbf{K}_{N-3} + \ldots + \mathbf{L}_{N-1}\mathbf{K}_0)
\end{aligned}
\tag{A.1.64}
$$

Obviously $\mathbf{L}_i = \mathbf{0}$, if $i > n_L$.

A very important feature of the shift matrices $\boldsymbol{S}_N^i(1) = \boldsymbol{S}^i$ is their interchangeability, which means that

$$
\boldsymbol{S}^i\boldsymbol{S}^j = \boldsymbol{S}^j\boldsymbol{S}^i = \boldsymbol{S}^{i+j} \tag{A.1.65}
$$

APPENDIX 2

State-Space Methods

A.2.1 SOLUTION OF STATE-SPACE EQUATIONS IN THE COMPLEX FREQUENCY DOMAIN

State equations can be transferred to the complex frequency domain by Laplace transformation (1.1.10). Note the transformed values of the time functions $\boldsymbol{x},u,y$ by $\boldsymbol{X}(s),U(s),Y(s)$. Taking the rules for transforming the differential quotient into account, we obtain

$$\begin{aligned} s\boldsymbol{X}(s) &= \boldsymbol{A}\boldsymbol{X}(s) + \boldsymbol{b}U(s) + \boldsymbol{x}(0) = \boldsymbol{A}\boldsymbol{X}(s) + \boldsymbol{b}U(s) + \boldsymbol{I}\boldsymbol{x}(0) \\ Y(s) &= \boldsymbol{c}^{\mathrm{T}}\boldsymbol{X}(s) + d_{\mathrm{c}}U(s) \end{aligned} \qquad \text{(A.2.1)}$$

In the first equation the vector $\boldsymbol{x}(0)$ of the initial conditions can be considered as an input signal that has effect on the system via an identity matrix $\boldsymbol{I}$. From the first equation, we obtain

$$\begin{aligned} \boldsymbol{X}(s) &= (s\boldsymbol{I} - \boldsymbol{A})^{-1}[\boldsymbol{b}U(s) + \boldsymbol{x}(0)] \\ &= (s\boldsymbol{I} - \boldsymbol{A})^{-1}\boldsymbol{b}U(s) + (s\boldsymbol{I} - \boldsymbol{A})^{-1}\boldsymbol{x}(0) \end{aligned} \qquad \text{(A.2.2)}$$

Applying the rules of matrix inversion, we obtain

$$(s\boldsymbol{I} - \boldsymbol{A})^{-1} = \frac{\mathbf{adj}(s\boldsymbol{I} - \boldsymbol{A})}{\det(s\boldsymbol{I} - \boldsymbol{A})} = \frac{\mathbf{adj}(s\boldsymbol{I} - \boldsymbol{A})}{\mathcal{A}(s)} = \frac{\boldsymbol{\Psi}(s)}{\mathcal{A}(s)} = \boldsymbol{\Phi}(s) \qquad \text{(A.2.3)}$$

Here $\boldsymbol{\Psi}(s) = \mathbf{adj}(s\boldsymbol{I} - \boldsymbol{A})$ is the transpose of a matrix whose elements are signed subdeterminants belonging to the corresponding elements of the matrix $(s\boldsymbol{I} - \boldsymbol{A})$. $\det(s\boldsymbol{I} - \boldsymbol{A})$ is the determinant of the matrix $(s\boldsymbol{I} - \boldsymbol{A})$, i.e., the denominator of the transfer function, which is n-order polynomial of s

$$\mathcal{A}(s) = s^n + k_1 s^{n-1} + \cdots + k_{n-1}s + k_n = \prod_{i-1}^{n} (s - \lambda_i) = \det(s\boldsymbol{I} - \boldsymbol{A}) \qquad \text{(A.2.4)}$$

$\mathcal{A}(s)$ is the so-called characteristic polynomial of matrix $\boldsymbol{A}$; the roots $\lambda_1,\ldots,\lambda_n$ of the characteristic equation are the eigenvalues of matrix $\boldsymbol{A}$, and are called poles of the system.

The elements of the matrix in the numerator of Equation (A.2.3) are polynomials of s, and since they are derived from $(n-1)$-order subdeterminants, their order can be $(n-1)$ at most; consequently, the quotient of the specific element and $\mathcal{A}(s)$ provides strictly proper transfer functions.

Because of Equation (A.2.2) the motion of the state vector is determined by the initial state vector $\boldsymbol{x}(0)$ and the input signal $U(s)$. The characteristic polynomial appears in the denominator of all elements depending on $\mathcal{A}(s)$, and therefore, because of the expansion theorem, their time functions are exclusively determined by the system poles. This part of the solution describes the free motion of the system, i.e., its response, when it was removed from the steady position and left on its own, and it exclusively depends on one of the system parameters, i.e., on the state matrix $\boldsymbol{A}$ both in the frequency and in the time domains.

The denominator of the elements in the part of the solution depending on $U(s)$ contains the polynomial in the denominator of $U(s)$ besides $\mathcal{A}(s)$, so the time functions depend not only on the poles of the system but also on the poles of the input signal. This part of the solution describes the excited motion of the system (forced response).

The output signal is obtained from Equations (A.2.1) and (A.2.2)

$$Y(s) = \boldsymbol{c}^{\mathrm{T}}(s\boldsymbol{I} - \boldsymbol{A})^{-1}[\boldsymbol{b}U(s) + \boldsymbol{x}(0)] + d_{\mathrm{c}}U(s) \tag{A.2.5}$$

The output of the excited motion is

$$Y(s) = \left[\boldsymbol{c}^{\mathrm{T}}(s\boldsymbol{I} - \boldsymbol{A})^{-1}\boldsymbol{b} + d_{\mathrm{c}}\right]U(s) \tag{A.2.6}$$

when $\boldsymbol{x}(0) = \boldsymbol{0}$.

Thus the transfer function of the system is

$$P(s) = \frac{Y(s)}{U(s)} = \boldsymbol{c}^{\mathrm{T}}(s\boldsymbol{I} - \boldsymbol{A})^{-1}\boldsymbol{b} + d_{\mathrm{c}}\bigg|_{d_{\mathrm{c}}=0} = \boldsymbol{c}^{\mathrm{T}}(s\boldsymbol{I} - \boldsymbol{A})^{-1}\boldsymbol{b} = \frac{\mathcal{B}(s)}{\mathcal{A}(s)} \tag{A.2.7}$$

The first term of $P(s)$ is strictly proper, since it consists of strictly proper elements (see (A.2.3): the adjugate always has a lower order than the determinant!). If $d_{\mathrm{c}} = 0$, then $P(s)$ is strictly proper; the order of its numerator is lower by one, at least, than the order of its denominator. If $d_{\mathrm{c}} \neq 0$, then $P(s)$ is proper, i.e., the order of the numerator is equal to the order of the denominator. The physical meaning of d_{c} is how the input signal takes effect on the output directly, without dynamics. Note that this effect does not disappear even at very high frequencies, and thus it can be stated that $P(j\omega \to \infty) = d_{\mathrm{c}}$. This means, at the same time, that the jump in the step response function in the time moment $t = 0$ is $v(t = 0) = d_{\mathrm{c}}$. In practice, the case $d_c \neq 0$ is traced back to the case $d_c = 0$ by introduction of a new output signal $\tilde{y} = y - d_c u$. The case $d_c \neq 0$ can be considered as a nonadequate linearization in the working point what should be corrected.

A.2.2 SOLUTION OF STATE-SPACE EQUATIONS IN THE TIME DOMAIN

The solution of state Equation (1.1.10) in the time domain can be given in the following closed form

$$\boldsymbol{x}(t) = e^{\boldsymbol{A}t}\boldsymbol{x}(0) + \int_0^t e^{\boldsymbol{A}(t-\tau)}\boldsymbol{b}u(\tau)\mathrm{d}\tau = e^{\boldsymbol{A}t}\boldsymbol{x}(0) + \left[\int_0^t e^{\boldsymbol{A}(t-\tau)}u(\tau)\mathrm{d}\tau\right]\boldsymbol{b} \tag{A.2.8}$$

The first term gives the motion of a system starting from the initial state $\boldsymbol{x}(0)$ and left on its own (free response), and the second term is the so-called convolution integral, i.e., the excited motion starting from the initial state $\boldsymbol{x}(0) = \boldsymbol{0}$.

To check Equation (A.2.8), let us derivate the above equation according to time

$$\frac{\mathrm{d}\boldsymbol{x}(t)}{\mathrm{d}t} = \boldsymbol{A}e^{\boldsymbol{A}t}\boldsymbol{x}(0) + \int_0^t \boldsymbol{A}e^{\boldsymbol{A}(t-\tau)}\boldsymbol{b}u(\tau)\mathrm{d}\tau + \boldsymbol{b}u(t) = \boldsymbol{A}\boldsymbol{x}(t) + \boldsymbol{b}u(t) \tag{A.2.9}$$

which proves the correctness of Equation (A.2.8). See the details of the derivation in A.2.5.1. Here $e^{\boldsymbol{A}t}$ is the fundamental matrix of the system defined by the Taylor series convergent for all t, i.e.,

$$e^{\boldsymbol{A}t} = \boldsymbol{I} + \boldsymbol{A}t + \frac{1}{2}(\boldsymbol{A}t)^2 + \cdots + \frac{1}{n!}(\boldsymbol{A}t)^n + \cdots \tag{A.2.10}$$

Differentiating the equation, we obtain an interesting and important feature of the fundamental matrix

$$\begin{aligned}\frac{\mathrm{d}e^{\boldsymbol{A}t}}{\mathrm{d}t} &= \boldsymbol{A} + \boldsymbol{A}^2 t + \frac{1}{2}\boldsymbol{A}^3 t^2 + \cdots + \frac{1}{(n-1)!}\boldsymbol{A}^n t^{n-1} + \cdots \\ &= \boldsymbol{A}\left(\boldsymbol{I} + \boldsymbol{A}t + \frac{1}{2}(\boldsymbol{A}t)^2 + \cdots + \frac{1}{n!}(\boldsymbol{A}t)^n + \cdots\right) = \boldsymbol{A}e^{\boldsymbol{A}t} = e^{\boldsymbol{A}t}\boldsymbol{A}\end{aligned} \tag{A.2.11}$$

From comparison of Equations (A.2.2) and (A.2.8), we obtain the Laplace transform of the fundamental matrix for the case $U(s) = 0$

$$\mathcal{L}\{e^{At}\} = (s\boldsymbol{I} - \boldsymbol{A})^{-1} = \boldsymbol{\Phi}(s) \tag{A.2.12}$$

and using this we can obtain a newer expression for the calculation of the fundamental matrix

$$e^{At} = \mathcal{L}^{-1}\left\{(s\boldsymbol{I} - \boldsymbol{A})^{-1}\right\} = \mathcal{L}^{-1}\left\{\boldsymbol{\Phi}(s)\right\} \tag{A.2.13}$$

Combining Equations (1.1.10) and (A.2.8), the output of the system becomes

$$y(t) = \boldsymbol{c}^{\mathrm{T}} e^{At}\boldsymbol{x}(0) + \boldsymbol{c}^{\mathrm{T}}\left[\int_0^t e^{A(t-\tau)}u(\tau)\mathrm{d}\tau\right]\boldsymbol{b} + d_c u(t) \tag{A.2.14}$$

From this latter equation the pulse response of the system can be simply obtained for zero initial conditions ($\boldsymbol{x}(0) = \boldsymbol{0}$) and case $d_c = 0$, and by applying the excitation $u(t) = \delta(t)$

$$\begin{aligned} w(t) &= \boldsymbol{c}^{\mathrm{T}} e^{At}\boldsymbol{b} = \boldsymbol{c}^{\mathrm{T}}\mathcal{L}^{-1}\left\{(s\boldsymbol{I} - \boldsymbol{A})^{-1}\right\}\boldsymbol{b} = \mathcal{L}^{-1}\left\{\boldsymbol{c}^{\mathrm{T}}(s\boldsymbol{I} - \boldsymbol{A})^{-1}\boldsymbol{b}\right\} \\ &= \mathcal{L}^{-1}\left\{\boldsymbol{c}^{\mathrm{T}}\boldsymbol{\Phi}(s)\boldsymbol{b}\right\} = \mathcal{L}^{-1}\left\{P(s)|_{d=0}\right\} \end{aligned} \tag{A.2.15}$$

The derivation of the pulse response (weighting function) in time domain can be seen in the A.2.5.2.

Because of the definition of the matrix function and the Cayley–Hamilton theorem, the fundamental matrix can also be calculated in the form of the final sum

$$e^{A\tau} = \alpha_0(\tau)\boldsymbol{I} + \alpha_1(\tau)\boldsymbol{A} + \ldots + \alpha_{n-1}(\tau)\boldsymbol{A}^{n-1} \tag{A.2.16}$$

since the state matrix $\boldsymbol{A}$ fulfills its characteristic equation, i.e.,

$$\mathcal{A}(\boldsymbol{A}) = 0 \tag{A.2.17}$$

(See the deduction in A.2.5.3.)

A.2.3 TRANSFORMATION OF STATE-SPACE EQUATIONS, CANONICAL FORMS

The input and output signals of a system are usually physical variables. State variables, however, depend on the chosen coordinate system. The parameter matrices $\boldsymbol{A}, \boldsymbol{b}, \boldsymbol{c}^{\mathrm{T}}$ also depend on the coordinate system. Introduce a new state vector $\boldsymbol{z}$ which can be derived from the original $\boldsymbol{x}$ by the linear transformation $\boldsymbol{z} = \boldsymbol{T}\boldsymbol{x}$, where $\boldsymbol{T}$ is regular. Using (1.1.10), the new state equation is

$$\begin{aligned} \frac{\mathrm{d}\boldsymbol{z}}{\mathrm{d}t} &= \boldsymbol{T}(\boldsymbol{A}\boldsymbol{x} + \boldsymbol{b}u) = \boldsymbol{T}\boldsymbol{A}\boldsymbol{T}^{-1}\boldsymbol{z} + \boldsymbol{T}\boldsymbol{b}u = \tilde{\boldsymbol{A}}\boldsymbol{z} + \tilde{\boldsymbol{b}}u \\ y &= \boldsymbol{c}^{\mathrm{T}}\boldsymbol{x} + d_{\mathrm{c}}u = \boldsymbol{c}^{\mathrm{T}}\boldsymbol{T}^{-1}\boldsymbol{z} + du = \tilde{\boldsymbol{c}}^{\mathrm{T}}\boldsymbol{z} + \tilde{d}_{\mathrm{c}}u \end{aligned} \tag{A.2.18}$$

where

$$\tilde{\boldsymbol{A}} = \boldsymbol{T}\boldsymbol{A}\boldsymbol{T}^{-1}; \quad \tilde{\boldsymbol{b}} = \boldsymbol{T}\boldsymbol{b}; \quad \tilde{\boldsymbol{c}}^{\mathrm{T}} = \boldsymbol{c}^{\mathrm{T}}\boldsymbol{T}^{-1}; \quad \tilde{d}_{\mathrm{c}} = d_{\mathrm{c}} \tag{A.2.19}$$

It is easy to check that the impulse response and the transfer function of the system are invariant for the linear transformation

$$w(t) = \tilde{\boldsymbol{c}}^{\mathrm{T}} e^{\tilde{\boldsymbol{A}}t}\tilde{\boldsymbol{b}} = \boldsymbol{c}^{\mathrm{T}}\boldsymbol{T}^{-1}e^{\boldsymbol{T}\boldsymbol{A}\boldsymbol{T}^{-1}t}\boldsymbol{T}\boldsymbol{b} = \boldsymbol{c}^{\mathrm{T}}e^{\boldsymbol{A}t}\boldsymbol{b} \tag{A.2.20}$$

$$H(s) = \tilde{\boldsymbol{c}}^{\mathrm{T}}\left(s\boldsymbol{I} - \tilde{\boldsymbol{A}}\right)^{-1}\tilde{\boldsymbol{b}} = \boldsymbol{c}^{\mathrm{T}}\boldsymbol{T}^{-1}\left(s\boldsymbol{I} - \boldsymbol{T}\boldsymbol{A}\boldsymbol{T}^{-1}\right)^{-1}\boldsymbol{T}\boldsymbol{b} = \boldsymbol{c}^{\mathrm{T}}(s\boldsymbol{I} - \boldsymbol{A})^{-1}\boldsymbol{b} \tag{A.2.21}$$

In Equation (A.2.20) the following simple identity

$$e^{\left(\boldsymbol{T}\boldsymbol{A}\boldsymbol{T}^{-1}\right)t} = \boldsymbol{T}e^{\boldsymbol{A}t}\boldsymbol{T}^{-1} \tag{A.2.22}$$

is taken into consideration.

It is known that the linear transformations have designated directions ($\boldsymbol{v}_i$) in which the vectors keep their direction, only their lengths are changed by λ_i times longer, i.e.,

$$\boldsymbol{A}\boldsymbol{v}_i = \lambda_i\boldsymbol{v}_i; \quad i = 1, \ldots, n \tag{A.2.23}$$

Here $\boldsymbol{v}_i$ is the eigenvector of $\boldsymbol{A}$, and λ_i is the corresponding eigenvalue. The eigenvalue problem can be written as a homogeneous equation system with n unknown variables:

$$(\lambda_i\boldsymbol{I} - \boldsymbol{A})\boldsymbol{v} = \boldsymbol{0} \tag{A.2.24}$$

where the unknown variables are the components of $\boldsymbol{v}$. The equation has a nontrivial solution only if the condition

$$\det(\lambda_i\boldsymbol{I} - \boldsymbol{A}) = \mathcal{A}(\lambda_i) = 0 \tag{A.2.25}$$

is fulfilled, i.e., if the eigenvalues λ_i are the roots of the characteristic polynomial. If the roots of the equation are single, then there are n eigenvalues, and for each eigenvalue only one eigenvector with unity length can be determined.

Diagonal Canonical Form

In this case of single eigenvalues the matrix $\boldsymbol{T}_\mathrm{d}\boldsymbol{A}(\boldsymbol{T}_\mathrm{d})^{-1}$ can be made diagonal by means of a special transformation matrix $\boldsymbol{T}_\mathrm{d}$

$$\tilde{\boldsymbol{A}}_\mathrm{d} = \boldsymbol{T}_\mathrm{d}\boldsymbol{A}(\boldsymbol{T}_\mathrm{d})^{-1} = \boldsymbol{\Lambda} = \begin{bmatrix} \lambda_1 & 0 & \dots & 0 \\ 0 & \lambda_2 & \dots & 0 \\ \vdots & \vdots & \ddots & \vdots \\ 0 & 0 & \dots & \lambda_n \end{bmatrix} = \boldsymbol{A}_\mathrm{d}$$

$$= \mathbf{diag}[\lambda_1, \lambda_2, \dots, \lambda_n] \tag{A.2.26}$$

The suitable transformation matrix $\boldsymbol{T}_\mathrm{d}$ is the inverse of the eigenvector matrix, i.e.,

$$\boldsymbol{T}_\mathrm{d} = [\boldsymbol{v}_1, \boldsymbol{v}_2, \dots, \boldsymbol{v}_n]^{-1} \tag{A.2.27}$$

The canonical state equation (diagonal form) obtained by the diagonal transformation is

$$\frac{\mathrm{d}\boldsymbol{z}}{\mathrm{d}t} = \begin{bmatrix} \lambda_1 & 0 & \dots & 0 \\ 0 & \lambda_2 & \dots & 0 \\ \vdots & \vdots & \ddots & \vdots \\ 0 & 0 & \dots & \lambda_n \end{bmatrix} \boldsymbol{z} + \begin{bmatrix} \beta_1 \\ \beta_2 \\ \vdots \\ \beta_n \end{bmatrix} u = \boldsymbol{\Lambda}\boldsymbol{z} + \boldsymbol{\beta} u$$
$$y = [\gamma_1 \quad \gamma_2 \quad \dots \quad \gamma_n]\boldsymbol{z} + du = \boldsymbol{\gamma}^\mathrm{T}\boldsymbol{z} + d_\mathrm{c}u \tag{A.2.28}$$

The transfer function of the transformed system is

$$P(s) = \sum_{i=1}^{n} \frac{\beta_i \gamma_i}{s - \lambda_i} + d_\mathrm{c} \tag{A.2.29}$$

Thus the transfer function is obtained from the canonical form in a partial fraction structure. Note that the eigenvalues of the matrix $\boldsymbol{A}$ are in the denominator. The transfer function remains unchanged if the product of β_i and γ_i remains constant. That is why there are an infinite number of canonical forms which have different $\boldsymbol{\beta}$ and $\boldsymbol{\gamma}^{\mathrm{T}}$ matrices but a common transfer function.

In the case of single poles the state equation system disintegrates to n independent first-order differential equations in the canonical coordinates. The state variables are separate and can be allocated to singular poles of the system.

If the characteristic equation has multiple roots, the matrix $\boldsymbol{A}_{\mathrm{d}}$ can be diagonized only in exceptional cases, but it can be rearranged into the Jordan form:

$$\boldsymbol{J} = \begin{bmatrix} \boldsymbol{J}_1 & \mathbf{0} & \dots & \mathbf{0} \\ \mathbf{0} & \boldsymbol{J}_2 & \dots & \mathbf{0} \\ \vdots & \vdots & \ddots & \vdots \\ \mathbf{0} & \mathbf{0} & \dots & \boldsymbol{J}_m \end{bmatrix} \tag{A.2.30}$$

Here $\boldsymbol{J}_i$ is a quadratic matrix assigned to the eigenvalue λ_i, whose order is equal to the multiplicity of the eigenvalue. Its main diagonal contains the eigenvalues, the right lateral diagonal has values of one, and all the other elements are zeros. If, e.g., λ_1 is a triple eigenvalue, the submatrix $\boldsymbol{J}_1$ has the structure

$$\boldsymbol{J}_1 = \begin{bmatrix} \lambda_1 & 1 & 0 \\ 0 & \lambda_1 & 1 \\ 0 & 0 & \lambda_1 \end{bmatrix} \tag{A.2.31}$$

(The number of values of one depends on how many linearly independent eigenvectors can be found for a multiple eigenvalue λ_1. If there is only one, this corresponds to the normal case (A.2.28), and all elements of the lateral diagonal have the value of one. If the number of independent eigenvectors increases by one, then the number of the values of one decreases by one. If the number of independent eigenvectors is equal to the multiplicity, then the Jordan matrix is diagonal. Disregarding from this case, the search for the transformation matrix requires special considerations different from the above; they are not discussed here.)

Controllable Canonical Form

Although in modeling practice, the state equations are usually based directly on the differential equations formulated for the physical variables, in many cases the initial information is a transfer function or the corresponding n-order differential equation. This procedure is usually called construction (or reconstruction) of the state equations. Assume that the operation of the system is given by the differential equation below [9]

$$\frac{\mathrm{d}^n y}{\mathrm{d}t^n} + a_1 \frac{\mathrm{d}^{n-1} y}{\mathrm{d}t^{n-1}} + \ldots + a_n y = b_1 \frac{\mathrm{d}^{n-1} u}{\mathrm{d}t^{n-1}} + \ldots + b_n u \tag{A.2.32}$$

The equation valid for the Laplace transforms is

$$Y(s) = \frac{b_1 s^{n-1} + \ldots + b_{n-1} s + b_n}{s^n + a_1 s^{n-1} + \ldots + a_{n-1} s + a_n} U(s) = \frac{\mathcal{B}(s)}{\mathcal{A}(s)} U(s) = P(s)\, U(s) \tag{A.2.33}$$

Introduce the following state variables with their Laplace transforms

$$\begin{aligned}
X_1(s) &= \frac{s^{n-1}}{\mathcal{A}(s)} U(s) \\
X_2(s) &= \frac{s^{n-2}}{\mathcal{A}(s)} U(s) = \frac{1}{s} X_1(s); \qquad \frac{\mathrm{d}x_2}{\mathrm{d}t} = x_1 \\
&\vdots \qquad\qquad\qquad\qquad\qquad\qquad \vdots \\
X_n(s) &= \frac{1}{\mathcal{A}(s)} U(s) = \frac{1}{s} X_{n-1}(s); \qquad \frac{\mathrm{d}x_n}{\mathrm{d}t} = x_{n-1}
\end{aligned} \tag{A.2.34}$$

according to which

$$\begin{aligned}
sX_1(s) &= -a_1 X_1(s) - \ldots - a_n X_n(s) + U(s); \qquad \frac{\mathrm{d}x_1}{\mathrm{d}t} = -a_1 x_1 - \ldots - a_n x_n + u \\
Y(s) &= b_1 X_1(s) + \ldots + b_n X_n(s); \qquad y = b_1 x_1 + \ldots + b_n x_n
\end{aligned} \tag{A.2.35}$$

Thus the resulting state equations are

$$\frac{\mathrm{d}\boldsymbol{x}}{\mathrm{d}t} = \begin{bmatrix} -a_1 & -a_2 & \ldots & -a_{n-1} & -a_n \\ 1 & 0 & \ldots & 0 & 0 \\ 0 & 1 & \ldots & 0 & 0 \\ \vdots & \vdots & \ddots & \vdots & \vdots \\ 0 & 0 & \ldots & 1 & 0 \end{bmatrix} \boldsymbol{x} + \begin{bmatrix} 1 \\ 0 \\ 0 \\ \vdots \\ 0 \end{bmatrix} u$$

$$y = \begin{bmatrix} b_1 & b_2 & \ldots & b_{n-1} & b_n \end{bmatrix} \boldsymbol{x} \tag{A.2.36}$$

This form with the corresponding state matrix of special structure

$$\boldsymbol{A}_c = \begin{bmatrix} -a_1 & -a_2 & \dots & -a_{n-1} & -a_n \\ 1 & 0 & \dots & 0 & 0 \\ 0 & 1 & \dots & 0 & 0 \\ \vdots & \vdots & \ddots & \vdots & \vdots \\ 0 & 0 & \dots & 1 & 0 \end{bmatrix}; \quad \boldsymbol{b}_c = \begin{bmatrix} 1 \\ 0 \\ 0 \\ \vdots \\ 0 \end{bmatrix}; \tag{A.2.37}$$

$$\boldsymbol{c}_c^{\mathrm{T}} = [b_1 \;\; b_2 \;\; \dots \;\; b_{n-1} \;\; b_n]$$

is a *controllable* canonical form or phase variable form. Its special feature is that all state variables, except the last one, are the derivated value of the state variable next in the leading direction and all are feedbacked to the first state variable. The feedback factors are the negative coefficients of the characteristic equation, so they appear in the first row of matrix $\boldsymbol{A}$. The input signal has effect only on x_1. The feedforward factors forming the output are the coefficients of the numerator of the transfer function.

If $P(s)$ is not strictly proper, i.e., $\mathcal{B}(s) = b'_o s^n + b'_1 s^{n-1} + \dots + b'_{n-1} s + b'_n$, the state equation contains also $d_c = b'_o$. Instead of the original coefficients b'_i, however, the coefficients b_i appear, which are obtained by the decomposition

$$\begin{aligned} P(s) &= \frac{\mathcal{B}(s)}{\mathcal{A}(s)} = \frac{b'_o s^n + b'_1 s^{n-1} + \dots + b'_{n-1} s + b'_n}{s^n + a_1 s^{n-1} + \dots + a_{n-1} s + a_n} \\ &= b_o + \frac{b_1 s^{n-1} + \dots + b_{n-1} s + b_n}{s^n + a_1 s^{n-1} + \dots + a_{n-1} s + a_n} \end{aligned} \tag{A.2.38}$$

Here the second term is already strictly proper and the coefficients of the numerator can be calculated by the expression $b_i = b'_i - b'_o a_i = b'_i - b_o a_i$, where $b_o = b'_o$.

The characteristic polynomial of the controllable canonical form is

$$\mathcal{A}(s) = \det \begin{bmatrix} s + a_1 & a_2 & \dots & a_{n-1} & a_n \\ -1 & s & \dots & 0 & 0 \\ 0 & -1 & \dots & 0 & 0 \\ \vdots & \vdots & \ddots & \vdots & \vdots \\ 0 & 0 & \dots & -1 & s \end{bmatrix} = \mathcal{A}_n(s) = s\mathcal{A}_{n-1}(s) + a_n \tag{A.2.39}$$

where the recursive relationship is obtained by the decomposition according to the last row. It is clear that the characteristic polynomial

$$\mathcal{A}_n(s) = s^n + a_1 s^{n-1} + \ldots + a_{n-1}s + a_n = \mathcal{A}(s) \tag{A.2.40}$$

is the denominator of the transfer function. Therefore the special matrix $\boldsymbol{A}_c$ is also called the companion matrix of the polynomial $\mathcal{A}(s)$ (see further details in A.2.5.10).

Note that the following parameter matrix

$$\overline{\boldsymbol{A}}_c = \begin{bmatrix} 0 & 1 & \ldots & 0 & 0 \\ \vdots & \vdots & \ddots & \vdots & \vdots \\ 0 & 0 & \ldots & 1 & 0 \\ 0 & 0 & \ldots & 0 & 1 \\ -a_n & -a_{n-1} & \ldots & -a_2 & -a_1 \end{bmatrix}; \quad \overline{\boldsymbol{b}}_c = \begin{bmatrix} 0 \\ 0 \\ 0 \\ \vdots \\ 1 \end{bmatrix}; \tag{A.2.41}$$

$$\overline{\boldsymbol{c}}_c^T = [b_n \ b_{n-1} \ \ldots \ b_2 \ b_1]$$

also provides a controllable canonical form, where the serial number of the state variables is opposite to the case of the previous form (A.2.37).

Observable Canonical Form

To construct this form, let us choose the state variables with their Laplace transforms according to the following recursive statement [9]

$$\begin{aligned}
&X_1(s) = Y(s) \\
&sX_1(s) = -a_1 X_1(s) + X_2(s) + b_1 U(s); && \frac{dx_1}{dt} = -a_1 x_1 + x_2 + b_1 u \\
&sX_2(s) = -a_2 X_2(s) + X_3(s) + b_2 U(s); && \frac{dx_2}{dt} = -a_2 x_2 + x_3 + b_2 u \\
&\quad \vdots \qquad\qquad \vdots \\
&sX_n(s) = -a_1 X_1(s) + b_n U(s); && \frac{dx_n}{dt} = -a_n x_n + b_n u
\end{aligned} \tag{A.2.42}$$

where $Y(s)$ is given by Equation (A.2.33). Then the following state equations can be formulated

$$\frac{\mathrm{d}\boldsymbol{x}}{\mathrm{d}t} = \begin{bmatrix} -a_1 & 1 & 0 & \dots & 0 \\ -a_2 & 0 & 1 & \dots & 0 \\ \vdots & \vdots & \vdots & \ddots & 0 \\ -a_{n-1} & 0 & 0 & \dots & 1 \\ -a_n & 0 & 0 & \dots & 0 \end{bmatrix} \boldsymbol{x} + \begin{bmatrix} b_1 \\ b_2 \\ \vdots \\ b_{n-1} \\ b_n \end{bmatrix} u \tag{A.2.43}$$

$$y = [1 \quad 0 \quad \dots \quad 0 \quad 0]\boldsymbol{x}$$

This form together with the system matrices of special structure

$$\boldsymbol{A}_\mathrm{o} = \begin{bmatrix} -a_1 & 1 & 0 & \dots & 0 \\ -a_2 & 0 & 1 & \dots & 0 \\ \vdots & \vdots & \vdots & \ddots & \vdots \\ -a_{n-1} & 0 & 0 & \dots & 1 \\ -a_n & 0 & 0 & \dots & 0 \end{bmatrix}; \quad \boldsymbol{b}_\mathrm{o} = \begin{bmatrix} b_1 \\ b_2 \\ \vdots \\ b_{n-1} \\ b_n \end{bmatrix}; \tag{A.2.44}$$

$$\boldsymbol{c}_\mathrm{o}^\mathrm{T} = [1 \ 0 \ \dots \ 0 \ 0]$$

is called an *observable* canonical form. Its feature is that the state variable x_1 itself is the output, which is feedbacked to the inputs of all state variables. The feedback factors are the negative coefficients of the characteristic equation, so they appear in the first column of matrix $\boldsymbol{A}_\mathrm{o}$.

Note that the parameter matrix

$$\overline{\boldsymbol{A}}_\mathrm{o} = \begin{bmatrix} 0 & \dots & 0 & 0 & -a_n \\ 1 & \dots & 0 & 0 & -a_{n-1} \\ \vdots & \ddots & \vdots & \vdots & \vdots \\ 0 & \dots & 1 & 0 & -a_2 \\ 0 & \dots & 0 & 1 & -a_1 \end{bmatrix}; \quad \overline{\boldsymbol{b}}_\mathrm{o} = \begin{bmatrix} b_n \\ b_{n-1} \\ \vdots \\ b_2 \\ b_1 \end{bmatrix}; \tag{A.2.45}$$

$$\overline{\boldsymbol{c}}_\mathrm{o}^\mathrm{T} = [0 \ 0 \ \dots \ 0 \ 1]$$

also provides an observable canonical form where the serial number of the state variables is opposite to the case above in Equation (A.2.43).

If $P(s)$ is not strictly proper, then $d_c = b_o$ also appears in the state equation, and the statements made in connection with the decomposition (A.2.38) above are still valid.

(If the poles and partial fractional form of the transfer function are known, then further canonical forms can be constructed.)

A.2.4 CONCEPTS OF CONTROLLABILITY AND OBSERVABILITY

One vital question regarding control is whether all state variables can be arbitrarily affected by the input signal. This question is answered by the controllability theorem introduced by Kalman [65].

The system is state controllable if, by the control u, its state vector can be moved from the arbitrary initial state $\boldsymbol{x}(t_o)$ to any prescribed state $\boldsymbol{x}(t_v)$ during a finite time $(t_v - t_o)$. If the above definition is fulfilled only for the output, then so-called output controllability exists. For *linear time-invariant* (*LTI*) systems the initial time $(t_o = 0)$ and the initial state $\boldsymbol{x}(0)$ can be chosen. In this case the concept of the controllability is regarded to the system. If it is valid for an initial state, then it remains valid for any initial state, since, e.g., starting from $\boldsymbol{x}(0)$ the state vector $\boldsymbol{x}(t_o)$ can be reached by a corresponding control signal.

The controllability can be best illustrated in the canonical coordinates. If in the canonical form (A.2.28) β_i is zero for a state variable, then this state cannot be controlled. This means that in this case there is no control which is parallel, but only perpendicular, with the eigenvector belonging to the eigenvalue λ_i; thus the effect of the control remains in the plane perpendicular to the eigenvector. (It follows from the basic condition of the canonical form that the system can only be state controllable if the poles of the canonical coordinates are different.)

If the system is not state controllable, it can be output controllable while at least one state variable is controllable and the corresponding γ_i is not zero (see (A.2.28)).

In the noncanonical coordinate system, the above statements cannot be applied directly because of the interconnected relations between the state variables, and therefore they must be substituted by more general criteria.

For simplicity, choose the initial state $\boldsymbol{x}(0) = \mathbf{0}$, so the solution of the state Equation (A.2.8) has the form

$$\boldsymbol{x}(t)=\left[\int_0^t e^{\boldsymbol{A}(t-\tau)}u(\tau)\mathrm{d}\tau\right]\boldsymbol{b}=\left[\int_0^t e^{\boldsymbol{A}\tau}u(t-\tau)\mathrm{d}\tau\right]\boldsymbol{b} \tag{A.2.46}$$

where using the final sum form of the basic matrix (see (A.2.16)), we obtain

$$e^{\boldsymbol{A}\tau}=\alpha_o(\tau)\boldsymbol{I}+\alpha_1(\tau)\boldsymbol{A}+\ldots+\alpha_{n-1}(\tau)\boldsymbol{A}^{n-1} \tag{A.2.47}$$

Thus the solution of the state equation is obtained in closed form

$$\boldsymbol{x}(t)=\boldsymbol{b}\int_0^t \alpha_o(\tau)u(\tau)\mathrm{d}\tau+\boldsymbol{A}\boldsymbol{b}\int_0^t \alpha_1(\tau)u(\tau)\mathrm{d}\tau+\ldots$$
$$+\boldsymbol{A}^{n-1}\boldsymbol{b}\int_0^t \alpha_{n-1}(\tau)u(\tau)\mathrm{d}\tau \tag{A.2.48}$$

Thus the right side of the equation is the linear combination of the columns of the *controllability* matrix

$$\boldsymbol{M}_c=\left[\boldsymbol{b}\quad \boldsymbol{A}\boldsymbol{b}\quad \ldots\quad \boldsymbol{A}^{n-1}\boldsymbol{b}\right] \tag{A.2.49}$$

Therefore the condition for reaching all points of the state space means that $\boldsymbol{M}_c$ must have n linearly independent columns, i.e., $\boldsymbol{M}_c$ must be invertible and regular. Since $\boldsymbol{M}_c$ depends on $\boldsymbol{A}$ and $\boldsymbol{b}$, it is also fair to say that it is $\boldsymbol{A};\boldsymbol{b}$ pair controllable.

Referring the above statements implicitly to the output, we see that the condition of the output controllability is that at least one element of

$$\boldsymbol{m}_c^{\mathrm{T}}=\left[\boldsymbol{c}^{\mathrm{T}}\boldsymbol{b}\quad \boldsymbol{c}^{\mathrm{T}}\boldsymbol{A}\boldsymbol{b}\quad \ldots\quad \boldsymbol{c}^{\mathrm{T}}\boldsymbol{A}^{n-1}\boldsymbol{b}\right] \tag{A.2.50}$$

is not zero.

The controllability matrix of the controllable canonical form (A.2.36) has special structure

$$\boldsymbol{M}_c^c=\begin{bmatrix} 1 & a_1 & a_2 & \ldots & a_{n-1} \\ 0 & 1 & a_1 & \ldots & a_{n-2} \\ \vdots & \vdots & \vdots & \ddots & \vdots \\ 0 & 0 & 0 & \ldots & a_1 \\ 0 & 0 & 0 & \ldots & 1 \end{bmatrix}^{-1} \tag{A.2.51}$$

which can be clearly seen if we generate the product $\boldsymbol{M}_c^c(\boldsymbol{M}_c^c)^{-1}$

$$
\begin{bmatrix} \boldsymbol{b}_c & \boldsymbol{A}_c\boldsymbol{b}_c & \dots & (\boldsymbol{A}_c)^{n-1}\boldsymbol{b}_c \end{bmatrix}
\begin{bmatrix} 1 & a_1 & a_2 & \dots & a_{n-1} \\ 0 & 1 & a_1 & \dots & a_{n-2} \\ \vdots & \vdots & \vdots & \ddots & \vdots \\ 0 & 0 & 0 & \dots & a_1 \\ 0 & 0 & 0 & \dots & 1 \end{bmatrix}
= \begin{bmatrix} \boldsymbol{w}_o & \boldsymbol{w}_1 & \boldsymbol{w}_2 & \dots & \boldsymbol{w}_{n-1} \end{bmatrix} \tag{A.2.52}
$$

Given the special structure of the system matrices $\boldsymbol{A}_c$ and $\boldsymbol{b}_c$ in Equation (A.2.37) we can write

$$
\begin{aligned}
\boldsymbol{w}_o &= \boldsymbol{b}_c \\
\boldsymbol{w}_1 &= a_1\boldsymbol{b}_c + \boldsymbol{A}_c\boldsymbol{b}_c \\
&\vdots \\
\boldsymbol{w}_{n-1} &= a_{n-1}\boldsymbol{b}_c + a_{n-2}\boldsymbol{A}_c\boldsymbol{b}_c + \dots + (\boldsymbol{A}_c)^{n-1}\boldsymbol{b}_c
\end{aligned} \tag{A.2.53}
$$

where the following recursive relationship exists

$$
\boldsymbol{w}_k = a_k\boldsymbol{b}_c + \boldsymbol{A}_c\boldsymbol{w}_{k-1} \tag{A.2.54}
$$

Using the above relationship, we obtain

$$
\begin{bmatrix} \boldsymbol{w}_o & \boldsymbol{w}_1 & \boldsymbol{w}_2 & \dots & \boldsymbol{w}_{n-1} \end{bmatrix} =
\begin{bmatrix} 1 & 0 & 0 & \dots & 0 \\ 0 & 1 & 0 & \dots & 0 \\ 0 & 0 & 1 & \dots & 0 \\ \vdots & \vdots & \vdots & \ddots & \vdots \\ 0 & 0 & 0 & \dots & 1 \end{bmatrix} = \boldsymbol{I} \tag{A.2.55}
$$

which proves the validity of Equation (A.2.51) [9].

The special $\boldsymbol{M}_c^c$ obtained by the *controllable canonical form* and always generated from the transfer function is always regular, since it is the inverse of a diagonal matrix (the determinant of a triangular matrix is the product of the elements in the main diagonal, which is now one), and thus it is *always controllable*. Only observability (see later) can be investigated by the pair $\boldsymbol{A}_o; \boldsymbol{c}_o^T$.

It is an interesting question how the linear transformation $\boldsymbol{z} = \boldsymbol{T}\boldsymbol{x}$ influences the controllability matrix. Using Equation (A.2.19) we can write

$$\begin{aligned}\tilde{\boldsymbol{b}} &= \boldsymbol{T}\boldsymbol{b} \\ \tilde{\boldsymbol{A}}\tilde{\boldsymbol{b}} &= \boldsymbol{T}\boldsymbol{A}\boldsymbol{T}^{-1}\boldsymbol{T}\boldsymbol{b} = \boldsymbol{T}\boldsymbol{A}\boldsymbol{b} \\ &\vdots \\ \tilde{\boldsymbol{A}}^{n-1}\tilde{\boldsymbol{b}} &= \boldsymbol{T}\boldsymbol{A}^{n-1}\boldsymbol{b}\end{aligned} \tag{A.2.56}$$

and

$$\tilde{\boldsymbol{M}}_{\mathrm{c}} = \begin{bmatrix}\tilde{\boldsymbol{b}} & \tilde{\boldsymbol{A}}\tilde{\boldsymbol{b}} & \ldots & \tilde{\boldsymbol{A}}^{n-1}\tilde{\boldsymbol{b}}\end{bmatrix} = \boldsymbol{T}\begin{bmatrix}\boldsymbol{b} & \boldsymbol{A}\boldsymbol{b} & \ldots & \boldsymbol{A}^{n-1}\boldsymbol{b}\end{bmatrix} = \boldsymbol{T}\boldsymbol{M}_{\mathrm{c}} \tag{A.2.57}$$

Given the above form of the controllability matrix and the transformation matrix $\boldsymbol{T}_{\mathrm{c}} = \boldsymbol{M}_{\mathrm{c}}^{\mathrm{c}}(\boldsymbol{M}_{\mathrm{c}})^{-1}$, all controllable systems can be given in a controllable canonical form.

The controllability matrix, however, is not always derived from the transfer function. Obviously, in this case, the controllability matrix $\boldsymbol{M}_{\mathrm{c}}$ needs to be directly investigated.

It is an important question in control engineering whether all state variables can be observed by perceiving the output. This question is answered by the observability theorem introduced by Kalman [65].

Observability, congenial to the controllability, tells us whether the initial state can be reconstructed after measurements of the input and output signals of a system of unknown state collected during a certain time. The system is observable if $\boldsymbol{x}(t_{\mathrm{o}})$ can be determined from the signals $y(t)$ and $u(t)$ observed in the time interval $t_{\mathrm{o}} < t < t_{\mathrm{V}}$.

It is sufficient to investigate the motion of the system generated by the initial value $u(t) \equiv 0$. Observability can be stated more simply again in the canonical coordinates. To this end, two criteria need to be fulfilled: the signal y must depend on all canonical state variables, and the poles of the system must be different. If any γ_i in Equation (A.2.28) is zero, then the output signal has no information concerning the given canonical state variable, so it cannot be reconstructed from the measurements. This means that none of the observations has a parallel, but only a perpendicular, component regarding the eigenvector belonging to the eigenvalue λ_i; thus the effect of the observation always remains in the plane perpendicular to the eigenvector.

In the noncanonical coordinate system the above conditions cannot be applied directly because of the interconnected relations between the state variables, and therefore they must be substituted by more general criteria.

When controllability was investigated as above then controllability of the state variables was considered and the output was disregarded. Now, with regard to observability, the input of the systems is neglected. Consider now the system below

$$\frac{d\boldsymbol{x}}{dt} = \boldsymbol{A}\boldsymbol{x} \qquad y = \boldsymbol{c}^{T}\boldsymbol{x} \tag{A.2.58}$$

By derivating the output we obtain the equation

$$\begin{bmatrix} y \\ \frac{dy}{dt} \\ \vdots \\ \frac{d^{n-1}y}{dt^{n-1}} \end{bmatrix} = \begin{bmatrix} \boldsymbol{c}^{T} \\ \boldsymbol{c}^{T}\boldsymbol{A} \\ \vdots \\ \boldsymbol{c}^{T}\boldsymbol{A}^{n-1} \end{bmatrix} \boldsymbol{x} \tag{A.2.59}$$

from which the state vector can be unambiguously determined from the observation of the output and its derivates, if the *observability* matrix

$$\boldsymbol{M}_{o} = \begin{bmatrix} \boldsymbol{c}^{T} \\ \boldsymbol{c}^{T}\boldsymbol{A} \\ \vdots \\ \boldsymbol{c}^{T}\boldsymbol{A}^{n-1} \end{bmatrix} \tag{A.2.60}$$

has n linearly independent rows. Thus $\boldsymbol{M}_{o}$ must be invertible and regular. Since $\boldsymbol{M}_{o}$ depends on $\boldsymbol{A}$ and $\boldsymbol{c}^{T}$, it is reasonable to call $\boldsymbol{A};\boldsymbol{c}^{T}$ pair observability. (Because of the Cayley–Hamilton theorem [38], there is no need to calculate the derivates of higher order than $(n-1)$ in Equation (A.2.59) (see A.2.5.3).)

The observability matrix of the observable canonical form has special structure

$$\boldsymbol{M}_{o}^{o} = \begin{bmatrix} 1 & 0 & \dots & 0 & 0 \\ a_1 & 1 & \dots & 0 & 0 \\ a_2 & a_1 & \ddots & 0 & 0 \\ \vdots & \vdots & \dots & 1 & 0 \\ a_{n-1} & a_{n-2} & \dots & a_1 & 1 \end{bmatrix}^{-1} \tag{A.2.61}$$

which can be easily proved by taking the product $(\boldsymbol{M}_{\mathrm{o}}^{\mathrm{o}})^{-1}\boldsymbol{M}_{\mathrm{o}}$

$$\begin{bmatrix} 1 & 0 & \dots & 0 & 0 \\ a_1 & 1 & \dots & 0 & 0 \\ a_2 & a_1 & \ddots & 0 & 0 \\ \vdots & \vdots & \dots & 1 & 0 \\ a_{n-1} & a_{n-2} & \dots & a_1 & 1 \end{bmatrix} \begin{bmatrix} \boldsymbol{c}_{\mathrm{o}}^{\mathrm{T}} \\ \boldsymbol{c}_{\mathrm{o}}^{\mathrm{T}}\boldsymbol{A}_{\mathrm{o}} \\ \vdots \\ \boldsymbol{c}_{\mathrm{o}}^{\mathrm{T}}(\boldsymbol{A}_{\mathrm{o}})^{n-1} \end{bmatrix} = \begin{bmatrix} \boldsymbol{w}_{\mathrm{o}}^{\mathrm{T}} \\ \boldsymbol{w}_{1}^{\mathrm{T}} \\ \vdots \\ \boldsymbol{w}_{n-1}^{\mathrm{T}} \end{bmatrix} \tag{A.2.62}$$

Given the special structure of the system matrices $\boldsymbol{A}_{\mathrm{o}}$ and $\boldsymbol{c}_{\mathrm{o}}^{\mathrm{T}}$ in Equation (A.2.44), we obtain

$$\begin{aligned} \boldsymbol{w}_{\mathrm{o}}^{\mathrm{T}} &= \boldsymbol{c}_{\mathrm{o}}^{\mathrm{T}} \\ \boldsymbol{w}_{1}^{\mathrm{T}} &= a_1\boldsymbol{c}_{\mathrm{o}}^{\mathrm{T}} + \boldsymbol{c}_{\mathrm{o}}^{\mathrm{T}}\boldsymbol{A}_{\mathrm{o}} \\ &\vdots \\ \boldsymbol{w}_{n-1}^{\mathrm{T}} &= a_{n-1}\boldsymbol{c}_{\mathrm{o}}^{\mathrm{T}} + a_{n-2}\boldsymbol{c}_{\mathrm{o}}^{\mathrm{T}}\boldsymbol{A}_{\mathrm{o}} + \ldots + \boldsymbol{c}_{\mathrm{o}}^{\mathrm{T}}(\boldsymbol{A}_{\mathrm{o}})^{n-1} \end{aligned} \tag{A.2.63}$$

where the following recursive relationship exists

$$\boldsymbol{w}_{k}^{\mathrm{T}} = a_k\boldsymbol{c}_{\mathrm{o}}^{\mathrm{T}} + \boldsymbol{w}_{k-1}^{\mathrm{T}}\boldsymbol{A}_{\mathrm{o}} \tag{A.2.64}$$

Using the above relationship we obtain

$$\begin{bmatrix} \boldsymbol{w}_{\mathrm{o}}^{\mathrm{T}} \\ \boldsymbol{w}_{1}^{\mathrm{T}} \\ \vdots \\ \boldsymbol{w}_{n-1}^{\mathrm{T}} \end{bmatrix} = \begin{bmatrix} 1 & 0 & 0 & \dots & 0 \\ 0 & 1 & 0 & \dots & 0 \\ 0 & 0 & 1 & \dots & 0 \\ \vdots & \vdots & \vdots & \ddots & \vdots \\ 0 & 0 & 0 & \dots & 1 \end{bmatrix} = \boldsymbol{I} \tag{A.2.65}$$

which proves the validity of Equation (A.2.61) [9].

The special $\boldsymbol{M}_{\mathrm{o}}^{\mathrm{o}}$ obtained by the *observable canonical form* and always generated from the transfer function is always regular, since it is the inverse of a regular matrix (the determinant of the triangular matrix is the product of the elements in the main diagonal, which is now one), and thus it is *always observable*. For this form only controllability can be investigated by the pair $\boldsymbol{A}_{\mathrm{c}}; \boldsymbol{c}_{\mathrm{c}}^{\mathrm{T}}$.

It is an interesting question how the linear transformation $\boldsymbol{z} = \boldsymbol{T}\boldsymbol{x}$ influences the observability matrix. Using Equation (A.2.19) we obtain

$$\begin{aligned}
\tilde{c}^{\mathrm{T}} &= c^{\mathrm{T}} T^{-1} \\
\tilde{c}^{\mathrm{T}} \tilde{A} &= c^{\mathrm{T}} T^{-1} T A T^{-1} = c^{\mathrm{T}} A T^{-1} \\
&\vdots \\
\tilde{c}^{\mathrm{T}} \tilde{A}^{n-1} &= c^{\mathrm{T}} A^{n-1} T^{-1}
\end{aligned} \tag{A.2.66}$$

and rearranging them in matrix form we obtain

$$\tilde{M}_{\mathrm{o}} = \begin{bmatrix} \tilde{c}^{\mathrm{T}} \\ \tilde{c}^{\mathrm{T}} \tilde{A} \\ \vdots \\ \tilde{c}^{\mathrm{T}} \tilde{A}^{n-1} \end{bmatrix} = \begin{bmatrix} c^{\mathrm{T}} \\ c^{\mathrm{T}} A \\ \vdots \\ c^{\mathrm{T}} A^{n-1} \end{bmatrix} T^{-1} = M_{\mathrm{o}} T^{-1} \tag{A.2.67}$$

Given the above transformation form of the observability matrix, all observable systems can be given in observable canonical form if the transformation matrix $T_{\mathrm{o}}^{-1} = (M_{\mathrm{o}})^{-1} M_{\mathrm{o}}^{\circ}$ (i.e., $T_{\mathrm{o}} = (M_{\mathrm{o}}^{\circ})^{-1} M_{\mathrm{o}}$) is applied.

The observability matrix, however, is not always derived from the transfer function. Obviously, in this case, the observability matrix M_{o} needs to be directly investigated.

Kalman Decomposition

The concepts of controllability and observability make it possible to understand the structure of a linear system. Remember that the space of the controllable states is the subspace spanned by the columns of the controllability matrix. If the dimension of this space is n, then the whole space is controllable. Introduce x_{c} and $x_{\bar{\mathrm{c}}}$ for the controllable and noncontrollable states, respectively. In this case the state equation has the form

$$\frac{\mathrm{d}}{\mathrm{d}t} \begin{bmatrix} x_{\mathrm{c}} \\ x_{\bar{\mathrm{c}}} \end{bmatrix} = \begin{bmatrix} A_{11} & A_{12} \\ 0 & A_{22} \end{bmatrix} \begin{bmatrix} x_{\mathrm{c}} \\ x_{\bar{\mathrm{c}}} \end{bmatrix} + \begin{bmatrix} b_1 \\ 0 \end{bmatrix} u \tag{A.2.68}$$

where it is clear from the structure that the states $x_{\bar{\mathrm{c}}}$ cannot be influenced by u. Similarly, let us introduce x_{o} and $x_{\bar{\mathrm{o}}}$ for the observable and nonobservable states, respectively. The state equation now is

$$\begin{aligned}
\frac{\mathrm{d}}{\mathrm{d}t} \begin{bmatrix} x_{\mathrm{o}} \\ x_{\bar{\mathrm{o}}} \end{bmatrix} &= \begin{bmatrix} A_{11} & 0 \\ A_{21} & A_{22} \end{bmatrix} \begin{bmatrix} x_{\mathrm{o}} \\ x_{\bar{\mathrm{o}}} \end{bmatrix} \\
y &= \begin{bmatrix} c_1^{\mathrm{T}} & 0^{\mathrm{T}} \end{bmatrix} \begin{bmatrix} x_{\mathrm{o}} \\ x_{\bar{\mathrm{o}}} \end{bmatrix}
\end{aligned} \tag{A.2.69}$$

where it is seen that no output component exists for the states $\boldsymbol{x}_{\bar{o}}$.

The linear systems can be decomposed to four subsystems:

- S_{co} controllable and observable $\boldsymbol{x}_{co}$
- $S_{c\bar{o}}$ controllable and nonobservable $\boldsymbol{x}_{c\bar{o}}$
- $S_{\bar{c}o}$ noncontrollable and observable $\boldsymbol{x}_{\bar{c}o}$
- $S_{\overline{co}}$ noncontrollable and nonobservable $\boldsymbol{x}_{\overline{co}}$

where the corresponding state variables are also presented (the input arrows mean the effect of the input and respecting state subspace). The complete Kalman decomposition of a linear system is [65]

$$\frac{\mathrm{d}}{\mathrm{d}t}\begin{bmatrix} \boldsymbol{x}_{co} \\ \boldsymbol{x}_{c\bar{o}} \\ \boldsymbol{x}_{\bar{c}o} \\ \boldsymbol{x}_{\overline{co}} \end{bmatrix} = \begin{bmatrix} \boldsymbol{A}_{11} & \boldsymbol{0} & \boldsymbol{A}_{13} & \boldsymbol{0} \\ \boldsymbol{A}_{21} & \boldsymbol{A}_{22} & \boldsymbol{A}_{23} & \boldsymbol{A}_{24} \\ \boldsymbol{0} & \boldsymbol{0} & \boldsymbol{A}_{33} & \boldsymbol{0} \\ \boldsymbol{0} & \boldsymbol{0} & \boldsymbol{A}_{43} & \boldsymbol{A}_{44} \end{bmatrix} \begin{bmatrix} \boldsymbol{x}_{co} \\ \boldsymbol{x}_{c\bar{o}} \\ \boldsymbol{x}_{\bar{c}o} \\ \boldsymbol{x}_{\overline{co}} \end{bmatrix} + \begin{bmatrix} \boldsymbol{b}_1 \\ \boldsymbol{b}_2 \\ \boldsymbol{0} \\ \boldsymbol{0} \end{bmatrix} u = \boldsymbol{A}\boldsymbol{x} + \boldsymbol{b}u$$

$$y = \begin{bmatrix} \boldsymbol{c}_1^{\mathrm{T}} & \boldsymbol{0}^{\mathrm{T}} & \boldsymbol{c}_2^{\mathrm{T}} & \boldsymbol{0}^{\mathrm{T}} \end{bmatrix} \boldsymbol{x} \tag{A.2.70}$$

The block scheme of the subsystems is shown in Figure A.2.1. Following the arrows on the figure we can see that the input influences the subsystems S_{co} and $S_{c\bar{o}}$, and the output depends only on subsystems S_{co} and $S_{\bar{c}o}$. The subsystem $S_{\overline{co}}$, however, is not connected either to the input or to the output.

The transfer function of the complete system can be obtained by simple calculation as

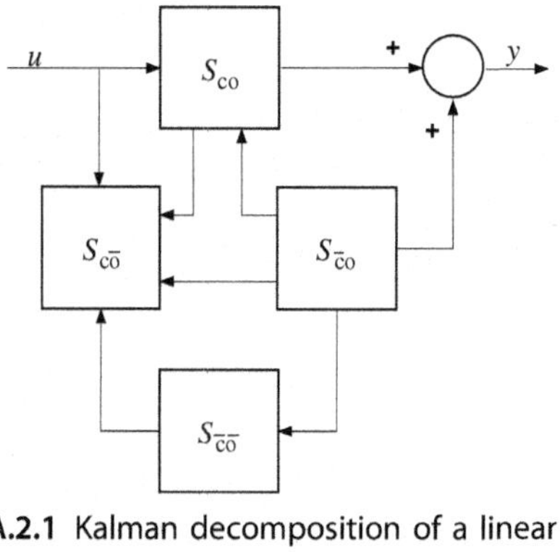

Figure A.2.1 Kalman decomposition of a linear system.

$$P(s) = \boldsymbol{c}_1^{\mathrm{T}}(s\boldsymbol{I} - \boldsymbol{A}_{11})^{-1}\boldsymbol{b}_1 \tag{A.2.71}$$

i.e., it is fully determined by the subsystem S_{co}. The reverse statement is also true, i.e., that only the controllable and observable subsystem S_{co} of the whole system can be given from the transfer function.

Effect of Common Poles and Zeros

The long-standing problem in control engineering of pole and zero cancellation can be explained by Kalman decomposition. To illustrate the problem, consider the following example [9]. Let the transfer function of the process be

$$P(s) = \frac{Y(s)}{U(s)} = \frac{s-1}{s-1} = 1 \tag{A.2.72}$$

i.e., the numerator and the denominator have a common root, and thus the pole and zero are equal. In this case, the common root means, at the same time, an unstable pole. It is clear that the following differential equation formally corresponds to the transfer function (A.2.72)

$$\frac{\mathrm{d}y}{\mathrm{d}t} - y = \frac{\mathrm{d}u}{\mathrm{d}t} - u \tag{A.2.73}$$

The solution obtained by integrating the differential equation is

$$y(t) = u(t) + ce^t \tag{A.2.74}$$

where c is a constant. Thus, never forget at the cancellation that the complete solution of the state equation is performed according to Equation (A.2.8), i.e., it also contains the initial condition, whose dynamics (the system is left to itself, free response) depend on the poles of the whole system, even on the possibly cancelled one, too. If this pole is unstable, then its unfavorable nonvanishing effect will appear in the solution.

The trivial system $y(t) = u(t)$ obtained after the pole cancellation from Equation (A.2.72) is obviously not equal to the system of Equation (A.2.74), which can be rearranged into the form with equivalent rewriting

$$P(s) = b_o + \frac{b_1}{s-1} = d + \frac{b_1}{s-1} = 1 + \frac{0}{s-1} \tag{A.2.75}$$

Given this a controllable canonical form can be simply written as

$$\frac{\mathrm{d}x_1}{\mathrm{d}t} = x_1 + u \;;\;\; y = u \tag{A.2.76}$$

which is not observable, and an observable canonical form can also be formulated as

$$\frac{dx_2}{dt} = x_2 \ ; \ \ y = x_2 + u \tag{A.2.77}$$

which is not controllable.

Using the system Equations (A.2.76) and (A.2.77), Figure A.2.2 shows the Kalman form of the whole system, which consists of the subsystems $S_{c\bar{o}}$, $S_{\bar{c}o}$, and S_{co}. S_{co} is a static system with transfer function $P(s) = 1$. $S_{c\bar{o}}$ is a nonobservable, but controllable subsystem; $S_{\bar{c}o}$, however, is a noncontrollable but observable subsystem.

Note that if the transfer function of the process is given, then first the investigation of the possible common divisor of the numerator and denominator needs to be performed. The common divisor can only be a common root. It is reasonable to reduce by the common roots while they exist. Such polynomials are called relative prime. A transfer function $P(s) = \mathcal{B}(s)/\mathcal{A}(s)$ is considered to be nonreducible, if the polynomials $\mathcal{A}(s)$ and $\mathcal{B}(s)$ are relative prime; the algebraic condition of this is that a special Diophantine (or Bezout) equation

$$\mathcal{A}(s)\mathcal{X}(s) + \mathcal{B}(s)\mathcal{Y}(s) = 1 \tag{A.2.78}$$

must have a solution, i.e., the relevant Sylvester matrix must be regular [38] (see further details in Chapter 9).

If the transfer function of the process is nonreducible, then the relevant state equation corresponds to the controllable and observable subsystem S_{co} of the Kalman form, and the other subsystems do not exist.

Equations (A.2.76) and (A.2.77) can be generalized for the case when the controllable and observable system $S_{co}\{\boldsymbol{A};\boldsymbol{b};\boldsymbol{c}^T;d\}$ cannot be reduced, and the numerator and the denominator of the transfer function $P(s)$ are extended by a common factor $(s-p)$ corresponding to a real pole. Thus for

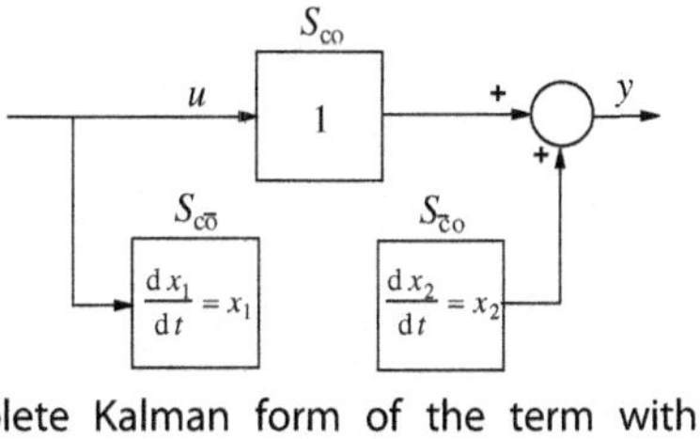

Figure A.2.2 The complete Kalman form of the term with the transfer function (A.2.72).

this general case the state equation of the redundant noncontrollable and nonobservable system is

$$\dot{\boldsymbol{x}}_r = \begin{bmatrix} \dot{\boldsymbol{x}} \\ \dot{x}_1 \\ \dot{x}_2 \end{bmatrix} = \begin{bmatrix} \boldsymbol{A} & 0 & 0 \\ 0^T & p & 0 \\ 0^T & 0 & p \end{bmatrix} \begin{bmatrix} \boldsymbol{x} \\ x_1 \\ x_2 \end{bmatrix} + \begin{bmatrix} \boldsymbol{b} \\ 1 \\ 0 \end{bmatrix} u = \boldsymbol{A}_r \boldsymbol{x}_r + \boldsymbol{b}_r u \tag{A.2.79}$$

$$y = \begin{bmatrix} \boldsymbol{c}^T & 0 & 1 \end{bmatrix} \boldsymbol{x}_r + d_c u = \boldsymbol{c}_r^T \boldsymbol{x}_r + d_c u$$

A.2.5 DEDUCTIONS/DERIVATIONS

A.2.5.1

The solution of the state equation is given by Equation (A.2.8). To prove this, let us derivate the equation

$$\frac{d\boldsymbol{x}(t)}{dt} = \frac{d}{dt}\left[e^{\boldsymbol{A}t}\boldsymbol{x}(0)\right] + \frac{d}{dt}\left[\int_0^t e^{\boldsymbol{A}(t-\tau)}\boldsymbol{b}u(\tau)d\tau\right] \tag{A.2.80}$$

where

$$\frac{d}{dt}\left[e^{\boldsymbol{A}t}\boldsymbol{x}(0)\right] = \boldsymbol{A}e^{\boldsymbol{A}t}\boldsymbol{x}(0) \tag{A.2.81}$$

and

$$\begin{aligned}\frac{d}{dt}\left[\int_0^t e^{\boldsymbol{A}(t-\tau)}\boldsymbol{b}u(\tau)d\tau\right] &= \int_0^t \frac{d}{dt}\left[e^{\boldsymbol{A}(t-\tau)}\boldsymbol{b}u(\tau)\right]d\tau + \frac{dt}{dt}\left[e^{\boldsymbol{A}(t-\tau)}\boldsymbol{b}u(\tau)\right]_{\tau=t} \\ &\quad - \frac{d0}{dt}\left[e^{\boldsymbol{A}(t-\tau)}\boldsymbol{b}u(\tau)\right]_{\tau=0} \\ &= \int_0^t \boldsymbol{A}e^{\boldsymbol{A}(t-\tau)}\boldsymbol{b}u(\tau)d\tau + \boldsymbol{b}u(\tau)\end{aligned} \tag{A.2.82}$$

where $dt/dt = 1$, $d0/dt = 0$, and $e^{\boldsymbol{A}(t-\tau)}\big|_{\tau=t} = 1$ are taken into account. Thus the derivate of Equation (A.2.8) is

$$\frac{\mathrm{d}\boldsymbol{x}(t)}{\mathrm{d}t} = \boldsymbol{A}e^{\boldsymbol{A}t}\boldsymbol{x}(0) + \int_0^t \boldsymbol{A}e^{\boldsymbol{A}(t-\tau)}\boldsymbol{b}u(\tau)\mathrm{d}\tau + \boldsymbol{b}u(\tau) = \boldsymbol{A}\boldsymbol{x}(t) + \boldsymbol{b}u(t) \tag{A.2.83}$$

A.2.5.2

The pulse response of the system can be calculated from Equation (A.2.8) under the zero initial condition $\boldsymbol{x}(0) = \boldsymbol{0}$, excitation $u(t) = \delta(t)$, and $d = 0$, as

$$\begin{aligned} \boldsymbol{x}(t) &= \int_0^t e^{\boldsymbol{A}(t-\tau)}\boldsymbol{b}\delta(\tau)\mathrm{d}\tau = e^{\boldsymbol{A}t}\left[\int_0^t e^{-\boldsymbol{A}\tau}\delta(\tau)\mathrm{d}\tau\right]\boldsymbol{b} = e^{\boldsymbol{A}t}\left[e^{-\boldsymbol{A}\tau}\delta(\tau)\right]_0^t\boldsymbol{b} \\ &= e^{\boldsymbol{A}t}\left[-e^{-\boldsymbol{A}t}\delta(t) + e^{-\boldsymbol{A}0}\delta(0)\right]\boldsymbol{b} = e^{\boldsymbol{A}t}\boldsymbol{b} \\ w(t) &= y(t) = \boldsymbol{c}^{\mathrm{T}}\boldsymbol{x}(t) = \boldsymbol{c}^{\mathrm{T}}e^{\boldsymbol{A}t}\boldsymbol{b} \end{aligned} \tag{A.2.84}$$

which is equal to Equation (A.2.15) obtained in the operator domain.

A.2.5.3

The Cayley–Hamilton theorem is considered vital in the matrix theory: any matrix $\boldsymbol{A}$ fulfills its own characteristic equation, i.e., the equation $\mathcal{A}(\boldsymbol{A}) = 0 = \det(s\boldsymbol{I} - \boldsymbol{A}) = 0$, which is formally the same in matrix form

$$\mathcal{A}(\boldsymbol{A}) = 0 \tag{A.2.85}$$

(see (A.2.17)). Equation (A.2.85) is also fulfilled by the matrix polynomial $\mathcal{P}(\boldsymbol{A})$ of matrix $\boldsymbol{A}$, and also by a matrix function $\boldsymbol{F}(\boldsymbol{A})$, and the corresponding function $f(s)$ is regular in a certain region around the origo of the complex plane of s. Let $\boldsymbol{F}(\boldsymbol{A}) = e^{\boldsymbol{A}\tau}$ be the fundamental matrix, and then because of the above statements

$$e^{\boldsymbol{A}\tau} = \alpha_0(\tau)\boldsymbol{I} + \alpha_1(\tau)\boldsymbol{A} + \ldots + \alpha_{n-1}(\tau)\boldsymbol{A}^{n-1} \tag{A.2.86}$$

A.2.5.4

To simplify the complex form $\boldsymbol{c}^{\mathrm{T}}\left(s\boldsymbol{I} - \boldsymbol{A} + \boldsymbol{b}\boldsymbol{k}^{\mathrm{T}}\right)^{-1}\boldsymbol{b}$, let us use the notation introduced in (A.2.3)

$$\boldsymbol{\Phi}(s) = (s\boldsymbol{I} - \boldsymbol{A})^{-1} = \frac{\mathbf{adj}(s\boldsymbol{I} - \boldsymbol{A})}{\det(s\boldsymbol{I} - \boldsymbol{A})} = \frac{\mathbf{adj}(s\boldsymbol{I} - \boldsymbol{A})}{\mathcal{A}(s)} = \frac{\boldsymbol{\Psi}(s)}{\mathcal{A}(s)} \tag{A.2.87}$$

and apply the matrix inversion lemma

$$\begin{aligned}\left(s\boldsymbol{I} - \boldsymbol{A} + \boldsymbol{b}\boldsymbol{k}^{\mathrm{T}}\right)^{-1} &= \left[\boldsymbol{\Phi}^{-1}(s) + \boldsymbol{b}\boldsymbol{k}^{\mathrm{T}}\right]^{-1} \\ &= \boldsymbol{\Phi}(s) - \boldsymbol{\Phi}(s)\boldsymbol{b}\left[1 + \boldsymbol{k}^{\mathrm{T}}\boldsymbol{\Phi}(s)\boldsymbol{b}\right]^{-1}\boldsymbol{k}^{\mathrm{T}}\boldsymbol{\Phi}(s)\end{aligned} \tag{A.2.88}$$

and

$$\boldsymbol{c}^{\mathrm{T}}\left(s\boldsymbol{I} - \boldsymbol{A} + \boldsymbol{b}\boldsymbol{k}^{\mathrm{T}}\right)^{-1}\boldsymbol{b} = \boldsymbol{c}^{\mathrm{T}}\boldsymbol{\Phi}(s)\boldsymbol{b} - \frac{\boldsymbol{c}^{\mathrm{T}}\boldsymbol{\Phi}(s)\boldsymbol{b}\boldsymbol{k}^{\mathrm{T}}\boldsymbol{\Phi}(s)\boldsymbol{b}}{1 + \boldsymbol{k}^{\mathrm{T}}\boldsymbol{\Phi}(s)\boldsymbol{b}} = \frac{\boldsymbol{c}^{\mathrm{T}}\boldsymbol{\Phi}(s)\boldsymbol{b}}{1 + \boldsymbol{k}^{\mathrm{T}}\boldsymbol{\Phi}(s)\boldsymbol{b}} \tag{A.2.89}$$

Thus (3.1.5) can be further modified as

$$T_{\mathrm{ry}}(s) = \frac{\boldsymbol{c}^{\mathrm{T}}\boldsymbol{\Phi}(s)\boldsymbol{b}k_{\mathrm{r}}}{1 + \boldsymbol{k}^{\mathrm{T}}\boldsymbol{\Phi}(s)\boldsymbol{b}} \tag{A.2.90}$$

Observe that here

$$\boldsymbol{c}^{\mathrm{T}}\boldsymbol{\Phi}(s)\boldsymbol{b} = \boldsymbol{c}^{\mathrm{T}}(s\boldsymbol{I} - \boldsymbol{A})^{-1}\boldsymbol{b} = H(s) = \frac{\mathcal{B}(s)}{\mathcal{A}(s)} \tag{A.2.91}$$

by means of which

$$\begin{aligned}T_{\mathrm{ry}}(s) &= \frac{\boldsymbol{c}^{\mathrm{T}}\boldsymbol{\Phi}(s)\boldsymbol{b}k_{\mathrm{r}}}{1 + \boldsymbol{k}^{\mathrm{T}}\boldsymbol{\Phi}(s)\boldsymbol{b}} = \frac{k_{\mathrm{r}}}{1 + \boldsymbol{k}^{\mathrm{T}}\frac{\boldsymbol{\Psi}(s)}{\mathcal{A}(s)}\boldsymbol{b}} \\ P(s) &= \frac{k_{\mathrm{r}}}{1 + \boldsymbol{k}^{\mathrm{T}}\frac{\boldsymbol{\Psi}(s)}{\mathcal{A}(s)}\boldsymbol{b}}\frac{\mathcal{B}(s)}{\mathcal{A}(s)} \\ &= \frac{k_{\mathrm{r}}\mathcal{B}(s)}{\mathcal{A}(s) + \boldsymbol{k}^{\mathrm{T}}\boldsymbol{\Psi}(s)\boldsymbol{b}}\end{aligned} \tag{A.2.92}$$

A.2.5.5

The calibration factor k_{r} can be used to ensure the unity static gain of the transfer function $T_{\mathrm{ry}}(s)$ of the closed system. From the condition

$$T_{\mathrm{ry}}(s)\big|_{s=0} = \boldsymbol{c}^{\mathrm{T}}\left(-\boldsymbol{A} + \boldsymbol{b}\boldsymbol{k}^{\mathrm{T}}\right)^{-1}\boldsymbol{b}k_{\mathrm{r}} = 1 \tag{A.2.93}$$

we obtain

$$k_r = -1 \Big/ c^T \left(A - bk^T\right)^{-1} b \tag{A.2.94}$$

Applying the matrix inversion lemma in the denominator, we obtain

$$\left(A - bk^T\right)^{-1} = A^{-1} + A^{-1}b\left[1 - k^T A^{-1} b\right]^{-1} k^T A^{-1} \tag{A.2.95}$$

and

$$\begin{aligned} c^T \left(A - bk^T\right)^{-1} b &= c^T A^{-1} b + \frac{c^T A^{-1} b k^T A^{-1} b}{1 - k^T A^{-1} b} \\ &= \frac{c^T A^{-1} b (1 + k^T A^{-1} b - k^T A^{-1} b)}{1 - k^T A^{-1} b} = \frac{c^T A^{-1} b}{1 - k^T A^{-1} b} \end{aligned} \tag{A.2.96}$$

Thus Equation (A.2.94) can be further modified as

$$k_r = \frac{-1}{c^T (A - bk^T)^{-1} b} = \frac{k^T A^{-1} b - 1}{c^T A^{-1} b} \tag{A.2.97}$$

A.2.5.6

As seen in the derivation of (3.1.10), the pole assignment state feedback vector $k^T = k_c^T$ can be very easily calculated from the controllable canonical form. It has already been mentioned in connection with Equation (A.2.57) that all controllable systems can be transferred to controllable canonical form by the transformation matrix $T_c = M_c^c (M_c)^{-1}$. Therefore we can obtain the similarity transformation (3.1.19) of the feedback vector as

$$k^T = k_c^T T_c = k_c^T M_c^c M_c^{-1} \tag{A.2.98}$$

There is a simpler calculation method instead of the complicated transformation matrix T_c. Let us seek the matrix T of similarity transformation in the form

$$A_c = TAT^{-1} \quad \text{and} \quad b_c = Tb \tag{A.2.99}$$

The similarity transformation of matrix A can also be given as

$$A_c T = TA \tag{A.2.100}$$

Introducing the notation $\boldsymbol{t}_i^{\mathrm{T}}$ for the row of the matrix $\boldsymbol{T}$ we obtain

$$\boldsymbol{A}_c\boldsymbol{T} = \begin{bmatrix} -a_1 & -a_2 & \dots & -a_{n-1} & -a_n \\ 1 & 0 & \dots & 0 & 0 \\ 0 & 1 & \dots & 0 & 0 \\ \vdots & \vdots & \ddots & \vdots & \vdots \\ 0 & 0 & 0 & 1 & 0 \end{bmatrix} \begin{bmatrix} \boldsymbol{t}_1^{\mathrm{T}} \\ \boldsymbol{t}_2^{\mathrm{T}} \\ \boldsymbol{t}_3^{\mathrm{T}} \\ \vdots \\ \boldsymbol{t}_n^{\mathrm{T}} \end{bmatrix} = \boldsymbol{T}\boldsymbol{A}$$

$$= \begin{bmatrix} \boldsymbol{t}_1^{\mathrm{T}} \\ \boldsymbol{t}_2^{\mathrm{T}} \\ \boldsymbol{t}_3^{\mathrm{T}} \\ \vdots \\ \boldsymbol{t}_n^{\mathrm{T}} \end{bmatrix} \boldsymbol{A} = \begin{bmatrix} \boldsymbol{t}_1^{\mathrm{T}}\boldsymbol{A} \\ \boldsymbol{t}_2^{\mathrm{T}}\boldsymbol{A} \\ \boldsymbol{t}_3^{\mathrm{T}}\boldsymbol{A} \\ \vdots \\ \boldsymbol{t}_n^{\mathrm{T}}\boldsymbol{A} \end{bmatrix} \tag{A.2.101}$$

and

$$\boldsymbol{A}_c\boldsymbol{T} = \begin{bmatrix} -a_1\boldsymbol{t}_1^{\mathrm{T}} - a_2\boldsymbol{t}_2^{\mathrm{T}} \dots - a_{n-1}\boldsymbol{t}_{n-1}^{\mathrm{T}} - a_n\boldsymbol{t}_n^{\mathrm{T}} \\ \boldsymbol{t}_1^{\mathrm{T}} \\ \boldsymbol{t}_2^{\mathrm{T}} \\ \vdots \\ \boldsymbol{t}_{n-1}^{\mathrm{T}} \end{bmatrix} = \begin{bmatrix} \boldsymbol{t}_1^{\mathrm{T}}\boldsymbol{A} \\ \boldsymbol{t}_2^{\mathrm{T}}\boldsymbol{A} \\ \boldsymbol{t}_3^{\mathrm{T}}\boldsymbol{A} \\ \vdots \\ \boldsymbol{t}_n^{\mathrm{T}}\boldsymbol{A} \end{bmatrix}$$

$$= \begin{bmatrix} \boldsymbol{t}_n^{\mathrm{T}}\boldsymbol{A}^{n-1} \\ \boldsymbol{t}_n^{\mathrm{T}}\boldsymbol{A}^{n-2} \\ \boldsymbol{t}_n^{\mathrm{T}}\boldsymbol{A}^{n-3} \\ \vdots \\ \boldsymbol{t}_n^{\mathrm{T}} \end{bmatrix} \boldsymbol{A} = \boldsymbol{T}\boldsymbol{A} \tag{A.2.102}$$

Given the identity of the two sides it can be concluded that the following recursive relationship exists for the row vectors $\boldsymbol{t}_i^{\mathrm{T}}$ if $\boldsymbol{t}_n^{\mathrm{T}}$ is known

$$\boldsymbol{t}_{i-1}^{\mathrm{T}} = \boldsymbol{t}_i^{\mathrm{T}}\boldsymbol{A}; \quad i = n, n-1, \dots, 2 \tag{A.2.103}$$

or

$$\boldsymbol{t}_{i-1}^{\mathrm{T}} = \boldsymbol{t}_n^{\mathrm{T}}\boldsymbol{A}^{n-i+1}; \quad i = n, n-1, \dots, 2 \tag{A.2.104}$$

which is also represented in the composition of the row vectors. Thus the transformation matrix for the controllable canonical form is

$$\boldsymbol{T}_{\mathrm{c}} = \boldsymbol{T} = \begin{bmatrix} \boldsymbol{t}_n^{\mathrm{T}}\boldsymbol{A}^{n-1} \\ \boldsymbol{t}_n^{\mathrm{T}}\boldsymbol{A}^{n-2} \\ \boldsymbol{t}_n^{\mathrm{T}}\boldsymbol{A}^{n-3} \\ \vdots \\ \boldsymbol{t}_n^{\mathrm{T}} \end{bmatrix} \tag{A.2.105}$$

Given (A.2.99) and (A.2.102), similarly we can obtain

$$\boldsymbol{b}_{\mathrm{c}} = \boldsymbol{T}\boldsymbol{b} = \begin{bmatrix} \boldsymbol{t}_1^{\mathrm{T}} \\ \boldsymbol{t}_2^{\mathrm{T}} \\ \boldsymbol{t}_3^{\mathrm{T}} \\ \vdots \\ \boldsymbol{t}_n^{\mathrm{T}} \end{bmatrix} \boldsymbol{b} = \begin{bmatrix} \boldsymbol{t}_n^{\mathrm{T}}\boldsymbol{A}^{n-1} \\ \boldsymbol{t}_n^{\mathrm{T}}\boldsymbol{A}^{n-2} \\ \boldsymbol{t}_n^{\mathrm{T}}\boldsymbol{A}^{n-3} \\ \vdots \\ \boldsymbol{t}_n^{\mathrm{T}} \end{bmatrix} \boldsymbol{b} = \begin{bmatrix} \boldsymbol{t}_n^{\mathrm{T}}\boldsymbol{A}^{n-1}\boldsymbol{b} \\ \boldsymbol{t}_n^{\mathrm{T}}\boldsymbol{A}^{n-2}\boldsymbol{b} \\ \boldsymbol{t}_n^{\mathrm{T}}\boldsymbol{A}^{n-3}\boldsymbol{b} \\ \vdots \\ \boldsymbol{t}_n^{\mathrm{T}}\boldsymbol{b} \end{bmatrix} \tag{A.2.106}$$

whose transposed form is

$$\boldsymbol{b}_{\mathrm{c}}^{\mathrm{T}} = \boldsymbol{t}_n^{\mathrm{T}}\left[\boldsymbol{b} \quad \boldsymbol{A}\boldsymbol{b} \quad \ldots \quad \boldsymbol{A}^{n-2}\boldsymbol{b} \quad \boldsymbol{A}^{n-1}\boldsymbol{b}\right] = \boldsymbol{t}_n^{\mathrm{T}}\boldsymbol{M}_{\mathrm{c}} \tag{A.2.107}$$

where $\boldsymbol{M}_{\mathrm{c}}$ is the controllability matrix. Hence

$$\boldsymbol{t}_n^{\mathrm{T}} = \boldsymbol{b}_{\mathrm{c}}^{\mathrm{T}}(\boldsymbol{M}_{\mathrm{c}})^{-1} \tag{A.2.108}$$

Thus $\boldsymbol{t}_n^{\mathrm{T}}$ is the first row of the inverse of the controllability matrix, since

$$\boldsymbol{b}_{\mathrm{c}}^{\mathrm{T}} = [1, 0, \ldots, 0] \tag{A.2.109}$$

Consider now the transpose (A.2.98) of the feedback vector

$$\boldsymbol{k}^{\mathrm{T}} = \boldsymbol{k}_{\mathrm{c}}^{\mathrm{T}}\boldsymbol{T}_{\mathrm{c}} = [r_1 - a_1, r_2 - a_2, \ldots, r_n - a_n] \begin{bmatrix} \boldsymbol{t}_n^{\mathrm{T}}\boldsymbol{A}^{n-1} \\ \boldsymbol{t}_n^{\mathrm{T}}\boldsymbol{A}^{n-2} \\ \boldsymbol{t}_n^{\mathrm{T}}\boldsymbol{A}^{n-3} \\ \vdots \\ \boldsymbol{t}_n^{\mathrm{T}} \end{bmatrix} \tag{A.2.110}$$

Executing the operations we obtain the equation

$$\boldsymbol{k}^{\mathrm{T}} = \boldsymbol{t}_n^{\mathrm{T}} \sum_{i=1}^{n} r_i \boldsymbol{A}^{n-i} - \boldsymbol{t}_n^{\mathrm{T}} \sum_{i=1}^{n} a_i \boldsymbol{A}^{n-i} \tag{A.2.111}$$

and adding $\boldsymbol{A}^n$ to both sides we obtain the following special form

$$\boldsymbol{k}^{\mathrm{T}} = \boldsymbol{t}_n^{\mathrm{T}} \mathcal{R}(\boldsymbol{A}) - \boldsymbol{t}_n^{\mathrm{T}} \mathcal{A}(\boldsymbol{A}) \tag{A.2.112}$$

Because of the Cayley–Hamilton theorem, all quadratic matrices fulfill their characteristic polynomial, and therefore $\mathcal{A}(\boldsymbol{A}) = \mathcal{R}(\boldsymbol{A}) = 0$. Thus the final form of Equation (A.2.112) is

$$\boldsymbol{k}^{\mathrm{T}} = \boldsymbol{t}_n^{\mathrm{T}} \mathcal{R}(\boldsymbol{A}) \tag{A.2.113}$$

This latter equation is called the Ackermann formula [1,2]. The relationship (A.2.98) can be calculated by computer methods much more easier than the expression (A.2.113).

A.2.5.7

Proceeding from the basic relationship (3.1.13) of the pole assignment and considering the diagonal canonical form we can obtain the following equation by identical manipulations

$$\begin{aligned}\mathcal{R}(s) - \mathcal{A}(s) &= \frac{\mathcal{B}(s)}{\boldsymbol{c}_{\mathrm{d}}^{\mathrm{T}}(s\boldsymbol{I} - \boldsymbol{A}_{\mathrm{d}})^{-1}\boldsymbol{b}^{\mathrm{d}}} \boldsymbol{k}_{\mathrm{d}}^{\mathrm{T}}(s\boldsymbol{I} - \boldsymbol{A}_{\mathrm{d}})^{-1}\boldsymbol{b}^{\mathrm{d}} \\ &= \mathcal{A}(s)\boldsymbol{k}_{\mathrm{d}}^{\mathrm{T}}(s\boldsymbol{I} - \boldsymbol{A}_{\mathrm{d}})^{-1}\boldsymbol{b}^{\mathrm{d}}\end{aligned} \tag{A.2.114}$$

from where

$$\frac{\mathcal{R}(s)}{\mathcal{A}(s)} = 1 + \boldsymbol{k}_{\mathrm{d}}^{\mathrm{T}}(s\boldsymbol{I} - \boldsymbol{A}_{\mathrm{d}})^{-1}\boldsymbol{b}^{\mathrm{d}} \tag{A.2.115}$$

Decomposing the left side to partial fractions and considering the diagonal feature of the system, we can write

$$\frac{\mathcal{R}(s)}{\mathcal{A}(s)} = 1 + \sum_{i=1}^{n} \frac{k_i^{\mathrm{d}} b_i^{\mathrm{d}}}{s - \lambda_i} = 1 + \sum_{i=1}^{n} \frac{k_i^{\mathrm{d}} \beta_i}{s - \lambda_i} \tag{A.2.116}$$

Applying the decomposition theorem to partial fractions valid for single poles, let us multiply both sides by $(s - \lambda_i)$ and substitute $s = \lambda_i$ so we obtain the relationship

$$k_i^{\mathrm{d}} b_i^{\mathrm{d}} = \prod_{j=1}^{n} (\lambda_i - \mu_j) \Big/ \prod_{\substack{j=1 \\ i \neq j}}^{n} (\lambda_i - \lambda_j) \tag{A.2.117}$$

This operation needs to be performed for each pole.

A.2.5.8

Given the matrix inversion identity of (A.1.17) in Appendix A.1 the following rearrangement seems to be quite obvious

$$T_{\mathrm{ry}}(s) = \frac{\left[\boldsymbol{c}^{\mathrm{T}}(s\boldsymbol{I}-\boldsymbol{A})^{-1}\boldsymbol{b}\right]\left[1-\boldsymbol{k}^{\mathrm{T}}\left(s\boldsymbol{I}-\boldsymbol{A}+\boldsymbol{b}\boldsymbol{k}^{\mathrm{T}}+\boldsymbol{l}\boldsymbol{c}^{\mathrm{T}}\right)^{-1}\boldsymbol{b}\right]k_{\mathrm{r}}}{1+\left[\boldsymbol{k}^{\mathrm{T}}\left(s\boldsymbol{I}-\boldsymbol{A}+\boldsymbol{b}\boldsymbol{k}^{\mathrm{T}}+\boldsymbol{l}\boldsymbol{c}^{\mathrm{T}}\right)^{-1}\boldsymbol{b}\right]\left[\boldsymbol{c}^{\mathrm{T}}(s\boldsymbol{I}-\boldsymbol{A})^{-1}\boldsymbol{b}\right]}$$
$$= \boldsymbol{c}^{\mathrm{T}}\left(s\boldsymbol{I}-\boldsymbol{A}+\boldsymbol{b}\boldsymbol{k}^{\mathrm{T}}\right)^{-1}\boldsymbol{b}k_{\mathrm{r}} = \frac{\boldsymbol{c}^{\mathrm{T}}(s\boldsymbol{I}-\boldsymbol{A})^{-1}\boldsymbol{b}k_{\mathrm{r}}}{1+\boldsymbol{k}^{\mathrm{T}}(s\boldsymbol{I}-\boldsymbol{A})^{-1}\boldsymbol{b}}$$
$$= \frac{k_{\mathrm{r}}P(s)}{1+\boldsymbol{k}^{\mathrm{T}}(s\boldsymbol{I}-\boldsymbol{A})^{-1}\boldsymbol{b}} = \frac{k_{\mathrm{r}}\mathcal{B}(s)}{\mathcal{A}(s)} \tag{A.2.118}$$

A.2.5.9

The so-called *linear quadratic* (*LQ*) regulator discussed in Section 3.2 can be considered as a special case of the generally defined optimization problem. During this general task a control $u(t)$ needs to be determined for the system given by the following state equation

$$\dot{\boldsymbol{x}}(t) = \frac{\mathrm{d}\boldsymbol{x}(t)}{\mathrm{d}t} = \boldsymbol{f}[\boldsymbol{x}(t), u(t)] \tag{A.2.119}$$

which minimizes the general integral criterion

$$I = \frac{1}{2}\int_0^{T_{\mathrm{f}}} F[\boldsymbol{x}(t), u(t)]\mathrm{d}t = I[u(t)] \tag{A.2.120}$$

The solution is provided by the so-called minimum principle [39], according to which the Hamilton function needs to be formulated as

$$H(t) = F[\boldsymbol{x}(t), u(t)] + \boldsymbol{\lambda}(t)^{\mathrm{T}}\boldsymbol{f}[\boldsymbol{x}(t), u(t)] \tag{A.2.121}$$

and the necessary conditions of the extremum must exist like

$$\frac{\mathrm{d}H(t)}{\mathrm{d}u(t)} = 0; \quad \frac{\mathrm{d}H(t)}{\mathrm{d}\boldsymbol{x}(t)} = -\frac{\mathrm{d}\boldsymbol{\lambda}(t)}{\mathrm{d}t} = -\dot{\boldsymbol{\lambda}}(t) \tag{A.2.122}$$

(The sufficient condition of the minimum is $\partial^2 H/\partial u^2 > 0$.) The Hamilton function [39] and the necessary condition (A.2.121) of the minimum formally correspond to the Lagrange method of the conditional optimum (thus $\boldsymbol{\lambda}$ is the associate vector of the method), since the minimum of $I[u(t)]$ must be reached under the condition (A.2.119). (Arbitrary motion in the state space cannot be allowed, only such kind which corresponds to (A.2.119).) It is usually assumed that $\boldsymbol{\lambda}(t) = \boldsymbol{P}(t)\boldsymbol{x}(t)$, i.e., it can be obtained from the state vector by linear transformation

$$\dot{\boldsymbol{\lambda}}(t) = \dot{\boldsymbol{P}}(t)\boldsymbol{x}(t) + \boldsymbol{P}\dot{\boldsymbol{x}}(t) \tag{A.2.123}$$

If the upper limit of the integral is infinite ($T_f = \infty$), then $\boldsymbol{P}(t) = \boldsymbol{P}$ is constant, and thus $\dot{\boldsymbol{P}} = \boldsymbol{0}$ and

$$\boldsymbol{\lambda}(t) = \boldsymbol{P}\boldsymbol{x}(t) \quad \text{and} \quad \dot{\boldsymbol{\lambda}}(t) = \boldsymbol{P}\dot{\boldsymbol{x}}(t) \tag{A.2.124}$$

The *LQ* regulator of the *LTI* process has to solve the task

$$I = \frac{1}{2}\int_0^\infty \left[\boldsymbol{x}^{\mathrm{T}}(t)\boldsymbol{W}_{\mathrm{x}}\boldsymbol{x}(t) + W_{\mathrm{u}}u^2(t)\right] \mathrm{d}t = \min_{u(t)} \tag{A.2.125}$$

on condition that the linear dynamic system is given as

$$\dot{\boldsymbol{x}}(t) = \boldsymbol{A}\boldsymbol{x}(t) + \boldsymbol{b}u(t) \tag{A.2.126}$$

The Hamilton function is now

$$H(t) = \frac{1}{2}\left[\boldsymbol{x}^{\mathrm{T}}(t)\boldsymbol{W}_{\mathrm{x}}\boldsymbol{x}(t) + W_{\mathrm{u}}u^2(t)\right] + \boldsymbol{\lambda}^{\mathrm{T}}[\boldsymbol{A}\boldsymbol{x}(t) + \boldsymbol{b}u(t)] \tag{A.2.127}$$

whose second derivative is $\partial^2 H/\partial u^2 = W_{\mathrm{u}} > 0$, so the necessary condition is, at the same time, the sufficient one. The necessary condition is

$$\frac{\mathrm{d}H(t)}{\mathrm{d}u(t)} = W_{\mathrm{u}}u(t) + \boldsymbol{\lambda}^{\mathrm{T}}\boldsymbol{b} = W_{\mathrm{u}}u(t) + \boldsymbol{b}^{\mathrm{T}}\boldsymbol{\lambda} = 0 \tag{A.2.128}$$

whence the optimal control is

$$u(t) = -\frac{1}{W_{\mathrm{u}}}\boldsymbol{b}^{\mathrm{T}}\boldsymbol{\lambda}(t) = -\frac{1}{W_{\mathrm{u}}}\boldsymbol{b}^{\mathrm{T}}\boldsymbol{P}\boldsymbol{x}(t) = -\boldsymbol{k}_{\mathrm{LQ}}^{\mathrm{T}}\boldsymbol{x}(t) \tag{A.2.129}$$

The second step is to determine matrix $\boldsymbol{P}$. Consider the overall state equation of the closed system

$$\dot{\boldsymbol{x}} = \boldsymbol{A}\boldsymbol{x} - \boldsymbol{b}\boldsymbol{k}_{\mathrm{LQ}}^{\mathrm{T}}\boldsymbol{x} = \left(\boldsymbol{A} - \boldsymbol{b}\boldsymbol{k}_{\mathrm{LQ}}^{\mathrm{T}}\right)\boldsymbol{x} = \left(\boldsymbol{A} - \frac{1}{W_{\mathrm{u}}}\boldsymbol{b}\boldsymbol{b}^{\mathrm{T}}\boldsymbol{P}\right)\boldsymbol{x} = \overline{\boldsymbol{A}}\boldsymbol{x} \tag{A.2.130}$$

which has the same form as in the state feedback. Thus the LQ regulator is a state feedback regulator. From Equations (A.2.122) and (A.2.130) we obtain the associate vector

$$\dot{\boldsymbol{\lambda}} = \boldsymbol{P}\dot{\boldsymbol{x}} = \left(\boldsymbol{PA} - \frac{1}{W_{\mathrm{u}}}\boldsymbol{Pbb}^{\mathrm{T}}\boldsymbol{P}\right)\boldsymbol{x} = \boldsymbol{P}\overline{\boldsymbol{A}}\boldsymbol{x} \tag{A.2.131}$$

which has to fulfill the equation

$$\begin{aligned}\dot{\boldsymbol{\lambda}}(t) &= -\frac{\mathrm{d}H(t)}{\mathrm{d}\boldsymbol{x}(t)} = -\boldsymbol{W}_{\mathrm{x}}\boldsymbol{x}(t) - \boldsymbol{A}^{\mathrm{T}}\boldsymbol{\lambda}(t) = -\boldsymbol{W}_{\mathrm{x}}\boldsymbol{x}(t) - \boldsymbol{A}^{\mathrm{T}}\boldsymbol{P}\boldsymbol{x}(t)\\ &= -\left(\boldsymbol{W}_{\mathrm{x}} + \boldsymbol{A}^{\mathrm{T}}\boldsymbol{P}\right)\boldsymbol{x}(t)\end{aligned} \tag{A.2.132}$$

obtained from the necessary condition (A.2.122). Comparing the last two equations we obtain the equation

$$\boldsymbol{PA} - \frac{1}{W_{\mathrm{u}}}\boldsymbol{Pbb}^{\mathrm{T}}\boldsymbol{P} = -\left(\boldsymbol{W}_{\mathrm{x}} + \boldsymbol{A}^{\mathrm{T}}\boldsymbol{P}\right) \tag{A.2.133}$$

to determine the symmetric matrix $\boldsymbol{P}$. By rearrangement we obtain the so-called algebraic Riccati matrix equation

$$\boldsymbol{PA} + \boldsymbol{A}^{\mathrm{T}}\boldsymbol{P} - \frac{1}{W_{\mathrm{u}}}\boldsymbol{Pbb}^{\mathrm{T}}\boldsymbol{P} = \boldsymbol{W}_{\mathrm{x}} \tag{A.2.134}$$

which being nonlinear has no explicit solution but there are, however, several numerical methods to solve the problem by computer.

Using Equations (A.2.130) and (A.2.132), we can write the combined state equation with the state vector of the closed system and the associate vector as

$$\begin{bmatrix}\dot{\boldsymbol{x}}\\ \dot{\boldsymbol{\lambda}}\end{bmatrix} = \begin{bmatrix}\boldsymbol{A} & \dfrac{1}{W_{\mathrm{u}}}\boldsymbol{bb}^{\mathrm{T}}\boldsymbol{P}\\ \boldsymbol{W}_{\mathrm{x}} & -\boldsymbol{A}^{\mathrm{T}}\end{bmatrix}\begin{bmatrix}\boldsymbol{x}\\ \boldsymbol{\lambda}\end{bmatrix} \tag{A.2.135}$$

Note that if the upper limit of the integral is finite ($T_{\mathrm{f}} < \infty$), then $\boldsymbol{P} = \boldsymbol{P}(t)$ depends on the time, and the Riccati matrix equation needs to be solved for the whole domain $0 \leq t \leq T_{\mathrm{f}}$ in advance.

Next it is shown that solution $\boldsymbol{P}$ of the Riccati matrix equation has special meaning. Substitute the equation of the optimal control (A.2.127) into the criterion (A.2.123)

$$\begin{aligned} I &= \frac{1}{2}\int_0^\infty \left[\boldsymbol{x}^{\mathrm{T}}(t)\boldsymbol{W}_{\mathrm{x}}\boldsymbol{x}(t) + W_{\mathrm{u}}u^2(t)\right]\mathrm{d}t \\ &= \frac{1}{2}\int_0^\infty \left\{\boldsymbol{x}^{\mathrm{T}}(t)\boldsymbol{W}_{\mathrm{x}}\boldsymbol{x}(t) + W_{\mathrm{u}}\left[-\boldsymbol{k}_{\mathrm{LQ}}^{\mathrm{T}}\boldsymbol{x}(t)\right]^2\right\}\mathrm{d}t \\ &= \frac{1}{2}\int_0^\infty \left[\boldsymbol{x}^{\mathrm{T}}(t)\boldsymbol{W}_{\mathrm{x}}\boldsymbol{x}(t) + \frac{1}{W_{\mathrm{u}}}\boldsymbol{x}^{\mathrm{T}}(t)\boldsymbol{P}^{\mathrm{T}}\boldsymbol{b}\boldsymbol{b}^{\mathrm{T}}\boldsymbol{P}\boldsymbol{x}(t)\right]\mathrm{d}t \\ &= \frac{1}{2}\int_0^\infty \left[\boldsymbol{x}^{\mathrm{T}}(t)\overline{\boldsymbol{W}}_{\mathrm{x}}\boldsymbol{x}(t)\right]\mathrm{d}t \end{aligned} \tag{A.2.136}$$

where

$$\overline{\boldsymbol{W}}_{\mathrm{x}} = \boldsymbol{W}_{\mathrm{x}} + \frac{1}{W_{\mathrm{u}}}\boldsymbol{P}^{\mathrm{T}}\boldsymbol{b}\boldsymbol{b}^{\mathrm{T}}\boldsymbol{P} \tag{A.2.137}$$

The solution for the nonexcited closed system (A.2.130) is

$$\boldsymbol{x}(t) = e^{\overline{\boldsymbol{A}}t}\boldsymbol{x}(0) \tag{A.2.138}$$

so the criterion (A.2.136) for the nonexcited case is

$$I = \frac{1}{2}\int_0^\infty \left[\boldsymbol{x}^{\mathrm{T}}(0)e^{\overline{\boldsymbol{A}}^{\mathrm{T}}t}\overline{\boldsymbol{W}}_{\mathrm{x}}e^{\overline{\boldsymbol{A}}t}\boldsymbol{x}(0)\right]\mathrm{d}t = \frac{1}{2}\boldsymbol{x}^{\mathrm{T}}(0)\boldsymbol{P}\boldsymbol{x}(0) \tag{A.2.139}$$

where it is assumed that

$$\boldsymbol{P} = \int_0^\infty e^{\overline{\boldsymbol{A}}^{\mathrm{T}}t}\overline{\boldsymbol{W}}_{\mathrm{x}}e^{\overline{\boldsymbol{A}}t}\,\mathrm{d}t \tag{A.2.140}$$

To prove this let us compute the integration

$$\begin{aligned} \boldsymbol{P} &= \int_0^\infty e^{\overline{\boldsymbol{A}}^{\mathrm{T}}t}\overline{\boldsymbol{W}}_{\mathrm{x}}e^{\overline{\boldsymbol{A}}t}\,\mathrm{d}t \\ &= \left[e^{\overline{\boldsymbol{A}}^{\mathrm{T}}t}\overline{\boldsymbol{W}}_{\mathrm{x}}\overline{\boldsymbol{A}}^{-1}e^{\overline{\boldsymbol{A}}t}\right]_0^\infty - \int_0^\infty \overline{\boldsymbol{A}}^{\mathrm{T}}e^{\overline{\boldsymbol{A}}^{\mathrm{T}}t}\overline{\boldsymbol{W}}_{\mathrm{x}}\overline{\boldsymbol{A}}^{-1}e^{\overline{\boldsymbol{A}}t}\,\mathrm{d}t \end{aligned} \tag{A.2.141}$$

Furthermore, if $\overline{\boldsymbol{A}}$ is stable, then it can be written that

$$\boldsymbol{P} = -\overline{\boldsymbol{W}}_{\mathrm{x}}\overline{\boldsymbol{A}}^{-1} - \overline{\boldsymbol{A}}^{\mathrm{T}}\left(\int_0^{\infty} e^{\overline{\boldsymbol{A}}^{\mathrm{T}}t}\overline{\boldsymbol{W}}_{\mathrm{x}}e^{\overline{\boldsymbol{A}}t}\,\mathrm{d}t\right)\overline{\boldsymbol{A}}^{-1}$$

$$= -\overline{\boldsymbol{W}}_{\mathrm{x}}\overline{\boldsymbol{A}}^{-1} - \overline{\boldsymbol{A}}^{\mathrm{T}}\boldsymbol{P}\overline{\boldsymbol{A}}^{-1} \tag{A.2.142}$$

Here the expression $\overline{\boldsymbol{A}}^{-1}e^{\overline{\boldsymbol{A}}t} = e^{\overline{\boldsymbol{A}}t}\overline{\boldsymbol{A}}^{-1}$ is used. Finally we obtain

$$\boldsymbol{P}\overline{\boldsymbol{A}} + \overline{\boldsymbol{A}}^{\mathrm{T}}\boldsymbol{P} = -\overline{\boldsymbol{W}}_{\mathrm{x}} \tag{A.2.143}$$

which is called the Lyapunov equation [39]. The equation is only virtually linear in $\boldsymbol{P}$, because $\overline{\boldsymbol{W}}_{\mathrm{x}}$ and also $\overline{\boldsymbol{A}}$ depend on $\boldsymbol{P}$. Further manipulating the equation we obtain again the algebraic Riccati matrix Equation (A.2.134). Consequently, the relationship (A.2.140) of $\boldsymbol{P}$ is proved and the meaning of $\boldsymbol{P}$ is also shown: it is the matrix of the quadratic cost function of the control ensuring the minimum of the criterion (A.2.123) in the case when there is no excitation.

It is also interesting to investigate how the Hamilton function changes in time. Formulate the derivative by time

$$\frac{\mathrm{d}H}{\mathrm{d}t} = \left[\frac{\mathrm{d}H}{\mathrm{d}\boldsymbol{x}}\right]^{\mathrm{T}}\frac{\mathrm{d}\boldsymbol{x}}{\mathrm{d}t} + \left[\frac{\mathrm{d}H}{\mathrm{d}u}\right]^{\mathrm{T}}\frac{\mathrm{d}u}{\mathrm{d}t} + \left[\frac{\mathrm{d}H}{\mathrm{d}\boldsymbol{\lambda}}\right]^{\mathrm{T}}\frac{\mathrm{d}\boldsymbol{\lambda}}{\mathrm{d}t} \tag{A.2.144}$$

Based on (A.2.119) and (A.2.121) the following expression obtains

$$\frac{\mathrm{d}H}{\mathrm{d}\boldsymbol{\lambda}} = \frac{\mathrm{d}H}{\mathrm{d}\boldsymbol{x}} = f \tag{A.2.145}$$

and if we take Equation (A.2.122) into account

$$\left[\frac{\mathrm{d}H}{\mathrm{d}\boldsymbol{x}}\right]^{\mathrm{T}}\frac{\mathrm{d}\boldsymbol{x}}{\mathrm{d}t} + \left[\frac{\mathrm{d}H}{\mathrm{d}\boldsymbol{\lambda}}\right]^{\mathrm{T}}\frac{\mathrm{d}\boldsymbol{\lambda}}{\mathrm{d}t} = 0 \tag{A.2.146}$$

Thus finally

$$\frac{\mathrm{d}H}{\mathrm{d}t} = 0 \tag{A.2.147}$$

i.e., the Hamilton function is constant (assuming that neither the control nor the state vector is under constraint). Thus, in the nonconstrained case the Hamilton function is time-invariant and also invariant regarding the input (see the necessary condition (A.2.122) for the extremum).

A.2.5.10

The system matrix $\boldsymbol{A}_c$ of the controllable canonical form

$$\boldsymbol{A}_c = \begin{bmatrix} -a_1 & -a_2 & \dots & -a_{n-1} & -a_n \\ 1 & 0 & \dots & 0 & 0 \\ 0 & 1 & \dots & 0 & 0 \\ \vdots & \vdots & \ddots & \vdots & \vdots \\ 0 & 0 & \dots & 1 & 0 \end{bmatrix} \tag{A.2.148}$$

is called the companion matrix of the characteristic polynomial

$$\mathcal{A}(s) = s^n + a_1 s^{n-1} + \ldots + a_{n-1}s + a_n = \det(s\boldsymbol{I} - \boldsymbol{A}_c) \tag{A.2.149}$$

In $\boldsymbol{A}_c$ the first and last rows and the first and last columns can also contain the parameter vector

$$-\begin{bmatrix} a_1 & a_2 & \dots & a_{n-1} & a_n \end{bmatrix}^T \tag{A.2.150}$$

The upper companion matrix $\boldsymbol{A}_c$ has the eigenvector

$$\boldsymbol{v}_i = \begin{bmatrix} \lambda_i^{n-1} & \lambda_i^{n-2} & \dots & \lambda_i & 1 \end{bmatrix}^T \tag{A.2.151}$$

belonging to the eigenvalue λ_i.

The rank of the matrix $(\lambda_i\boldsymbol{I} - \boldsymbol{A})$ is $n - 1$. Matrix $\boldsymbol{A}_c$ is nonsingular if and only if $a_n \neq 0$. The inverse of $\boldsymbol{A}_c$ is a so-called lower companion matrix, where the last row is

$$-\frac{1}{a_n}\begin{bmatrix} a_1 & a_2 & \dots & a_{n-1} & a_n \end{bmatrix}^T \tag{A.2.152}$$

i.e.,

$$\boldsymbol{A}_c^{-1} = \begin{bmatrix} 0 & 1 & \dots & 0 & 0 \\ \vdots & \vdots & \ddots & \vdots & \vdots \\ 0 & 0 & \dots & 0 & 0 \\ 0 & 0 & \ddots & 1 & 0 \\ -\dfrac{a_1}{a_n} & -\dfrac{a_2}{a_n} & \dots & -\dfrac{-a_{n-1}}{a_n} & 1 \end{bmatrix} \tag{A.2.153}$$

For the controllable canonical form, when $\boldsymbol{b}_c = \begin{bmatrix} 1 & 0 & \ldots & 0 \end{bmatrix}^T$, the factor $(s\boldsymbol{I} - \boldsymbol{A}_c)^{-1}\boldsymbol{b}_c$ can be easily computed by

$$(s\boldsymbol{I} - \boldsymbol{A}_c)^{-1}\boldsymbol{b}_c = \frac{1}{\mathcal{A}(s)}\begin{bmatrix} s^{n-1} & s^{n-2} & \ldots & s & 1 \end{bmatrix}^T \tag{A.2.154}$$

Similar formulas can be derived for the system matrix $\boldsymbol{A}_o$ of the dual observable canonical form.

APPENDIX 3

Sampled Data Systems

A.3.1 SAMPLING

It is common practice today that control systems are implemented by digital equipment with real-time services. Depending on the complexity of the task to be solved, digital equipment ranges from a microcontroller with a single chip to midi- or mini-computers of higher power. The development of the networking of digital devices makes it possible to establish distributed control systems.

The digital solution of a control function characterizes the control engineering culture applied in the industry. Namely, practical control engineering requires the integrated, obviously digital, implementation of a combination of different guiding and control tasks.

The basic block scheme of a closed sampled control loop is illustrated in Figure A.3.1. It is assumed in the figure that the process to be controlled is a continuous-time (CT) process. Contrary to the CT input and output signals of the CT process, the digital system works in discrete-time (DT), since the control program is running on a control device driven by a CPU of given working frequency. It is obvious that the signals of the analogous CT process and the digital controller need to be fitted, and this task is solved by the converters in the control loop. The task of the holding element is to produce the CT input signal from the DT signal series. The elements that

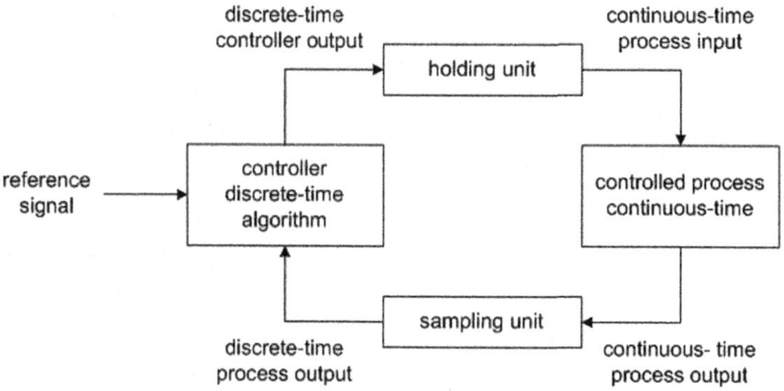

Figure A.3.1 Basic block scheme of a closed sampled control loop.

convert the signals from analogous to digital (A/D) are called *sampling* devices, and in the opposite direction, from digital to analogous (D/A), they are called *holding* devices. The sampling device in practically all cases is an A/D converter, well known in the field, whose holding element is typically, but not inevitably, a D/A converter. Since in the closed-loop control CT and DT signals are present and analogous and digital information can be available from the same signal, it is reasonable to introduce an unambiguous notation system:

$-x(t)$: CT signal

$-x[k] = x(kT_s)$: DT signal, where the *sampling time* is noted by T_s

$k = 0, 1, 2, \ldots$means the serial numbers of the sample

Information on the operation of a CT process can be obtained by observing the CT signals. If the information on the signals is processed by a digital device, then these CT signals are represented by the samples taken in the sampling times. Assuming equidistant sampling, we can use the very important theoretical result of the signals and systems subject area to decide whether the CT signal (assumed to be band-constrained) can be unambiguously reconstructed from its sampled data. The condition for the reproducibility is formulated by the Shannon sampling theorem, according to which, the sampling must be performed by a minimum frequency which obtains at least two samples from the CT signal component of the highest frequency in the sampling period. The importance of the sampling theorem is that it constitutes the basic principles of signal processing (e.g., visualization, frequency analysis) in a flexible programmable digital environment.

As already mentioned, the sampling is physically performed by A/D converters. These converters are guided by the real-time clock signal of the digital system, and they directly provide the signal series $x[k]$ in coded digital form from the CT analogous signal $x(t)$. The relevant operation of the A/D converter is symbolized by a switch closed periodically for a short time (see Figure A.3.2).

The A/D converters have several important parameters for use in different applications. Among them the conversion time, characterizing the speed of the conversion, can be of a very low-order magnitude, *microsecundum* (μs). Other important aspects are the disturbance filtering ability and the resolution of the converter, i.e., the bit width of the digitized signal. The usual value region for the latter is between 8 and 16 bits.

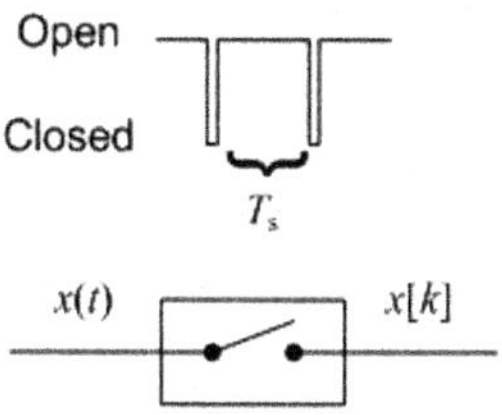

Figure A.3.2 Switch symbolizing the operation of the A/D converter.

Application of the notations introduced above in sampled systems $f[k] = f(kT_s)$ is called *mathematical sampling*.

Considering the signal set, the closed sampled control loop is a hybrid system, where analogous and digital signals can occur simultaneously. This means that the operation time domain of the CT process and the DT controller (or regulator) needs to be fitted. The sampling is realized by a CT/DT converter and to close the loop, a DT/CT converter, i.e., a holding element, needs to be applied. The holding element is also a guided device, and its task is to decode the digital signal and ensure a CT output between two samples. There are no prescribed constrains for the holding character between two samples (constant, linear, second order, etc.) but because of its simplicity the zero-order holding element which maintains a constant value between two sampling times is preferred in practice. The application of a zero-order holding results in the *step-like* curve as shown in Figure A.3.3.

The mathematical description of a *zero-order hold* (ZOH) is illustrated in Figure A.3.3. Note that the pulse response of a ZOH is $1(t) - 1(t - T_s)$.

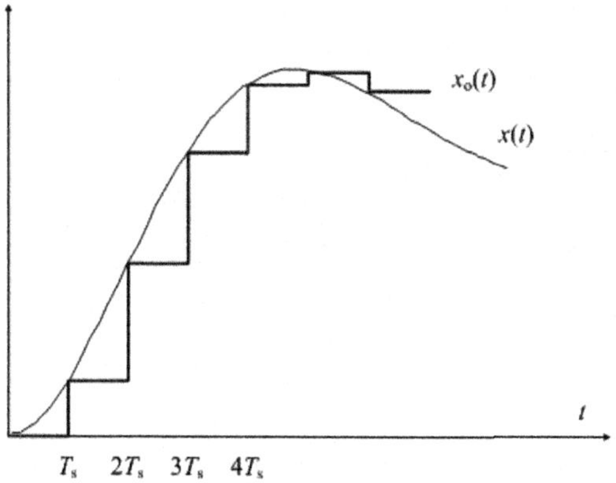

Figure A.3.3 Application of a zero-order hold: $x_o(t)$ is the output of the unit.

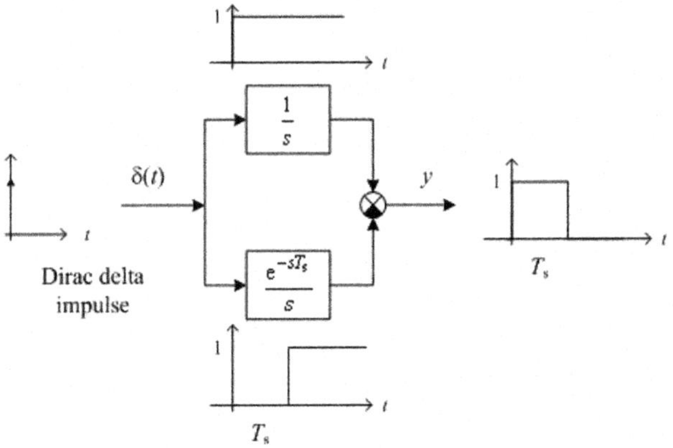

Figure A.3.4 Components of the impulse response of the *ZOH*.

Since the step response is derived from a Dirac-impulse by an integrator and the dead-time T_s is represented by the transfer function e^{-sT_s}, then from Figure A.3.4 the transfer function of a zero-order holding is obtained as

$$W_{\mathrm{ZOH}}(s) = \frac{1}{s} - \frac{e^{-sT_s}}{s} = \frac{1 - e^{-sT_s}}{s} \tag{A.3.1}$$

The basic features of the *ZOH* can be discovered without detailed investigations in the frequency domain. Consider the following first-order Taylor series

$$\begin{aligned} W_{\mathrm{ZOH}}(s) &= \frac{1 - e^{-sT_s}}{s} \cong \frac{1 - \left(1 - sT_s + \dfrac{s^2T_s^2}{2} - \ldots\right)}{s} \\ &\approx T_s\left(1 - \frac{sT_s}{2} + \ldots\right) \approx T_s e^{-sT_s/2} \end{aligned} \tag{A.3.2}$$

Let $f(t)$ be a CT signal, whose spectrum is $F(j\omega)$. It is known from the theory of signals and systems that the frequency spectrum of the sampled signal series $f^*(t)$ is

$$F^*(j\omega) = \frac{1}{T_s}\sum_{k=-\infty}^{\infty} F(j\omega + jk\omega_s) \tag{A.3.3}$$

where $\omega_s = 2\pi/T_s$ is the sampling frequency. Thus if a *ZOH* is serially connected to a CT system of transfer function $H(s)$, then a good approximation of the overall CT system transfer frequency characteristic can be found as

$$H_{\mathrm{d}}(j\omega) = \frac{1}{T_{\mathrm{s}}}\tilde{H}(j\omega) \approx \frac{1}{T_{\mathrm{s}}}T_{\mathrm{s}}e^{-j\omega T_{\mathrm{s}}/2}H(j\omega) = e^{-j\omega T_{\mathrm{s}}/2}H(j\omega) \qquad \text{(A.3.4)}$$

A.3.2 DESCRIPTION OF THE SAMPLED SYSTEMS IN TIME AND OPERATOR DOMAINS

As seen in the discussion of CT systems, both the analysis and design of closed-control systems require abstract system description. If we assume *ZOH* in the sampled data systems, the value of the input of the controlled element and its sampled output change only in the sampling time-points. Therefore, it seems reasonable to choose a mathematical description, i.e., a model that directly describes the behavior of the CT process in the sampling points, avoiding repetition of the physical sampling, i.e., the reconstruction of the CT time functions and their sampling. Hereafter the description of the DT behavior of the CT process will be called *discretization*. Of the several approaches, first consider the state-space description and then derive two further description modes from this, namely the pulse transfer function and the difference equation forms. These triple description modes are in accordance with the concepts used in discussion of CT systems (state-space, differential equation, transfer function).

State-Space Models

Let the *linear time-invariant* state-space model of the CT system (see (1.1.10)) be discretized with sampling time T_{s},

$$\begin{aligned} \dot{\boldsymbol{x}}(t) &= \boldsymbol{A}\boldsymbol{x}(t) + \boldsymbol{b}u(t) \\ y(t) &= \boldsymbol{c}^{\mathrm{T}}\boldsymbol{x}(t) + d_{\mathrm{c}}u(t) \end{aligned} \qquad \text{(A.3.5)}$$

According to expression (A.2.8), the solution of the state equation for the initial time-point t_{o} and the initial state vector $\boldsymbol{x}(t_{\mathrm{o}})$ is

$$\begin{aligned} \boldsymbol{x}(t) &= e^{\boldsymbol{A}(t-t_{\mathrm{o}})}\boldsymbol{x}(t_{\mathrm{o}}) + \int_{t_{\mathrm{o}}}^{t} e^{\boldsymbol{A}(t-\tau)}\boldsymbol{b}u(\tau)\mathrm{d}\tau \\ &= e^{\boldsymbol{A}(t-t_{\mathrm{o}})}\boldsymbol{x}(t_{\mathrm{o}}) + \left[\int_{t_{\mathrm{o}}}^{t} e^{\boldsymbol{A}(t-\tau)}u(\tau)\mathrm{d}\tau\right]\boldsymbol{b} \end{aligned} \qquad \text{(A.3.6)}$$

Assuming that the discretized model is preceded by a *ZOH* (i.e., the input of the process model is a staged function, see later on *step response equivalent* (*SRE*) transformation (A.3.34)), let us perform the integration for the time domain between *k*th and $(k+1)$th sampling time-points (the initial time-point is $t_o = kT_s$)

$$\boldsymbol{x}(kT_s + T_s) = e^{\boldsymbol{A}T_s}\boldsymbol{x}(kT_s) + \int_{kT_s}^{kT_s+T_s} e^{\boldsymbol{A}(kT_s+T_s-\tau)}\boldsymbol{b}u(\tau)\mathrm{d}\tau$$

$$= e^{\boldsymbol{A}T_s}\boldsymbol{x}(kT_s) + \boldsymbol{b}u(kT_s)\int_{kT_s}^{kT_s+T_s} e^{\boldsymbol{A}(kT_s+T_s-\tau)}\mathrm{d}\tau \tag{A.3.7}$$

$$= e^{\boldsymbol{A}T_s}\boldsymbol{x}(kT_s) + \left[u(kT_s)\int_0^{T_s} e^{\boldsymbol{A}\lambda}\mathrm{d}\lambda\right]\boldsymbol{b}$$

where it was taken into account that $u(\tau) =$ constant during the sampling period $[kT_s \leq \tau < (k+1)T_s]$, namely $u(\tau) = u(kT_s)$, and furthermore, the variable $\lambda = kT_s + T_s - \tau$ is introduced to facilitate the calculation of the integral. Applying the former notations $\boldsymbol{x}(kT_s + T_s) = \boldsymbol{x}[k+1]$, $\boldsymbol{x}(kT_s) = \boldsymbol{x}[k]$, and $u(kT_s) = u[k]$, we obtain

$$\boldsymbol{x}[k+1] = e^{\boldsymbol{A}T_s}\boldsymbol{x}[k] + \left[\int_0^{T_s} e^{\boldsymbol{A}\lambda}\mathrm{d}\lambda\right]\boldsymbol{b}u[k] \tag{A.3.8}$$

and, introducing parameter matrices of the discretized model

$$\boldsymbol{F} = e^{\boldsymbol{A}T_s} \quad \text{and} \quad \boldsymbol{g} = \int_0^{T_s} e^{\boldsymbol{A}\lambda}\mathrm{d}\lambda\boldsymbol{b} \tag{A.3.9}$$

we obtain the state equation of the discretized system as $\boldsymbol{x}[k+1] = \boldsymbol{F}\boldsymbol{x}[k] + \boldsymbol{g}u[k]$. Note that if matrix $\boldsymbol{A}$ is invertible, then the integral in the expression of the vector $\boldsymbol{g}$ can be executed and we obtain the closed form $\boldsymbol{g} = \boldsymbol{A}^{-1}(e^{\boldsymbol{A}T_s} - \boldsymbol{I})\boldsymbol{b}$. Using the substitution $t = kT_s$, the discretized form of the output equation $y(t) = \boldsymbol{c}^{\mathrm{T}}\boldsymbol{x}(t) + d_c u(t)$ of the CT state model can be directly obtained as

$$y(kT_s) = \boldsymbol{c}^{\mathrm{T}}\boldsymbol{x}(kT_s) + d_c u(kT_s) \tag{A.3.10}$$

or, with the formalism used for discrete systems,

$$y[k] = \boldsymbol{c}^{\mathrm{T}}\boldsymbol{x}[k] + d_{\mathrm{c}}u[k] \tag{A.3.11}$$

Thus, the DT state model

$$\boldsymbol{x}[k+1] = \boldsymbol{F}\boldsymbol{x}[k] + \boldsymbol{g}u[k] \tag{A.3.12}$$

is determined by the state difference equation and the output Equation (A.3.11) for the values $k = 0,1,2,\ldots$ Comparing the obtained model with the CT state model, we can state that the parameter matrices of the state difference equation must be derived from the parameter matrices of the CT system according to Equation (A.3.9), whereas the parameters ($\boldsymbol{c}^{\mathrm{T}}$ and d_{c}) of the output equation are the same in both CT and DT systems. Matrix $\boldsymbol{F}$ is called a state-transition matrix, since in the case of zero input the state transition of the system is guided by the elements of matrix $\boldsymbol{F}$ between the kth and $(k+1)$ steps

$$\boldsymbol{x}[k+1] = \boldsymbol{F}\boldsymbol{x}[k] \tag{A.3.13}$$

Input–Output Models Based on the Shift-Operator

It was seen in the discussion of CT systems, that several so-called input–output equivalent state models can belong to an input–output model given by a differential equation or transfer function, but only one input–output model can be derived from a state model. This feature is also valid for DT systems. To demonstrate this, first introduce a *shift-operator*

$$q\boldsymbol{x}[k] = \boldsymbol{x}[k+1] \quad (k = \ldots, -1, 0, 1, \ldots) \tag{A.3.14}$$

The operator q is a shift-operator, whose argument is a DT signal series of scalar or vector value. Application of the operator q for a signal series results in a one-step-ahead-shifted signal series. By the repeated application of the shift-operator, a signal series can be shifted ahead by more steps in the direction of positive time points, so for m step it can be written that

$$q^{m}\boldsymbol{x}[k] = \boldsymbol{x}[k+m] \quad (k = \ldots, -1, 0, 1, \ldots) \tag{A.3.15}$$

Delay can be performed by the inverse of the operator q

$$q^{-1}\boldsymbol{x}[k] = \boldsymbol{x}[k-1] \quad (k = \ldots, -1, 0, 1, \ldots) \tag{A.3.16}$$

or

$$q^{-m}\boldsymbol{x}[k] = \boldsymbol{x}[k-m] \quad (k = \ldots, -1, 0, 1, \ldots) \tag{A.3.17}$$

The operator q^{-1} is called a shift-operator. Applying the operator q, we can write the discretized state model as

$$\begin{aligned} \boldsymbol{x}[k+1] &= q\boldsymbol{x}[k] = \boldsymbol{F}\boldsymbol{x}[k] + \boldsymbol{g}u[k] \\ y[k] &= \boldsymbol{c}^{\mathrm{T}}\boldsymbol{x}[k] + d_{\mathrm{c}}u[k] \end{aligned} \tag{A.3.18}$$

Expressing $\boldsymbol{x}[k]$ from the state difference equation we obtain

$$\boldsymbol{x}[k] = (q\boldsymbol{I} - \boldsymbol{F})^{-1}\boldsymbol{g}u[k] \tag{A.3.19}$$

by means of which the output equation has the form

$$y[k] = \boldsymbol{c}^{\mathrm{T}}(q\boldsymbol{I} - \boldsymbol{F})^{-1}\boldsymbol{g}u[k] + d_{\mathrm{c}}u[k] = \left[\boldsymbol{c}^{\mathrm{T}}(q\boldsymbol{I} - \boldsymbol{F})^{-1}\boldsymbol{g} + d_{\mathrm{c}}\right]u[k] \tag{A.3.20}$$

Introduce the dependence of the output series from the input series according to the following expression

$$y[k] = G(q)u[k] = \left[\boldsymbol{c}^{\mathrm{T}}(q\boldsymbol{I} - \boldsymbol{F})^{-1}\boldsymbol{g} + d_{\mathrm{c}}\right]u[k] \tag{A.3.21}$$

Here the rational fractional function $G(q)$ is called the *pulse transfer operator*. $G(q)$ is a proper rational fractional function, since

$$G(q) = \boldsymbol{c}^{\mathrm{T}}(q\boldsymbol{I} - \boldsymbol{F})^{-1}\boldsymbol{g} + d_{\mathrm{c}} = \boldsymbol{c}^{\mathrm{T}}\frac{\mathbf{adj}(q\boldsymbol{I} - \boldsymbol{F})}{\det(q\boldsymbol{I} - \boldsymbol{F})}\boldsymbol{g} + d_{\mathrm{c}} = \frac{\mathcal{G}(q)}{\mathcal{F}(q)} \tag{A.3.22}$$

where both $\det(q\boldsymbol{I} - \boldsymbol{F})$ and $\boldsymbol{c}^{\mathrm{T}}\mathbf{adj}(q\boldsymbol{I} - \boldsymbol{F})\boldsymbol{g}$ are polynomials of q with real coefficients, so $G(q)$ can be written as the quotient of two polynomials

$$\begin{aligned} \mathcal{F}(q) &= \det(q\boldsymbol{I} - \boldsymbol{F}) = q^n + f_1q^{n-1} + \ldots + f_n \\ \mathcal{G}(q) &= g_0q^{n_{\mathrm{G}}} + g_1q^{n_{\mathrm{G}}-1} + \ldots + g_{n_{\mathrm{G}}} \end{aligned} \tag{A.3.23}$$

Considering the order of the introduced polynomials $\mathcal{F}(q)$ and $\mathcal{G}(q)$, we see that n is the number of state variables, n_{G} is the order of the polynomial $\mathcal{G}(q)$, and the relation $n_{\mathrm{G}} \leq n$ exists. If $d_{\mathrm{c}} = 0$, which is true in the majority of cases, Equation (A.3.22) is a strictly proper rational fractional function, so it is reasonable to use the notations

$$\begin{aligned} \mathcal{F}(q) &= \det(q\boldsymbol{I} - \boldsymbol{F}) = q^n + f_1q^{n-1} + \ldots + f_n \\ &= q^n\left(1 + f_1q^{-1} + \ldots + f_nq^{-n}\right) = q^n\mathcal{F}\left(q^{-1}\right) \\ \mathcal{G}(q) &= g_1q^{n-1} + g_2q^{n-2} + \ldots + g_n \\ &= q^n\left(g_1q^{-1} + g_2q^{-2} + \ldots + g_nq^{-n}\right) = q^n\mathcal{G}\left(q^{-1}\right) \end{aligned} \tag{A.3.24}$$

instead of Equation (A.3.23). Similarly to the terminology introduced for CT systems, here the roots of the equation $\mathcal{F}(q) = 0$ are called poles and the roots of the equation $\mathcal{G}(q) = 0$ are called zeros.

Based on the introduced polynomials and the direct time-domain meaning of the shift-operator, a further model, i.e., the difference equation, of the DT systems can be simply derived. Because of the above we can write

$$y[k] = G(q)u[k] = \frac{\mathcal{G}(q)}{\mathcal{F}(q)} u[k] \tag{A.3.25}$$

whence

$$\mathcal{F}(q)y[k] = \mathcal{G}(q)u[k] \tag{A.3.26}$$

In details

$$\left(q^n + f_1 q^{n-1} + \ldots + f_n\right) y[k] = \left(g_1 q^{n-1} + g_2 q^{n-2} + \ldots + g_n\right) u[k] \tag{A.3.27}$$

we obtain the input–output difference equation as

$$\begin{aligned} &y[k+n] + f_1 y[k+n-1] + \ldots + f_n y[k] \\ &\quad = g_1 u[k+n-1] + g_1 u[k+n-2] + \ldots + g_n u[k] \end{aligned}$$

Dividing both sides of Equation (A.3.27) by q^n we obtain the operator equation

$$\left(1 + f_1 q^{-1} + \ldots + f_n q^{-n}\right) y[k] = \left(g_1 q^{-1} + g_2 q^{-2} + \ldots + g_n q^{-n}\right) u[k] \tag{A.3.28}$$

which corresponds to the input–output difference equation

$$\begin{aligned} y[k] = &\, g_1 u[k-1] + g_2 u[k-2] + \ldots + g_n u[k-n] - f_1 y[k-1] - \ldots \\ &- f_n y[k-n] \end{aligned} \tag{A.3.29}$$

This relationship offers a further important interpretation of Equation (A.3.25), since this equation is a recursive formula, which produces the DT output. It can be seen that the recursive form makes direct use of the polynomials of Equation (A.3.24) ordered by q^{-1}. Therefore the equivalent form of $G(q)$

$$G\left(q^{-1}\right) = \frac{\mathcal{G}(q^{-1})}{\mathcal{F}(q^{-1})} = \frac{\mathcal{G}(q^{-1})}{1 + \tilde{\mathcal{F}}(q^{-1})} \tag{A.3.30}$$

is also called the *filter form*. Based on Equation (A.3.30), the recursive Equation (A.3.29) can always be written very simply as

$$\begin{aligned} y[k] &= \mathcal{G}(q^{-1})u[k] - \tilde{\mathcal{F}}(q^{-1})y[k] \\ &= g_1 u[k-1] + g_2 u[k-2] + \ldots + g_n u[k-n] \\ &\quad - f_1 y[k-1] - \ldots - f_n y[k-n] \end{aligned} \tag{A.3.31}$$

which is linear in parameters $\{f_i; g_i\}$ and can be realized easily in computer-aided environment.

Input–Output Models Based on z-Transformation

The pulse transfer function representing the discretized system is (under zero initial conditions)

$$G(z) = \frac{\mathcal{Z}\{y[k]\}}{\mathcal{Z}\{u[k]\}} \tag{A.3.32}$$

which is the quotient of the z-transform of the input and output signals. The sampled values of the output for any input can be calculated as

$$y[k] = \mathcal{Z}\left\{ y(t)|_{t=kT_s} \right\} = \mathcal{Z}\left\{ \mathcal{L}^{-1}[Y(s)]_{t=kT_s} \right\} = \mathcal{Z}\left\{ \mathcal{L}^{-1}[U(s)H(s)]_{t=kT_s} \right\} \tag{A.3.33}$$

It can be seen that the expression depends on the input, in other words, instead of a general relationship, we can only obtain a DT model equivalent to certain kinds of input classes. Consider the *step response equivalent* (*SRE*) DT *model*. Here $u[k]$ is the DT and $u(t)$ is the CT unit step input signal, and thus

$$y[k] = \mathcal{Z}\left\{ \mathcal{L}^{-1}\left[\frac{H(s)}{s}\right]_{t=kT_s} \right\} \tag{A.3.34}$$

i.e., it is necessary to have the same values in the sampling times for both CT and DT model under step signal excitation. Notice that the inverse Laplace transformation provides the transfer function (the step response) of the CT process as

$$\mathcal{L}^{-1}\left\{\frac{H(s)}{s}\right\} = v(t) \tag{A.3.35}$$

whose sampled values are

$$v[k] = \mathcal{L}^{-1}\left[\frac{H(s)}{s}\right]_{t=kT_s} \tag{A.3.36}$$

so the DT model becomes

$$G(z) = \frac{\mathcal{Z}\{v[k]\}}{\mathcal{Z}\{u[k]\}} = \frac{\mathcal{Z}\{v[k]\}}{\dfrac{z}{z-1}} = \frac{z-1}{z}\mathcal{Z}\{v[k]\} = \left(1 - z^{-1}\right)\mathcal{Z}\{v[k]\} \tag{A.3.37}$$

Note that the above important relationship often found in the non-precise form

$$G(z) = \left(1 - z^{-1}\right)\mathcal{Z}\left\{\frac{H(s)}{s}\right\} \tag{A.3.38}$$

in many publications. Though the expression is compact and formally expressive, it is not correct and cannot be interpreted mathematically, since a Laplace transform cannot have a direct z-transform. The correct procedure is to go back to the CT domain via the Laplace transformation, after which the obtained CT function needs to be sampled by substitution $t = kT_s$. The *SRE* transformation, in principle, is equal to the case when a *ZOH* is applied in the closed loop. This means a step excitation between the sampling times.

Observe the formal identity of the results obtained by the pulse transfer operator $G(q)$ and the z-transformation technique. Since in the z-transformation expressions the multiplication by z and z^{-1} mean shifting ahead and delaying, respectively, it is generally true that when we determine the pulse transfer function $G(z)$ the pulse transfer operator $G(q)$ can be obtained by the substitution $z = q$. In spite of this formal identity, however, it needs to be taken into account that the shift-operator and the variable of the z-transformation are different quantities!

Repeating the deduction shown at the discussion on the shift-operator the pulse transfer function $G(z)$ can also be written in the form $G(z) = \frac{\mathcal{G}(z)}{\mathcal{F}(z)} = \boldsymbol{c}^{\mathrm{T}} \frac{\mathbf{adj}(z\boldsymbol{I}-\boldsymbol{F})}{\det(z\boldsymbol{I}-\boldsymbol{F})}\boldsymbol{g} + d_c$, where the *characteristic equation* of the discretized system is $\det(z\boldsymbol{I} - \boldsymbol{F}) = 0$, and the condition $|z_i| < 1 \quad \mathrm{i} = 1, 2, .., \mathrm{n}$, relating to the eigenvalues, is the condition of the stability of the discretized system.

In discrete systems, it is quite usual to use the notation z^{-1} and q^{-1}, instead of z and q, in the argument of the pulse transfer function $G(z)$ and pulse transfer operator $G(q)$, respectively. The filter forms $G(z^{-1})$ and $G(q^{-1})$ can be obtained by dividing $G(z)$ and $G(q)$ with the higher-order term in z and q of their denominator. These forms are

$$G(z^{-1}) = \frac{\mathcal{G}(z^{-1})}{\mathcal{F}(z^{-1})} = \frac{g_1 z^{-1} + g_2 z^{-2} + \ldots + g_n z^{-n}}{1 + f_1 z^{-1} + f_2 z^{-2} + \ldots + f_n z^{-n}} = \frac{\left(g_1 + g_2 z^{-1} + \ldots + g_n z^{-(n-1)}\right) z^{-1}}{1 + f_1 z^{-1} + f_2 z^{-2} + \ldots + f_n z^{-n}} \tag{A.3.39}$$

and

$$G(q^{-1}) = \frac{\mathcal{G}(q^{-1})}{\mathcal{F}(q^{-1})} = \frac{g_1 q^{-1} + g_2 q^{-2} + \ldots + g_n q^{-n}}{1 + f_1 q^{-1} + f_2 q^{-2} + \ldots + f_n q^{-n}} = \frac{\left(g_1 + g_2 q^{-1} + \ldots + g_n q^{-(n-1)}\right) q^{-1}}{1 + f_1 q^{-1} + f_2 q^{-2} + \ldots + f_n q^{-n}} \tag{A.3.40}$$

Observe that if the transfer function of the CT system is proper, then the order of $G(z)$ or $G(q)$ is lower by one than the order of the denominator, and thus the pole access is one, independent of how the dead-time e^{-sT_d} of the CT process is represented

$$z^{-d}; \quad q^{-d} \tag{A.3.41}$$

where

$$d = \text{int}\left\{\frac{T_d}{T_s}\right\} \tag{A.3.42}$$

Here the notation int{...} means the integer value. z^{-d} corresponds to the general shifting expression (A.3.17).

The z^{-1} and q^{-1} in the numerator of $G(z^{-1})$ and $G(q^{-1})$, shown above, do not belong to the factors z^{-d} and q^{-d} representing the time-delays. Assume the following pulse transfer function

$$G(z^{-1}) = \frac{g_1 + g_2 z^{-1} + \ldots + g_n z^{-(n-1)}}{1 + f_1 z^{-1} + f_2 z^{-2} + \ldots + f_n z^{-n}} z^{-(d+1)} \tag{A.3.43}$$

for a real process with time-delay, which is the *SRE* transformation of the transfer function of the general CT system

$$H(s) = k\frac{\prod_{i=1}^{m}\left(s - z_i^{\mathrm{c}}\right)}{\prod_{i=1}^{n}\left(s - p_i^{\mathrm{c}}\right)}e^{-sT_{\mathrm{d}}}; \quad m < n \tag{A.3.44}$$

In the DT and CT state-space form $d = 0$.

SRE Transformation of First-Order Lag

Introduce the analytical relationship $H(s) = K/(1 + sT)$ for the *SRE* discrete model of the first-order lag. Using the simplified notation $G(z) = (1 - z^{-1})\mathcal{Z}\{H(s)/s\}$ we obtain

$$\begin{aligned}
G(z) &= (1 - z^{-1})\mathcal{Z}\left\{\frac{H(s)}{s}\right\} = (1 - z^{-1})\mathcal{Z}\left\{\frac{K}{s(1 + sT)}\right\} \\
&= (1 - z^{-1})\mathcal{Z}\left\{\frac{K}{s} - \frac{KT}{1 + sT}\right\} = \left(1 - z^{-1}\right)\frac{Kz}{z - 1} - \left(1 - z^{-1}\right)\frac{Kz}{z - e^{-T_s/T}} \\
&= K\left(1 - \frac{z - 1}{z - e^{-T_s/T}}\right) = K\frac{1 - e^{-T_s/T}}{z - e^{-T_s/T}} = \frac{K\left(1 - e^{-T_s/T}\right)z^{-1}}{1 - e^{-T_s/T}z^{-1}} = \frac{g_1 z^{-1}}{1 + f_1 z^{-1}}
\end{aligned} \tag{A.3.45}$$

Investigate how the coefficients g_1 and f_1 in the numerator and denominator of $G(z)$ depend on T and T_s, respectively. Introduce the relative sampling time $x = T_s/T$

$$\begin{aligned}
g_1 &= K\left(1 - e^{-x}\right) \\
f_1 &= -e^{-x}
\end{aligned} \tag{A.3.46}$$

The static gain can be easily checked: $G(z = 1) = K$. The pole of $G(z)$ is

$$p_1^{\mathrm{d}} = -f_1 = e^{-x}$$

Bilinear Transformation

The concept of bilinear transformation is very simple since the integrators of the CT system are substituted by DT integrators formed according to the trapezoid rule. This means the substitution

$$\frac{1}{s} = \frac{T_s}{2}\frac{1 + z^{-1}}{1 - z^{-1}} \tag{A.3.47}$$

in the transfer function of the CT system. The inverse z-transformation means the substitution

$$z^{-1} = \frac{1 - sT_s/2}{1 + sT_s/2} \tag{A.3.48}$$

in the pulse transfer function. The only advantage of bilinear transformation is that it does not require complicated calculations so it can be applied without hardware arithmetic in physical systems. Though its frequency characteristic suits the transfer function for delay-free systems, it does not include delaying because of the *ZOH*, and it gives noninterpretable distortion (proportional channel) on very high frequency. Vajk recommended the modification of the original bilinear transformation to apply an integrator as the first serial integrator at the input of the system which uses a rectangular approach instead of the trapezoid one [152]

$$\frac{1}{s} = T_s \frac{z^{-1}}{1 - z^{-1}} \tag{A.3.49}$$

Thus the formal summary of the modified bilinear transformation is

$$G(z^{-1}) = \frac{2z^{-1}}{1 - z^{-1}} H(s)\Big|_{s=\frac{2}{T_s}\frac{1-z^{-1}}{1+z^{-1}}} \tag{A.3.50}$$

and

$$H(s) = \frac{1 + z^{-1}}{2z^{-1}} G(z^{-1})\Big|_{z^{-1}=\frac{1-sT_s/2}{1+sT_s/2}} \tag{A.3.51}$$

Delta Transformation

Delta transformation [52] is based on the application of numerical integration by the left-side rectangular sums. The delta operator is the differential operator

$$\delta = \frac{q - 1}{T_s} \approx s \tag{A.3.52}$$

obtained from the integration by the rectangular rule. Applying it to the DT signal series $f[kT_s]$ we obtain

$$\delta f[k] = \delta f[kT_s] = \frac{f(kT_s + T_s) - f(kT_s)}{T_s} = \frac{f[k+1] - f[k]}{T_s} \tag{A.3.53}$$

The transfer function of a system represented by the delta operator (or delta transform) can be formally derived both from the CT and the DT forms. This seems to be obvious since the substitutions shown in Equation (A.3.52) and $q = \delta T_s + 1$ can be made, but it needs to be noted that one of the substitutions results in an approximate solution and the other in an exact solution.

Starting with the DT form and substitution $q = \delta T_s + 1$, we also obtain a rational fraction function for the delta operator form

$$G(q) = \frac{\mathcal{B}(q)}{\mathcal{A}(q)} = \frac{\mathcal{B}(\delta T_s + 1)}{\mathcal{A}(\delta T_s + 1)} = \frac{\overline{\mathcal{B}}(\delta)}{\overline{\mathcal{A}}(\delta)} = \overline{G}(\delta) \tag{A.3.54}$$

The delta transform form can be obtained from the CT transfer function $H(s)$, where $G(\delta) = H(s)|_{s=\delta}$, and thus it keeps the structure (e.g., the pole access) of $H(s)$.

The theoretical importance of the delta transform is that it provides a direct relation between the CT and DT systems. It is true for a CT system given by $H(s)$ that

$$\lim_{T_s \to 0} G(\delta) = H(s) \tag{A.3.55}$$

Lately delta transformation has become the preferred method in practical applications. It can be explained by the relationship (A.3.55), since, for small sampling times, the poles and zeros of the delta transfer function and the CT system are very close to each other. Returning to *SRE* form and for the transfer function

$$P(s) = \frac{1}{(1 + 5s)(1 + 10s)} \tag{A.3.56}$$

we obtain the following discretized model under $T_s = 1$ [s]

$$P_d(z) = \frac{0.0091(z + 0.9048)}{(z - 0.9048)(z - 0.8187)} \tag{A.3.57}$$

where the information regarding the poles, because of the mapping $z = e^{sT_s}$, appears numerically very unfavorable, i.e., just slightly different from one. The situation is deteriorated by a decrease in the sampling time, e.g., in the case $T_s = 0.1$ [s]

$$P_d(z) = \frac{9.9006(z + 0.99)10^{-5}}{(z - 0.9900)(z - 0.9802)} \tag{A.3.58}$$

The development of the hardware platforms for digital regulators makes lower and lower sampling times possible, and therefore, the application of delta transforms is at the forefront of application in discretization. Since in the case of fast sampling delta transformation is quite accurate, it is possible to avoid the use of bilinear transformation and the problem of frequency distortion.

State-Space Canonical Forms of Sampled Data Systems

Given the controllable canonical form (A.2.37) of the CT systems, the controllable canonical form of the DT systems can be given by the following system matrices

$$\boldsymbol{F}_{\mathrm{c}} = \begin{bmatrix} -f_1 & -f_2 & \dots & -f_{n-1} & -f_n \\ 1 & 0 & \dots & 0 & 0 \\ 0 & 1 & \dots & 0 & 0 \\ \vdots & \vdots & \ddots & \vdots & \vdots \\ 0 & 0 & \dots & 1 & 0 \end{bmatrix}; \quad \boldsymbol{g}_{\mathrm{c}} = \begin{bmatrix} 1 \\ 0 \\ 0 \\ \vdots \\ 0 \end{bmatrix}; \tag{A.3.59}$$

$$\boldsymbol{c}_{\mathrm{c}}^{\mathrm{T}} = \begin{bmatrix} g_1 & g_2 & \dots & g_{n-1} & g_n \end{bmatrix}$$

The system matrices $\overline{\boldsymbol{F}}_{\mathrm{c}}$, $\overline{\boldsymbol{g}}_{\mathrm{c}}$, and $\overline{\boldsymbol{c}}_{\mathrm{c}}^{\mathrm{T}}$ of the structure (A.2.41) can be similarly derived. The controllable canonical forms are always controllable; observability, however, needs to be investigated.

Given the observability canonical form (A.2.44) of CT systems, the observability canonical form of DT systems can be given by the following system matrices

$$\boldsymbol{F}_{\mathrm{o}} = \begin{bmatrix} -f_1 & 1 & 0 & \dots & 0 \\ -f_2 & 0 & 1 & \dots & 0 \\ \vdots & \vdots & \vdots & \ddots & \vdots \\ -f_{n-1} & 0 & 0 & \dots & 1 \\ -f_n & 0 & 0 & \dots & 0 \end{bmatrix}; \quad \boldsymbol{g}_{\mathrm{o}} = \begin{bmatrix} g_1 \\ g_2 \\ \vdots \\ g_{n-1} \\ g_n \end{bmatrix}; \tag{A.3.60}$$

$$\boldsymbol{c}_{\mathrm{o}}^{\mathrm{T}} = \begin{bmatrix} 1 & 0 & \dots & 0 & 0 \end{bmatrix}$$

The system matrices $\overline{\boldsymbol{F}}_{\mathrm{o}}$, $\overline{\boldsymbol{g}}_{\mathrm{o}}$, and $\overline{\boldsymbol{c}}_{\mathrm{o}}^{\mathrm{T}}$ of the structure (A.2.45) can be similarly derived. The observable canonical forms are always observable; controllability, however, needs to be investigated.

The DT version of the CT diagonal form (A.2.28) can be derived by the transformation $\boldsymbol{T}_{\mathrm{d}}\boldsymbol{F}(\boldsymbol{T}_{\mathrm{d}})^{-1}$.

APPENDIX 4

Optimization Problems

A.4.1 THE WIENER PROBLEM

Wiener sought a method to reconstruct the original signal from its noisy observation. The block scheme of this problem is shown in Figure A.4.1, where y_x is the unknown ideal signal, and it can be measured as signal y with noise y_n. Our goal is to construct signal $\hat{y}$ from the signal measured by the filter H_W which supremely approaches the ideal signal, i.e., the error ε_x must be the smallest possible.

The value of the error is measured by its mean square (*mean square error, MSE*), i.e.,

$$J^2 = \bar{\varepsilon_x^2} = MSE = \lim_{T\to\infty} \frac{1}{2T} \int_{-\infty}^{\infty} \varepsilon_x^2(t)\mathrm{d}t = \frac{1}{2\pi} \int_{-\infty}^{\infty} \Phi_{\varepsilon\varepsilon}(j\omega)\mathrm{d}\omega$$

$$= \frac{1}{2\pi j} \int_{-j\infty}^{j\infty} \bar{\Phi}_{\varepsilon\varepsilon}(s)\mathrm{d}s \qquad \text{(A.4.1)}$$

The function J^2 given in the time domain can also be expressed by the power density spectrum $\Phi_{\varepsilon\varepsilon}(j\omega) = \bar{\Phi}_{\varepsilon\varepsilon}(s)\big|_{s=j\omega}$ of the error signal $\varepsilon_x = y_x - \hat{y}$ (see Parseval theorem). Here $\bar{\Phi}_{\varepsilon\varepsilon}(s)$ is the two-sided Laplace transform of $\varepsilon_x(t)$. In the complex operator domain

$$\varepsilon_x(s) = Y_x(s) - H_W(s)Y(s) \qquad \text{(A.4.2)}$$

The autocorrelation function of the error signal is

$$\varphi_{\varepsilon\varepsilon}(\tau) = \overline{\varepsilon_x(t)\varepsilon_x(t+\tau)} = \varphi_{y_x y_x}(\tau) - \varphi_{y_x\hat{y}}(\tau) - \varphi_{\hat{y}y_x}(\tau) + \varphi_{\hat{y}\hat{y}}(\tau) \qquad \text{(A.4.3)}$$

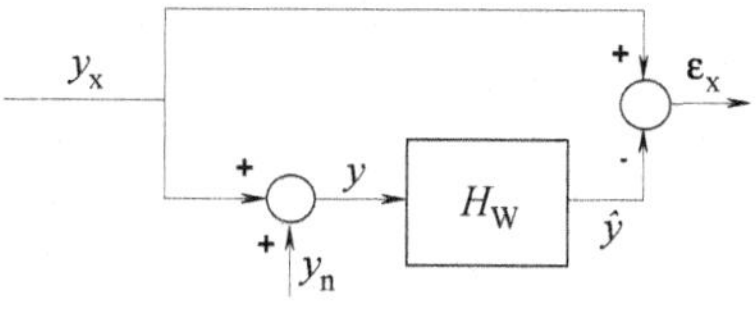

Figure A.4.1 Block scheme of the Wiener problem.

The power density spectrum can be simply calculated from the above as

$$\bar{\Phi}_{\varepsilon\varepsilon}(s) = \bar{\Phi}_{y_x y_x}(s) - H_W(s)\bar{\Phi}_{y_x y}(s) - H_W(-s)\bar{\Phi}_{y y_x}(s) + \bar{\Phi}_{yy}(s) \qquad (A.4.4)$$

where the filter equation $\hat{Y}(s) = H_W(s)Y(s)$ and the well-known index-changing rules are used.

The so-called Wiener filter minimizes the *MSE* (A.4.1), i.e.,

$$H_W = \arg\left\{ \min_{H_W} \left[J^2 \right] \right\} = \arg\left\{ J^2_{\min} \right\} \qquad (A.4.5)$$

The minimum seeking task is solved by variational computation to find a stable, realizable filter

$$H(s) = H_W(s) + \delta\, \Delta H(s); \quad J^2 = J^2_{\min} + \Delta J^2 \qquad (A.4.6)$$

Here the minimum of (A.4.5) is ensured by H_W, and δ is a positive real variable and increases the value of $J^2_{\min}$ by ΔJ^2. Because of the stability of $H(s)$, all poles of $\Delta H(s)$ must be on the left side plane. The necessary condition of the extremum is that the first derivate of ΔJ^2 by δ must be zero at $\delta = 0$, i.e.,

$$\frac{dJ}{d\delta} = \frac{d}{d\delta}\left(J^2_{\min} + \Delta J^2\right) = \left.\frac{d\Delta J^2}{d\delta}\right|_{\delta=0} = 0 \qquad (A.4.7)$$

since $J^2_{\min}$ does not depend on δ. Introduce the following simple notations: $H_W^- = H_W(-s)$, $H_W^+ = H_W(s)$ and $\Delta^+ = \Delta H(s)$, $\Delta^- = \Delta H(-s)$. Combining Equations (A.4.1), (A.4.4), and (A.4.6) we obtain

$$J^2_{\min} = \frac{1}{2\pi j} \int_{-j\infty}^{j\infty} \left[\bar{\Phi}_{y_x y_x}(s) - H_W^+ \bar{\Phi}_{y_x y}(s) - H_W^- \bar{\Phi}_{y y_x}(s) + H_W^- H_W^+ \bar{\Phi}_{yy}(s) \right] ds \qquad (A.4.8)$$

and

$$\Delta J^2 = \frac{1}{2\pi j} \int_{-j\infty}^{j\infty} \left[-\delta\Delta^+ \bar{\Phi}_{y_x y} - \delta\Delta^- \bar{\Phi}_{y y_x} + \delta\Delta^- H_W^+ \bar{\Phi}_{yy} + \delta\Delta^+ H_W^- \bar{\Phi}_{yy} + \delta^2 H_W^- H_W^+ \bar{\Phi}_{yy} \right] ds \qquad (A.4.9)$$

From here the condition of the extremum is

$$\left.\frac{d\Delta J}{d\delta}\right|_{\delta=0} = \frac{1}{2\pi j}\int_{-j\infty}^{j\infty}\left[-\Delta^{+}\bar{\Phi}_{y_{x}y} - \Delta^{-}\bar{\Phi}_{yy_{x}} + \Delta^{-}H_{W}^{+}\bar{\Phi}_{yy} + \Delta^{+}H_{W}^{-}\bar{\Phi}_{yy}\right]ds$$
$$= J_1 + J_2 = 0 \tag{A.4.10}$$

The extremum is minimum, since the second derivate is

$$\left.\frac{d^2\Delta J^2}{d\delta^2}\right|_{\delta=0} = \frac{1}{\pi j}\int_{-j\infty}^{j\infty}\left[H_{W}^{-}H_{W}^{+}\bar{\Phi}_{yy}\right]ds = \frac{1}{\pi j}\int_{-j\infty}^{j\infty}|H_{W}(j\omega)|^2\Phi_{yy}(j\omega)ds$$
$$\geq 0 \tag{A.4.11}$$

positive. Equation (A.4.9) is the sum of two integrals, where

$$J_1^2 = \frac{1}{2\pi j}\int_{-j\infty}^{j\infty}\Delta H^{-}\left[H_{W}^{+}\bar{\Phi}_{yy} - \bar{\Phi}_{yy_{x}}\right]ds \quad \text{and}$$
$$J_2^2 = \frac{1}{2\pi j}\int_{-j\infty}^{j\infty}\Delta H^{+}\left[H_{W}^{-}\bar{\Phi}_{yy} - \bar{\Phi}_{y_{x}y}\right]ds \tag{A.4.12}$$

Apply the index-changing rules and substitute $s = -s'$ in J_2^2, and consider $\bar{\Phi}_{yy}(-s) = \bar{\Phi}_{yy}(s)$, $\bar{\Phi}_{y_{x}y}(-s) = \bar{\Phi}_{yy_{x}}(s)$, so

$$J_2^2 = \frac{1}{2\pi j}\int_{-j\infty}^{j\infty}\Delta H(-s')\left[H_W(s')\bar{\Phi}_{yy}(s') - \bar{\Phi}_{y_{x}y}(s')\right]ds = J_1 \tag{A.4.13}$$

Thus the condition of the extremum is simplified for the equation

$$J_1^2 = \frac{1}{2\pi j}\int_{-j\infty}^{j\infty}\Delta H(-s)\left[H_W(s)\bar{\Phi}_{yy}(s) - \bar{\Phi}_{yy_{x}}(s)\right]ds = 0 \tag{A.4.14}$$

The integral J_1^2 is evaluated by the residuum theorem. To this end, the imaginary axis must be completed as a closed curve by a half curve in the left side half plane. The method is the same as in the case of the inverse Laplace transformation. Decompose the function $\bar{\Phi}_{yy}(s)$ as a product

$$\bar{\Phi}_{yy}(s) = \bar{\Phi}^{+}_{yy}(s)\bar{\Phi}^{-}_{yy}(s) \tag{A.4.15}$$

where all poles and zeros of $\bar{\Phi}^{+}_{yy}(s)$ have negative real values, so all poles and zeros of $\bar{\Phi}^{-}_{yy}(s)$ have positive real values. The superscripts refer to the fact that, after the inverse transformation, the value of the autocorrelation function $\varphi^{+}_{yy}(\tau)$, corresponding to $\bar{\Phi}^{+}_{yy}(s)$, can be different from zero only for positive τ, i.e., $\varphi^{+}_{yy}(\tau) = 0$, if $\tau < 0$ and $\varphi^{-}_{yy}(\tau) = 0$, if $\tau > 0$. In the time domain (now in the shift domain) the decomposition of (A.4.14) into a product means that $\varphi_{yy}(\tau)$ can be derived as the convolution of $\varphi^{+}_{yy}(\tau)$ and $\varphi^{-}_{yy}(\tau)$. Since $\bar{\Phi}_{yy}(s)$ is even/conjugated, the following identity exists

$$\bar{\Phi}^{-}_{yy}(s) = \bar{\Phi}^{+}_{yy}(-s) \quad \text{or} \quad \bar{\Phi}^{+}_{yy}(s) = \bar{\Phi}^{-}_{yy}(-s) \tag{A.4.16}$$

and using it in (A.4.13) we obtain

$$J_1^2 = \frac{1}{2\pi j}\int_{-j\infty}^{j\infty} \Delta(-s)\bar{\Phi}^{-}_{yy}(s)\left[H_W(s)\bar{\Phi}^{+}_{yy}(s) - \frac{\bar{\Phi}_{yy_x}(s)}{\bar{\Phi}^{-}_{yy}(s)}\right] ds = 0 \tag{A.4.17}$$

It is evident that the factors $\Delta(-s)$ and $\bar{\Phi}^{-}_{yy}(s)$ do not have poles on the left half plane. Therefore the integral is zero if the expression in the bracket does not have a pole on the left half plane, i.e., if

$$\left[H_W(s)\bar{\Phi}^{+}_{yy}(s) - \frac{\bar{\Phi}_{yy_x}(s)}{\bar{\Phi}^{-}_{yy}(s)}\right]_+ = H_W(s)\bar{\Phi}^{+}_{yy}(s) - \left[\frac{\bar{\Phi}_{yy_x}(s)}{\bar{\Phi}^{-}_{yy}(s)}\right]_+ = 0 \tag{A.4.18}$$

Here the first term has poles only on the left half plane, so $[\ldots]_+$ is identical to itself. Finally the optimal and realizable Wiener filter

$$H_W(s) = \frac{1}{\bar{\Phi}^{+}_{yy}(s)}\left[\frac{\bar{\Phi}_{yy_x}(s)}{\bar{\Phi}^{-}_{yy}(s)}\right]_+ \tag{A.4.19}$$

can be calculated in closed form. The solution does not depend on the arbitrarily chosen error $\Delta H(s)$. The operation $[\ldots]_+$ is defined by the well-known decomposition of the two-sided Laplace transform ($\bar{\mathcal{L}}$)

$$\begin{aligned}\bar{F}(s) &= \bar{\mathcal{L}}\{f\} = F_+(s) + F_-(s) = \mathcal{L}\{f_+\} + \mathcal{L}\{f_-\};\\ f(t) &= f_+\{t\} + f_-\{t\}\end{aligned} \tag{A.4.20}$$

where $f_+\{t\}$ and $f_-\{t\}$ are positive and negative time functions, respectively. The two-sided Laplace transform is now a power density spectrum $\bar{\Phi}(s)$, and therefore the decomposition (A.4.18) becomes

$$\begin{aligned}\bar{\Phi}(s) &= \bar{\mathcal{L}}\{\varphi\} = \Phi_+(s) + \Phi_-(s) = \mathcal{L}\{\varphi_+\} + \mathcal{L}\{\varphi_-\};\\ \varphi(\tau) &= \varphi_+\{\tau\} + \varphi_-\{\tau\}\end{aligned} \tag{A.4.21}$$

where $\varphi_+\{\tau\}$ and $\varphi_-\{\tau\}$ are the correlation functions for positive and negative shifts, respectively. In the decompositions for sums (A.4.20) and (A.4.21), all poles of the one-sided Laplace transforms ($\mathcal{L}$) (so $[\ldots]_+$) are on the left half plane, whereas all poles of $F_-(s)$ and $\Phi_-(s)$ (so $[\ldots]_-$) are on the right half plane, but there is no restriction on the zeros. Thus the condition of the Wiener optimum (A.4.18) means that the correlation function

$$H_W(s)\bar{\Phi}^+_{yy}(s) - \frac{\bar{\Phi}_{yy_x}(s)}{\bar{\Phi}^-_{yy}(s)} \tag{A.4.22}$$

belonging to the power density spectrum can be only zero for positive shifts. This interpretation is known as the Wiener orthogonality condition. The orthogonality condition is illustrated in the signal domain in Figure A.4.2, where the signal is mapped perpendicularly to the subspace of the observations. In this case the chosen *MSE* criterion is minimum.

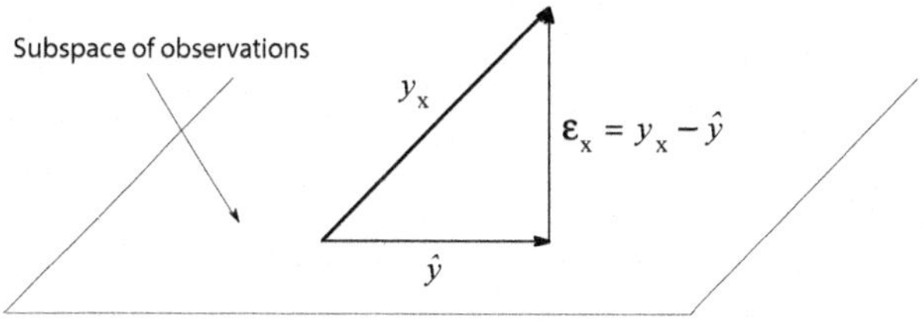

Figure A.4.2 Geometric interpretation of the orthogonality principle in the signal domain.

A.4.2 MINIMIZATION OF THE $\mathcal{H}_2$ NORM

The Wiener filter makes possible the best approximation of a prescribed ideal stationary stochastic signal from a measured stationary stochastic signal, which is the input of the filter. This filter minimizes the square mean of the error between the ideal signal and the output of the filter (see (A.4.1)). It is a typical application when the measured signal has an additional noise to the ideal signal.

Consider the block scheme in Figure A.4.3, where the excitation y_x is an impulse function (Dirac delta), whose Laplace transform is equal to one ($Y_x(s) = 1$). The task is to determine an optimal compensator K_2, which provides the best approach to the stable reference model R_x when it is serially connected to the process $P = P_+P_-$. The ideal signal is now the output of the R_x. The square of the $\mathcal{H}_2$ norm of the error $\varepsilon_x(t)$ is used as a cost function of the approach

$$J_2^2 = \int_{-\infty}^{\infty} \varepsilon_x^2(t)\mathrm{d}t = \frac{1}{2\pi j}\int_{-j\infty}^{j\infty} \varepsilon_x(s)\varepsilon_x(-s)\mathrm{d}s = \frac{1}{2\pi}\int_{-\infty}^{\infty} |\varepsilon_x(j\omega)|^2\mathrm{d}\omega$$

$$= \|\varepsilon_x(j\omega)\|_2^2 \qquad \text{(A.4.23)}$$

Thus the criterion now is the energy of the signal $\varepsilon_x(t)$ (and not the power, as in the Wiener filter). Thus the square of the $\mathcal{H}_2$ norm is the energy of the impulse function of the error. In the complex operator domain

$$\varepsilon_x(s) = R_x(s) - K_2(s)P(s) \qquad \text{(A.4.24)}$$

Thus the optimization task means the determination of the compensator

$$K_2 = \arg\left\{\min_{K_2}\left[J^2\right]\right\} = \arg\left\{J_{\min}^2\right\} \qquad \text{(A.4.25)}$$

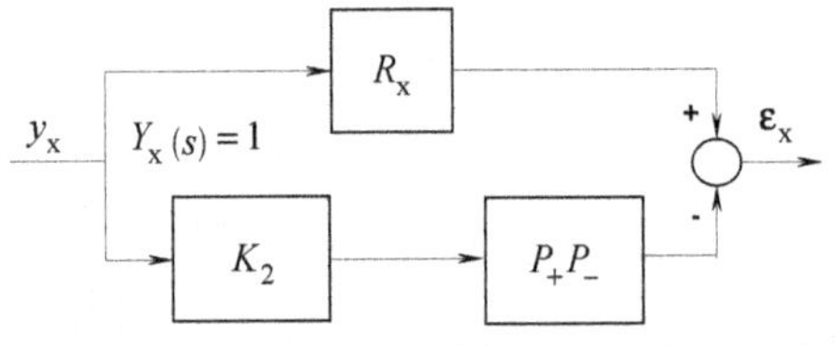

Figure A.4.3 $\mathcal{H}_2$ optimal approach of the prescribed reference model.

which minimizes the $\mathcal{H}_2$ norm (A.4.23). The task is solved by variational computation in the search for a stable, realizable compensator

$$K(s) = K_2(s) + \delta\Delta K(s); \quad J^2 = J^2_{\min} + \Delta J^2 \tag{A.4.26}$$

Here the minimum of (A.4.25) is ensured by K_2, and δ is a positive variable, by means of which the minimum cost function $J^2_{\min}$ is increased by ΔJ^2. Because of the stability requirement concerning $K(s)$, it is assumed that all poles of $\Delta K(s)$ are on the left half plane. The necessary condition of the extreme is that the first derivate of ΔJ^2 by δ must be zero at $\delta = 0$, i.e.,

$$\frac{dJ^2}{d\delta} = \frac{d}{d\delta}\left(J^2_{\min} + \Delta J^2\right) = \left.\frac{d\Delta J^2}{d\delta}\right|_{\delta=0} = 0 \tag{A.4.27}$$

since $J^2_{\min}$ does not depend on δ.

It is assumed at the decomposition $P(s) = P_+(s)P_-(s)$ of the process that all poles and zeros of $P_+(s)$ have negative real value so all poles and zeros of $P_-(s)$ have positive real value. Introduce the following simplified notations

$$\begin{aligned} &P_+(s) = P_+; \quad P_+(-s) = P_+^-; \quad P_-(s) = P_-; \quad P_-(-s) = P_-^- \\ &R(s) = R_+; \quad R(-s) = R_-; \quad K(s) = K_+; \quad K(-s) = K_- \\ &\Delta K(s) = \Delta^+; \quad \Delta K(-s) = \Delta^-; \quad K_2(s) = K_2^+; \quad K_2(-s) = K_2^- \end{aligned} \tag{A.4.28}$$

By combining Equations (A.4.23), (A.4.24), and (A.4.26), we obtain

$$\begin{aligned} J^2_{\min} &= \frac{1}{2\pi j}\int_{-j\infty}^{j\infty}\left[R_+R_- - K_2^+P_+P_-R_- - K_2^-P_+^-P_-^-R_+ \right. \\ &\qquad \left. + K_2^+K_2^-\left(P_+P_-^-\right)\left(P_-P_+^-\right)\right]ds \\ &= \frac{1}{2\pi j}\int_{-j\infty}^{j\infty}\left(R_+ - K_2^+P_+P_-\right)\left(R_- - K_2^-P_+^-P_-^-\right)ds \\ &= \|R(s) - K_2(s)P(s)\|_2^2 \end{aligned} \tag{A.4.29}$$

and

$$\Delta J^2 = \frac{1}{2\pi j}\int_{-j\infty}^{j\infty}\left[-\Delta^+ P_+ P_- R_- - \Delta^- P_+^- P_-^- R_+ + \Delta^+ K_2^-\left(P_+ P_-^-\right)\left(P_- P_+^-\right)\right.$$
$$\left. + \Delta^- K_2^+\left(P_+ P_-^-\right)\left(P_- P_+^-\right)\right]\delta$$
$$+ \left[\Delta^+\Delta^-\left(P_+ P_-^-\right)\left(P_- P_+^-\right)\delta^2\right]\mathrm{d}s \tag{A.4.30}$$

The condition of the extremum is

$$\left.\frac{\mathrm{d}\Delta J^2}{\mathrm{d}\delta}\right|_{\delta=0} = \frac{1}{2\pi j}\int_{-j\infty}^{j\infty}\left[-\Delta^+ P_+ P_- R_- - \Delta^- P_+^- P_-^- R_+ + \Delta^+ K_2^-\left(P_+ P_-^-\right)\right.$$
$$\left.\times\left(P_- P_+^-\right) + \Delta^- K_2^+\left(P_+ P_-^-\right)\left(P_- P_+^-\right)\right]\mathrm{d}s$$
$$= J_1^2 + J_2^2 = 0 \tag{A.4.31}$$

The extremum is minimum, since the second-order derivate is

$$\left.\frac{\mathrm{d}^2\Delta J}{\mathrm{d}\delta^2}\right|_{\delta=0} = \frac{1}{\pi j}\int_{-j\infty}^{j\infty}\left[\Delta K(s)\Delta K(-s)P(s)P(-s)\right]\mathrm{d}s \geq 0 \tag{A.4.32}$$

positive. Equation (A.4.31) is the sum of two integrals, where

$$J_1^2 = \frac{1}{2\pi j}\int_{-j\infty}^{j\infty} -\left(\Delta^- P_- P_+^-\right)\left[\frac{P_-^- R_+}{P_-} - K_2^+\left(P_+ P_-^-\right)\right]\mathrm{d}s$$
$$J_2^2 = \frac{1}{2\pi j}\int_{-j\infty}^{j\infty} -\left(\Delta^+ P_+ P_-^-\right)\left[\frac{P_- R_-}{P_-^-} - K_2^-\left(P_- P_+^-\right)\right]\mathrm{d}s \tag{A.4.33}$$

Apply the substitution $s = -s'$ in J_2^2, so

$$J_2^2 = \frac{1}{2\pi j}\int_{-j\infty}^{j\infty} -\left(\Delta^- P_- P_+^-\right)\left[\frac{P_-^- R_+}{P_-} - K_2^+\left(P_+ P_-^-\right)\right]\mathrm{d}s = J_1^2 \tag{A.4.34}$$

Thus the condition of the extremum becomes

$$J_1^2 = \frac{1}{2\pi j}\int_{-j\infty}^{j\infty} -\left(\Delta^- P_- P_+^-\right)\left[\frac{P_-^- R_+}{P_-} - K_2^+\left(P_+ P_-^-\right)\right] \mathrm{d}s = 0 \tag{A.4.35}$$

The integral J_1^2 is evaluated by the residual theorem, and to this the imaginary axis needs to be extended to a closed curve by a half circle on the left half plane. The method is the same as in the case of the inverse Laplace transformation. The factors $\Delta^- P_- P_+^-$ definitely do not have poles on the left half plane, i.e., if

$$\left[\frac{P_-^- R_+}{P_-} - K_2^+\left(P_+ P_-^-\right)\right]_+ = \left[\frac{P_-^- R_+}{P_-}\right]_+ - K_2^+\left(P_+ P_-^-\right) = 0 \tag{A.4.36}$$

Here the second term has poles only on the left half plane, so the term $[\ldots]_+$ is equal to itself. Finally the optimal and realizable compensator can be given by the closed formula

$$K_2^+ = K_2 = \frac{1}{P_+ P_-^-}\left[\frac{P_-^- R_+}{P_-}\right]_+ \tag{A.4.37}$$

The solution does not depend on the arbitrarily chosen error $\Delta K(s)$. The operation $[\ldots]_+$ is defined by the following well-known decomposition of the two-sided Laplace transform ($\bar{\mathcal{L}}$)

$$\begin{aligned} \bar{F}(s) &= \bar{\mathcal{L}}\{f\} = F_+(s) + F_-(s) = \mathcal{L}\{f_+\} + \mathcal{L}\{f_-\}; \\ f(t) &= f_+\{t\} + f_-\{t\} \end{aligned} \tag{A.4.38}$$

where $f_+\{t\}$ and $f_-\{t\}$ are positive and negative time functions, respectively. Because of this fact, $[\ldots]_+$ has all poles on the left half plane, and $[\ldots]_-$ has all poles on the right half plane and there are no specifications for the zeros.

Let $R_x = \mathcal{B}_x/\mathcal{A}_x$ and $P_- = \mathcal{B}_-$ be, or $P_-^- = \mathcal{B}_-^-$. The $\mathcal{H}_2$ optimal compensator is

$$K_2 = \frac{1}{P_+ \mathcal{B}_-^-}\left[\frac{\mathcal{B}_-^- \mathcal{B}_x}{\mathcal{A}_x \mathcal{B}_-}\right]_+ \tag{A.4.39}$$

To determine $[\ldots]_+$, consider the decomposition

$$\left[\frac{\mathcal{B}_-^- \mathcal{B}_x}{\mathcal{A}_x \mathcal{B}_-}\right]_+ = \left[\frac{\mathcal{R}}{\mathcal{B}_-} + \frac{\mathcal{K}}{\mathcal{A}_x}\right]_+ = \left[\frac{\mathcal{R}\mathcal{A}_x + \mathcal{K}\mathcal{B}_-}{\mathcal{A}_x \mathcal{B}_-}\right]_+ = \frac{\mathcal{K}}{\mathcal{A}_x} \tag{A.4.40}$$

where $\mathcal{R}$ and $\mathcal{K}$ are obtained from the solution of the *Diophantine equation (DE)* of special form

$$\mathcal{B}_-^- \mathcal{B}_x = \mathcal{R}\mathcal{A}_x + \mathcal{K}\mathcal{B}_- \tag{A.4.41}$$

Using the result we obtain

$$K_2 = \frac{\mathcal{K}}{P_+ \mathcal{B}_-^- \mathcal{A}_x} = \frac{\mathcal{B}_x}{\mathcal{A}_x} \frac{\mathcal{K}}{P_+ \mathcal{B}_-^- \mathcal{B}_x} = R_x \frac{\mathcal{K}}{P_+ \mathcal{B}_-^- \mathcal{B}_x} = R_x K_2' \tag{A.4.42}$$

It is evident that R_x is serially connected in the optimal compensator, and therefore the optimization of the block scheme shown in Figure A.4.4 is completely the same as that in Figure A.4.3. Remember that the optimal compensator contains the factor P_+^{-1} which cancels the invertible part of the process P_+.

Using the optimal compensator and (A.4.28), we obtain the minimum of the cost function as

$$J_{\min}^2 = \|R(s) - K_2(s)P(s)\|_2^2 = \left\| \frac{\mathcal{B}_x}{\mathcal{A}_x} - \frac{\mathcal{K}}{P_+ \mathcal{B}_-^- \mathcal{A}_x} P_+ \mathcal{B}_- \right\|_2^2$$

$$= \left\| \frac{\mathcal{B}_-^- \mathcal{B}_x - \mathcal{K}\mathcal{B}_-}{\mathcal{A}_x \mathcal{B}_-^-} \right\|_2^2 = \left\| \frac{\mathcal{R}\mathcal{A}_x}{\mathcal{A}_x \mathcal{B}_-^-} \right\|_2^2 = \left\| \frac{\mathcal{R}}{\mathcal{B}_-^-} \right\|_2^2 \tag{A.4.43}$$

In the above methodology, the general optimum is determined by variational calculus. In the literature of control engineering, it is common practice to use a simpler, more informal, procedure, the so-called orthogonality principle, which has no consistent verification. This method is discussed below. From the expression of the $\mathcal{H}_2$ norm we obtain

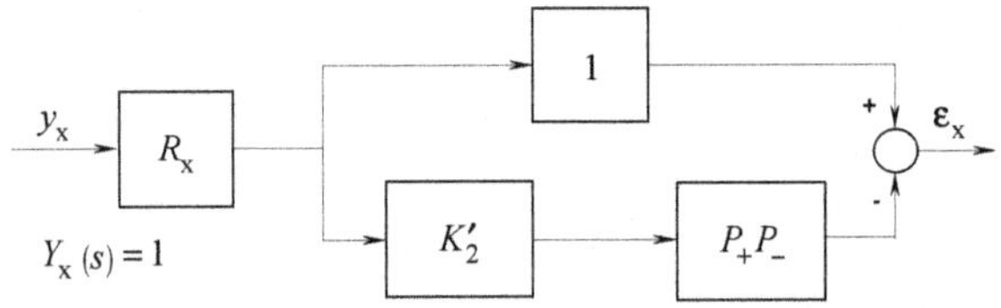

Figure A.4.4 Identical scheme for the $\mathcal{H}_2$ optimal approach.

$$J = \|R_x - K_2P_+P_-\|_2 = \left\|\frac{\mathcal{B}_x}{\mathcal{A}_x} - K_2P_+\mathcal{B}_-\right\|_2 \tag{A.4.44}$$

The expression (A.4.44) can be further modified by identical rearrangements

$$J = \|R_x - K_2P_+\mathcal{B}_-\|_2 = \left\|\frac{\mathcal{B}_-}{\mathcal{B}_-^-}\right\|_2 \left\|\frac{\mathcal{B}_-^-\mathcal{B}_x}{\mathcal{B}_-\mathcal{A}_x} - K_2P_+\mathcal{B}_-^-\right\|_2$$
$$= \left\|\frac{\mathcal{B}_-^-\mathcal{B}_x}{\mathcal{B}_-\mathcal{A}_x} - K_2P_+\mathcal{B}_-^-\right\|_2 \tag{A.4.45}$$

The invariant factor of the process is the previously introduced $P_- = \mathcal{B}_-$, where $\mathcal{B}_-(0) = 1$ and $\mathcal{B}_-(s) = \prod_{i=1}^{m_-}(1 - s/z_i)$. z_i are the zeros of $\mathcal{B}_-$ and m_- is the number of these *inverse unstable* (*IU*) zeros. $\mathcal{B}_-^-$ is a polynomial, derived by mirroring the zeros of $\mathcal{B}_-$ on the imaginary axis, thus for the obtained zeros z_i^*, it is fulfilled that $\mathrm{Re}\{z_i^*\} = -\mathrm{Re}\{z_i\}$ and $\mathrm{Im}\{z_i^*\} = \mathrm{Im}\{z_i\}$. In this way $\mathcal{B}_-^-$ becomes a stable polynomial, and the factor B_-/B_-^- is called the norm-keeping inner function (i.e., all-pass filter concerning signal forming; see (1.1.18)), and thus $\|B_-/B_-^-\|_2 = 1$. All factors of the second term have only left poles; this component is called causal. The first term is decomposed for causal and noncausal parts by partial fractional decomposition

$$\frac{\mathcal{B}_-^-\mathcal{B}_x}{\mathcal{B}_-\mathcal{A}_x} = \frac{\mathcal{R}}{\mathcal{B}_-} + \frac{\mathcal{K}}{\mathcal{A}_x} = \frac{\mathcal{R}\mathcal{A}_x + \mathcal{K}\mathcal{B}_-}{\mathcal{B}_-\mathcal{A}_x} \tag{A.4.46}$$

Here $\mathcal{R}/\mathcal{B}_-$ is the noncausal component (it has only right side poles) and $\mathcal{K}/\mathcal{A}_x$ is the causal component (it has only left side poles). Because of the orthogonality theorem, the $\mathcal{H}_2$ norm (A.4.45) gives the minimum in the form of the following inequality

$$J = \left\|\frac{\mathcal{B}_-^-\mathcal{B}_x}{\mathcal{B}_-\mathcal{A}_x} - K_2P_+\mathcal{B}_-^-\right\|_2 = \left\|\frac{\mathcal{R}}{\mathcal{B}_-} + \frac{\mathcal{K}}{\mathcal{A}_x} - K_2P_+\mathcal{B}_-^-\right\|_2 \geq \left\|\frac{\mathcal{R}}{\mathcal{B}_-}\right\|_2$$
$$= \left\|\frac{\mathcal{B}_-}{\mathcal{B}_-^-}\right\|_2 \left\|\frac{\mathcal{R}}{\mathcal{B}_-}\right\|_2 = \left\|\frac{\mathcal{R}}{\mathcal{B}_-^-}\right\|_2 = J_{\min} \tag{A.4.47}$$

The minimum can be reached if the sum of the causal parts is zero, i.e.,

$$\frac{\mathcal{K}}{\mathcal{A}_x} - K_2 P_+ \mathcal{B}_-^- = 0 \tag{A.4.48}$$

From the above equation the optimality of the $\mathcal{H}_2$ norm is ensured by the compensator

$$K_2 = \frac{\mathcal{K}}{P_+ \mathcal{B}_-^- \mathcal{A}_x} \tag{A.4.49}$$

which is the same as (A.4.42) obtained by the variational method. The minimum of J is now

$$J_{\min} = \min_{K_2} [J] = \left\| \frac{\mathcal{R}}{\mathcal{B}_-^-} \right\|_2 \tag{A.4.50}$$

which is the same as (A.4.43).

Instead of a very serious verification, the validity of the above orthogonal decomposition for the computation of the $\mathcal{H}_2$ norm is illustrated by a very simple method. Assume that the two-sided Laplace transform ($\bar{\mathcal{L}}$) of the error signal $\boldsymbol{\varepsilon}_x(s) = \bar{\boldsymbol{\varepsilon}}(s)$ (see (A.4.38)) can be decomposed as follows

$$\bar{\boldsymbol{\varepsilon}}(s) = \bar{\mathcal{L}}\{\varepsilon(t)\} = \boldsymbol{\varepsilon}_+(s) + \boldsymbol{\varepsilon}_-(s) = \mathcal{L}\{\varepsilon_+(t)\} + \mathcal{L}\{\varepsilon_-(t)\} \tag{A.4.51}$$

where $\varepsilon_+(t)$ and $\varepsilon_-(t)$ are positive and negative time functions, respectively. Here $\boldsymbol{\varepsilon}_+(s)$ is a causal component (all poles are on the left half plane), and $\boldsymbol{\varepsilon}_-(s)$ is a noncausal component (all poles are on the right half plane). By definition $\varepsilon_+(t)\varepsilon_-(t) \equiv 0$. Thus the energy J_2^2 for the signal $\varepsilon(t)$ is

$$J_2^2 = \int_{-\infty}^{\infty} \varepsilon^2(t)\mathrm{d}t = \int_{-\infty}^{\infty} [\varepsilon_+(t) + \varepsilon_-(t)]^2 \mathrm{d}t = \int_{-\infty}^{\infty} \varepsilon_+^2(t)\mathrm{d}t + \int_{-\infty}^{\infty} \varepsilon_-^2(t)\mathrm{d}t \tag{A.4.52}$$

In the complex operator and frequency domain we obtain

$$J_2^2 = \frac{1}{2\pi j} \int_{-j\infty}^{j\infty} \boldsymbol{\varepsilon}(s)\boldsymbol{\varepsilon}(-s)\mathrm{d}s = \frac{1}{2\pi} \int_{-\infty}^{\infty} |\boldsymbol{\varepsilon}(j\omega)|^2 \mathrm{d}\omega$$
$$= \|\boldsymbol{\varepsilon}_+(j\omega)\|_2^2 + \|\boldsymbol{\varepsilon}_-(j\omega)\|_2^2 \tag{A.4.53}$$

Thus in the time domain, the positive and negative time function components are orthogonal, therefore the total energy of the signal is the sum of the two signals. In the optimization, the component $\varepsilon_+(s)$ or the corresponding signal $\varepsilon_+(t)$ is made equal to zero by the K_2 compensator calculated from the $\mathcal{H}_2$ norm.

A.4.3 MINIMIZATION OF THE $\mathcal{H}_\infty$ NORM

Consider the block scheme in Figure A.4.5, and let the excitation y_x be an impulse function (Dirac delta), whose Laplace transform is one ($Y_x(s)=1$). The task is to find an optimal compensator K_∞, which best approaches the stable reference model R_x when it is serially connected to the process $P(s)$. The ideal signal is now the output of R_x. In the decomposition $P(s) = P_+(s)P_-(s)$ of the process it is assumed that all poles and zeros of $P_+(s)$ have negative real values, so all poles and zeros of $P_-(s)$ have positive real values. The cost function of the approach is the $\mathcal{H}_\infty$ norm of the frequency function $\boldsymbol{\varepsilon}_x(j\omega)$ of the error signal $\varepsilon_x(t)$.

$$J_\infty = \max_{\omega}|\boldsymbol{\varepsilon}_x(j\omega)| = \|\boldsymbol{\varepsilon}_x(j\omega)\|_\infty = \|R_x(s) - K_x(s)P(s)\|_\infty \qquad \text{(A.4.54)}$$

and in the complex operator domain

$$\boldsymbol{\varepsilon}_x(s) = R_x(s) - K_x(s)P(s) \qquad \text{(A.4.55)}$$

The optimization task is to determine the compensator

$$K_\infty = \arg\left\{\min_{K_\infty}[J_\infty]\right\} = \arg\left\{\min_{K_\infty}\left[\|R_x(s) - K_x(s)P(s)\|_\infty\right]\right\} \qquad \text{(A.4.56)}$$

by minimizing the $\mathcal{H}_\infty$ norm (A.4.54). The minimization task (A.4.56) is solved by the optimal interpolation theorem, which is known as the Nevanlinna–Pick problem [134], [135] in the literature for this case.

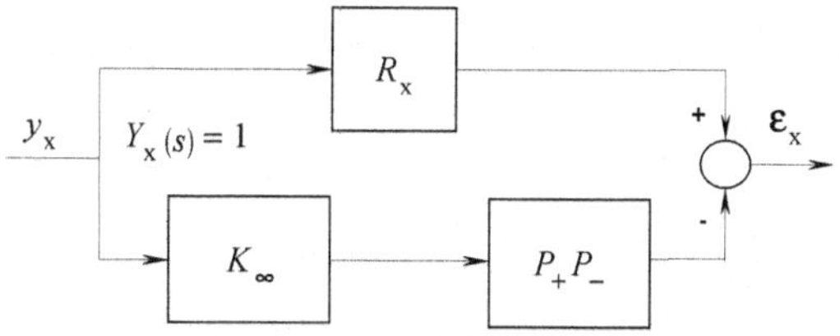

Figure A.4.5 $\mathcal{H}_\infty$ optimal approach of the prescribed reference model.

If P_- has m_- zeros, then the minimum W^{o} of the general supreme $\|W\|_\infty$ needs to be found in the form

$$W^{\mathrm{o}} = \begin{cases} \mu \dfrac{\mathcal{H}^\#}{\mathcal{H}}, & \text{if } m_- \geq 1 \\ 0, & \text{if } m_- = 0 \end{cases} \tag{A.4.57}$$

Here $\mathcal{H}$ is a Hurwitz polynomial of order $(m_- - 1)$. The polynomial $\mathcal{H}^\#$ is obtained by mirroring the zeros of $\mathcal{H}$ on the imaginary axis (see the derivation of $\mathcal{B}_-^-$ above). The constant μ and the coefficients of $\mathcal{H}$ are unambiguously determined by the following m_-, the so-called, interpolation restrictions

$$W(z_i) = \mu \frac{\mathcal{H}^\#(z_i)}{\mathcal{H}(z_i)} = (R_x - K_\infty P_+ P_-)\big|_{z_i} = R_x(z_i) = r_i; \quad i = 1, 2, \ldots, m_- \tag{A.4.58}$$

where $z_1, z_2, \ldots, z_{m_-}$ mean the disjunct zeros of $P_- = \mathbf{B}_-$. (The multiplicity of the zeros can be simply handled by taking further restrictions into account.) The $(m_- - 1)$ coefficients of the polynomial $\mathcal{H}$ and μ, together, mean m_- unknown parameters, which can be unambiguously determined from the m_- equations of (A.4.58). Note that (A.4.58) is a nonlinear equation system. With the rearrangement of (A.4.58) it can easily be seen that P_- is a divisor of the expression $R_x - W$, and thus it can be written that

$$R_x - W = \frac{\mathcal{B}_x}{\mathcal{A}_x} - \mu \frac{\mathcal{H}^\#}{\mathcal{H}} = \frac{\mathcal{N}}{\mathcal{D}} P_- = \frac{\mathcal{N}}{\mathcal{D}} \mathcal{B}_- = K_\infty P_+ \mathcal{B}_- \tag{A.4.59}$$

Hence the $\mathcal{H}_\infty$ optimal compensator is

$$K_\infty = \frac{\mathcal{N}}{\mathcal{D} P_+} \tag{A.4.60}$$

The polynomial $\mathcal{D}$ in the denominator can be obtained by comparing the two sides of (A.4.59)

$$\mathcal{D} = \mathcal{A}_x \mathcal{H} \tag{A.4.61}$$

The polynomial $\mathcal{N}$ in the numerator can also be calculated from (A.4.59) as

$$\mathcal{N} = \frac{\mathcal{B}_x \mathcal{H} - \mathcal{A}_x \mu \mathcal{H}^\#}{\mathcal{B}_-} \tag{A.4.62}$$

Because of the above mentioned features the division must be performed without residuals. The final form of the $\mathcal{H}_\infty$ optimal compensator is

$$K_\infty = \frac{\mathcal{B}_x\mathcal{H} - \mathcal{A}_x\mu\mathcal{H}^\#}{\mathcal{B}_-\mathcal{A}_x\mathcal{H}P_+} \tag{A.4.63}$$

The minimum of the loss function can be simply obtained from (A.4.57), since

$$\min\{\|W(j\omega)\|_\infty\} = \|W^\circ(j\omega)\|_\infty = \mu \tag{A.4.64}$$

A.4.4 THE NEHARI PROBLEM

Several problems in telecommunication, control, modeling, etc., can be formulated by seeking the so-called nearest inverse. The simplest description of this, often named the Nehari problem [132], is shown in Figure A.4.6. The question is how to determine the serial compensator K_x in order to minimize a norm of the error signal ε_x while only the inverse of P_+ from the components P_+ and P_- of the process is realizable. Therefore the product $K_xP_+P_-$ must be as close as possible to one. Let the optimality criterion be the $\mathcal{H}_2$ norm

$$J_x = \|R_x(1 - K_xP_+P_-)\|_2 = \left\| \frac{\mathcal{B}_x}{\mathcal{A}_x} - \frac{\mathcal{B}_xK_xP_+P_-}{\mathcal{A}_x} \right\|_2 \tag{A.4.65}$$

i.e., the task is to determine

$$K_x = \arg\left\{ \min_{K_x} [J_x] \right\} = \arg\left\{ \min_{K_x} \left[\|R_x(1 - K_xP_+P_-)\|_2 \right] \right\} \tag{A.4.66}$$

Here the filter $R_x = \mathcal{B}_x/\mathcal{A}_x$ means the weighting of the distance to one in a given frequency domain. In the case of the $\mathcal{H}_2$ norm, the excitation y_x is an impulse function (Dirac delta), whose Laplace transform is one ($Y_x(s) = 1$). The square of the $\mathcal{H}_2$ norm is J_x^2, and thus it is the energy of the impulse response of the error.

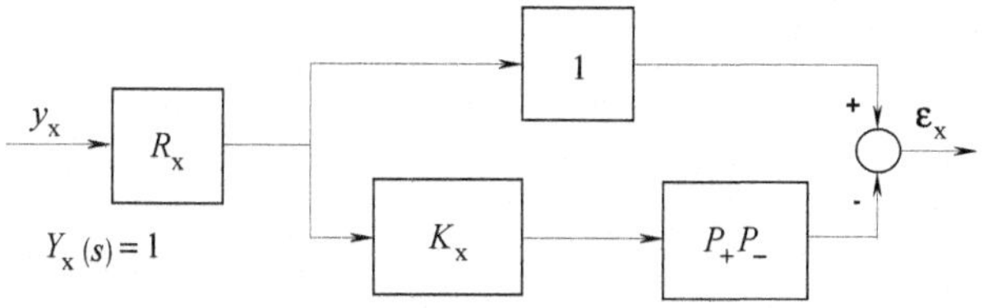

Figure A.4.6 Optimal scheme of the Nehari problem.

Let the invariant factor of the process be $P_- = \mathcal{B}_-$, where $\mathcal{B}_-(0) = 1$ and $\mathcal{B}_-(s) = \prod_{i=1}^{m_-}(1 - s/z_i)$. Here z_i are the zeros of $\mathcal{B}_-$ and m_- is number of these *IU* zeros. It can be written for the $\mathcal{H}_2$ norm (A.4.65) that

$$J_x = \|R_x(1 - K_x P_+ \mathcal{B}_-)\|_2 = \left\|\frac{\mathcal{B}_-}{\mathcal{B}_-^-}\right\|_2 \left\|\frac{\mathcal{B}_-^-\mathcal{B}_x}{\mathcal{B}_-\mathcal{A}_x} - \frac{\mathcal{B}_x K_x P_+ \mathcal{B}_-^-}{\mathcal{A}_x}\right\|_2$$
$$= \left\|\frac{\mathcal{B}_-^-\mathcal{B}_x}{\mathcal{B}_-\mathcal{A}_x} - \frac{\mathcal{B}_x K_x P_+ \mathcal{B}_-^-}{\mathcal{A}_x}\right\|_2 \tag{A.4.67}$$

$\mathcal{B}_-^-$ is a polynomial obtained by mirroring the zeros of $\mathcal{B}_-$ on the imaginary axis; thus the conditions $\mathrm{Re}\{z_i^*\} = -\mathrm{Re}\{z_i\}$ and $\mathrm{Im}\{z_i^*\} = \mathrm{Im}\{z_i\}$ must be fulfilled for the new zeros z_i^*. Therefore $\mathcal{B}_-^-$ becomes stable polynomial and the factor $\mathcal{B}_-/\mathcal{B}_-^-$ is called the norm-keeping inner function (the so-called all-pass filter in the context of signal forming; see (1.1.16)), and thus $\left\|\mathcal{B}_-/\mathcal{B}_-^-\right\|_2 = 1$.

At minimizing (A.4.66), the classical orthogonality method of the $\mathcal{H}_2$ paradigm is applied, i.e., the argument of (A.4.67) is decomposed to causal and noncausal parts, thus the first term needs to be decomposed as

$$\frac{\mathcal{B}_-^-\mathcal{B}_x}{\mathcal{B}_-\mathcal{A}_x} = \frac{\mathcal{R}}{\mathcal{B}_-} + \frac{\mathcal{K}}{\mathcal{A}_x} = \frac{\mathcal{R}\mathcal{A}_x + \mathcal{K}\mathcal{B}_-}{\mathcal{B}_-\mathcal{A}_x} \tag{A.4.68}$$

Here $\mathcal{R}/\mathcal{B}_-$ and $\mathcal{K}/\mathcal{A}_x$ are the noncausal and causal components, respectively (the polynomials $\mathcal{R}$ and $\mathcal{K}$ introduced here have no connection with the ones applied in Section 3.1 with the same notations). The new form of (A.4.67) is now

$$J_x = \|R_x(1 - K_x P_+ \mathcal{B}_-)\|_2 = \left\|\frac{\mathcal{R}}{\mathcal{B}_-} + \frac{\mathcal{K}}{\mathcal{A}_x} - \frac{\mathcal{B}_x K_x P_+ \mathcal{B}_-^-}{\mathcal{A}_x}\right\|_2 \tag{A.4.69}$$

Because of the orthogonality theorem discussed above, the minimum can be reached by making the causal part equal to zero, i.e., the condition of the minimum is

$$\frac{\mathcal{K}}{\mathcal{A}_x} - \frac{\mathcal{B}_x K_x P_+ \mathcal{B}_-^-}{\mathcal{A}_x} = 0 \tag{A.4.70}$$

From the above equation the optimality of the $\mathcal{H}_2$ norm can be ensured by the following inner filter

$$K_x = \frac{\mathcal{K}}{\mathcal{B}_x \mathcal{B}_-^- P_+} = P_+^{-1} G_x; \quad G_x = \frac{\mathcal{K}}{\mathcal{B}_x \mathcal{B}_-^-} \tag{A.4.71}$$

$\mathcal{R}$ and $\mathcal{K}$ can be calculated by solving the *DE* of a special form, which is derived from the comparison of the two sides of (A.4.68)

$$\mathcal{B}_{-}^{-}\mathcal{B}_{x} = \mathcal{R}\mathcal{A}_{x} + \mathcal{K}\mathcal{B}_{-} \tag{A.4.72}$$

It is clear that the solutions can correspond if the above error signals $\varepsilon_{X} = R_{x}(1 - K_{x}P_{+}\mathcal{B}_{-})$ or $\varepsilon'_{X} = (R_{x} - K'_{x}P_{+}\mathcal{B}_{-})$ are optimized. The G_{X}, introduced in (A.4.71), can be interpreted as the only chance to compensate the effect of $P_{-} = \mathcal{B}_{-}$ if P_{+}^{-1} is applied.

Thus the optimization cancels all stable factors. Therefore we have to accept the old-established practice of the classical regulators whereby if the dominant part of the process (usually when the poles correspond to the biggest time constants) is invertible then we can apply this solution. The minimum of J_{X} becomes

$$\min_{K_{x}} [J_{x}] = \left\| \frac{\mathcal{R}}{\mathcal{B}_{-}^{-}} \right\|_{2} \tag{A.4.73}$$

Note that the same result is obtained if G_{X} is optimized in the reduced task

$$G_{x} = \arg\left\{ \min_{G_{x}} [J'_{x}] \right\} = \arg\left\{ \min_{G_{x}} [\|R_{x}(1 - G_{x}P_{-})\|_{2}] \right\} \tag{A.4.74}$$

The minimum of J'_{x} is equal to the minimum of J_{X}.

Sampled Data Systems

The above results can be implicitly repeated for the discrete-time control loops, if the transfer functions are replaced by pulse transfer functions. In sampled systems the assumption of zero delay time is not required, because the form z^{-d} provides rational fractional function in all cases. When an integrator appears, the root $z = 1$ needs to be used in the denominator. To derive G_{-}^{-}, the unstable zeros in the numerator of the process need to be mirrored on the unit circle instead of the imaginary axis, and thus the simple transformation $z_{i}^{*} = 1/z_{i}$ needs to be applied.

A.4.5 GENERALIZED MINIMUM VARIANCE REGULATOR

The *generalized minimum variance (GMV)* uses the following loss function as the optimality criterion

$$V = E\left\{ \left(W_{y}(z)y[k+d] \right)^{2} + \left(W_{u}(z)u[k] \right)^{2} \middle| k \right\} \tag{A.4.75}$$

and thus penalizes the variance of the input. The stable filters $W_y(z)$ and $W_u(z)$ weight (penalize [34,35]) the corresponding frequency region of the output and input, respectively.

To determine the optimal regulator, let us introduce the output variable filtered by $W_y(z)$

$$y^F[k] = W_y(z)y[k] = \frac{\mathcal{W}_n^y}{\mathcal{W}_d^y}y[k] = \frac{w_{no}^y + w_{n1}^y z^{-1} + \ldots}{1 + w_{d1}^y z^{-1} + \ldots}y[k] \tag{A.4.76}$$

Here $\mathcal{W}_n^y$ and $\mathcal{W}_d^y$ are the polynomials of the numerator and denominator of the rational fractional function filter, respectively. Similarly, let us define the filter of the input

$$W_u(z) = \frac{\mathcal{W}_n^u}{\mathcal{W}_d^u} = \frac{w_{no}^u + w_{n1}^u z^{-1} + \ldots}{1 + w_{d1}^u z^{-1} + \ldots} \tag{A.4.77}$$

where $\mathcal{W}_n^u$ and $\mathcal{W}_d^u$ are the polynomials of the numerator and denominator of the rational fractional function filter, respectively.

If the d-step ahead prediction of $y^F[k]$ is determined (see (6.4)), the expected value in (A.4.75) can be given as

$$\begin{aligned} V &= E\left\{\left(y^F[k+d|k]\right)^2 + \left(W_u(z)u[k]\right)^2\right\} + E\left\{\varepsilon^F[k+d|k]\right\} \\ &= \left(y^F[k+d|k]\right)^2 + \left(W_u(z)u[k]\right)^2 + E\left\{\varepsilon^F[k+d|k]\right\} \\ &= Q + E\left\{\varepsilon^F[k+d|k]\right\} \end{aligned} \tag{A.4.78}$$

Here the independence of the d-step prediction $y^F[k+d|k]$ of the prediction error $\varepsilon^F[k+d|k]$ is considered; furthermore, it is also taken into account that the expected value of the first term in the kth sampling time is equal to itself. The minimum of V is searched by $u[k]$ and this task is equivalent to the determination of the minimum of Q. The necessary condition of the optimum is

$$\frac{\partial Q}{\partial u[k]} = 2y^F[k+d|k]\frac{\partial y^F[k+d|k]}{\partial u[k]} + 2\frac{\partial u^F[k]}{\partial u[k]}W_u(z)u[k] = 0 \tag{A.4.79}$$

where the filtered value $u^F[k] = W_u(z)u[k]$ of the input is introduced.

Given the polynomial forms of $W_y(z)$ and $W_u(z)$, and the Equation (6.1) of the process, where the polynomial in the numerator is $\mathcal{G} = g_o + g_1 z^1 + \ldots$, then the partial derivates in (A.4.79) can be obtained by simple calculations as

$$\frac{\partial y^{\mathrm{F}}[k+d|k]}{\partial u[k]} = w_{\mathrm{no}}^{\mathrm{y}} g_{\mathrm{o}}; \quad \frac{\partial u^{\mathrm{F}}[k]}{\partial u[k]} = w_{\mathrm{no}}^{\mathrm{u}} \tag{A.4.80}$$

The condition of the minimum gives the equality

$$y^{\mathrm{F}}[k+d|k] + \frac{w_{\mathrm{no}}^{\mathrm{u}}}{w_{\mathrm{no}}^{\mathrm{y}} g_{\mathrm{o}}} W_{\mathrm{u}}(z)u[k] = y^{\mathrm{F}}[k+d|k] + W_{\mathrm{u}}'(z)u[k] = 0 \tag{A.4.81}$$

Now determine the expressions for the independent d-step predictor $y^{\mathrm{F}}[k+d|k]$. To do this, let us convert Equation (6.1) for the case of the general noise model $\mathcal{D} \neq \mathcal{F}$

$$y^{\mathrm{F}}[k+d] = \frac{\mathcal{W}_{\mathrm{n}}^{\mathrm{y}}}{\mathcal{W}_{\mathrm{d}}^{\mathrm{y}}} \frac{\mathcal{G}z^{-d}}{\mathcal{F}} u[k+d] + \frac{\mathcal{W}_{\mathrm{n}}^{\mathrm{y}}}{\mathcal{W}_{\mathrm{d}}^{\mathrm{y}}} \frac{\mathcal{C}}{\mathcal{D}} e[k+d] = \frac{\mathcal{G}'}{\mathcal{F}'} u[k] + \frac{\mathcal{C}'}{\mathcal{D}'} e[k+d] \tag{A.4.82}$$

Introducing the polynomials

$$\mathcal{G}' = \mathcal{W}_{\mathrm{n}}^{\mathrm{y}} \mathcal{G}; \quad \mathcal{F}' = \mathcal{W}_{\mathrm{d}}^{\mathrm{y}} \mathcal{F}; \quad \mathcal{C}' = \mathcal{W}_{\mathrm{n}}^{\mathrm{y}} \mathcal{C}; \quad \mathcal{D}' = \mathcal{W}_{\mathrm{d}}^{\mathrm{y}} \mathcal{D} \tag{A.4.83}$$

the new form of the process precisely corresponds to the original equation. Introduce the polynomial equations

$$\begin{aligned} \mathcal{C}' &= \mathcal{D}'\mathcal{T}' + \mathcal{P}'z^{-d} \\ (\mathcal{W}_{\mathrm{n}}^{\mathrm{y}} \mathcal{C}) &= (\mathcal{W}_{\mathrm{d}}^{\mathrm{y}} \mathcal{D})\mathcal{T}' + \mathcal{P}'z^{-d} \end{aligned} \tag{A.4.84}$$

whose solution is unambiguous seeking the $(d-1)$-order polynomial $\mathcal{T}' = 1 + t_1' z^{-1} + \cdots + t_{d-1}' z^{-(d-1)}$ and $(r-1)$-order polynomial $\mathcal{P}' = p_{\mathrm{o}}' + p_1' z^{-1} + \cdots + p_{r-1}' z^{-(r-1)}$, if the order of the polynomials $\mathcal{C}'$ and $\mathcal{F}'$ is r. Here r is the resulting order of the product polynomials of $\mathcal{C}'$ and $\mathcal{F}'$. Then (A.4.82) can be translated into the form

$$y^{\mathrm{F}}[k+d] = y^{\mathrm{F}}[k+d|k] + \varepsilon^{\mathrm{F}}[k+d|k] \tag{A.4.85}$$

where $\varepsilon^{\mathrm{F}}[k+d|k] = \mathcal{T}'e[k+d]$ and

$$y^{\mathrm{F}}[k+d|k] = \frac{\mathcal{G}'\mathcal{D}'\mathcal{T}'}{\mathcal{F}'\mathcal{C}'} u[k] + \frac{\mathcal{P}'}{\mathcal{C}'} y[k] = \frac{\mathcal{G}\mathcal{D}\mathcal{T}'}{\mathcal{F}\mathcal{C}} u[k] + \frac{\mathcal{P}'}{\mathcal{C}\mathcal{W}_{\mathrm{n}}^{\mathrm{y}}} y[k] \tag{A.4.86}$$

Given the above expressions, the minimum condition of (A.4.81) results in the optimal input signal

$$u[k] = -\frac{\mathcal{FP}'}{(\mathcal{GDT}' + \mathcal{FC}W'_{\text{u}})W^{\text{y}}_{\text{n}}}y[k], \quad \text{i.e.,}$$
$$C_{\text{GMV}} = \frac{\mathcal{FP}'}{(\mathcal{GDT}' + \mathcal{FC}W'_{\text{u}})W^{\text{y}}_{\text{n}}} \tag{A.4.87}$$

Using C_{GMV}, determine the characteristic equation of the closed system. Do not forget that the output of the closed loop is now the virtual filtered variable $y^{\text{F}}[k] = W_{\text{y}}(z)y[k]$. Based on (A.4.74) and (A.4.81), the controlled process is now $G' = W_{\text{y}}G = \mathcal{G}'z^{-d}/\mathcal{F}'$, so the complementary sensitivity function of the closed loop is

$$T' = \frac{C_{\text{GMV}}G'}{1 + C_{\text{GMV}}G'} \tag{A.4.88}$$

Apply the form of the regulator C_{GMV} equivalent to (A.4.87)

$$C_{\text{GMV}} = \frac{\mathcal{F}'\mathcal{P}'}{\mathcal{G}'\mathcal{D}'\mathcal{T}' + \mathcal{F}'\mathcal{C}'W'_{\text{u}}} \tag{A.4.89}$$

and using this form, we obtain

$$T' = \frac{\dfrac{\mathcal{F}'\mathcal{P}'}{\mathcal{G}'\mathcal{D}'\mathcal{T}' + \mathcal{F}'\mathcal{C}'W'_{\text{u}}}\dfrac{\mathcal{G}'z^{-d}}{\mathcal{F}'}}{1 + \dfrac{\mathcal{F}'\mathcal{P}'}{\mathcal{G}'\mathcal{D}'\mathcal{T}' + \mathcal{F}'\mathcal{C}'W'_{\text{u}}}\dfrac{\mathcal{G}'z^{-d}}{\mathcal{F}'}} = \frac{\mathcal{P}'\mathcal{G}'z^{-d}}{\mathcal{G}'\mathcal{D}'\mathcal{T}' + \mathcal{F}'\mathcal{C}'W'_{\text{u}} + \mathcal{P}'\mathcal{G}'z^{-d}}$$
$$= \frac{\mathcal{P}'\mathcal{G}'z^{-d}}{\mathcal{G}'(\mathcal{D}'\mathcal{T}' + \mathcal{P}'z^{-d}) + \mathcal{F}'\mathcal{C}'W'_{\text{u}}} = \frac{\mathcal{P}'\mathcal{G}'z^{-d}}{\mathcal{G}'\mathcal{C}' + \mathcal{F}'\mathcal{C}'W'_{\text{u}}} = \frac{\mathcal{P}'\mathcal{G}'z^{-d}}{(\mathcal{G}' + \mathcal{F}'W'_{\text{u}})\mathcal{C}'}$$
$$= \frac{\mathcal{P}'W^{\text{y}}_{\text{n}}\mathcal{G}z^{-d}}{(W^{\text{y}}_{\text{n}}\mathcal{G} + W^{\text{y}}_{\text{d}}\mathcal{F}W'_{\text{u}})W^{\text{y}}_{\text{n}}\mathcal{C}} = \frac{\mathcal{P}'\mathcal{G}z^{-d}}{(W^{\text{y}}_{\text{n}}\mathcal{G} + W^{\text{y}}_{\text{d}}\mathcal{F}W'_{\text{u}})\mathcal{C}} \tag{A.4.90}$$

Thus the characteristic equation of the closed system is

$$(W^{\text{y}}_{\text{n}}\mathcal{G} + W^{\text{y}}_{\text{d}}\mathcal{F}W'_{\text{u}})\mathcal{C} = 0 \tag{A.4.91}$$

whence it follows that the stability of $\mathcal{C}$ needs to be acquired. It can be seen in (A.4.88) that the regulator C_{GMV} leaves the zeros of the process untouched.

A.4.6 SOLUTION OF THE DIOPHANTINE EQUATION (*DE*)

The general polynomial design of a regulator requires the solution of the *DE*

$$\mathcal{A}(s)\mathcal{X}(s) + \mathcal{B}(s)\mathcal{Y}(s) = \mathcal{AX} + \mathcal{BY} = \mathcal{R} = \mathcal{R}(s) \tag{A.4.92}$$

where $\mathcal{A}$, $\mathcal{B}$, and $\mathcal{R}$ are known, and the unknown parameters to be determined are in the polynomials $\mathcal{X}$ and $\mathcal{Y}$. The *DE* has a solution [99], if and only if all common divisors of $\mathcal{A}$ and $\mathcal{B}$ are, at the same time, also the common divisors of $\mathcal{R}$. If $\mathcal{A}$ and $\mathcal{B}$ are relative primes (they do not have a common divisor), then *DE* always has solution for any $\mathcal{R}$; the number of solutions is infinite. If $\mathcal{A}$ is a polynomial of order n then the order of $\mathcal{X}$ and $\mathcal{Y}$ can always be searched with the order of $(n-1)$, since in this case *DE* always has a solution. Observe that several optimization tasks in this appendix can be solved using similar *DE*.

Define the following matrix and vector variables

$$\begin{aligned} \boldsymbol{x}_{\text{xy}} &= \left[x_o, x_1, \dots, x_{n-1}; y_o, y_1, \dots, y_{n-1}\right]^{\text{T}} \quad \text{and} \\ \boldsymbol{r} &= \left[r_o, r_1, \dots, r_n, 0, \dots 0\right]^{\text{T}} \end{aligned} \tag{A.4.93}$$

$$\boldsymbol{M}_{\text{AB}} = \begin{bmatrix} a_o & 0 & \dots & 0 & b_o & 0 & \dots & 0 \\ a_1 & a_o & \dots & 0 & b_1 & b_o & \dots & 0 \\ a_2 & a_1 & \dots & 0 & b_2 & b_1 & \dots & 0 \\ \vdots & a_2 & \dots & 0 & \vdots & b_2 & \dots & 0 \\ a_{n-1} & \vdots & \ddots & \vdots & b_{n-1} & \vdots & \ddots & \vdots \\ a_n & a_{n-1} & \dots & a_{n-2} & b_n & b_{n-1} & \dots & b_{n-2} \\ 0 & a_n & \dots & a_{n-1} & 0 & b_n & \dots & b_{n-1} \\ 0 & 0 & \dots & a_n & 0 & 0 & \dots & b_n \end{bmatrix} \tag{A.4.94}$$

Then the *DE* corresponds to a linear matrix equation

$$\boldsymbol{M}_{\text{AB}}\, \boldsymbol{x}_{\text{xy}} = \boldsymbol{r} \tag{A.4.95}$$

Hence the vector $\boldsymbol{x}_{\text{xy}}$ containing the unknown parameters of the $\mathcal{X}$ and $\mathcal{Y}$ polynomials can be obtained by the explicit solution

$$\boldsymbol{x}_{\text{xy}} = \boldsymbol{M}_{\text{AB}}^{-1}\boldsymbol{r} \tag{A.4.96}$$

The number of parameters in $\boldsymbol{x}_{\text{xy}}$ is $2(n-1)+2=2n$, so the matrix $\boldsymbol{M}_{\text{AB}}$ is a $(2n \times 2n)$ matrix. Selecting the normalized $a_o = x_o = r_o = 1$ values, we can decrease the number of parameters. In the case of other *DE*, the structure of the unknown parameter vector and therefore the parameter matrix change accordingly.

APPENDIX 5

Norms of Signals and Operators

A norm can be defined in the complex linear space and the concept of a norm is somewhat similar to that of the absolute value. A norm is a function that assigns to every vector $\boldsymbol{x}$ in a given vector space a real number denoted by $\|\boldsymbol{x}\|$ that satisfies the following relationships:

$\|\boldsymbol{x}\| > 0$, if $\boldsymbol{x} \neq 0$, and $\|0\| = 0$

$\|a\boldsymbol{x}\| = |a|\|\boldsymbol{x}\|$ for arbitrary complex number a

$\|\boldsymbol{x} + \boldsymbol{y}\| \leq \|\boldsymbol{x}\| + \|\boldsymbol{y}\|$ which is the so-called triangle inequality.

The same concept is valid for the n-dimensional linear vector spaces and it can be similarly formulated for the functions, too.

The quality/performance of the control, as seen in previous sections, is related to the error signal or to the sensitivity function. The error signal is a time function and the sensitivity function is a complex frequency function. A measure needs to be found for the functions because its value in a time point or a given frequency does not characterize the whole function, especially its magnitude. The norm concept can be used for this purpose. Some basic norms are discussed below.

A.5.1 NORMS OF SIGNALS

$$\mathcal{L}_1 \text{ norm:} \quad \|u(t)\|_1 = \int_{-\infty}^{\infty} |u(t)|\mathrm{d}t \tag{A.5.1}$$

$$\mathcal{L}_2 \text{ norm:} \quad \|u(t)\|_2 = \sqrt{\int_{-\infty}^{\infty} |u(t)|^2 \mathrm{d}t} \tag{A.5.2}$$

$$\mathcal{L}_\infty \text{ norm:} \quad \|u(t)\|_\infty = \max_t |u(t)| \tag{A.5.3}$$

In practice, admitting signals are usually investigated ($u(t) \equiv 0$, if $t < 0$), where the lower limit of the integrals is zero.

Nonfinite time signals have so-called power, whose definition is

$$\mathrm{pow}[u(t)] = \sqrt{\lim_{T \to \infty} \frac{1}{2T} \int_{-T}^{T} |u(t)|^2 \mathrm{d}t} \tag{A.5.4}$$

Note that the finite time bounded signals have only energy and their power is zero. Thus if $\|u(t)\|_2 < \infty$ then pow[$u(t)$] = 0.

The simplest inequalities referring to the norm of the signals are:

$$\text{pow}[u(t)] \le \|u(t)\|_\infty; \quad \text{if } \|u(t)\|_\infty < \infty \tag{A.5.5}$$

$$\|u(t)\|_2 \le \sqrt{\|u(t)\|_\infty \|u(t)\|_1}; \quad \text{if } \|u(t)\|_\infty < \infty \text{ and } \|u(t)\|_1 < \infty \tag{A.5.6}$$

A.5.2 NORMS OF OPERATORS

Using the frequency function $H(j\omega)$ of the *linear time-invariant* system with the stable transfer function $H(s)$, we obtain

$$\mathcal{H}_2 \text{ norm:} \quad \|H(j\omega)\|_2 = \sqrt{\frac{1}{2\pi}\int_{-\infty}^{\infty} |H(j\omega)|^2 \mathrm{d}\omega} \tag{A.5.7}$$

$$\mathcal{H}_\infty \text{ norm:} \quad \|H(j\omega)\|_\infty = \max_\omega |H(j\omega)| \tag{A.5.8}$$

The operator norms are called system norms, too.

The $\mathcal{H}_2$ norm can be computed by the Parseval theorem

$$\begin{aligned}\|H(j\omega)\|_2^2 &= \frac{1}{2\pi}\int_{-\infty}^{\infty} |H(j\omega)|^2 \mathrm{d}\omega = \frac{1}{2\pi j}\oint H(-s)H(s)\mathrm{d}s \\ &= \sum \mathrm{Res}[H(-s)H(s)]\end{aligned} \tag{A.5.9}$$

where the residues of $[H(-s)H(s)]$ on the left half plane need to be considered. The expression (A.5.9) can be used (i.e., $\mathcal{H}_2$ is finite) if and only if $H(s)$ is strictly proper and has no poles on the imaginary axis. It is worth noting that

$$\|H(j\omega)\|_2 = \sqrt{\int_{-\infty}^{\infty} |w(t)|^2 \mathrm{d}t} = \sqrt{\int_0^{\infty} |w(t)|^2 \mathrm{d}t} \tag{A.5.10}$$

where $w(t)$ is the impulse response of the system with the transfer function $H(s)$.

If the system $H(s)$ is given in the state-space form $(\boldsymbol{A}, \boldsymbol{b}, \boldsymbol{c}^{\mathrm{T}})$, then the $\mathcal{H}_2$ norm can be computed as

$$\|H(j\omega)\|_2 = \sqrt{\boldsymbol{c}^{\mathrm{T}}\boldsymbol{L}\boldsymbol{c}} \tag{A.5.11}$$

where

$$L = \int_0^\infty e^{At} b b^{\mathrm{T}} e^{A^{\mathrm{T}} t} \mathrm{d}t \tag{A.5.12}$$

L can be computed much more simply by solving the following equation system linear in L instead of (A.5.12)

$$AL + LA^{\mathrm{T}} = -bb^{\mathrm{T}} \tag{A.5.13}$$

(see A.5.5). Equation (A.5.13) can be solved conventionally by arranging the unknown column vectors of L into a single column vector.

The calculation of the $\mathcal{H}_\infty$ norm is not easy, but its geometric interpretation is very simple: it is the distance between the origin and the farthest point from the origin of the $H(j\omega)$ Nyquist diagram. Since $H(s)$ and $H(j\omega)$ are usually rational fractional functions, the feasible places of the extreme of the absolute value (necessary condition) come from the zero allocations of the first-order derivate. This equation, however, even in the case of a low-order system, leads to a high-order polynomial equation, whose solution needs a numerical technique. Therefore, instead of an analytical solution, a numerical method is used to find the maximum of $|H(j\omega)|$. The $\mathcal{H}_\infty$ norm is finite if $H(s)$ is proper and has no pole on the imaginary axis and on the right half plane.

The Nevanlinna–Pick approximation procedure calculating the $\mathcal{H}_\infty$ norm, discussed in Chapter 4, can be applied for the error function operators [134,135].

One of the most important inequalities concerning the $\mathcal{H}_\infty$ norm is

$$\|H_1(j\omega)H_2(j\omega)\|_\infty \le \|H_1(j\omega)\|_\infty \|H_2(j\omega)\|_\infty \tag{A.5.14}$$

Keeping the above notation conventions, let the input of the system of transfer function $H(s)$ be $u(t)$ and the output $y(t)$. For stable processes the most important relationships of the signals and norms of the system are

$$\|y(t)\|_2 \le \|H(j\omega)\|_\infty \|u(t)\|_2 \tag{A.5.15}$$

therefore it can be said that the $\mathcal{H}_\infty$ norm is the upper limit of the gain factor of the $\mathcal{L}_2$ norm. Given the inequality

$$\|y(t)\|_\infty \le \|w(t)\|_1 \|u(t)\|_\infty \tag{A.5.16}$$

it can be realized that the $\mathcal{L}_1$ norm of the impulse response is the upper limit of the gain factor of the $\mathcal{L}_\infty$ norms. Thus the upper limit of the maximum of the step response function $y(t) = v(t)$, (if $u(t) = 1(t)$), is equal to the integral of the absolute value of the impulse response. Similarly, the following inequality can also be admitted

$$\|y(t)\|_\infty \le \|H(j\omega)\|_2 \|u(t)\|_2 \tag{A.5.17}$$

Comparing Equations (A.5.16) and (A.5.17) we obtain

$$\|y(t)\|_\infty \le \min\left\{\|w(t)\|_1 \|u(t)\|_\infty ; \|H(j\omega)\|_2 \|u(t)\|_2\right\} \tag{A.5.18}$$

where the stricter condition is applied. Thus the $\mathcal{H}_\infty$, $\mathcal{H}_2$, and $\mathcal{L}_1$ norms can be considered as the upper limits of the transfer factors in relation to the signal norms.

Similar relationships can be formulated for the power of the input and output signals:

$$\mathrm{pow}[y(t)] \le \|H(j\omega)\|_\infty \mathrm{pow}[u(t)] \tag{A.5.19}$$

whence we can obtain

$$\mathrm{pow}[y(t)] \le \|H(j\omega)\|_\infty \|u(t)\|_\infty \tag{A.5.20}$$

Comparing the latter inequalities we obtain

$$\begin{aligned}\mathrm{pow}[y(t)] &\le \min\left\{\|H(j\omega)\|_\infty \mathrm{pow}[u(t)]; \|H(j\omega)\|_\infty \|u(t)\|_\infty\right\} \\ &= \|H(j\omega)\|_\infty \min\left\{\mathrm{pow}[u(t)]; \|u(t)\|_\infty\right\}\end{aligned} \tag{A.5.21}$$

where the stricter condition is applied.

A.5.3 ENERGY NORM OF SIGNALS

In the evaluation of the dynamic transient processes of control loops, an index is of major importance in measuring the energy used. A generalized energy norm is defined which can easily be formulated from $\mathcal{L}$ and $\mathcal{H}$ norms. The definition of the energy norm can be given by the frequency characteristic of the signal for a stable transient as

$$\mathcal{E}_2 \text{ norm:} \quad \|u(t)\|_{\mathcal{E}_2} = \sqrt{\frac{1}{2\pi}\int_{-\infty}^{\infty} |U(j\omega)|^2 d\omega} \tag{A.5.22}$$

The reason for the special notation is that the $\mathcal{L}_2$ norm of Equation (A.5.2) measures the energy of the signal $u(t)$ in the same way as $\mathcal{E}_2$ is defined by the frequency characteristic $U(j\omega)$ of the signal. The frequency functions $H(j\omega)$ used in $\mathcal{H}$ norm computations of the popular modern optimization methods of control systems, however, provide the energy content of a given signal only in the case of a special Dirac-delta excitation.

A.5.4 SUPREMUM NORM OF FREQUENCY CHARACTERISTICS OF SIGNALS

In the evaluation of the dynamic transient processes of control loops, an index can be introduced to measure the maximum absolute value of the frequency characteristic of the signal which appears during the transient. Therefore a generalized supremum norm is defined next, which can be formulated easily by $\mathcal{L}$ and $\mathcal{H}$ norms. Its definition is given by the frequency characteristic of the signal for the stable transient as

$$\mathcal{E}_\infty \text{ norm:} \quad \|u(t)\|_{\mathcal{E}_\infty} = \|U(j\omega)\|_{\mathcal{E}_\infty} = \max_{\omega}|U(j\omega)| \tag{A.5.23}$$

A.5.5 INTEGRAL CRITERIA

The key to design in control engineering is reaching a compromise between high overexcitation and long control time. This compromise can be formulated by the distance from the stability limit $(-1 + j0)$ (see Chapter 9) and an index whose minimum (optimum) balances the dynamic behavior between two extreme transients. These indices were the integral criteria applied before the present modern norms. In this case the quality of the control is judged by the value of the integral criteria formulated for the error signal $e(t)$ of the closed control loop. The parameters of the regulator are chosen to minimize an integral error.

Integral errors are as follows:

$$I_1 = \int_0^\infty e(t)\mathrm{d}t \quad \text{linear error area (control area),} \tag{A.5.24}$$

(it can be applied only for aperiodic systems, and can be analytically computed)

$$I_2 = \int_0^\infty e^2(t)\mathrm{d}t \quad \text{square area} \tag{A.5.25}$$

(it can be analytically computed)

$$I_3 = \int_0^\infty |e(t)|\mathrm{d}t = IAE \quad \text{absolute value area} \tag{A.5.26}$$

(IAE—integral of absolute value error)

$$I_4 = \int_0^\infty t|e(t)|\mathrm{d}t = ITAE \quad \text{absolute value area weighted by time} \tag{A.5.27}$$

(*ITAE—integral of time multiplied absolute value error*). Criteria I_3 and I_4 can be evaluated only by simulation. The integral criterion $I_3 = IAE = \|e(t)\|_1$, and thus it is the $\mathcal{L}_1$ norm of the error signal $e(t)$, and $I_2 = \|e(t)\|_2^2$ is the square of the $\mathcal{L}_2$ norm. Although the above expressions are apparent, the integral criteria can be considered only as engineering qualitative measures, whereas the norms have strict mathematical definitions.

A.5.6 DEDUCTIONS/DERIVATIONS

The definition of the $\mathcal{H}_2$ norm is

$$\|H(j\omega)\|_2 = \sqrt{\int_0^\infty \|w(t)\|^2 \mathrm{d}t} = \sqrt{\int_0^\infty \boldsymbol{c}^{\mathrm{T}} e^{\boldsymbol{A}t} \boldsymbol{b}\boldsymbol{b}^{\mathrm{T}} e^{\boldsymbol{A}^{\mathrm{T}}t} \boldsymbol{c}\mathrm{d}t} \tag{A.5.28}$$

according to the Parseval theorem, and thus

$$\|H(j\omega)\|_2^2 = \boldsymbol{c}^{\mathrm{T}} \left[\int_0^\infty e^{\boldsymbol{A}t} \boldsymbol{b}\boldsymbol{b}^{\mathrm{T}} e^{\boldsymbol{A}^{\mathrm{T}}t} \mathrm{d}t \right] \boldsymbol{c} \tag{A.5.29}$$

Introduce the notation

$$\boldsymbol{L} = \int_0^\infty e^{\boldsymbol{A}t} \boldsymbol{b}\boldsymbol{b}^{\mathrm{T}} e^{\boldsymbol{A}^{\mathrm{T}}t} \mathrm{d}t \tag{A.5.30}$$

so finally the $\mathcal{H}_2$ norm can be computed as

$$\|H(j\omega)\|_2 = \sqrt{\boldsymbol{c}^{\mathrm{T}} \boldsymbol{L} \boldsymbol{c}} \tag{A.5.31}$$

Differentiate the Equation (A.5.30)

$$\frac{\mathrm{d}}{\mathrm{d}t} e^{\boldsymbol{A}t} \boldsymbol{b}\boldsymbol{b}^{\mathrm{T}} e^{\boldsymbol{A}^{\mathrm{T}}t} = \boldsymbol{A} e^{\boldsymbol{A}t} \boldsymbol{b}\boldsymbol{b}^{\mathrm{T}} e^{\boldsymbol{A}^{\mathrm{T}}t} + e^{\boldsymbol{A}t} \boldsymbol{b}\boldsymbol{b}^{\mathrm{T}} e^{\boldsymbol{A}^{\mathrm{T}}t} \boldsymbol{A}^{\mathrm{T}} \tag{A.5.32}$$

then integrate both sides of the equation in the region $[0, \infty]$

$$\left[e^{\boldsymbol{A}t} \boldsymbol{b}\boldsymbol{b}^{\mathrm{T}} e^{\boldsymbol{A}^{\mathrm{T}}t} \right]_0^\infty = \boldsymbol{A} \left[\int_0^\infty e^{\boldsymbol{A}t} \boldsymbol{b}\boldsymbol{b}^{\mathrm{T}} e^{\boldsymbol{A}^{\mathrm{T}}t} \mathrm{d}t \right] + \left[\int_0^\infty e^{\boldsymbol{A}t} \boldsymbol{b}\boldsymbol{b}^{\mathrm{T}} e^{\boldsymbol{A}^{\mathrm{T}}t} \mathrm{d}t \right] \boldsymbol{A}^{\mathrm{T}} \tag{A.5.33}$$

whence, by simple computation and taking Equation (A.5.30) into account, we obtain the following linear equation system for $\boldsymbol{L}$

$$\boldsymbol{A}\boldsymbol{L} + \boldsymbol{L}\boldsymbol{A}^{\mathrm{T}} = -\boldsymbol{b}\boldsymbol{b}^{\mathrm{T}} \tag{A.5.34}$$

Linear Error Area

By simple means we obtain

$$I_1 = \lim_{t\to\infty}\int_0^t e(\tau)\mathrm{d}\tau = \lim_{s\to 0} s\frac{E(s)}{s} = E(0) \tag{A.5.35}$$

where $E(s) = \mathcal{L}\{e(t)\}$ is the Laplace transform of the error signal. For a system, given by aperiodic and transfer function $T(s)$ $(T(0) = A)$ (see Figure A.5.1), compute the linear control area.

$$T(s) = A\frac{\prod_{k=1}^{m}(1+s\tau_k)}{\prod_{j=1}^{n}(1+sT_j)} = AT'(s), \tag{A.5.36}$$

Given (A.5.35) we obtain

$$\begin{aligned} I_1 &= \int_0^\infty (A - v(t))\mathrm{d}t = \left[(A - T(s))\frac{1}{s}\right]_{s=0} = A\left[\frac{1-T'(s)}{s}\right]_{s=0} \\ &= A\left[\frac{\prod_{j=1}^{n}(1+sT_j) - \prod_{k=1}^{m}(1+s\tau_k)}{s\prod_{j=1}^{n}(1+sT_j)}\right]_{s=0} = A\left(\sum_{j=1}^{n}T_j - \sum_{k=1}^{m}\tau_k\right) \end{aligned} \tag{A.5.37}$$

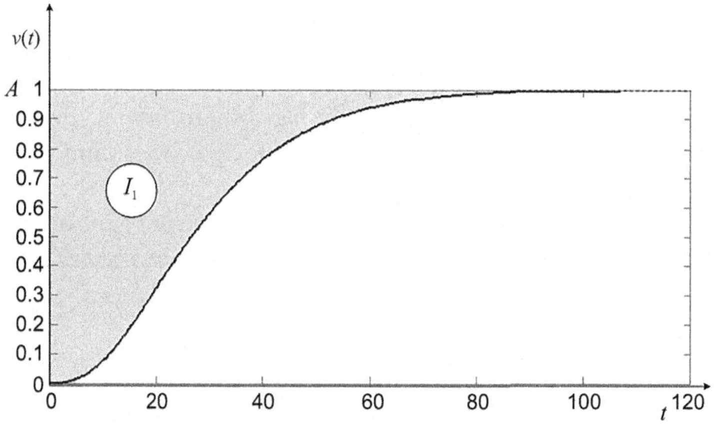

Figure A.5.1 Linear control area of an aperiodic process.

The time constants of the denominator of the transfer function increase, whereas the time constants of the numerator decrease the control region. Thus, by introducing zeros we can accelerate the system.

For aperiodic transients an equivalent time shift T_e or dead time can be defined which is the time shift of the step response with amplitude A and measured from $(t = 0)$, with the same control region as the actual step response

$$T_e = \frac{I_1}{A} = \sum_{j=1}^{n} T_j - \sum_{k=1}^{m} \tau_k \tag{A.5.38}$$

Square Error Area

The square integral criterion can also be evaluated by the Parseval theorem in the frequency region.

$$I_2 = \int_0^{\infty} e^2(t)\mathrm{d}t = \frac{1}{2\pi j}\int_{-\infty}^{\infty} E(-s)E(s)\mathrm{d}s = \frac{1}{\pi}\int_0^{\infty} |E(j\omega)|^2 \mathrm{d}\omega. \tag{A.5.39}$$

The square error region can also be calculated analytically. For low-order cases the following expression can be used to calculate the Laplace transform of the error signal when it is strictly proper $(m < n)$

$$E(s) = \frac{\sum_{i=0}^{m} c_i s^i}{\sum_{i=0}^{n} d_i s^i} \tag{A.5.40}$$

and gives the value of I_2 in closed form, in parameters of c_i and d_i for a given order [37]. It is not difficult to formulate a special recursive algorithm, too. Note that minimizing the error region by a parameter of the regulator means the resulting minimum is usually flat. Unfortunately, the optimal transient has relatively high overexcitation (20–25%), so it cannot be used in quality control.

Observe that the physical meaning of the square error region and the $\mathcal{H}_2$ norm is the same. If $H(j\omega) = E(j\omega)$, i.e., $E(s)$ is the Laplace transform of the error signal $e(t)$ that measures the energy content of the signal.

Absolute Value Criteria

The absolute value criteria cannot be evaluated analytically, so the minimum can be determined by simulation or optimum seeking procedures.

The minimum of the cost function, however, is characteristic of a given parameter value. The absolute value criterion weighted by time penalizes the high values in the error signal at the beginning, while penalizes more the latter ones. The optimum of this criterion ensures the nicest over-excitation ~5%.

Solution of a Special Matrix Equation

The special matrix equation (see e.g., (A.5.34))

$$\boldsymbol{AX} + \boldsymbol{XB} = \boldsymbol{C} \tag{A.5.41}$$

has a unique solution if $\boldsymbol{A}$ and $-\boldsymbol{B}$ have no common eigenvalues. The equation can be rewritten in classical form as

$$(\boldsymbol{A} \otimes \boldsymbol{I}_n + \boldsymbol{I}_m \otimes \boldsymbol{B})\boldsymbol{\Sigma} = \boldsymbol{\Psi} \tag{A.5.42}$$

where $\boldsymbol{\Sigma}$ and $\boldsymbol{\Psi}$ contain the rows of $\boldsymbol{X}$ and $\boldsymbol{C}$ in column vector form, i.e.,

$$\boldsymbol{\Sigma} = [x_{11}, x_{12}, \ldots, x_{1n}, \ldots, x_{n1}, x_{n2}, \ldots, x_{nn}]^{\mathrm{T}} = \mathbf{vec}(\boldsymbol{X}) \tag{A.5.43}$$

and

$$\boldsymbol{\Psi} = [c_{11}, c_{12}, \ldots, c_{1n}, \ldots, c_{n1}, c_{n2}, \ldots, c_{nn}]^{\mathrm{T}} = \mathbf{vec}(\boldsymbol{C}) \tag{A.5.44}$$

If $\boldsymbol{A}$ and $\boldsymbol{B}$ have no common eigenvalues which are single ones, furthermore

$$\mathrm{Re}[\lambda_i(\boldsymbol{A})] + \mathrm{Re}[\lambda_i(\boldsymbol{B})] \geq 0 \quad \text{for all } i \text{ and } j \tag{A.5.45}$$

then the unique solution can be obtained as

$$\boldsymbol{X} = \int_0^\infty e^{\boldsymbol{A}t}\boldsymbol{C}e^{\boldsymbol{B}t}\mathrm{d}t \tag{A.5.46}$$

APPENDIX 6

Derivations of Process Identification Methods

A.6.1

Given $\boldsymbol{Z} = \lambda^2 \boldsymbol{I}$ in (10.7), i.e., in the case of uncorrelated error, the detailed form of the likelihood function is

$$L\left(\boldsymbol{p}, \hat{\lambda}\right) = -\frac{N}{2}\ln 2\pi - \frac{N}{2}\ln\hat{\lambda}^2 - \frac{1}{2\hat{\lambda}^2}\hat{\boldsymbol{e}}^{\mathrm{T}}\hat{\boldsymbol{e}} \tag{A.6.1}$$

where it is indicated that $L(\boldsymbol{p}, \hat{\lambda})$ is the function of $\boldsymbol{p}$ and $\hat{\lambda}$, too. The task (10.7) was obtained by assuming λ was known. It is shown now that (10.7) is also valid for unknown λ and it is possible to estimate $\hat{\lambda}$. Formulate the partial derivate of Equation (A.6.1) by $\hat{\lambda}$ and make it equal to zero, so we obtain the identity

$$\frac{\partial L\left(\boldsymbol{p}, \hat{\lambda}\right)}{\partial \hat{\lambda}} = -\frac{N}{2\hat{\lambda}} + \frac{\hat{\boldsymbol{e}}^{\mathrm{T}}\hat{\boldsymbol{e}}}{\hat{\lambda}^3} = 0 \tag{A.6.2}$$

whence the estimation of $\hat{\lambda}^2$ is

$$\hat{\lambda}^2 = \hat{\lambda}^2(\hat{\boldsymbol{p}}) = \frac{1}{N}\hat{\boldsymbol{e}}^{\mathrm{T}}\hat{\boldsymbol{e}}\Big|_{\hat{\boldsymbol{p}}} = \frac{2}{N}V(\hat{\boldsymbol{p}}) \tag{A.6.3}$$

Here the notation $\hat{\boldsymbol{p}}$ means that the calculations have to be performed by a $\boldsymbol{p}$ that ensures the maximum of L, thus by *minimum variance* (*MV*) estimated parameter vector $\hat{\boldsymbol{p}}$. Since now $\hat{\lambda}$ depends on $\hat{\boldsymbol{p}}$, after substituting it in Equation (A.6.1) we obtain

$$\begin{aligned} L\left(\boldsymbol{p}, \hat{\lambda}\right)\Big|_{\hat{\lambda}(\hat{\boldsymbol{p}})} = L(\hat{\boldsymbol{p}}) &= -\frac{N}{2}\ln 2\pi - \frac{N}{2}\ln\frac{\hat{\boldsymbol{e}}^{\mathrm{T}}\hat{\boldsymbol{e}}}{N} - \frac{1}{2}\frac{N}{\hat{\boldsymbol{e}}^{\mathrm{T}}\hat{\boldsymbol{e}}}\hat{\boldsymbol{e}}^{\mathrm{T}}\hat{\boldsymbol{e}} \\ &= -\frac{N}{2}\ln(2\pi + 1) - \frac{N}{2}\ln\frac{\hat{\boldsymbol{e}}^{\mathrm{T}}\hat{\boldsymbol{e}}}{N} \end{aligned} \tag{A.6.4}$$

Since the logarithm function is monotonous, the maximum of $L(\hat{\boldsymbol{p}})$ is ensured by the minimum of $\hat{\boldsymbol{e}}^{\mathrm{T}}\hat{\boldsymbol{e}}$, and thus it is equivalent to the minimization of the loss function, namely

$$\min_{\hat{\boldsymbol{p}}}\{V(\hat{\boldsymbol{p}})\} = \min_{\hat{\boldsymbol{p}}}\left\{\frac{1}{2}\hat{\boldsymbol{e}}^{\mathrm{T}}\hat{\boldsymbol{e}}\right\} \tag{A.6.5}$$

as seen for other methods.

In the case of the likelihood function (10.14) written for *multiple input multiple output* (*MIMO*) systems, the solution is not so plausible as in the above derivation. If we consider only that for a given $\boldsymbol{\Lambda}$ the maximum of L can be obtained by minimizing $\sum_{j=1}^{N}\boldsymbol{e}^{\mathrm{T}}[j]\boldsymbol{\Lambda}^{-1}\boldsymbol{e}[j]$ then the result will be completely different from that obtained by the detailed investigation of L. The main difficulty is that the (co)variance is matrix ($\boldsymbol{\Lambda}$). The likelihood function now depends on $\boldsymbol{P}$ and $\hat{\boldsymbol{\Lambda}}$, i.e.,

$$L(\boldsymbol{P},\hat{\boldsymbol{\Lambda}}) = -\frac{pN}{2}\ln 2\pi - \frac{N}{2}\ln|\hat{\boldsymbol{\Lambda}}| - \frac{1}{2}\sum_{j=1}^{N}\hat{\boldsymbol{e}}^{\mathrm{T}}[j]\hat{\boldsymbol{\Lambda}}^{-1}\hat{\boldsymbol{e}}[j] \tag{A.6.6}$$

To determine the extremum of L by $\hat{\boldsymbol{\Lambda}}$, for simplicity, make the matrix derived by $\hat{\boldsymbol{\Lambda}}^{-1}$ equal to zero

$$\frac{\partial L(\boldsymbol{P},\hat{\boldsymbol{\Lambda}})}{\partial\hat{\boldsymbol{\Lambda}}^{-1}} = \frac{N}{2}\hat{\boldsymbol{\Lambda}} - \frac{1}{2}\sum_{j=1}^{N}\hat{\boldsymbol{e}}[j]\hat{\boldsymbol{e}}^{\mathrm{T}}[j] = \boldsymbol{0} \tag{A.6.7}$$

whence the estimation of $\hat{\boldsymbol{\Lambda}}$ is obtained as

$$\hat{\boldsymbol{\Lambda}} = \hat{\boldsymbol{\Lambda}}(\hat{\boldsymbol{P}}) = \frac{1}{N}\sum_{j=1}^{N}\hat{\boldsymbol{e}}[j]\hat{\boldsymbol{e}}^{\mathrm{T}}[j]\Bigg|_{\hat{\boldsymbol{P}}} = \frac{1}{N}\hat{\boldsymbol{E}}\hat{\boldsymbol{E}}^{\mathrm{T}} \tag{A.6.8}$$

where

$$\hat{\boldsymbol{E}} = [\hat{\boldsymbol{e}}[1],\ldots,\hat{\boldsymbol{e}}[N]] \tag{A.6.9}$$

The notation $\hat{\boldsymbol{P}}$ means that the calculations have to be performed at such $\boldsymbol{P}$ that ensures the maximum of L, and thus at the *MV* estimated parameter matrix. Since now $\hat{\boldsymbol{\Lambda}}$ depends on $\hat{\boldsymbol{P}}$, substituting it for (A.5.6) we obtain

$$L(\boldsymbol{P},\hat{\boldsymbol{\Lambda}})\big|_{\hat{\boldsymbol{\Lambda}}(\hat{\boldsymbol{P}})} = L(\hat{\boldsymbol{P}}) = -\frac{pN}{2}(\ln 2\pi + 1) - \frac{N}{2}\ln|\hat{\boldsymbol{\Lambda}}| \tag{A.6.10}$$

This expression cannot be easily proved. Using the identities valid for the trace of matrices, Equation (A.6.6) can also be written in the following form to take account of (A.6.9)

$$L(\mathbf{P},\hat{\mathbf{\Lambda}}) = -\frac{pN}{2}\ln 2\pi - \frac{N}{2}\ln\left|\hat{\mathbf{\Lambda}}\right| - \frac{1}{2}\sum_{j=1}^{N}\hat{\mathbf{e}}^{\mathrm{T}}[j]\hat{\mathbf{\Lambda}}^{-1}\hat{\mathbf{e}}[j]$$
$$= -\frac{pN}{2}\ln 2\pi - \frac{N}{2}\ln\left|\hat{\mathbf{\Lambda}}\right| - \frac{1}{2}\mathrm{tr}\left(\mathbf{E}^{\mathrm{T}}\hat{\mathbf{\Lambda}}^{-1}\mathbf{E}\right) \qquad \text{(A.6.11)}$$

Taking the $\hat{\mathbf{\Lambda}}$ of Equation (A.6.8) the rewritings are given step by step

$$L(\mathbf{P},\hat{\mathbf{\Lambda}}) = -\frac{pN}{2}\ln 2\pi - \frac{N}{2}\ln\left|\hat{\mathbf{\Lambda}}(\hat{\mathbf{P}})\right| - \frac{N}{2}\mathrm{tr}\left(\mathbf{E}^{\mathrm{T}}\left(\mathbf{E}^{\mathrm{T}}\mathbf{E}\right)\mathbf{E}\right)$$
$$= -\frac{pN}{2}\ln 2\pi - \frac{N}{2}\ln\left|\hat{\mathbf{\Lambda}}(\hat{\mathbf{P}})\right| - \frac{N}{2}\mathrm{tr}\left(\left(\mathbf{E}^{\mathrm{T}}\mathbf{E}\right)^{-1}\mathbf{E}^{\mathrm{T}}\mathbf{E}\right)$$
$$= -\frac{pN}{2}\ln 2\pi - \frac{N}{2}\ln\left|\hat{\mathbf{\Lambda}}(\hat{\mathbf{P}})\right| - \frac{N}{2}\mathrm{tr}\left(\mathbf{I}_p\right)$$
$$= -\frac{pN}{2}(\ln 2\pi + 1) - \frac{N}{2}\ln\left|\hat{\mathbf{\Lambda}}(\hat{\mathbf{P}})\right| \qquad \text{(A.6.12)}$$

i.e., $\mathbf{I}_p$ is a $p\times p$ identity matrix.

Thus for *MIMO* systems the loss function is to be minimized which ensures the maximum of the likelihood function is

$$V(\hat{\mathbf{P}}) = \left|\hat{\mathbf{\Lambda}}(\hat{\mathbf{P}})\right| = \left|\frac{1}{N}\hat{\mathbf{E}}\hat{\mathbf{E}}^{\mathrm{T}}\right| = \frac{1}{N}\left|\sum_{j=1}^{N}\hat{\mathbf{e}}^{\mathrm{T}}[j]\hat{\mathbf{e}}[j]\right| \qquad \text{(A.6.13)}$$

i.e., it is the determinant of a matrix (the covariance matrix of the source noise).

A.6.2

Unfolding the criterion (10.9) in detail we obtain

$$V(\hat{\mathbf{p}}) = V(\hat{\mathbf{p}},N) = \frac{1}{2}\left[\mathbf{y}^{\mathrm{T}}\mathbf{y} - 2\mathbf{y}^{\mathrm{T}}\mathbf{F}_{\mathrm{u}}(\mathbf{u})\hat{\mathbf{p}} + \hat{\mathbf{p}}^{\mathrm{T}}\mathbf{F}_{\mathrm{u}}^{\mathrm{T}}(\mathbf{u})\mathbf{F}_{\mathrm{u}}(\mathbf{u})\hat{\mathbf{p}}\right] \qquad \text{(A.6.14)}$$

and making its gradient equal to zero

$$\frac{\mathrm{d}V(\hat{\mathbf{p}})}{\mathrm{d}\hat{\mathbf{p}}} = -\mathbf{F}_{\mathrm{u}}^{\mathrm{T}}(\mathbf{u})\mathbf{y} + \mathbf{F}_{\mathrm{u}}^{\mathrm{T}}(\mathbf{u})\mathbf{F}_{\mathrm{u}}(\mathbf{u})\hat{\mathbf{p}} = 0 \qquad \text{(A.6.15)}$$

and solving the equation for $\hat{\boldsymbol{p}}$, the best parameter estimation we obtain is

$$\hat{\boldsymbol{p}} = \left[\boldsymbol{F}_{\mathrm{u}}^{\mathrm{T}}(\boldsymbol{u})\boldsymbol{F}_{\mathrm{u}}(\boldsymbol{u})\right]^{-1}\boldsymbol{F}_{\mathrm{u}}^{\mathrm{T}}(\boldsymbol{u})\boldsymbol{y} \tag{A.6.16}$$

The Hesse-matrix of the second derivates is the Gram-matrix (or Gramian) $\boldsymbol{F}_{\mathrm{u}}^{\mathrm{T}}(\boldsymbol{u})\boldsymbol{F}_{\mathrm{u}}(\boldsymbol{u})$, i.e., it is nonnegative definite. So the solution (A.6.16) is a minimum point if this matrix is nonsingular (i.e., it has full rank). The necessary condition of the regularity is that the number of rows (i.e., the number of measured points) of $\boldsymbol{F}_{\mathrm{u}}(\boldsymbol{u})$ must be equal to or greater than the number of columns (i.e., the number of parameters) and the elements of the function component vectors $\boldsymbol{f}(u)$ must be linearly independent.

A.6.3

Investigate the system Equation (10.1.53)

$$\boldsymbol{Y} = \boldsymbol{P}_{\mathrm{BA}}\boldsymbol{F} + \boldsymbol{E} \tag{A.6.17}$$

i.e., the *least squares* (*LS*) estimation of the parameter matrix $\boldsymbol{P}_{\mathrm{BA}}$. The minimization of the following loss function can be considered as the direct generalization of the *LS* method discussed for *single-input single-output* (*SISO*) systems

$$\boldsymbol{V}\left(\hat{\boldsymbol{P}}_{\mathrm{BA}}, N\right) = \frac{1}{2}\sum_{j=1}^{N}\hat{\boldsymbol{e}}[j]\hat{\boldsymbol{e}}^{\mathrm{T}}[j] = \frac{1}{2}\hat{\boldsymbol{E}}\hat{\boldsymbol{E}}^{\mathrm{T}} \tag{A.6.18}$$

where

$$\hat{\boldsymbol{E}} = \boldsymbol{Y} - \hat{\boldsymbol{P}}_{\mathrm{BA}}\boldsymbol{F} \tag{A.6.19}$$

So, in fact, $\boldsymbol{V}$ is proportional to the matrix of moments of the computed residuals $\hat{\boldsymbol{e}}[j]$. Since $\boldsymbol{V}$ is a matrix, different kinds of scalar measures can be used for the minimization and this is a vital difference compared with the *SISO* systems. After a suitable scalar measure has been chosen, *MIMO* systems require the application of a derivate of a scalar by vector to the minimization and avoid derivation by matrix (though it might be possible, see next, the first method). Using different approaches closed explicit solutions are given next to the *LS* identification of the *MIMO* systems.

Choose first the logarithm of the determinant of $\boldsymbol{V}$ as a scalar measure (loss function) [123]. Here the condition for minimization is

$$\frac{\mathrm{d}\ln|\boldsymbol{V}|}{\mathrm{d}\hat{\boldsymbol{P}}_{\mathrm{BA}}} = \left(\boldsymbol{Y}\boldsymbol{F}^{\mathrm{T}} - \hat{\boldsymbol{P}}_{\mathrm{BA}}\boldsymbol{F}\boldsymbol{F}^{\mathrm{T}}\right)\boldsymbol{V}^{-1} = \boldsymbol{0} \tag{A.6.20}$$

Expressing $\hat{\boldsymbol{P}}_{\mathrm{BA}}$ from the matrix equation, the *LS* estimation of the *MIMO* system parameters is

$$\hat{\boldsymbol{P}}_{\mathrm{BA}} = \boldsymbol{Y}\boldsymbol{F}^{\mathrm{T}}\left(\boldsymbol{F}\boldsymbol{F}^{\mathrm{T}}\right)^{-1} \tag{A.6.21}$$

In the expression (A.6.20) the derivation rules of the determinant by matrix are used as

$$\frac{\mathrm{d}\ln\boldsymbol{V}}{\mathrm{d}\hat{\boldsymbol{P}}_{\mathrm{BA}}} = \mathrm{tr}\left[\boldsymbol{V}^{-1}\frac{\mathrm{d}\boldsymbol{V}}{\mathrm{d}\hat{\boldsymbol{P}}_{\mathrm{BA}}}\right] = \mathrm{tr}\left[\frac{\mathrm{d}\boldsymbol{V}}{\mathrm{d}\hat{\boldsymbol{P}}_{\mathrm{BA}}}\boldsymbol{V}^{-1}\right] \tag{A.6.22}$$

and

$$\begin{aligned}\frac{\mathrm{d}\boldsymbol{V}}{\mathrm{d}\hat{\boldsymbol{P}}_{\mathrm{BA}}} &= \hat{\boldsymbol{E}}\frac{\mathrm{d}\hat{\boldsymbol{E}}}{\mathrm{d}\hat{\boldsymbol{P}}_{\mathrm{BA}}} = \hat{\boldsymbol{E}}\left[\frac{\mathrm{d}}{\mathrm{d}\hat{\boldsymbol{P}}_{\mathrm{BA}}}\left(\boldsymbol{Y} - \hat{\boldsymbol{P}}_{\mathrm{BA}}\boldsymbol{F}\right)\right] \\ &= \left(\boldsymbol{Y} - \hat{\boldsymbol{P}}_{\mathrm{BA}}\boldsymbol{F}\right)\boldsymbol{F}^{\mathrm{T}} = \boldsymbol{Y}\boldsymbol{F}^{\mathrm{T}} - \hat{\boldsymbol{P}}_{\mathrm{BA}}\boldsymbol{F}\boldsymbol{F}^{\mathrm{T}}\end{aligned} \tag{A.6.23}$$

A.6.4

Suppose that after processing of N data pairs the result of the off-line *LS* parameter estimation is available as

$$\hat{\boldsymbol{p}}[N] = \left[\boldsymbol{F}^{\mathrm{T}}[N]\boldsymbol{F}[N]\right]^{-1}\boldsymbol{F}^{\mathrm{T}}[N]\boldsymbol{y}_N \tag{A.6.24}$$

and calculate the new *LS* estimation valid for $N+1$ point

$$\begin{aligned}\hat{\boldsymbol{p}}[N+1] &= \left[\boldsymbol{F}^{\mathrm{T}}[N+1]\boldsymbol{F}[N+1]\right]^{-1}\boldsymbol{F}^{\mathrm{T}}[N+1]\boldsymbol{y}_{N+1} \\ &= \left\{\begin{bmatrix}\boldsymbol{F}[N] \\ \boldsymbol{f}^{\mathrm{T}}[N+1]\end{bmatrix}^{\mathrm{T}}\begin{bmatrix}\boldsymbol{F}[N] \\ \boldsymbol{f}^{\mathrm{T}}[N+1]\end{bmatrix}\right\}^{-1}\begin{bmatrix}\boldsymbol{F}[N] \\ \boldsymbol{f}^{\mathrm{T}}[N+1]\end{bmatrix}^{\mathrm{T}}\begin{bmatrix}\boldsymbol{y}_N \\ y[N+1]\end{bmatrix} \\ &= \left\{\boldsymbol{F}^{\mathrm{T}}[N]\boldsymbol{F}^{\mathrm{T}}[N] + \boldsymbol{f}[N+1]\boldsymbol{f}^{\mathrm{T}}[N+1]\right\}^{-1} \\ &\quad\times\left\{\boldsymbol{F}^{\mathrm{T}}[N]\boldsymbol{y}_N + \boldsymbol{f}[N+1]y[N+1]\right\}\end{aligned} \tag{A.6.25}$$

It is essential to calculate the matrix inverse extended by the dyadic product

$$\boldsymbol{R}[N+1] = \left\{\boldsymbol{F}^{\mathrm{T}}[N+1]\boldsymbol{F}^{\mathrm{T}}[N+1]\right\}^{-1}$$
$$= \left\{\boldsymbol{F}^{\mathrm{T}}[N]\boldsymbol{F}^{\mathrm{T}}[N] + \boldsymbol{f}[N+1]\boldsymbol{f}^{\mathrm{T}}[N+1]\right\}^{-1} \tag{A.6.26}$$

Because of the expression A.1.17 of the Appendix A.1 we can write

$$\left\{\boldsymbol{F}^{\mathrm{T}}[N+1]\boldsymbol{F}^{\mathrm{T}}[N+1]\right\}^{-1} = \left\{\boldsymbol{F}^{\mathrm{T}}[N]\boldsymbol{F}^{\mathrm{T}}[N]\right\}^{-1}$$
$$-\frac{\left\{\boldsymbol{F}^{\mathrm{T}}[N]\boldsymbol{F}^{\mathrm{T}}[N]\right\}^{-1}\boldsymbol{f}[N+1]\boldsymbol{f}^{\mathrm{T}}[N+1]\left\{\boldsymbol{F}^{\mathrm{T}}[N]\boldsymbol{F}^{\mathrm{T}}[N]\right\}^{-1}}{1+\boldsymbol{f}^{\mathrm{T}}[N+1]\left\{\boldsymbol{F}^{\mathrm{T}}[N]\boldsymbol{F}^{\mathrm{T}}[N]\right\}^{-1}\boldsymbol{f}[N+1]} \tag{A.6.27}$$

The resulting equation is the so-called Sherman–Morrison expression [64]. Using the definition of $\boldsymbol{R}[N]$ in (10.2.4) we obtain

$$\boldsymbol{R}[N+1] = \boldsymbol{R}[N] - \frac{\boldsymbol{R}[N]\boldsymbol{f}(N+1)\boldsymbol{f}^{\mathrm{T}}(N+1)\boldsymbol{R}[N]}{1+\boldsymbol{f}^{\mathrm{T}}(N+1)\boldsymbol{R}[N]\boldsymbol{f}(N+1)} \tag{A.6.28}$$

Substituting the above recursive equation of the convergence matrix for (A.6.27), we obtain the recursive equation of the parameter estimation

$$\hat{\boldsymbol{p}}[N+1] = \hat{\boldsymbol{p}}[N] + \boldsymbol{R}[N+1]\boldsymbol{f}(N+1)\left\{y[N+1] - \boldsymbol{f}^{\mathrm{T}}(N+1)\hat{\boldsymbol{p}}[N]\right\} \tag{A.6.29}$$

"Recursive" derives from the fact that the renewing equations of both $\boldsymbol{R}[N]$ and $\hat{\boldsymbol{p}}[N]$ can be calculated from the previous values by the addition of an additive term.

It is clear that the vector $\boldsymbol{R}[N+1]\boldsymbol{f}(N+1)$ in the Equation (A.6.29) of the recursive parameter estimation can be expressed in another form as

$$\boldsymbol{R}[N+1]\boldsymbol{f}(N+1) = \boldsymbol{R}[N]\boldsymbol{f}(N+1) - \frac{\boldsymbol{R}[N]\boldsymbol{f}(N+1)\boldsymbol{f}^{\mathrm{T}}(N+1)\boldsymbol{R}[N]}{1+\boldsymbol{f}^{\mathrm{T}}(N+1)\boldsymbol{R}[N]\boldsymbol{f}(N+1)}\boldsymbol{f}(N+1)$$
$$= \frac{\boldsymbol{R}[N]\boldsymbol{f}(N+1)}{1+\boldsymbol{f}^{\mathrm{T}}(N+1)\boldsymbol{R}[N]\boldsymbol{f}(N+1)} \tag{A.6.30}$$

Because of this Equation (A.6.29) can also be written as

$$\hat{\boldsymbol{p}}[N+1] = \hat{\boldsymbol{p}}[N] + \frac{\boldsymbol{R}[N]\boldsymbol{f}[N+1]}{1+\boldsymbol{f}^{\mathrm{T}}[N+1]\boldsymbol{R}[N]\boldsymbol{f}[N+1]} \times \left\{y[N+1] - \boldsymbol{f}^{\mathrm{T}}[N+1]\hat{\boldsymbol{p}}[N]\right\} \tag{A.6.31}$$

The basic equation of the convergence matrix of the recursive estimation is the following dyadic adjunct

$$\begin{aligned}\{\boldsymbol{R}[N+1]\}^{-1} &= \boldsymbol{F}^{\mathrm{T}}[N]\boldsymbol{F}[N] + \boldsymbol{f}[N+1]\boldsymbol{f}^{\mathrm{T}}[N+1] \\ &= \{\boldsymbol{R}[N]\}^{-1} + \boldsymbol{f}[N+1]\boldsymbol{f}^{\mathrm{T}}[N+1]\end{aligned} \tag{A.6.32}$$

which will have the following form if the forgetting factor λ of (10.2.7) is applied

$$\begin{aligned}\{\boldsymbol{R}[N+1]\}^{-1} &= \lambda^2\boldsymbol{F}^{\mathrm{T}}[N]\boldsymbol{F}[N] + \boldsymbol{f}[N+1]\boldsymbol{f}^{\mathrm{T}}[N+1] \\ &= \lambda^2\{\boldsymbol{R}[N]\}^{-1} + \boldsymbol{f}[N+1]\boldsymbol{f}^{\mathrm{T}}[N+1]\end{aligned} \tag{A.6.33}$$

Using (A.1.17) and based on Equation (A.6.27) the Sherman–Morrison formula becomes

$$\boldsymbol{R}[N+1] = \frac{1}{\lambda^2}\left\{\boldsymbol{R}[N] - \frac{\boldsymbol{R}[N]\boldsymbol{f}[N+1]\boldsymbol{f}^{\mathrm{T}}[N+1]\boldsymbol{R}[N]}{\lambda^2 + \boldsymbol{f}^{\mathrm{T}}[N+1]\boldsymbol{R}[N]\boldsymbol{f}[N+1]}\right\} \tag{A.6.34}$$

It is interesting to calculate how the determinant and trace of the matrix $\boldsymbol{R}[N] = \{\boldsymbol{F}^{\mathrm{T}}[N]\boldsymbol{F}^{\mathrm{T}}[N]\}^{-1}$ change during the recursive algorithm. This happens because $\boldsymbol{R}[N]$ is proportional to the covariance matrix (10.1.24) of the estimation, and thus these are scalar measures with simple geometrical meanings which provide information regarding the quality of the estimation. Based on Equation (A.6.32) of the dyadic adjunct, the recursive calculation of the determinant can be made according to the following expression using (A.1.18)

$$\det\{\boldsymbol{R}[N+1]\} = \frac{\det\{\boldsymbol{R}[N]\}}{1 + \boldsymbol{f}^{\mathrm{T}}[N+1]\boldsymbol{R}[N]\boldsymbol{f}[N+1]} \tag{A.6.35}$$

Since the trace of the dyadic products can be simply computed as $\mathrm{tr}(\boldsymbol{a}\boldsymbol{b}^{\mathrm{T}}) = \boldsymbol{b}^{\mathrm{T}}\boldsymbol{a}$, and $\mathrm{tr}(\boldsymbol{A}+\boldsymbol{B}) = \mathrm{tr}(\boldsymbol{A}) + \mathrm{tr}(\boldsymbol{B})$, using Equation (A.6.28) it can be written that

$$\mathrm{tr}\{\boldsymbol{R}[N+1]\} = \mathrm{tr}\{\boldsymbol{R}[N]\} - \frac{\boldsymbol{f}^{\mathrm{T}}[N+1]\boldsymbol{R}^{\mathrm{T}}[N]\boldsymbol{R}[N]\boldsymbol{f}[N+1]}{1 + \boldsymbol{f}^{\mathrm{T}}[N+1]\boldsymbol{R}[N]\boldsymbol{f}[N+1]} \tag{A.6.36}$$

REFERENCES

[1] J. Ackermann, Der Entwurf linearer Regelungssysteme im Zustandsraum, vol. 20, Regelungstechnik, 1972, pp. 297–300.

[2] J. Ackermann, Parameter space design of robust control systems, IEEE Trans. Aut. Control AC-25 (6) (1980) 1058–1072.

[3] D. Allerton, Principles of Flight Simulation, John Wiley & Sons, 2009, p. 123.

[4] B.D.O. Anderson, R.L. Kosut, Adaptive robust control: on-line learning, in: 30th IEEE Conf. Decision and Control CDC'91, Brighton, UK, 1991.

[5] B.D.O. Anderson, From Youla-Kučera to identification, adaptive and nonlinear control, Automatica 34 (12) (1998) 1485–1506.

[6] K.J. Åström, T. Bohlin, S. Wensmark, Automatic Construction of Linear Stochastic Dynamic Models for Stationary Industrial Processes with Random Disturbances Using Operating Records, IBM Nordic Laboratory, Sweden, 1965. TP 18.150, Technical paper.

[7] K.J. Åström, T. Bohlin, Numerical identification of linear dynamic systems from normal operating records, in: IFAC Symposium on the Theory of Self-adaptive Systems, Teddington, UK, 1965.

[8] K.J. Åström, U. Borisson, Self-tuning Regulators – Industrial Applications and Multivariable Theory, Internal report, Lund Institute of Technology, Lund, 1975.

[9] K.J. Åström, P. Hagander, J. Sternby, Zeros of sampled systems, in: 19th IEEE Conf. Decision and Control CDC'80, Albuquerque, NM, USA, 1980, pp. 1077–1081.

[10] K.J. Åström, Matching criteria for control and identification, in: 2nd European Control Conference ECC'93, Groningen, NL, 1993, pp. 248–251.

[11] K.J. Åström, B. Wittenmark, Computer-controlled Systems, Theory and Design, third ed., Prentice Hall, 1997.

[12] K.J. Åström, Control System Design, Lecture Notes, University of California, Santa Barbara, USA, 2002.

[13] Cs. Bányász, L. Keviczky, Direct methods for self-tuning PID regulators, in: 6th IFAC Symposium on Ident. and Syst. Par. Est., Washington DC, USA, 1982, pp. 1249–1254.

[14] Cs. Bányász, L. Keviczky, Convergence and robustness properties of a generic regulator refinement scheme, in: 2nd IFAC Symposium on Robust Control Design ROCOND'97, Budapest, Hongrie, 1997, pp. 495–500.

[15] Cs. Bányász, L. Keviczky, Designing high performance two-degree of freedom controllers, in: Instrumentation and Measurement Technology Conference IMTC'98, St. Paul, MN, USA, 1998, pp. 1359–1366.

[16] Cs. Bányász, L. Keviczky, I. Vajk, A novel adaptive control system for raw material blending, IEEE Control Syst. Mag. 23 (1) (2003) 87–96.

[17] Cs. Bányász, L. Keviczky, A new PID regulator design scheme based on a K-B parameterized closed-loop, in: IFAC Workshop on Digital Control — Past, Present and Future of PID Control, Terrassa, ES, 2000, pp. 521–526.

[18] Cs. Bányász, L. Keviczky, A simple PID regulator applicable for a class of factorable nonlinear plants, in: American Control Conference, Anchorage, AL, USA, 2002, pp. 2354–2359.

[19] Cs. Bányász, L. Keviczky, CLCR optimal input design for IS and IU plants, IIGSS Symp. on Control Theory and Engineering, Pittsburg, PA, USA, Kybernetes: Int. J. Syst. Cybern. 9 (9/10) (2002) 1220–1235.

[20] Cs. Bányász, L. Keviczky, State-feedback solutions via transfer function representations, J. Syst. Sci. 30 (2) (2004) 21–34.

[21] R.W. Bass, I. Gura, High-order system design via state-space considerations, in: Joint Aut. Control Conf., Troy, NY, USA, 1965, pp. 311–318.

[22] G.J. Bierman, Factorization Methods for Discrete Sequential Estimation, Academic Press, New York, 1976.

[23] G.J. Bierman, Efficient time propagation of U-D covariance factors, IEEE Trans. Aut. Control AC-26 (4) (1981) 890–894.

[24] R.R. Bitmead, M. Gevers, V. Wertz, Adaptive optimal control, in: M.J. Grimble (Ed.), The Thinking Man's GPC, Prentice Hall, 1990.

[25] R. Bitmead, Iterative control design approaches, in: 12th IFAC Congres, vol. 9, Sydney, Australia, 1993, pp. 381–384.

[26] U. Borisson, Self-tuning Regulators – Industrial Applications and Multivariable Theory, Report 7513, Dept. of Aut. Control, Lund Institute of Technology, Lund, Sweden, 1975.

[27] U. Borisson, Self-tuning regulators for a class of multivariable systems, Automatica 15 (2) (1979) 209–215.

[28] J. Bokor, L. Keviczky, Structural properties and structure determination of vector difference equations, Int. J. Control 36 (3) (1982) 461–475.

[29] J. Bokor, L. Keviczky, Structure and parameter estimation of MIMO systems using elementary subsystem representation, Int. J. Control 39 (5) (1984) 965–986.

[30] J. Bokor, L. Keviczky, ARMA canonical forms obtained from constructibility invariants, Int. J. Control 45 (3) (1987) 861–873.

[31] O.H. Bosgra, H. Kwakernaak, G. Meinsma, Design Methods for Control Systems, Winter Course, Dutch Institute of Systems and Control, NL, 2004.

[32] S.P. Boyd, C.H. Barrett, Linear Controller Design: The Limits of Performance, Prentice-Hall, Englewood Cliffs, NJ, 1991.

[33] R. de Callafon, Feedback oriented identification for enhanced and robust control, a fractional approach applied to a wafer stage (thesis), Delft University Press, NL, 1998.

[34] D.W. Clarke, Generalized least-squares estimation of the parameters of a dynamic model, in: IFAC Symposium on Identification and Automatic Control Systems, Prague, CZ, 1967, pp. 1–11.

[35] D.W. Clarke, P.J. Gawthrop, Self-tuning controller, Proc. IEE 122 (1975) 929–934.

[36] A.O. Cordero, D.Q. Mayne, Deterministic convergence of a self-tuning regulator with variable forgetting factor, IEE Proc.-D 128 (1) (1981) 19–23.

[37] F. Csáki, Szabályozások Dinamikája, Akadémiai Kiadó, Budapest, 1966.

[38] F. Csáki, Fejezetek a Szabályozástechnikából. Állapotegyenletek, Műszaki Könyvkiadó, Budapest, 1973.

[39] F. Csáki, Korszerű Szabályozáselmélet, Akadémiai Kiadó, Budapest, 1973.

[40] E.B. Dahlin, Designing and tuning digital controllers, Instrum. Control Syst. 42 (6) (1968) 77–83.

[41] R.C. Dorf, R.H. Bishop, Modern Control Systems, Prentice Hall Intern, 2001.

[42] J.C. Doyle, B.A. Francis, A. Tannenbaum, Feedback Control Theory, Macmillan, New York, 1992.

[43] B. Etkin, L.D. Reid, Dynamics of Flight Stability and Control, John Wiley & Sons, 1996.

[44] P. Eykhoff, System Identification, Parameter and State Estimation, John Wiley & Sons, 1974.

[45] A.A. Feldbaum, Theoretical Foundations of Automatic Systems (in Russian), State Publishing House in Physics and Mathematics, Moscow, 1963.

[46] T.R. Fortescue, L.S. Kershenbaum, B.E. Ydstie, Implementation of self-tuning regulators with variable forgetting factors, Automatica 17 (6) (1985) 29–36.

[47] W.M. Gentleman, Least squares computations of givens transformations without square roots, J. Inst. Math. Appl. 12 (1973) 329–336.

[48] M. Gevers, Connecting identification and robust control: a new challenge, in: 9th IFAC/IFORS Symposium on Identification and System Parameter Estimation, Budapest, Hungary, 1991, pp. 1–10.
[49] M. Gevers, Learning from identification to control design, in: European Science Foundation Symp on Complex Systems ESF-cosy 49, Valencia, ES, 1996, pp. 1–60.
[50] M. Gevers, L. Ljung, P.M.J. van den Hof, Asymptotic variance expressions for closed-loop identification and their relevance in identification and control, in: 11th IFAC/IFORS Symp. SYSID'97, Fukuoka, Japan, 1997, pp. 1449–1454.
[51] G.C. Goodwin, R.I. Payne, Dynamic System Identification: Experiment Design and Data Analysis, Academic Press, 1977.
[52] G.C. Goodwin, K.S. Sin, Adaptive Filtering, Prediction and Control, Prentice Hall, 1984.
[53] G.C. Goodwin, Model Identification and Adaptive Control, Springer, 2001.
[54] R. Haber, L. Keviczky, Identification of nonlinear dynamic systems (Survey Paper), in: 4th IFAC Symposium on Ident. and Syst. Par. Est., Tbilisi, USSR, vol. 1, 1976, pp. 62–112.
[55] R. Haber, J. Hetthéssy, L. Keviczky, I. Vajk, A. Fehér, N. Czeiner, Z. Császár, A. Turi, Identification and adaptive control of a glass furnace, Automatica 17 (1) (1981) 175–185.
[56] R. Haber, L. Keviczky, Nonlinear System Identification - Input-output Modelling Approach, Vol. 1: Nonlinear System Parameter Identification, Vol. 2: Nonlinear System Structure Identification, Kluwer Academic Publishers, 1999.
[57] L.W. Harold, Inverted decoupling – a neglected technique, IEEE Trans. Aut. Control AC-36 (1) (1997) 3–10.
[58] J. Hetthéssy, L. Keviczky, M. Hilger, Számítógépes Folyamatirányítási Algoritmusok Sztochasztikus Zavarásoknak Kitett Rendszerekre, SZIKKTI Report, Budapest, 1975.
[59] J. Hetthéssy, L. Keviczky, Cs. Bányász, On a class of adaptive PID regulators, in: IFAC Workshop on Adaptive Systems in Control and Signal Processing, San Francisco, CA, USA, 1983.
[60] I.M. Horowitz, Synthesis of Feedback Systems, Academic Press, New York, 1963.
[61] R. Isermann, Digital Control Systems, in: Fundamentals, Deterministic Control, second ed., vol. I, Springer-Verlag, 1989.
[62] R. Isermann, Digital Control Systems, in: Stochastic Control, Multivariable Control, Adaptive Control, Applications, second ed., vol. II, Springer-Verlag, 1989.
[63] R. Isermann, K.H. Lachmann, D. Matko, Adaptive Control Systems, Prentice Hall, 1992.
[64] T. Kailath, Linear Systems, Prentice Hall, 1980.
[65] R.E. Kalman, On the general theory of control systems, in: 1st IFAC Congress, Moscow, USSR, 1960, pp. 481–492.
[66] P.G. Kaminski, A.E. Bryson, S.F. Schmidt, Discrete square root filtering: a survey of current techniques, IEEE Trans. Aut. Control AC-16 (7) (1971) 727–736.
[67] L. Keviczky, Cs. Bányász, On input signal synthesis for linear discrete-time systems, in: 3rd IFAC Symposium on Ident. and Syst. Par. Est., Hague/Delft, NL, 1973.
[68] L. Keviczky, "Design of Experiments" for the identification of linear dynamic systems, Technometrics 17 (3) (1975) 303–308.
[69] L. Keviczky, J. Hetthéssy, Self-tuning minimum variance control of MIMO discrete time systems, J. Autom. Control Theory Appl. 5 (1) (1977) 11–17.
[70] L. Keviczky, J. Hetthéssy, M. Hilger, J. Kolostori, Self-tuning adaptive control of cement raw material blending, Automatica 14 (6) (1978) 525–532.
[71] L. Keviczky, J. Bokor, Cs. Bányász, A new identification method with special parametrization for model structure determination, in: 5th IFAC Symposium on Ident. and Syst. Par. Est., Darmstadt, D, 1979.

[72] L. Keviczky, On the Transfer Functions of Sampled Continuous Systems, Technical Report, University of Minnesota, Department of Electrical Engineering, Minneapolis, MN, USA, 1979.

[73] L. Keviczky, K.S.P. Kumar, The Multivariable Extension of Clarke's Self-tuning Controller, Technical Report, University of Minnesota, Department of Electrical Engineering, Minneapolis, MN, USA, 1979.

[74] L. Keviczky, K.S.P. Kumar, Multivariable self-tuning regulator with generalized cost-function, Int. J. Control 33 (1981) 913–921.

[75] L. Keviczky, Control in cement production (Invited plenary paper), in: 4th IFAC Symp. On Automation in Mining, Mineral and Metal Processing, Helsinki, SF, 1983.

[76] L. Keviczky, J. Hetthéssy, Cement industry: adaptive control, in: M.G. Singh (Ed.), Systems and Control Encyclopedia: Theory, Technology, Applications, Pergamon Press, 1987, pp. 561–564.

[77] L. Keviczky, R. Haber, Glass industry: adaptive control, in: M.G. Singh (Ed.), Systems and Control Encyclopedia: Theory, Technology, Applications, Pergamon Press, 1987, pp. 2008–2011.

[78] L. Keviczky, I. Vajk, M. Vajta, Load-frequency regulation, in: M.G. Singh (Ed.), Adaptive Control in: Systems and Control Encyclopedia: Theory, Technology, Applications, Pergamon Press, 1987, pp. 2889–2892.

[79] L. Keviczky, Cs. Bányász, A new structure to design optimal control systems, in: IFAC Workshop on New Trends in Design of Control Systems, Smolenice, SK, 1994, pp. 102–105.

[80] L. Keviczky, Combined identification and control: another way (Plenary paper), in: 5th IFAC Symp. On Adaptive Control and Signal Processing, ACASP'95, Budapest, Hungary, 1995, pp. 13–30.

[81] L. Keviczky, Combined identification and control: another way, IFAC J. Control Engineering Practice 4 (5) (1996) 685–698.

[82] L. Keviczky, Cs. Bányász, On the dialectics of identification and control in iterative learning schemes (A Heisenberg-type uncertainty law in combined identification and control), in: DYCOMANS Workshop, Algarve, Portugal, 1996, pp. 77–82.

[83] L. Keviczky, Cs. Bányász, Extension of dual control: the concept of triple control, in: 13th IFAC Congress'96, San Francisco, CA, USA, vol. K., 1996, pp. 265–270.

[84] L. Keviczky, Cs. Bányász, Robustness and reachable bandwidth versus control and identification performances, in: 2nd IFAC Workshop on New Trends in Design of Control Systems, NTDCS'97, Smolenice, SK, 1997, pp. 89–95.

[85] L. Keviczky, Cs. Bányász, Influence of amplitude constraints: an iterative redesign technique of reference model, in: European Control Conference ECC'97, Brussels, Belgium, 1997. TH-M-F1.

[86] L. Keviczky, Cs. Bányász, An iterative reference model redesign technique for intelligent regulator tuners, in: Y. Lin, Y. Ma (Eds.), 2nd IIGSS Workshop, San Marcos, USA, Advances in Systems Science & Applications, an Official Journal of the International Institute for General Systems Studies, 1997, pp. 87–100.

[87] L. Keviczky, Cs. Bányász, Dialectics of identification and control: product inequalities in an iterative scheme, in: World Automation Congress ISIAC'98, Anchorage, AL, USA, 1998.

[88] L. Keviczky, Cs. Bányász, Optimal two-degree of freedom controllers: iterative refinement, robustness and error properties, in: American Control Conference, Philadelphia, PA, USA, vol. 1, 1998, pp. 188–189.

[89] L. Keviczky, Cs. Bányász, Optimality of two-degree of freedom controllers in H_2- and H_∞-norm space, their robustness and minimal sensitivity, in: 14th IFAC Congress, Beijing, PRC, 1999.

[90] L. Keviczky, Cs. Bányász, Asymptotic variances in closed-loop identification using K-B parametrization, in: IASTED International Conference on Control and Applications, CA'99, Banff, Canada, 1999.
[91] L. Keviczky, Cs. Bányász, Dialectics of identification and control (New paradigms - new solutions), in: European Science Foundation Symp. on Complex Systems ESF COSY Course, Valencia, ES, 1999, p. 109.
[92] L. Keviczky, Cs. Bányász, Comparison of different closed-loop parametrization schemes, in: 12th IFAC Symposium on System Identification SYSID'2000, Santa Barbara, CA, USA, 2000 (CD printing).
[93] L. Keviczky, Cs. Bányász, Closed-loop parametrization schemes: identification of the K-B and d-Y parameters, in: 3rd IFAC Symposium Robust Control Design ROCOND'00, Prague, CZ, 2000 (CD printing).
[94] L. Keviczky, Cs. Bányász, Iterative identification and control using K-B parametrization, in: K.J. Aström, P. Albertos, M. Blanke, A. Isidori, W. Schaufelberger, R. Sanz (Eds.), Lecture Note: Control of Complex Systems, Springer, 2000, pp. 101–121.
[95] Keviczky, A new gap metric for robustness measure and regulator design, in: 9th Mediterranean Conf. Control and Automation MED'01, Dubrovnik, CR, 2001 (CD printing).
[96] L. Keviczky, Cs. Bányász, Adaptive iterative refinement of an optimal 2DF nonlinear controller, in: IFAC Workshop Adaptation and Learning in Control and Signal Processing ALCOSP'01, Como, Italy, vol. I, 2001, pp. 425–430.
[97] L. Keviczky, Cs. Bányász, Generic two-degree of freedom control systems for linear and nonlinear processes, J. Syst. Sci. 26 (4) (2001) 5–24.
[98] L. Keviczky, Cs. Bányász, Theoretical performance/robustness limits for linear SISO control systems, in: 4th Asian Control Conference, Singapore, 2002, pp. 984–989 (CD printing).
[99] L. Keviczky, Cs. Bányász, Direct relationships of performance, robustness measures and amplitude constraint, in: IEEE Conf. Decision and Control CDC'02, Las Vegas, CA, USA, 2002, pp. 4160–4161.
[100] L. Keviczky, Cs. Bányász, Influence of time-delay mismatch on robustness and performance, in: European Control Conference ECC'03, Cambridge, UK, 2003 (CD printing).
[101] L. Keviczky, Cs. Bányász, Optimal structure for model predictive control, in: IFAC Workshop on Adaptation and Learning in Control and Signal Processing ALCOSP'04, Yokohama, Japan, 2004, pp. 621–626 (CD printing).
[102] L. Keviczky, Cs. Bányász, New structural considerations for model predictive and iterative control, J. Syst. Sci. 30 (1) (2004) 23–36.
[103] L. Keviczky, Cs. Bányász, On the equivalence of a minimal order MPC and a GTDOF control of time-delay plants, in: 16th IFAC Congress, Prague, CZ, 2005 (CD printing).
[104] L. Keviczky, Cs. Bányász, Decomposition of error based control system optimization, in: IEEE Int Conf on Systems, Man and Cybernetics, SMC'05, Waikoloa, Hawaii, 2005, pp. 2174–2181.
[105] L. Keviczky, R. Bars, J. Hetthéssy, A. Barta, Cs. Bányász, Control Engineering, ISBN 9819, Széchenyi University Press, Győr, 2011. Original in Hungarian: L. Keviczky, R. Bars, J. Hetthéssy, A. Barta és, Cs. Bányász (2006). Szabályozástechnika, Egyetemi jegyzet, ISBN 55079, Műegyetemi Kiadó, Budapest, Hungary.
[106] L. Keviczky, Cs. Bányász, Robust stability and performance of time-delay control systems, ISA Trans. J. Sci. Eng. Measure. Autom., Am. Inst. Phys. 46 (2007) 233–237.
[107] L. Keviczky, Cs. Bányász, Future of the Smith predictor based regulators comparing to Youla parametrization, T19–001, in: 15th Mediterranean Conf. On Control and Automation MED'07, Athens, GR, 2007.

[108] L. Keviczky, Cs. Bányász, On the H2, L2 and H-infinity, L-infinity optimality of some two-degree of freedom control systems, J. Syst. Sci. 33 (3) (2008) 39–49.
[109] L. Keviczky, Cs. Bányász, A formal generalization of the Youla-parametrization for multivariable plants, in: European Control Conference ECC09, Budapest, Hungary, 2009, pp. 779–783 (CD printing).
[110] L. Keviczky, Cs. Bányász, MIMO controller design for stable multivariable processes, in: 17th Int Conf. On Systems Science, Wroclaw, PL, 2010, pp. 43–52.
[111] L. Keviczky, Cs. Bányász, Model error in observer based state feedback and Youla-parametrized regulator, in: 19th Mediterranean Conf on Control and Automation MED2011, Corfu, GR, 2011, pp. 219–224.
[112] L. Keviczky, Cs. Bányász, MIMO controller design for decoupling aircraft lateral dynamics, in: 9th IEEE Conf. On Control and Automation ICCA'11, Santiago, CL, 2011, pp. 1079–1084.
[113] L. Keviczky, Cs. Bányász, On an unknown property of the LQ control problem, in: Systemics and Informatics World Congress SIWN'08 Int Conf. Industrial Informatics and System Engineering IISE'08, Glasgow, UK, vol. 3, 2008, pp. 125–129.
[114] Cs. Bányász, L. Keviczky, Pole-placement constraints using LQ controller design, 3, in: Int Conf on Modeling, Simulation and Applied Optimization, Sharjah, UAE, 2009 (CD printing).
[115] L. Keviczky, J. Bokor, On the solution of an LQ control problem anomaly, in: Proc. of the Automation and Applied Computer Science Workshop AACS'11 (50th Anniversary of the Dept. of Automation and Applied Informatics), Budapest, Hungary, 2011, pp. 22–37.
[116] L. Keviczky, J. Bokor, On an LQ problem anomaly and a possible solution, in: 14th IASTED Int. Conf. on Control and Applications CA 2012, Crete, GR, 2012 (CD printing).
[117] R. Kulhavy, M. Karny, Tracking of slowly varying parameters by directional forgetting, in: 9th IFAC Congress, Budapest, Hungary, 1984.
[118] V. Kučera, Stability of discrete linear feedback systems, in: 6th IFAC Congress, Boston, MA, USA, 1975.
[119] V. Kučera, Exact model matching, polynomial equation approach, Int. J. Syst. Sci. 12 (1981) 1477–1484.
[120] V. Kučera, Analysis and Design of Discrete Linear Control Systems, Prentice Hall, 1991.
[121] I.D. Landau, System Identification and Control Design, Prentice Hall, 1990.
[122] L. Ljung, T. Söderström, Theory and Practice of Recursive Identification, MIT Press, 1985.
[123] L. Ljung, in: T. Kailath (Ed.), System Identification, Theory for the User, second ed., Prentice Hall, 1999.
[124] D.G. Luenberger, Observing the state of a linear system, IEEE Trans. Mil Electron. MIL-8 (1963) 74–80.
[125] D. Luenberger, Observers for multivariable systems, IEEE Trans. Aut. Control AC-11 (1966) 190–197.
[126] D.G. Luenberger, An introduction to observers, IEEE Trans. Aut. Control AC-16 (1971) 596–603.
[127] J.M. Maciejowski, Multivariable Feedback Design, Addison Wesley, 1989.
[128] D. McLean, Automatic Flight Control Systems, Prentice Hall, 1990, 125.
[129] R. Middleton, G.C. Goodwin, Digital Control and Estimation: A Unified Approach, Prentice Hall, 1990.
[130] J. Mikles, M. Fikar, Process Modelling, Identification and Control (Identification and Optimal Control), STU Press, Bratislava, SK, 2004.
[131] M. Morari, E. Zafiriou, Robust Process Control, Prentice-Hall, London, 1989.

[132] Z. Nehari, On bounded bilinear forms, Ann. Math. 65 (1) (1957) 153–162.
[133] R.C. Nelson, Flight Stability and Automatic Control, WCB/McGraw-Hill, 1998.
[134] R. Nevanlinna, Uber beschränkte funktionen, die in gegebenen punkten vorgeschrieben werte annehmen, Ann. Acad. Sci. Fenn. A 13 (1) (1919) 1–72.
[135] G. Pick, Uber die beschränkungen analytischer funktionen, welche durch vorgegebene funktionswerte bewirkt werden, Math. Ann. 77 (1916) 7–23.
[136] K. Ogata, Discrete-time Control Systems, Prentice-Hall, 1987.
[137] V. Peterka, Adaptive digital regulation of noisy systems, in: IFAC Symp. On Syst. Identification, Prague, CZ, 1970.
[138] V. Peterka, On a steady-state minimum variance control strategy, Kybernetika 3 (1972).
[139] Q.-G. Wang, Decoupling Control, Springer, 2003.
[140] Z.V. Rekasius, A general performance index for analytica of control systems, IRE Trans. Autom. Control (1960) 217.
[141] E. Rosenwasser, R. Yusupov, Sensitivity of Automatic Control Systems, CRC Press, 2000.
[142] M.G. Safonov, B.S. Chen, Multivariable stability-margin optimization with decoupling and output regulation, IEE Proc. D 129 (6) (1982) 276–283.
[143] R. Schrama, Approximate identification and control design with application to a mechanical system (thesis), Delft University, 1992.
[144] S. Skogestad, I. Postlethwaite, Multivariable Feedback Control: Analysis and Design, second ed., Wiley, 2005, pp. 297–300.
[145] O.J.M. Smith, Feedback Control Systems, McGraw-Hill, NewYork, 1958.
[146] O.J.M. Smith, Closer control of loops with dead time, Chem. Eng. Prog. 53 (1957) 217–219.
[147] T. Söderström, P. Stoica, System Identification, Prentice Hall, 1989.
[148] J.G. Truxal, Automatic Feedback Control System Synthesis, McGraw-Hill, New York, 1955.
[149] Y.Z. Tsypkin, Adaptation and Learning in Automatic Systems, 1968. Original in Russian: Цыпкин, Я.З. (1968). Адаптация и обучение в автоматических системах, Изд. Наука, Москва.
[150] Y.Z. Tsypkin, Fundamental Theories in Learning Systems, 1970. Original in Russian: Цыпкин, Я.З. (1970). Основы теории обучающихся систем. Изд. Наука, Москва.
[151] I. Vajk, M. Vajta, L. Keviczky, R. Haber, J. Hetthéssy, K. Kovács, Adaptive load-frequency control of the Hungarian power system, Automatica 21 (2) (1985) 129–137.
[152] I. Vajk, Implementation Questions of Adaptive Regulators, 1988 (Original in Hungarian: Adaptív szabályozók implementálásának kérdései. Kandidátusi értekezés, Magyar Tudományos Akadémia).
[153] P.M.J. Van den Hof, R.J.P. Schrama, R.A. de Callafon, O.H. Bosgra, Identification of normalized coprime plant factors from closed-loop experimental data, Eur. J. Control 1 (1995) 62–74.
[154] M. Vidyasagar, Control System Synthesis: A Factorization Approach, MIT Press, 1985.
[155] G. Vinnicombe, Frequency domain uncertainty and the graph topology, CA-38, IEEE Trans. Aut. Control 6 (1993) 1371–1383.
[156] G. Vinnicombe, On IQCs and the ν-gap metric, in: 37th IEEE Conf. Decision and Control CDC'98, Tampa, FL, USA, 1998.
[157] G. Vinnicombe, Approximating uncertainty representations using the ν-gap metric, in: 5th European Control Conference ECC'99, Karlsruhe, D, 1999.
[158] S. Wang, B. Chen, Optimal model matching control for time-delay systems, Int. J. Control 47 (3) (1988) 883–894.

[159] N. Wiener, Extrapolation, Interpolation and Smoothing of Stationary Time Series, Wiley and Sons, 1949.
[160] B.E. Ydstie, Adaptive control and estimation with forgetting factors, in: IFAC Symp. On Identification and System Parameter Estimation, York, UK, 1985.
[161] D.C. Youla, H.A. Jabri, J.J. Bongiorno, Modern Wiener-Hopf design of optimal controllers: part II: the multivariable case, IEEE Trans. Aut. Control AC-21 (1976) 319–338.
[162] D.C. Youla, J.J. Bongiorno, C.N. Lu, Single-loop feedback stabilization of linear multivariable dynamical plants, Automatica 10 (2) (1974) 159–173.
[163] P.C. Young, An instrumental variable method for real-time identification of a noisy process, Automatica 6 (1970).
[164] M.B. Zarrop, Variable forgetting factors in parameter estimation, Automatica 19 (3) (1983) 295–298.

AUTHOR INDEX

Note: Page numbers followed by "f" and "t" indicate figures and tables respectively.

SUBJECT INDEX

Note: Page numbers followed by "f" and "t" indicate figures and tables respectively.